Advances in Carbohydrate Chemistry and Biochemistry

Volume 56

Advances in Carbohydrate Chemistry and Biochemistry

Editor
DEREK HORTON
The American University
Washington, DC

Board of Advisors

LAURENS ANDERSON
STEPHEN J. ANGYAL
HANS H. BAER
JOHN S. BRIMACOMBE
J. GRANT BUCHANAN
DAVID R. BUNDLE

STEPHEN HANESSIAN
BENGT LINDBERG
HANS PAULSEN
NATHAN SHARON
J. F. G. VLIEGENTHART
ROY L. WHISTLER

Volume 56

ACADEMIC PRESS
A Harcourt Science and Technology Company

San Diego San Francisco New York Boston
London Sydney Tokyo Toronto

This book is printed on acid-free paper.

Copyright © 2001 by ACADEMIC PRESS

All Rights Reserved.
No part of this publication may be reproduced or transmitted in any form or by any means, electronic or mechanical, including photocopy, recording, or any information storage and retrieval system, without permission in writing from the Publisher.

The appearance of the code at the bottom of the first page of a chapter in this book indicates the Publisher's consent that copies of the chapter may be made for personal or internal use of specific clients. This consent is given on the condition, however, that the copier pay the stated per copy fee through the Copyright Clearance Center, Inc. (222 Rosewood Drive, Danvers, Massachusetts 01923), for copying beyond that permitted by Sections 107 or 108 of the U.S. Copyright Law. This consent does not extend to other kinds of copying, such as copying for general distribution, for advertising or promotional purposes, for creating new collective works, or for resale. Copy fees for pre-2001 chapters are as shown on the title pages. If no fee code appears on the title page, the copy fee is the same as for current chapters.
0065-2318/01 $35.00

Explicit permission from Academic Press is not required to reproduce a maximum of two figures or tables from an Academic Press chapter in another scientific or research publication provided that the material has not been credited to another source and that full credit to the Academic Press chapter is given.

Academic Press
A Harcourt Science and Technology Company
525 B Street, Suite 1900, San Diego, California 92101-4495, USA
http://www.academicpress.com

Academic Press
Harcourt Place, 32 Jamestown Road, London NW1 7BY, UK
http://www.academicpress.com

International Standard Book Number: 0-12-007256-4

PRINTED IN THE UNITED STATES OF AMERICA
00 01 02 03 04 05 QW 9 8 7 6 5 4 3 2 1

CONTENTS

PREFACE ix

Guy Gordon Studdy Dutton, 1923–1997
HARALAMBOS PAROLIS

Text 1

Iodine in Carbohydrate Chemistry
ANDREW R. VAINO AND WALTER A. SZAREK

I.	Introduction	9
II.	Glycosylation	10
	1. Uses of Iodine	10
	2. Glycosyl Iodides	10
	3. Activation of *n*-Pentenyl Glycosides	13
	4. Activation of Thioglycosides	15
III.	Activation of Glycals	19
IV.	Pseudohalogens	26
V.	Hypervalent Iodine	30
	1. Dess–Martin Reagent	36
VI.	Reactions Involving Phosphorus	38
	1. Rydon Reagents	38
	2. The Garegg–Samuelsson Reagent	39
VII.	Installation and Removal of Protecting Groups Using Iodine	44
VIII.	Iodocyclizations	50
IX.	Miscellaneous Reactions	55
	References	57

Structure of Anomeric Glycosyl Radicals and Their Transformations under Reductive Conditions
JEAN-PIERRE PRALY

I.	Introduction	65
II.	Structure of Sugar-Derived Radicals	67
	1. Structure of Carbon-Centered Acetal Radicals	68
	2. ESR Studies of Carbohydrate-Derived Radicals	69

		3. Interpretation in the Framework of the Frontier Molecular Orbital (FMO) Theory	85
III.		Radical Methods to Anhydroalditols from Glycosyl Derivatives	86
	1.	Radical Reduction with Tin and Silicon Hydrides	87
	2.	Synthesis of Anhydroalditols by Single-Electron Transfer	100
	3.	Synthesis of Ring-Fused and C-Branched Anhydroalditols	102
	4.	Stereoselectivity of Hydrogen-Atom Transfer at Anomeric Radicals	103
IV.		Diastereoselective Reductions of Substituted Anomeric Radicals	106
	1.	Stereocontrolled Synthesis of O-Glycosides and Analogs	107
	2.	Stereocontrolled Synthesis of C-Glycosyl Compounds and Analogues	115
	3.	Radical Routes to Nucleosides	127
	4.	Reduction Diastereoselectivity of Substituted Anomeric Radicals	129
V.		Synthesis of 2-Deoxy Sugars by Radical-Induced Rearrangement	130
	1.	Synthesis of 2-Deoxy Sugars by 2,1-Acyloxy Group Rearrangements	131
	2.	Synthesis of 2-Deoxy Sugars by 2,1-Phosphonyloxy Group Rearrangement	135
	3.	Mechanism of Radical-Induced 2,1-Migrations of Ester Groups	138
VI.		Radical-Mediated Eliminations: Formation of Glycals	139
VII.		Conclusion	142
		References	143

Oligosaccharide–Protein Conjugates as Vaccine Candidates against Bacteria

VINCE POZSGAY

I.		Overview of Polysaccharide-Based Antibacterial Vaccines	153
II.		Synthesis and Immunological Evaluation of Protein Conjugates of Oligosaccharide Components of Bacterial Polysaccharides	157
	1.	Studies on Dextran-Related Oligosaccharide–Protein Conjugates	157
	2.	Protein Conjugates of Oligosaccharide Fragments of Capsular Polysaccharides	161
	3.	Protein Conjugates of Oligosaccharides Related to Lipopolysaccharides	179
III.		Conclusion	193
		References	194

Molecular Structure of the Carbohydrate–Protein Linkage Region Fragments from Connective-Tissue Proteoglycans

N. RAMA KRISHNA AND PAWAN K. AGRAWAL

I.	Introduction	201
II.	Isolation of Linkage-Region Oligosaccharides	205
III.	Identification of Carbohydrate Spin Systems	206
IV.	Identification of the Peptide Spin Systems	209
V.	Conformational Analysis of the Interglycosidic Linkages	210
VI.	Conformation of the Glycopeptide Linkage	211

VII.	Influence of Glycosylation on Proteoglycan Core-Peptide Backbone Conformation	213
VIII.	Structural Studies on the Carbohydrate–Protein Linkage Region	215
	1. Dermatan Sulfate	215
	2. Chondroitin Sulfate	216
	3. Heparan Sulfate	219
	4. Heparin	223
IX.	NMR Spectral Properties of Linkage-Region Oligosaccharides	224
X.	Biosynthesis of Carbohydrate–Protein Linkage-Region Oligosaccharides	228
	References	229

The Conformation of *C*-Glycosyl Compounds

JESUS JIMÉNEZ-BARBERO, JUAN FELIX ESPINOSA, JUAL LUIS ASENSIO,
FRANCISCO JAVIER CAÑADA, AND ANA POVEDA

I.	Introduction	235
II.	Nomenclature	236
III.	*C*-Glycosyl Compounds in Chemistry and Biochemistry	237
IV.	Conformation of Saccharides: An Overview on the Relative Role of the Exoanomeric Effect and Steric Interactions	239
V.	The Use of NMR and Molecular Modeling as Tools in Conformational Analysis of *C*-Glycosyl Compounds in Solution	243
VI.	Conformation of *C*-Glycosyl Compounds	246
	1. Conformation in the Solid State	246
	2. Conformation in Solution	247
VII.	Conclusions	279
	References	280

AUTHOR INDEX 285
SUBJECT INDEX 311

PREFACE

This fifty-sixth volume of *Advances* features two articles devoted to synthetic aspects. Vaino (La Jolla, CA) and Szarek (Kingston, Ontario) survey the versatility of the heaviest accessible elemental halogen, iodine, for effecting a wide variety of useful transformations in the carbohydrate field under particularly mild conditions. Reactions involving free radicals, long underutilized in the carbohydrate field, have received considerable attention in recent years, notably through the work of the late D. H. R. Barton and his coworkers, and by the group of B. Giese. Here Praly (Lyon, France) contributes an extensive survey of radicals at the anomeric center of sugars, from both the structural viewpoint and with respect to their chemical transformations, especially under reductive conditions.

Although broad-spectrum antibiotics have long held center stage in the clinical treatment of bacterial infections, the use of carbohydrate vaccines for immunizing subjects against specific pathogens dates back still further, and is once more coming to the forefront. High current interest is focused on the use of specific oligosaccharide determinants chemically conjugated to proteins as vaccine candidates, an area here discussed by Pozsgay (Bethesda, MD).

The complex structures of mammalian proteoglycans continues to challenge researchers using even the most sophisticated current methodology. The glycosaminoglycan and protein subunits in these glycoconjugates are connected through a specific oligosaccharide linkage-region, and the molecular structure of this region is the subject of the article by Rama Krishna and Agrawal (Birmingham, Alabama).

The final chapter in this volume, by Jiménez-Barbero, Espinosa, Asiento, Cañada, and Poveda (Madrid, Spain), is devoted to the conformations of C-glycosyl compounds, and focuses on the complementary use of computational methodology and experimental studies, largely by NMR, to probe the conformational properties of this class of glycoside analogues, which are of high interest in their behavior with glycoside hydrolases and transferases.

The passing is noted with regret of three major figures in the carbohydrate world, George A. Jeffrey on February 13, 2000, Sumio Umezawa on March 30, 2000, and Raymond U. Lemieux on July 22, 2000. All three were authors of important articles in this Series, and detailed obituaries are scheduled to appear in a future volume.

An appreciation of the life and work of Guy G. S. Dutton, a frequent contributor to this Series and a member of its Advisory Board, is contributed by Parolis (Constantia, South Africa).

Washington, DC
May, 2000

DEREK HORTON

GUY GORDON STUDDY DUTTON

1923–1997

A very rich and rewarding life ended in Vancouver in the early hours of the morning of 16 April 1997, when Guy Dutton passed away at the age of 74 after losing his battle with cancer. In spite of the threat, which hung over him for four years, Guy Dutton never stopped looking forward. He simply got on with his life the way he always had and was planning a summer sailing trip to Alaska with a friend when, sadly, the progress of his illness accelerated and his plans came to naught. Guy was born in London on 26 February 1923, the only child of Henry and Charlotte Dutton. He received his early education at Rokeby Preparatory School in Wimbledon, and then went to Eastbourne College, where his interest in chemistry was stimulated by the way that his teacher, Mr, T. E. Rodd, taught the subject. After leaving Eastbourne, Dutton went to Queen's College, Cambridge, where he read for a degree in Chemistry, Physics, and Mineralogy. On completing the B.A. in 1943, he was drafted into the Ministry of Aircraft Production and was posted initially to RAE Farnborough, where he worked on a team whose task was to develop for night fighters a black paint having a reflectance of less than one percent. Some time later, he was transferred to the branch dealing with the development of smoke and gas weapons. Products developed at this branch were used in the laying of a smokescreen at sea before the D-Day invasion.

After the war, Guy Dutton was appointed lecturer in Organic Chemistry at Sir John Cass College, London, a position he held until his departure for Canada in mid-1949. During this period, he also commenced research toward a University of London M.Sc. on the synthesis and toxicity of dinitrophenols under the guidance of Professor E. de B. Barnett. The period at Sir John Cass College expanded Dutton's chemical knowledge significantly and contributed to his development as an academic. While running practical classes where, in one laboratory students at different levels would be conducting inorganic, analytical, or organic experiments, he acquired a mental dexterity in fielding and dealing with a great variety of questions. During the spring of 1949, Dutton applied for a teaching position at the University of British Columbia (UBC) in Vancouver, Canada. Much to his surprise, he was offered a lectureship, which he accepted. He arrived in Vancouver on 1 September 1949, having sailed from Southampton to New York on the *Mauretania*, and then travelled across Canada on the Canadian Pacific Railway.

His first lecturing duties were to teach two fourth-year chemistry courses to the last class of war veterans.

Dutton initially continued his studies on nitrophenols and completed the M.Sc. in 1952. His first publication, "The characterisation of dinitrophenols," appeared in the *Canadian Journal of Chemistry* in 1953. In all, Dutton published six papers on aspects of phenol chemistry. He was introduced to carbohydrate chemistry by E.V. White, who had also joined the staff of the University of British Columbia in 1949. White's interest was studying the structure of plant gum exudates. Guy Dutton's first publication on a carbohydrate topic appeared in the *Canadian Journal of Chemistry* in 1953. By this time, he had realized that he would not progress very far in academia if he did not obtain the Ph.D. degree. Armed with a University Fellowship and two years of unpaid study leave, which was to commence from September 1953, Dutton sought an institution offering an appropriate program in carbohydrate chemistry. At that time there were very few places in North America offering postgraduate studies in carbohydrate chemistry. Eventually, Dutton chose to go to the University of Minnesota, where Fred Smith had established himself in the Department of Agricultural Chemistry. Smith had studied at the University of Birmingham (UK) under the direction of Professor W.N. (later Sir Norman) Haworth. Guy Dutton's doctoral research focused on the structure of hemicelluloses, which were known to be important in the manufacture of paper and, in particular, of rayon. He was able to complete his Ph.D. in two years as, having scored virtually 100% on the organic placement examination, he was excused from taking the usual collection of graduate courses. The results of his doctoral research, which established the first structure of a coniferous hemicellulose, that of Western hemlock (*Tsuga heterophylla*), were published in the *Journal of the American Chemical Society* in 1956. On his return to the University of British Columbia in 1955 and for the next fifteen years, his research programs were directed toward the study of hemicelluloses from several coniferous and deciduous species. Papers were published on the hemicelluloses of Sitka Spruce, Black Spruce, Ponderosa Pine, Douglas Fir, cherry wood, apple wood, and corn stalks and leaves. During this period his research program was generously funded by grants from the National Science and Engineering Council (NSERC) and from MacMillan Bloedel and the Institute of Paper Making. Grants from the U.S. National Institutes of Health from 1960 to 1962 allowed Dutton to expand his research program to include structural studies of synthetic polysaccharides. Most of this work was reported in the *Canadian Journal of Chemistry* from 1962 to 1965.

One of the strategies employed to obtain sequence information in the study of plant polysaccharides is to hydrolyze the polysaccharide partially and elucidate the structures of the separated oligosaccharide fragments. Frequently, only small quantities of oligosaccharides were obtained in such studies, and identification was accomplished on the basis of comparisons with authentic reference samples. As the latter were not always readily available, Dutton's group embarked on a program of oligosaccharide synthesis. Most of the results from these studies appear in the *Canadian Journal of Chemistry*.

During the early 1960s the relatively new technique of gas chromatography was beginning to replace paper and thin-layer chromatography as the method of choice in carbohydrate analysis, and it was quickly incorporated into Dutton's research program. Methodologies were developed for a variety of applications in carbohydrate chemistry, including the determination of oligo- and poly-saccharide structures, and these were reported in the *Journal of Chromatography, Carbohydrate Research*, and *Analytical Letters*. Dutton's keen interest in and extensive knowledge of the application of gas chromatography to carbohydrate chemistry resulted in an invitation to write a comprehensive review entitled, "Applications for Gas–Liquid Chromatography to Carbohydrates," which was published in two parts in *Advances in Carbohydrate Chemistry*, volumes 28 and 30. He was pleased to be invited to the Board of Advisors for *Advances* in 1978 and remained a member until his death.

Dutton's work in structural polysaccharide chemistry, particularly on hemicelluloses, resulted in an invitation to speak at a Symposium on the Chemistry and Biochemistry of Lignin, Cellulose, and Hemicellulose in Grenoble, France, in 1964. This was followed by an invitation to spend a sabbatical year in Grenoble at the École Francaise de Papeterie, which he took up the following year as a Visiting Professor. His period in Grenoble (1965–1966) formed the basis of a long and fruitful association with several groups there, some members of which later came to study in his laboratory in Vancouver. During his stay in Grenoble, Dutton introduced his co-workers to the chemistry of carbohydrates, particularly the analysis and characterization of hemicelluloses. He also developed and improved his knowledge of the French language to such an extent that he was able to deliver a course for students and a series of research lectures in French.

In 1967, Guy Dutton received news that was to have a profound influence on his scientific career. He had been nominated for a NATO Fellowship that would permit him to spend two periods of three months each at any two institutions in NATO countries. Guy Dutton chose the Technical University in Lyngby, Denmark, and the Max Planck Institute for Immunology in Freiburg, Germany. Some years previously, he had attended a series of lectures on the relationship between the chemical composition of the bacterial cell surface and the immunological differences that exist between different strains. He was fascinated by this field and wished to get involved. He now had the opportunity to study at firsthand at the Max Planck Institute the necessary experimental techniques. He took up the NATO Fellowship during the first part of 1968. The three months in the Chemistry Department of the Technical University in Lyngby were spent doing the spadework for the two-part review on gas–liquid chromatography mentioned earlier. The time at the Max Planck Institute was spent acquainting himself with the various chemical and microbiological techniques used in the structural elucidation of bacterial polysaccharides. While in Freiburg, he was encouraged by Drs. O. Westphal and S. Stirm to participate with other laboratories, in a collaborative program, in the study of the capsular polysaccharides (K-antigens) of the more than seventy capsular strains of *Klebsiella*. On his return to Vancouver, he set about changing his research direction from the study of hemicelluloses to the study of bacterial cell-surface polysaccharides. From 1968 until his retirement in 1988, Guy Dutton's research concentrated mainly on the structural elucidation of the capsular polysaccharides of *Klebsiella* and *Escherichia coli* strains. In spite of their high molecular weights (10^5–10^7 daltons), the structures of the capsular polysaccharides could readily be established because they are composed of regular repeating oligosaccharide units (usually between three and seven monosaccharides). This was not the case with hemicelluloses, which have far more complex structures. It is thus not surprising that Guy Dutton found his research on capsular polysaccharides more absorbing than the research on hemicelluloses and the results infinitely more satisfying. His group was the first to demonstrate that, despite their high molecular weights, capsular polysaccharides have reasonably well resolved ^{1}H NMR spectra at elevated temperatures. From these spectra, it was possible to determine the number of sugar residues in the oligosaccharide repeat-units and the identity of any non-carbohydrate substituents.

The award of a Senior Killam Fellowship in 1974 enabled Dutton to return to the Max Planck Institute on sabbatical in order to work with Dr. Stephan Stirm and to learn about bacteriophages. These bacterial viruses had been shown to contain specific endoglycanases that are capable of depolymerizing the capsular polysaccharides of the host bacterial strains into oligosaccharides representing the repeating units. Members of Dutton's group immediately appreciated the potential of bacteriophage depolymerization in structural studies of capsular polysaccharides. Methods were soon developed for rapidly preparing and purifying bacteriophages and applying them as reagents in structural studies of specific *Klebsiella* and *E. coli* strains. A host of publications emanated from his laboratory, featuring bacteriophage depolymerization as a key method in the production of an oligosaccharide-repeat unit with such labile substituents as *O*-acetyl or 1-carboxyethylidene intact. Dutton's studies on bacterial polysaccharides are reported in nearly one hundred publications. The John Labatt Award in 1989 recognized his innovative contributions to the field of bacterial antigens.

Guy Dutton's organizational skills and his desire to serve his profession were evident early in his academic career. In 1952, he served as conference secretary of the British Columbia Academy of Sciences and in 1953 was chairman for the 7th Annual Conference. In the same year, he was chairman of the Division of Organic Chemistry, American Chemical Society Pacific Northwest Regional Meeting in Bellingham, Washington. These were the first of several conferences that Guy Dutton was to organize. On his return to the University of British Columbia, after completing the Ph.D. at Minnesota, Guy Dutton re-established his involvement with the Canadian Institute of Chemistry (CIC). During 1956–1958, he was elected to the executive of the Vancouver Section of the CIC, serving as its chairman for the year 1957–1958. In 1956, he was awarded the Fellowship of both the CIC and the Royal Institute of Chemistry. From 1962–1965, Guy Dutton served as CIC Provincial Counsellor. He served again in this capacity from 1982–1985, and in 1988 he was elected to a two-year term as president of the CIC. Guy Dutton served as chairman of the 62nd CIC Conference in Vancouver in 1979. His attention to detail and thorough planning of both the scientific and social events ensured a successful conference. With this success to his credit, it was quite understandable that he should be drafted as chairman of the 28th IUPAC Congress in Vancouver in 1981. At that time, this was the largest conference ever held at UBC. The following year he chaired the 11th International Carbohydrate Symposium, also held in Vancouver. This conference was also an outstanding success and bore the now-familiar Dutton trademark. During the week of the symposium, each delegate was invited to attend at least one function on the lawns of the Dutton's beautiful home. Such attention to detail and concern for the comfort and well-being of the delegates made this a most memorable conference. Guy served a three-year term as secretary of the International Carbohydrate Organization. After he retired from UBC, Guy Dutton assisted (as Vice-Chairman) with the organization of the 1989 Pacific Chemical Conference in Hawaii. In recognition of his outstanding contributions to and significant leadership in the profession of chemistry, Guy Dutton was awarded the 1986 Montreal Medal of the Montreal CIC Section.

Guy Dutton remained a faculty member of the Department of Chemistry at UBC from the time of his appointment as a Lecturer in 1949 until his retirement in 1988. He was promoted to Assistant Professor in 1950, Associate Professor in 1959, and Full Professor in 1964. After his retirement he continued his association with the Department as an Honorary Professor. Dutton taught undergraduate courses in general chemistry, organic chemistry, polymer chemistry, and industrial organic chemistry. He was an excellent teacher at all levels. He thought deeply about his subject and how best to present it to his audience. The same thorough preparation went into all his lectures, whether he was presenting a formal lecture to an undergraduate class or a research paper to a sophisticated international audience. Understandably, he received numerous invitations to lecture in Canada and in many parts of the world. He was fully bilingual and relished speaking French.

In 1951, Guy Dutton met his future wife, Marise Atkinson, at a cocktail party, and after a short courtship of a few months they were married. Marise was a very important factor in Guy's career and supported him staunchly throughout. She accompanied Guy on many of his sabbatical trips and to most of the scientific and professional conferences he attended. Marise was always a warm and gracious hostess at the many gatherings in the Dutton home. Their hospitality earned them a very special place in the hearts of many people who visited them in Vancouver. Their first child, Anthony, was born in 1957, and their twin sons, Peter and Robin, four years later. The arrival of the twins precipitated a need for more living space, and in order to accomplish this within their financial means, Guy decided to purchase power tools and add to the house himself. He did a very professional job and also established a lovely garden. In 1968, the Dutton family purchased a home on the picturesque University Endowment Lands within walking distance of the university campus. Their beautiful home overlooked the Burrard Inlet and had breathtaking views of the sea and the mountains of West and North Vancouver. Guy Dutton, by now an accomplished gardener, established and maintained a magnificent garden that boasted beautiful roses and eye-catching rhododendrons.

During his career at UBC, Guy Dutton directed the research programs of more than thirty M.Sc. and Ph.D. students and of numerous postdoctoral fellows from different parts of the world. Those who chose to work with Guy, whether as postgraduate student or visiting postdoctoral fellow, were rapidly made to feel at home. Guy and Marise took an enduring interest in those who came to work in his laboratory. Frequently, they would assist in the search for accommodation, often putting people up in their home until suitable lodgings were found. Frequent invitations to dine at the Dutton home resulted in relaxed and productive postgraduate students. Guy would promote the interest of his postgraduates and co-workers at every opportunity. He invariably assisted them in the search for employment and made a point of keeping in touch. Many of his students secured positions at academic institutions and in industry, and went on to make important contributions in their professions.

Dutton's laboratory was also frequently the destination for academics on sabbatical leave, many returning for a second or even a third sabbatical period. Guy Dutton, in turn, spent several periods abroad. In addition to the Fellowships and sabbaticals that were mentioned earlier, he spent three further periods as a visiting professor at the University of Grenoble in 1975, 1984, and 1989. He was a visiting professor at the University of Cape Town (1980), the Technical University of Chile (1980), and was appointed to the Hugh Kelly Fellowship of Rhodes University (Grahamstown) in 1985.

Although chemistry and science consumed most of Guy's time, there were other activities that gave him much pleasure. He and Marise enjoyed listening to classical music and opera; they loved to travel, which they did extensively, making many friends all over the world; and they enjoyed good food and wine. Guy was very knowledgeable on wines, particularly those of France, and his interest in wines extended beyond imbibing to wine making. He had a vast range of wines in his cellar, the majority of which he had made from grapes purchased in California and brought to Vancouver in a refrigerated vehicle. Guy had a great passion for the sea and sailing. He was an accomplished yachtsman and was a long-time member of the Royal Vancouver Yacht Club. Although Guy sailed in many parts of the world, his greatest sailing accomplishment was probably his first ocean crossing from Hawaii to Vancouver in 1970. He repeated the crossing a few years later, but the first time was very special, as he had a great crew and the weather was perfect. Guy had hoped to spend his retirement cruising around the world, but circumstances prevented that dream. He did, however, occupy himself in retirement with teaching sailing and navigation, including celestial navigation, which he found particularly rewarding.

Guy Dutton was a true gentleman and a scholar. Tall and distinguished in appearance, and with a superb command of the English language he will be remembered for many different reasons. Professionally, he will be remembered as an inspiring teacher, for his contributions to carbohydrate chemistry, for his service to Canadian chemistry, and his exceptional organizational talents. Very many people worldwide will remember his career with affection and honor. Guy is survived by his wife, Marise, and their three sons, Anthony, Peter, and Robin.

HARALAMBOS PAROLIS

IODINE IN CARBOHYDRATE CHEMISTRY

By Andrew R. Vaino and Walter A. Szarek

Department of Chemistry, Queen's University, Kingston, Ontario K7L 3N6, Canada

I. Introduction	9
II. Glycosylation	10
1. Uses of Iodine	10
2. Glycosyl Iodides	10
3. Activation of *n*-Pentenyl Glycosides	13
4. Activation of Thioglycosides	15
III. Activation of Glycals	19
IV. Pseudohalogens	26
V. Hypervalent Iodine	30
1. Dess–Martin Reagent	36
VI. Reactions Involving Phosphorus	38
1. Rydon Reagents	38
2. The Garegg–Samuelsson Reagent	39
VII. Installation and Removal of Protecting Groups Using Iodine	44
VIII. Iodocyclizations	50
IX. Miscellaneous Reactions	55
References	57

I. Introduction

Iodine, the heaviest of the commonly encountered halogens, has a long and storied history in organic chemistry. Iodine was discovered in 1811, a full 75 years before the isolation of its lighter chemical relative fluorine.[1] Despite its longevity and widespread use in carbohydrate chemistry, far more attention is given to fluorine. Although the topic of fluorine in carbohydrate chemistry has been reviewed extensively, in this series[2,3] and elsewhere,[4,5] as have halogenated sugars in general,[6] no review has yet appeared devoted solely to iodine in carbohydrate chemistry. A chapter with a large section devoted to starch–iodine complexes has, however, appeared in this series.[7] The importance of iodine to carbohydrate chemistry, and indeed to organic chemistry as a whole, is astounding—from the simplest incarnation as a leaving group to the elegant glycal addition sequences employed in oligosaccharide synthesis—the range of applications of iodine to carbohydrate chemistry is tremendous. Iodine's ability to act as a leaving group is well known, and is not explicitly covered in the present chapter. Instead, an attempt has been made to introduce the reader to what may be less well-known applications of iodine in carbohydrate synthesis, for example, the number of different types of reaction that require the presence of molecular iodine, the number of different ways iodine participates in oxidation or initiates radical reactions, and its use as a simple and efficient reagent for the introduction and removal of protecting groups.

II. Glycosylation

1. Uses of Iodine

Iodine has been used to increase yields in the Königs–Knorr glycosylation. Simple addition of iodine to the reaction mixture was reported to increase yields in the synthesis of β,β-trehalose octaacetate by condensation of 2,3,4,6-tetra-O-acetyl-β-D-glucopyranose and 2,3,4,6-tetra-O-acetyl-α-D-glucopyranosyl bromide,[8] and in the preparation of various mannosyl disaccharides.[9] Iodine was rationalized to impart increased polarity to the reaction medium, enhancing the stability of the intermediate oxacarbenium cation, leading to an increase in rate.[10] However, given that the use of iodine in the synthesis of α- and β-gentiobiose octaacetates actually decreased the rate of reaction, but was essential in their synthesis,[11] the possibility exists that iodine was acting to suppress undesired side reactions between the glycosyl halide and the metal promoter.

Iodine has been used as a glycosylation promoter on its own. Treating D-glucose, D-fructose, or inulin with a catalytic amount of iodine in a boiling alcohol afforded the glycoside corresponding to the alcohol used.[12] In the case of inulin, a preponderance of the fructofuranoside was obtained. Iodine is known to oxidize methanol to produce HI along with formaldehyde,[13] and, thus, this is another example of the classic acid-catalyzed Fischer glycosylation.

2. Glycosyl Iodides

The use of glycosyl chlorides, bromides, and even fluorides are all commonplace in the formation of glycosidic bonds. Glycosyl iodides, owing to their inherent instability, have not been used as frequently; nevertheless, they are powerful glycosyl acceptors. Glycosyl iodides have been exploited synthetically for the very reason that makes them difficult to handle—namely, the high reactivity they possess.

The synthesis of 2,3,4,6-tetra-O-acetyl-α-D-glucopyranosyl iodide, obtained by treatment the corresponding bromide with sodium iodide in acetone, was reported[14] in 1929. Similarly, Kronzer and Schuerch reported the *in situ* generation of benzylated glucosyl and galactosyl iodides from the corresponding chlorides using an excess of sodium iodide in the presence of lutidine and acetonitrile.[15] Treating the highly reactive glycosyl iodides with a slight excess of a variety of alcohols resulted in glycoside formation, with a preponderance of the thermodynamically favored α anomer. When a larger excess of alcohol was used, up to 40-fold, an increase of the proportion of β anomer was noted—consistent with the decrease in anomeric stabilization known to occur as solvent polarity increases.[16] When acetonitrile was used as the solvent, the possibility of the intermediacy of an α-nitrilium cation could account for the increased amount of β anomer formed.[17,18] It should be noted that the so-called α-nitrilium cation effect is not universally accepted.[19] A preponderance of β anomer formation in polar solvents was noted in a 1950 report of the formation, in low yield, of methyl 2,3,4,6-tetra-O-benzoyl-β-D-glucopyranoside by subjecting the corresponding iodide to refluxing methanol[20]; in this instance, anchimeric assistance from the ester protecting group at C-2 favors formation of the β anomer.

Glycosyl iodides have been successfully prepared in the same manner as the more common chlorides and bromides[21]—namely by treatment of the peracetylated sugar with hydrogen iodide in glacial acetic acid.[20] The use of less forcing conditions has also been explored. Thiem and Meyer[22] reported the synthesis of glycosyl iodides by treating fully protected sugars with iodotrimethylsilane. Owing to the ability of the trimethylsilyl cation to act as a powerful Lewis acid, polarization of the π electrons of oxygen in a cyclic acetal allowed for the direct introduction of iodine, with concomitant introduction of a trimethylsilyl group at the newly liberated hydroxyl group (Scheme 1).

The Lewis acidity of iodotrimethylsilane has also been used to convert ethereal protecting groups into the corresponding iodide.[23] Replacement of the ether occurs only if the possibility of stabilizing the resultant cation exists; thus, it has best been applied to the conversion of a primary trityl ether group into an iodide. Acetic esters, with no possibility of cationic stabilization, were not converted into iodides, except when at the anomeric position. In this manner, the synthesis of methyl 2,3,4-tri-O-acetyl-6-deoxy-6-iodo-α-D-glucopyranoside was achieved by treating the corresponding 6-O-trityl compound with iodotrimethylsilane. With

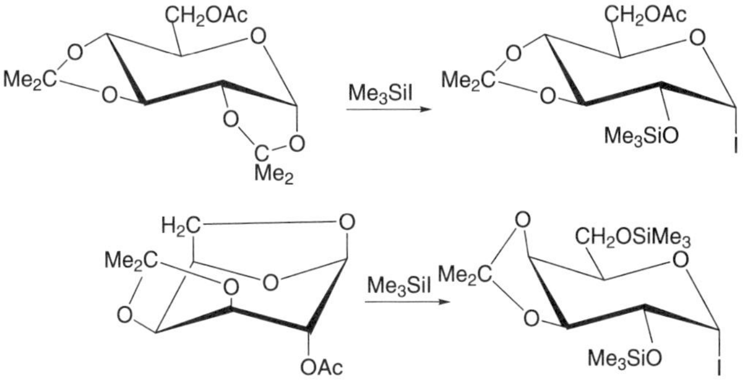

SCHEME 1

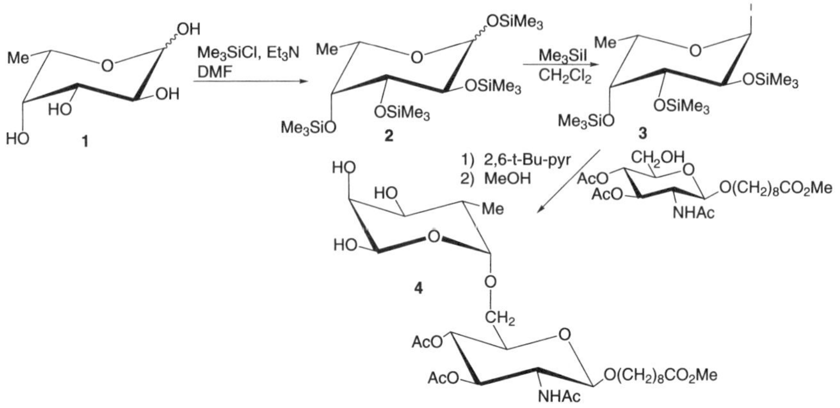

SCHEME 2

an acetic ester at the anomeric position, iodine was introduced both at C-1 and C-6 on treatment with iodotrimethylsilane. The preparation of glycosyl iodides has also been achieved using 1-iodo-N,N-2-trimethylpropenylamine[24]; although this method suffers somewhat from the necessity of preparing the unstable haloenamine.

Given the instability of glycosyl iodides, their generation and immediate use *in situ* is of benefit in oligosaccharide synthesis. Uchiyama and Hindsgaul[25] reported an effective means for the glycosylation of L-fucose (**1**, Scheme 2) by exposing the per-O-trimethylsilylated species (**2**) to iodotrimethylsilane, followed by treatment of the mixture with a nucleophile. Yields obtained in this manner are generally high and, in most cases, none of the β anomer was detected.[25] During processing of the reaction mixture—specifically, treatment with methanol—the remaining trimethylsilyl ethers were cleaved. Addition of tetrabutylammonium iodide had no effect on the reaction, implying that the means of iodine introduction was not simply an intermolecular nucleophilic displacement of iodide.

The conversion of glycosyl phosphates into glycosyl iodides has been achieved using 1 molar $LiClO_4$ in CH_2Cl_2. The use of the perchlorate salt, which imparts a high degree of polarity to the reaction medium, had previously been reported to aid in the glycosylation of less-reactive glycosyl acceptors, such as glycosyl fluorides and phosphates, in the absence of any glycosylation promoter.[26,27] By treating 2,3,4,6-tetra-O-benzyl-α-D-glucopyranosyl diphenyl phosphate (**5**, Scheme 3) with lithium iodide in a 1 molar solution of $LiClO_4$ in CH_2Cl_2, conversion into the highly reactive β-D-glycosyl iodide (**6**) was achieved; this unstable intermediate has been characterized by nuclear magnetic resonance (NMR) spectroscopy. The phenyl phosphate was found to afford higher yields than other phosphates examined owing to the greater ability of the aromatic system to stabilize the anion. Introduction of a nucleophile resulted in formation of a glycosidic bond,

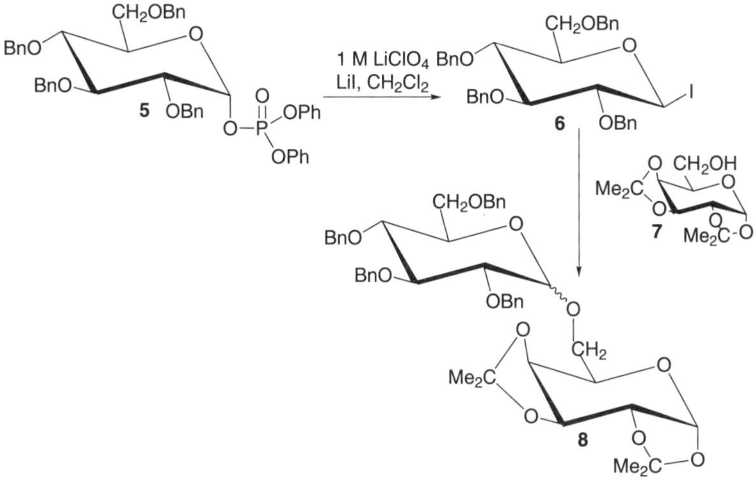

SCHEME 3

in moderate yields, and with a slight preference for the α anomer.[28] Although the use of glycosyl phosphates to generate glycosyl iodides is of interest, that glycosyl phosphates are required, necessitating extra steps, as the starting material detracts from the value of this approach.

Reaction of peracetylated sugars with iodine in the presence of hexamethyl-disilazane has been reported to afford the corresponding glycosyl iodide.[29] Glycosyl iodides have also been prepared by treatment of glycosyl chlorides or bromides with substoichiometric amounts of iodine and DDQ.[30] In this instance, DDQ acts to oxidize iodide remaining after the initial attack on the electrophilic iodine. The benefit of this approach is to limit the amount of free iodide is solution, as recycling of iodine at the expense of DDQ is clearly not cost-effective.

3. Activation of *n*-Pentenyl Glycosides

The elegant technique of exploiting *n*-pentenyl glycosides (NPGs) in glycoside synthesis[31,32] has been enhanced by the use of electrophilic sources of iodine. Using *N*-bromosuccinimide, it was possible to activate an *n*-pentenyl glycoside (**9,** Scheme 4) by transforming it into a cationic, substituted furan ring (**11**), which, in turn, is an effective leaving group. The generation of an oxacarbenium cation (**12**), common to most, if not all, glycosylation techniques,[33–35] is succeeded by formation of a new glycosidic bond by reaction with a nucleophile in solution (**13**).

The choice of protecting group at C-2 was found to have a profound effect on the activation of NPGs, and fomented a new means of selectivity.[36] The use of an electron-withdrawing protecting group at C-2—for example, an ester or halogen—resulted in an inability to activate the *n*-pentenyl glycoside toward

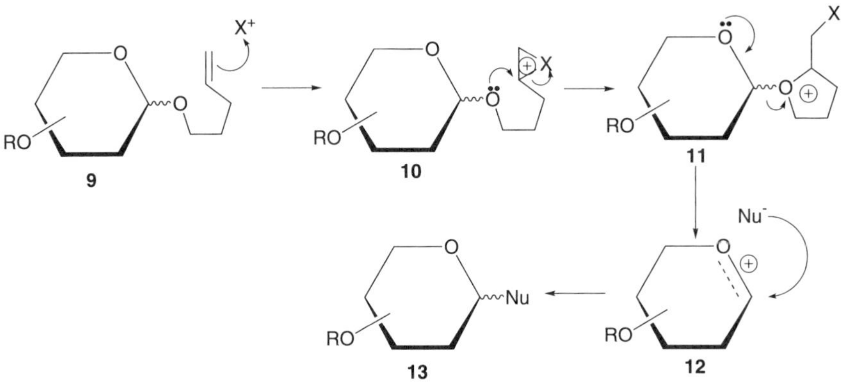

SCHEME 4

reaction. Conversely, the use of an ethereal protecting group allows for ready activation (Scheme 5). Thus, a simple change in protecting group allows the glycosyl acceptor to be either "armed" or "disarmed." This phenomenon was extended with the discovery that what was thought to be a disarmed glycosyl donor could in fact act as an armed glycosyl donor simply by using a more powerful electrophile. Using either NIS coupled with TFA,[37] or NIS and Et$_3$SiOTf,[38] a disarmed glycosyl donor was found to react to form a glycosidic bond. An example of this selectivity is depicted in Scheme 5. Reaction of **14** with **15** is performed with a less active bromine electrophile, which activates only the *n*-pentenyl glycoside of **14.** The disarming ester protecting group at C-2 of **15** prohibits undesired self-condensation,

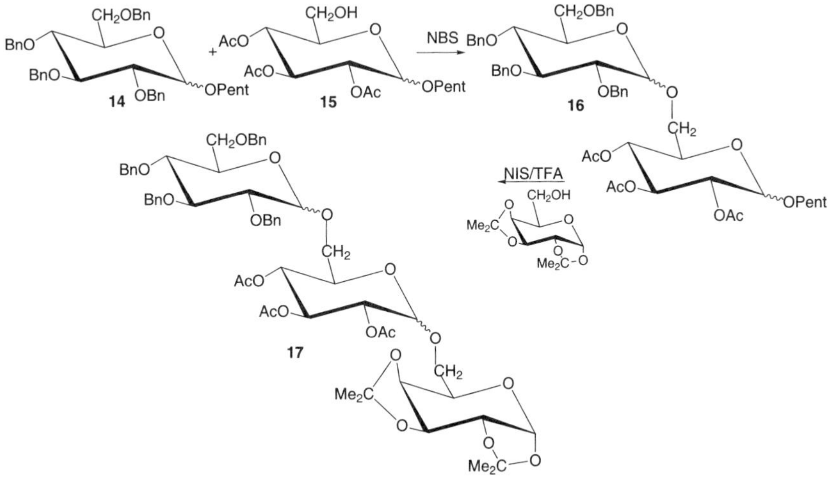

SCHEME 5

allowing formation of **16**. Further coupling to form trisaccharide **17**, simply by choosing the more electrophilic NIS/TFA, is now readily achieved. Even when used with armed glycosyl donors, the NIS systems afforded higher reaction rates. The ability to arm or disarm a glycoside donor simply by a change in protecting group or activator has been readily exploited in oligosaccharide synthesis.[39–41]

The use glycosides of *n*-pentenyl esters as glycosyl donors has also been examined.[42] *n*-Pentenyl esters behave in much the same manner as NPG glycosides; differences include that they are less affected by an electron-withdrawing group at C-2, and formation of the activated cationic furan ring occurs via the carbonyl oxygen, not the glycosidic oxygen. *n*-Pentenyl ester glycosides are not as commonly used in glycosylation reactions as their ethereal counterparts.

The activation of vinyl glycosides using NIS–trimethylsilyl triflate has been reported by Boons and co-workers.[43] Treatment of (2-buten-2-yl) 2,3,4,6-tetra-*O*-benzyl-β-D-galactopyranoside with a mixture of NIS, trimethylsilyl triflate, and dibenzyl phosphate afforded the corresponding glycosyl phosphate; anomeric selectivity, however, was poor. The same reaction, with a participating protecting group at C-2, namely a benzoic ester, afforded 66% of the β-D-galactosyl phosphate.

4. Activation of Thioglycosides

The use of thioglycosides as glycosyl donors has become an increasingly common technique.[44–46] Thioglycosides are advantageous in that they are easy to prepare (once prepared, the stench associated with the starting thiol is absent), are readily activated towards glycosylation, and are readily stored—unlike the more reactive glycosyl halides. Sulfur, a large, highly polarizable, nucleophilic atom,[47] readily accepts a positive charge, and thus becomes activated. Iodine is an excellent electrophile for activating sulfur, as it too is large, highly polarizable, and can act as a powerful electrophile. Thus, that sulfur should have such an affinity for iodine is no surprise when rationalized in terms of matching diffuseness of frontier molecular orbitals.[48]

Molecular iodine, the simplest possible source of electrophilic iodine, has been reported to activate thioglycosides.[49] Nucleophilic attack of the soft sulfur on the equally soft iodine affords a cationic trivalent sulfur, an effective leaving group, which readily departs to afford the intermediate oxacarbenium cation. Reaction with a nucleophile affords the desired glycoside. Nucleophiles ranging from simple alcohols to suitably protected monosaccharides have been submitted to this treatment. The anomeric selectivity is not spectacular, although the yields are high. The simplicity of this reaction is appealing.

The ability to "arm" and "disarm" thioglycosides in a manner similar to the NPG glycosyl acceptors has been reported. As with *n*-pentenyl glycosyl acceptors, the use of an electron-withdrawing substituent at C-2 was reported to inactivate the anomeric position toward glycosylation using iodonium dicollidine perchlorate (IDCP),[50] a switch to the more electrophilic NIS/TfOH system allowed for activation of the thioglycoside.[51] Similarly, the use of an electron-withdrawing 4-nitrophenyl thioglycoside acted as a "disarmed" glycosyl donor toward IDCP or NIS/TfOH, but was activated by using dimethyl(methylthio)sulfonium triflate.[52] The electron-withdrawing ability of the nitro group leads to a decrease in nucleophilicity of the glycosidic sulfur, requiring a more powerful electrophile. The steric bulk of the aglycone has also been demonstrated to disarm thioglycosyl donors; the use of a dicyclohexylmethyl thioglycoside precluded activation by either IDCP or NIS/TfOH.

Constraint of the pyranose ring of a carbohydrate has been exploited as a means of activating/deactivating glycosyl donors. By appending a dispiroketal group on a sugar (**19,** Scheme 6), it was discovered that the corresponding thioglycoside could no longer be activated toward glycosylation using IDCP. Using the better electrophile NIS/TfOH under carefully controlled conditions, **20** acted as a glycosyl acceptor preferentially over glycosylation with a substrate having an ester group at C-2 to afford[53] **21**. As such, this effect can be thought of as a "semi-disarmed" glycoside. Presumably, the difficulty in attaining the requisite half-chair[54] geometry, required to form an oxacarbenium cation in a bicyclic system, is the cause of this intermediate reactivity.

Iodine has been used to activate methyl thioglycosides for the synthesis of disaccharides.[49] Similarly, it has been reported that the use of IBr allowed for the

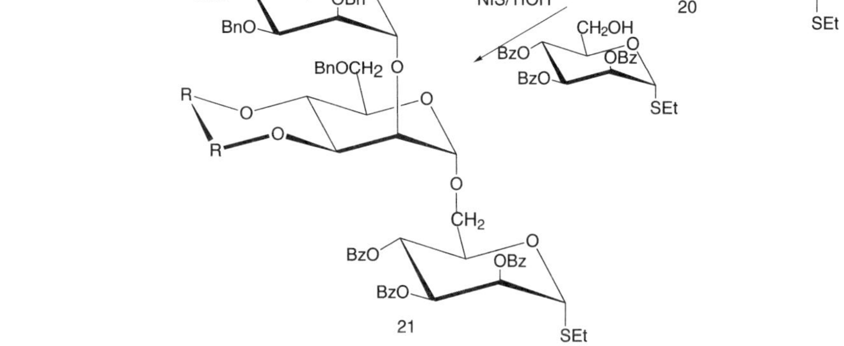

SCHEME 6

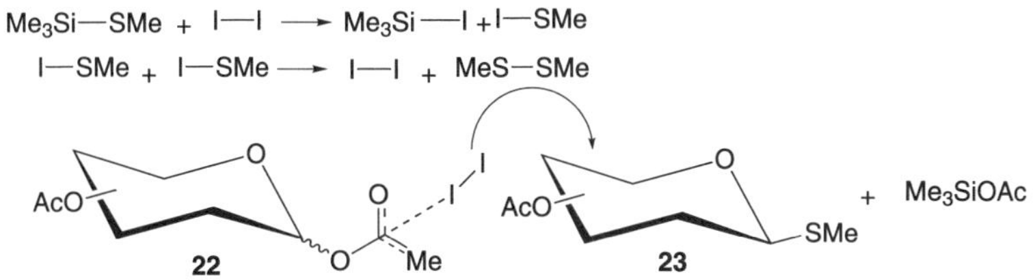

```
Me₃Si—SMe  +  I—I   →   Me₃Si—I  +  I—SMe
I—SMe  +  I—SMe   →   I—I  +  MeS—SMe
```

Scheme 7

coupling of disarmed thioglycosides by direct conversion into the more reactive glycosyl bromide.[55]

An interesting report from Kartha and Field[56] on the synthesis of thioglycosides from peracetylated sugars has appeared. Treatment of a peracetylated sugar (**22**, Scheme 7) with a two-fold excess of trimethylmethylthiosilane and an equivalent, or slightly less than an equivalent, of iodine in acetonitrile afforded the thioglycoside in high yield. A mechanism has been proposed wherein iodotrimethylsilane, created *in situ* by the action of iodine on methylthiotrimethylsilane, acts as a Lewis acid to polarize the anomeric acetoxy group, thus prompting formation of an oxacarbenium intermediate. Reaction can then occur to form the highly reactive glycosyl iodide, which will be quenched by the powerful sulfur nucleophile.

The use a dimethylphosphinothioate glycoside as a stable glycoside acceptor has been explored, and many agents have been used to activate these donors.[57] Among the most active has been found to be the use of iodine in the presence of triphenylmethyl perchlorate (TrClO₄), which presumably acts to generate iodonium perchlorate *in situ*. By varying the amount of TrClO₄ used in the glycosylation of mannose derivatives, a surprisingly high degree of β selectivity was noted when 5% TrClO₄ was used in the reaction—the same selectivity was not observed using either more or less of the promoter.[58] This result, the opposite of what would be expected on steric and electronic grounds, is rationalized in terms of the generation of a relatively stable α iodo glycoside. The expected α mannoside, generated from an oxacarbenium cation, reacts in an S_N2-like displacement, likely through a tightion-pair, in the same manner as reported by Lemieux and co-workers in their halide-exchange method.[59]

Another popular promoter for glycosylation with thioglycosides is the reagent dimethyl(methylthio)sulfonium triflate (DMTST).[60] A comparison between DMTST and NIS/TfOH for the formation of sialoglycoconjugates demonstrated that the iodine system provided superior yields and anomeric selectivity in nearly all cases.[61] The yields and stereoselectivity also increased with an increase in solvent polarity, suggesting that enhanced stabilization of the oxacarbenium intermediate, the rate-limiting step in glycosylation, was responsible for the increased rate. A similar comparison of the two activators for the activation of thiofructofuranosides provided the same conclusion.[62]

An attempt has been made to repeat Lemieux's halide-ion glycosylation method using thioglycosides. By treating a variety of thioglycosides with a catalytic amount of IDCP, Boons and Stauch[63] observed the degree of anomerization of thioglycosides of 2,3,4,6-tetra-O-benzyl-1-thio-α/β-D-glucose. It was hoped that varying the nature of the glycoside would permit specific glycosylation. No control was noted, and the process was found to be independent of the configuration of the starting thioglycoside; that is, using the opposite anomer gave the same degree of anomerization. As mentioned earlier, the use of bulky thioglycosides gave no anomerization at all. Given that glycosylation of the oxacarbenium intermediate occurs after the thioglycoside has left, the lack of stereocontrol is not surprising given the tightness of the ion-pair responsible for the stereoselectivity in halide exchange approaches to glycosylation is not operational.

The use of iodine to activate thioglycosides or acetylated glycosyl bromides with iodine has been applied to the synthesis of glycoamino acids.[64] Activation of either the ester or thioglycoside with iodine proceeds as already described to permit addition of the hydroxyl group of suitably protected serine. The conditions are mild enough that nitrogen protecting Fmoc groups are unaffected. Owing to the greater nucleophilicity of the hydroxyl group versus a carboxylic acid, the glycosylation occurs regardless of whether the carboxyl group of serine is protected as an ester. In the absence of any base, there is a preference for the thermodynamically preferred α anomer (**26,** Scheme 8). With a mild base—for example, sodium hydrogencarbonate (which is not basic enough to affect the Fmoc group)—the β anomer (**27**) is favored.

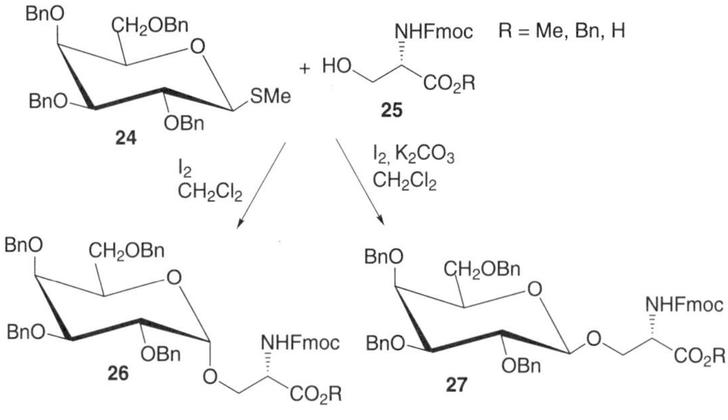

SCHEME 8

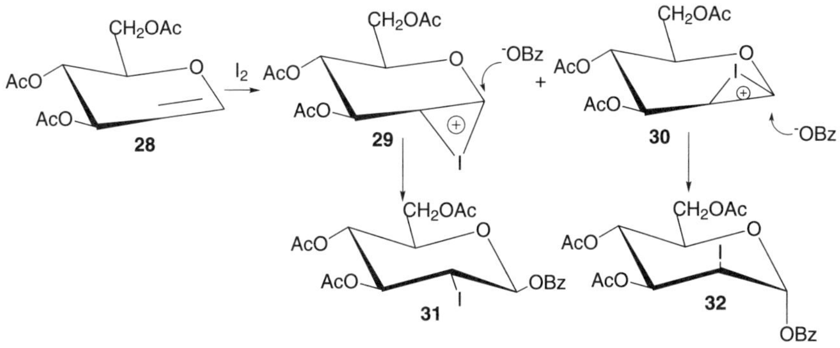

SCHEME 9

III. ACTIVATION OF GLYCALS

Addition of iodine to a glycal was reported by Staněk and Schwarz[65] in 1955, in the formation of 3,4,6-tri-*O*-acetyl-1-*O*-benzoyl-2-deoxy-2-iodo-D-glucopyranose, by treatment of 3,4,6-tri-*O*-acetyl-D-glucal (**28,** Scheme 9) with iodine and silver benzoate in anhydrous benzene, the Prévost reaction.[66] A reexamination of this sequence by Lemieux and Levine[67] resulted in the formation of a crystalline material in a high yield. The product, however, melted over a wide range of 123–148 °C. Characterization, using the then novel technique of ^{1}H NMR spectroscopy, revealed that in fact a 1:1 mixture of 3,4,6-tri-*O*-acetyl-1-*O*-benzoyl-2-deoxy-2-iodo-β-D-glucopyranose (**31**) and 3,4,6-tri-*O*-acetyl-1-*O*-benzoyl-2-deoxy-2-iodo-α-D-mannopyranose (**32**) had been formed.

The nature of iodine addition to glycals was further examined by Lemieux and Fraser-Reid.[68] The commonly accepted first step of this type of reaction is nucleophilic attack of the π electrons of the double bond on the electrophilic halogen, followed by attack of an incoming nucleophile at the carbon atom having the greatest cationic character. The anomeric position is the most reactive, as the adjacent oxygen readily stabilizes the incipient positive charge. It was recognized that although the possibility of further resonance structures for the intermediate cationic species exist for all the halogens (Scheme 10), the greater bulk of iodine[69]

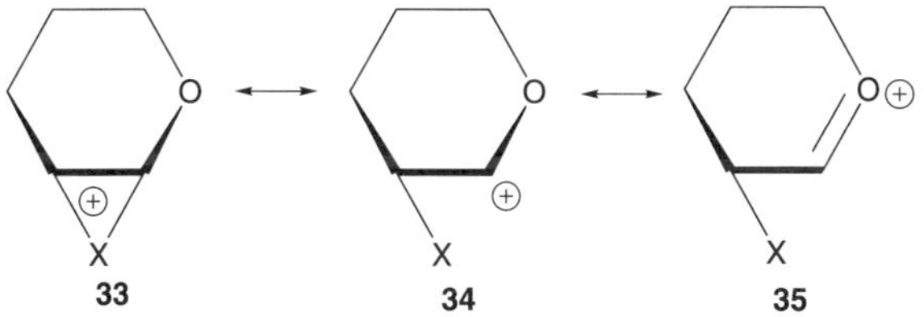

SCHEME 10

results in a preponderance of **33**, while for bromine and chlorine **34** and **35** are both possible. Conversely, the analogous halogenation with the much smaller chlorine results in chlorination at both C-1 and C-2 to afford a 4:1 ratio of products having the α-D-*gluco* and α-D-*manno* configuration.[70]

The formation of glycosidic bonds through iodine-promoted glycal addition was extended to the formation of methyl, cyclohexyl, and *tert*-butyl glycosides.[71] Formation of the respective glycosides was accomplished by treating 3,4,6-tri-*O*-acetyl-D-glucal with stoichiometric amounts of silver perchlorate, *sym*-collidine, and iodine, together with the corresponding alcohol in benzene. It was noted that the proportion of the product having the α-D-*manno* configuration increased with the steric bulk of the aglycone—this was rationalized in that formation of the intermediate iodonium species was reversible (as has been clearly demonstrated for bromination[72]), and that the diequatorial ring-opening of the 1,2-α-D-*gluco* iodonium ion was less favorable than diaxial opening of the 1,2-α-D-*manno* iodonium ion, consistent with the Fürst–Plattner rule.[73] This route is useful in the preparation of 2-deoxy sugars.

A disadvantage of silver perchlorate is its low solubility in nonpolar solvents. The introduction of iodonium dicollidine perchlorate (IDCP) by Lemieux and Morgan[74] in 1965 circumvented this difficulty. The steric bulk of the methyl groups ortho to the nitrogen of the pyridine ring were found to be essential for successful O-glycoside synthesis, less hindered pyridinium or picolinium salts resulted in the formation of the corresponding *N*-glycosylpyridines. The use of IDCP as a source of electrophilic iodine was successful in the formation of glycosides by promoting reaction between the glycal and simple alcohols.

With the success of the IDCP method in the formation of simple glycosides, the next logical step was to attempt the synthesis of a disaccharide. Beginning with 3,4,6-tri-*O*-acetyl-D-glucal, treatment with IDCP in chloroform together with an equimolar amount of 2,3,4,6-tetra-*O*-acetyl-β-D-glucose (**36**, Scheme 11) as the glycosyl donor, resulted in the formation of 2,3,4,6-tetra-*O*-acetyl-β-D-glucosyl-3,4,6-tri-*O*-acetyl-2-deoxy-2-iodo-α-D-mannopyranoside (**37**) in moderate yield. Catalytic hydrogenation of **37** provided the expected 2,3,4,6-tetra-*O*-acetyl-β-D-glucosyl-3,4,6-tri-*O*-acetyl-2-deoxy-α-D-*arabino*-hexopyranoside (**38**).

An analogous route to the formation of α-D-mannosides starting with 3,4,6-tri-*O*-acetyl-D-glucal using *N*-iodosuccinimide (NIS) as a source of electrophilic iodine has been reported by Thiem and co-workers.[75] Reaction of 3,4,6-tri-*O*-acetyl-D-glucal with cyclohexanol and NIS in acetone gave mainly cyclohexyl 3,4,6-tri-*O*-acetyl-2-deoxy-2-iodo-α-D-mannopyranoside together with a small amount of the corresponding β-D-glucopyranoside.

In a study of NIS-promoted additions to glycals it was noted that when the nucleophilicity of the alcohol was low, a competing reaction occurred whereby the succinimide anion itself acted as a nucleophile, attacking the anomeric carbon.[76,77]

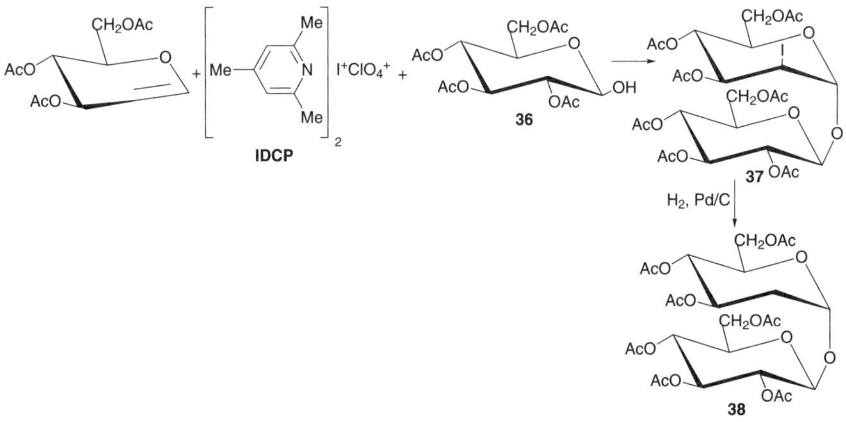

SCHEME 11

Furthermore, in the absence of competing nucleophiles and light stereoselective addition of the succinimide, in the α-D-*manno* configuration, was observed. This has been exploited in the synthesis of biologically significant 2-deoxy-α-*N*-glycopeptides.[78] Performing the NIS reaction on the 3,4,6-tri-*O*-acetyl-D-glucal followed by removal of the halogen at C-2 afforded **40** (Scheme 12). Nucleophilic ring-opening of the succinimide forms the open chain **41**. The resultant ester was then easily saponified to the acid. Formation of the amide bond using EEDQ to give **42** was readily accomplished.

Thiem and Ossowski[79] considered that the conformation of the starting glycal was critical to the configuration of the product obtained. It was reasoned that **45**

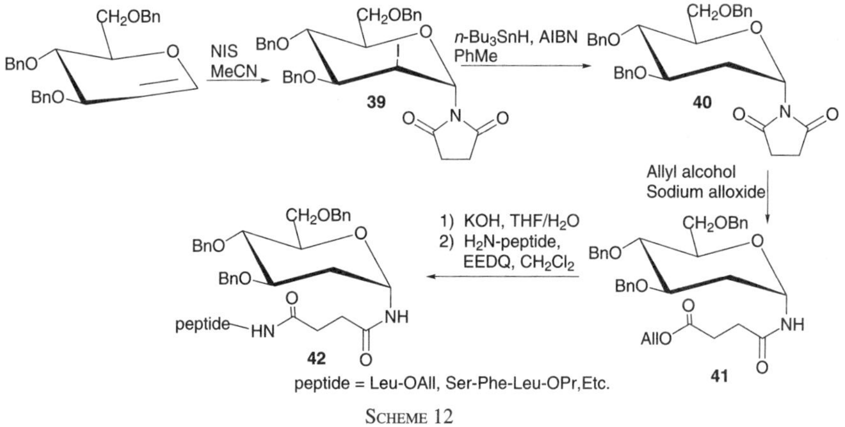

SCHEME 12

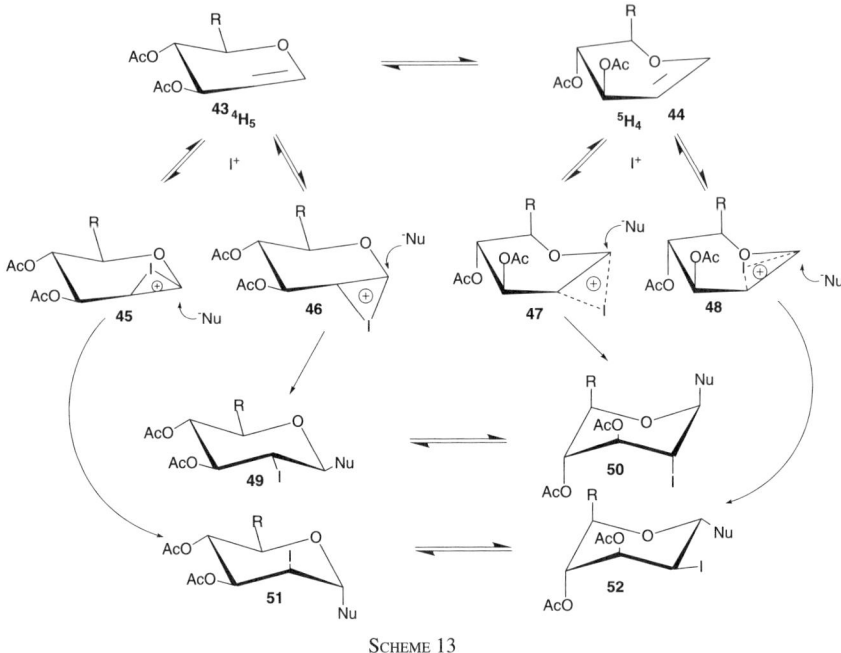

SCHEME 13

and **48** (Scheme 13) should be the more probable, as a result of their cationic substituents giving rise to a reverse anomeric effect (RAE).[80] Arguments based on the RAE should be regarded with some caution, as the very existence of the RAE is somewhat controversial,[81,82] and indeed has been postulated to not exist at all for hexose sugars.[83–85] Nucleophilic "ring-opening" should then proceed to give the sterically favored product, having both substituents in *trans*-diaxial orientations—in agreement with the Fürst–Plattner rule. An increased ratio of the $^5H_4$ half-chair conformation for the glycal led to an increase in the proportion of the β anomer obtained, evidenced by reactions of glycals of glucuronic acid methyl ester derivatives, which are more likely to exist in this conformation, up to 40% of the β anomer was obtained in this manner.

The mechanism of glycal addition was further examined by Horton and co-workers,[86] who have used this method to form (7S,9S)-4-demethoxy-7-O-(2,6-dideoxy-2-iodo-α-L-mannopyranosyl)-adriamycinone[87] and -daunomycinone,[88] which are iodinated analogs of natural antitumor compounds. The nature of the solvent was found to be critical. In particular, they were able to generalize that in non-coordinating polar solvents, where no possibility for interaction with iodine exists—that is, no lone-pair interactions—the formation of the iodonium intermediate was irreversible, and the resultant stereochemistry reflects the electronic stability of the product. In coordinating solvents such as THF or MeOH, where the aforementioned interactions with the iodonium species are possible, the iodination step was considered to be reversible and, thus, controlled by steric factors to afford a preponderance of *trans*-diaxial products. Similar explanations have been proposed for the bromination of cyclohexene.[89,90]

After many years of moderate use, the technique of activating glycals with electrophilic iodine has undergone somewhat of a recrudescence.[91] By exploiting the ease of activation of glycals, as pioneered by Lemieux,[67] Friesen and Danishefsky[92] reported the application of the iodine–glycal method to the iterative synthesis of oligosaccharides. Beginning with the introduction of a glycal as an aglycone to form a disaccharide (**53,** Scheme 14), activation of the glycal is

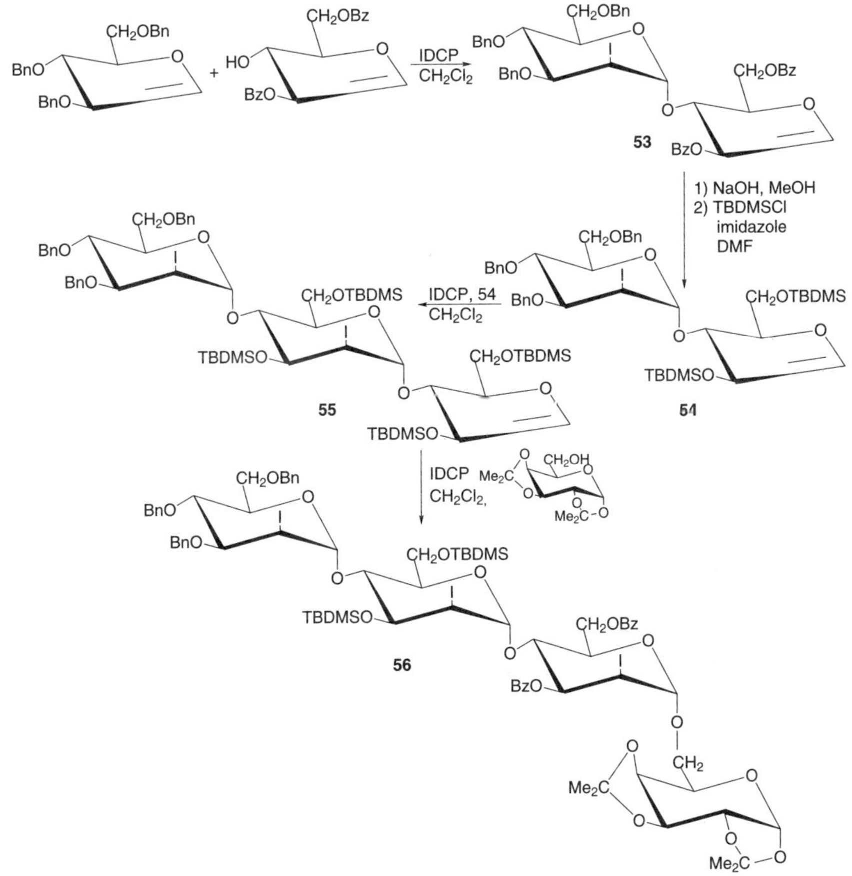

SCHEME 14

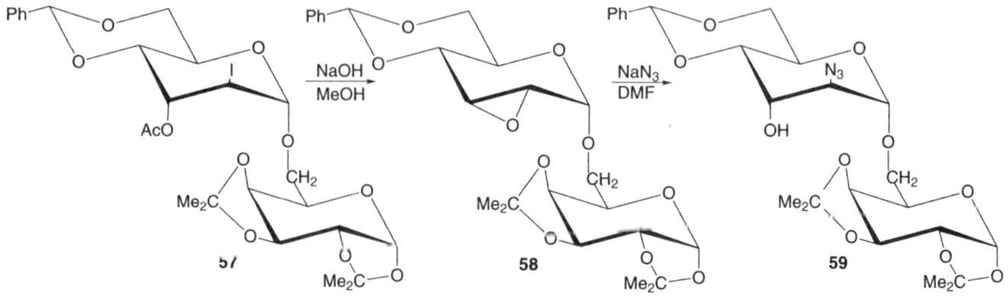

SCHEME 15

accomplished with IDCP, followed by introduction of the next unit. By continually adding and then activating suitably protected glycals, large oligo-saccharides are potentially accessible. The main limitation of this approach is the strict demand imposed on the stereochemical outcome of the reaction, which affords sugars in preponderantly the α-D-*manno* configuration.

One of the difficulties of the NIS-activation method is the limitation placed on the moiety bearing iodine at C-2—nucleophilic displacement at C-2 of pyranose rings is notoriously difficult for stereoelectronic reasons.[93] For the most part the only further elaboration of 2-deoxy-2-iodo sugars have been reduction to the 2-deoxy sugar, either through catalytic hydrogenation or via tinhydride reduction. An interesting approach that allows for a wider range of substituents has been the generation of the 2,3-anhydro sugar, followed by nucleophilic substitution of the reactive strained three-membered ring. Opening of the oxirane again tends to favor a *trans*-diaxial product. Friesen and Danishefsky[94] have taken 1,2:3,4-di-*O*-isopropylidene-6-*O*-(3-*O*-acetyl-4,6-*O*-benzylidene-2-deoxy-2-iodo-α-D-altropyranosyl)-α-D-galactopyranose (**57**, Scheme 15), prepared using the NIS-mediated glycal addition method, and formed the epoxide **58** simply by exposure to methoxide anion to remove the C-3 acetyl protecting group, followed by cyclization. Treatment with the strongly nucleophilic azide afforded the 2-azido sugar, in the D-*altro* configuration (**59**), which is amenable to further modification. This method suffers from the limitation of requiring an axially oriented hydroxyl at C-3.

Carbohydrates bearing amino groups at C-2 have been prepared by glycal addition, using benzenesulfonamide as the nucleophile.[95–97] This method makes use of an intramolecular migration of sulfonamide from the anomeric position to C-2 upon treatment with a nucleophile. Beginning with tri-*O*-benzyl-D-glucal, treatment with NIS and benzenesulfonamide in CH_2Cl_2 afforded the expected product **60** (Scheme 16). Further treatment with sodium azide resulted in displacement at the anomeric position, with concomitant migration of the sulfonamide to displace the iodine at C-2 in an intramolecular fashion. Removal of the sulfonamide protecting group, to afford the free amine at C-2, in the β-D-*gluco* configuration, is

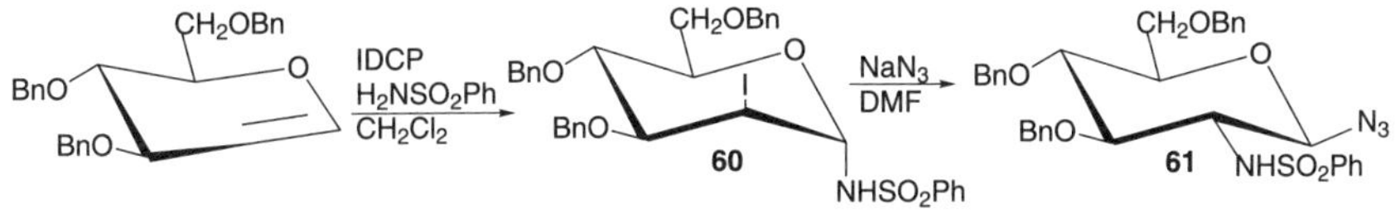

SCHEME 16

readily achieved using a photolytic electron-transfer desulfonylation.[98] The use of this technique has been applied successfully to the synthesis of such complex and biologically significant molecules as sialyl Lewis X.[99]

Whereas solid-phase organic synthesis is now commonplace,[100] effective solid-phase oligosaccharide synthesis remains a challenge. Although work in this area has been underway for some time,[101,102] as yet it has not become a universal technique has as the solid-phase synthesis of peptides or oligonucleotides. The ability to add glycosyl units in an iterative fashion, using glycals has been examined in conjunction with a solid support, using either an oxygen electrophile as activator[103] or NIS.[104] One promising method entails attaching the glycal to a solid support via a silyl linker; the formation of glycosidic bonds is then effected by repetitive activation and addition of suitably protected glycals. Once the desired number of saccharide units has been installed, the oligosaccharide is cleaved from its support by treatment with fluoride, a standard method for the cleavage of silyl ethers.[105]

The Ferrier reaction[106] has long been a means of forming a glycosidic bond, which entails activating a glycal with a Lewis acid to form a 2,3-unsaturated glycoside. The use of iodine as a Lewis acid in THF has been reported to afford the glycoside, with either complete selectivity for the α anomer[107] or a large preponderance of the α anomer.[108] Using 4,6-O-isopropylidene-3-O-n-pentenoyl-D-glucal (**62,** Scheme 17), the pentenoyl ester, upon activation by IDCP, was converted into

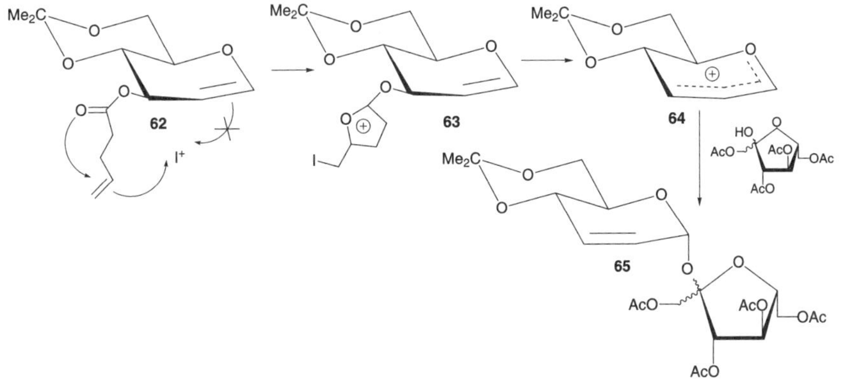

SCHEME 17

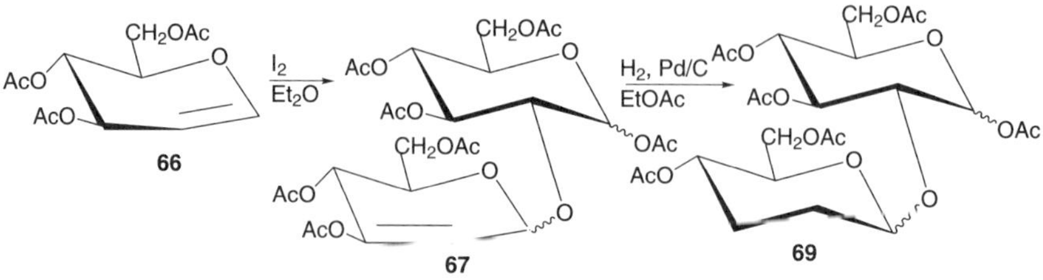

SCHEME 18

a leaving group after cyclization (**63**). That no effect on the unsaturation between C-1 and C-2 is noted is of interest, as terminal alkenes are notoriously unreactive, and implies participation of the carbonyl oxygen. Addition of 1,3,4,6-tetra-*O*-acetyl-D-fructofuranose allowed for the formation of 1′,3′,4′,6′-tetra-*O*-acetyl-α- and β-D-fructofuranosyl-2,3-dideoxy-4,6-*O*-isopropylidene-α-D-*erythro*-hex-2-enopyranoside **65**,[109] an intermediate that had been used 13 years prior in the synthesis of sucrose by Fraser-Reid and Iley.[110]

A report on attempts to add nitryl iodide to a glycal yielded an unexpected outcome. Reaction of 3,4,6-tri-*O*-acetyl-D-glucal with INO_2 afforded a complex mixture of products, none of which, upon treatment to bring about elimination, showed any sign of a nitro group.[111] Performing the reaction in the absence of silver nitrite afforded the same mixture of products. Careful processing of this complex mixture revealed a low yield of a dimer (**69**, Scheme 18); the same result was obtained by Ferrier and Prasad[112] by treating the glucal with $BF_3 \cdot Et_2O$. Yields of up to 30% of this dimer, of which a preponderance of the β anomer was formed, could be achieved simply by employing iodine as the Lewis acid. Examination of the NMR spectrum after hydrogenation permitted an accurate determination of a 5:2 ratio of anomers in favor of the β form.

IV. Pseudohalogens

The addition of a halogen across a double bond is one of the best known reactions in organic chemistry. The remarkable ability of the diatomic iodine molecule—the ease of polarization, the stability imparted to the halide anion by a full octet of electrons, that allows it to act both as an electrophile and a nucleophile in the same reaction, is unparalleled. The sequence of events leading to the addition of iodine to an alkene are simple, nucleophilic π electrons of the double bond attack one atom of the halogen dimer, donating an electron to the other, thereby stabilizing it. The newly created anion now attacks the highly electron-deficient halonium species. In the presence of a nucleophile more powerful that the halogen anion, anisotropic addition across the double bond occurs. The class of compounds known as *pseudohalogens* behaves in this manner. Pseudohalogens

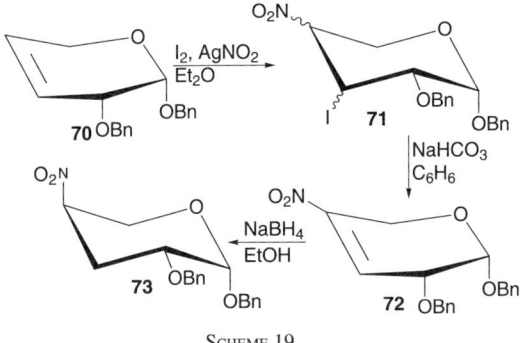

SCHEME 19

consist of a halogen as an electrophilic component bound to another molecule, the nucleophilic component. Pseudohalogens have been known for some time, iodine azide was reported by Hantsch in 1900,[113] and nitryl iodide was reported in 1932.[114]

The use of the pseudohalogen nitryl iodide, prepared *in situ* from iodine and silver nitrite, has been found to add to an alkene in what is strictly an anti-Markownikov fashion. The explanation for this lies in that nitryl iodide adds in a radical manner, initially forming the more stable secondary radical after addition of NO_2.[115] Treatment of 3-*O*-acetyl-5,6-dideoxy-1,2-*O*-isopropylidene-α-D-*xylo*-hex-5-enofuranose with nitryl iodide was found to afford an unstable adduct, with the nitro group appended to C-6, and iodine attached to the more substituted C-5.[116–118] Similarly, treatment of benzyl 2-*O*-benzyl-3,4-dideoxy-α-D-*glycero*-pent-3-enopyranoside (**70**, Scheme 19) with nitryl iodide afforded the unstable adduct **71**, which, upon exposure to mild base ($NaHCO_3$), afforded the eliminated product, namely benzyl 2-*O*-benzyl-3,4-dideoxy-4-nitro-α-D-*glycero*-pent-3-enopyranoside (**72**). The eliminated product was then readily converted into benzyl 2-*O*-benzyl-3,4-dideoxy-β-L-*threo*-pentopyranoside (**73**) by reduction with sodium borohydride. Addition of deuteride using $NaBD_4$ led to axial deuteration at C-3.

The pseudohalogen iodine trifluoroacetate has similarly been exploited in carbohydrate synthesis. Reaction of the unsaturated sugar 5,6-dideoxy-1,2-*O*-isopropylidene-α-D-*xylo*-hex-5-enofuranose (**74**, Scheme 20) with iodine trifluoroacetate afforded roughly a 1:1 mixture of **76** and **77**.[119] The extreme lability of the trifluoroacetate ester[120] resulted in its cleavage prior to isolation; in the case of **76**, this resulted in cyclization, and for **77** simply deprotection. Performing the reaction with the hydroxyl group at C-3 blocked to avoid the possibility of cyclization produced only **77**. The configuration of the iodine on C-5 of **77** could not be determined owing to its instability. (The compound decomposed to a reddish-brown syrup even on standing in the dark at 5 °C.) Hydrogenation to the stable 5-deoxy

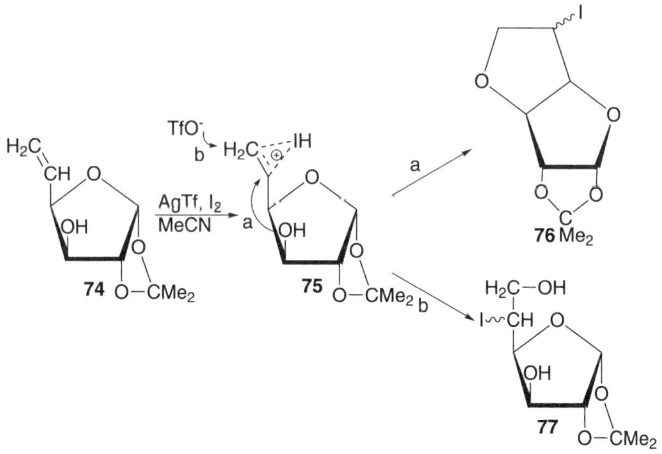

SCHEME 20

sugar permitted the unequivocal determination that iodination had indeed occurred at C-5. Iodine azide, prepared in a similar manner, also adds to a double bond, save that an azido group is appended at the less substituted position.[121,122] Iodine azide is particularly useful in the creation of amino sugars.

Treatment of an epoxide, namely methyl 2,3-anhydro-4,6-O-benzylidene-α-D-allopyranoside (**78,** Scheme 21), with nitryl iodide using methanol as the solvent has been reported.[123] Rather than the expected nitration, opening of the epoxide occurred to afford 30% of methyl 4,6-O-benzylidene-3-O-methyl-α-D-glucopyranoside (**79**), 10% of methyl 4,6-O-benzylidene-2-O-methyl-α-D-altropyranoside (**80**), and a water-soluble material. The water-soluble material was later determined to be a mixture of the 2-O- (**81**) and 3-O-methyl (**82**) products, with the benzylidene acetal removed, a result consistent with a later study.[124] Performing the reaction in the absence of the silver salt resulted solely in formation of the water-soluble product. The preponderance of the diequatorial product **81** is in violation of the Fürst–Plattner rule, which predicts more of the diaxial product. The authors explanation for the anti-Fürst–Plattner addition is as follows.

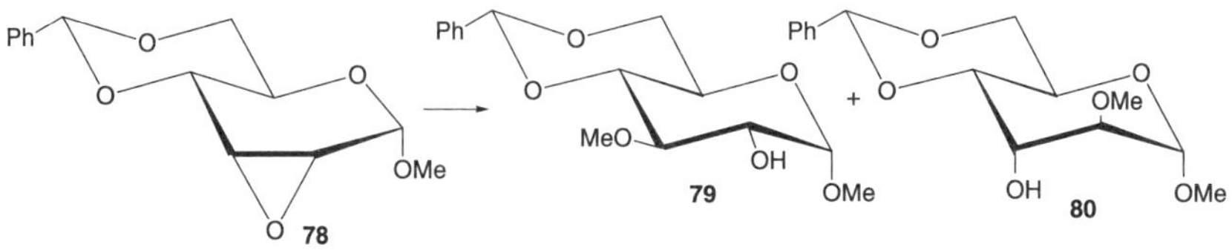

SCHEME 21

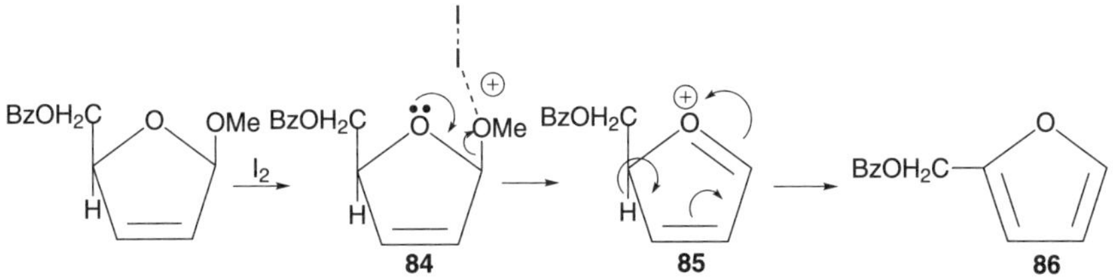

SCHEME 22

Ring opening occurs in the presence of iodine in an S$_N$1 process, the oxirane oxygen being initially complexed with the iodine. The two oxygen atoms bound to C-1 serve to destabilize positive charge formation at C-2, rendering nucleophilic attack to C-3 more favorable. This accounts for the diequatorial product observed.

The use of iodine-containing pseudohalogens has been demonstrated to cause a variety of side reactions. Treatment of methyl 5-O-benzoyl-2,3-dideoxy-β-D-glycero-pent-2-enofuranoside (**83**, Scheme 22) with nitryl iodide afforded none of the desired addition product, but gave a quantitative yield of furfuryl benzoate (**86**). The same result was obtained by treating **83** with iodine nitrate, or with iodine alone.[125] The rationale for this undesired result lies in the ability of iodine to act as a Lewis acid, polarizing the glycosidic bond, generating an oxacarbenium cation, which then decomposes to the stable aromatic **86**.

Iodine azide has been used to form 1-azido-2-deoxy sugars directly from glycals. Further reaction with trimethyl phosphite afforded the corresponding 2-deoxyiodoglycosyl phosphoramidates.[126] An interesting report on the action of iodine azide on a 2,3-unsaturated sugar has appeared.[127] Rather than just introduction of the iodine and azido groups, the groups were added, but the unsaturation remained. An explanation for this was that the azide of **90** (Scheme 23) was

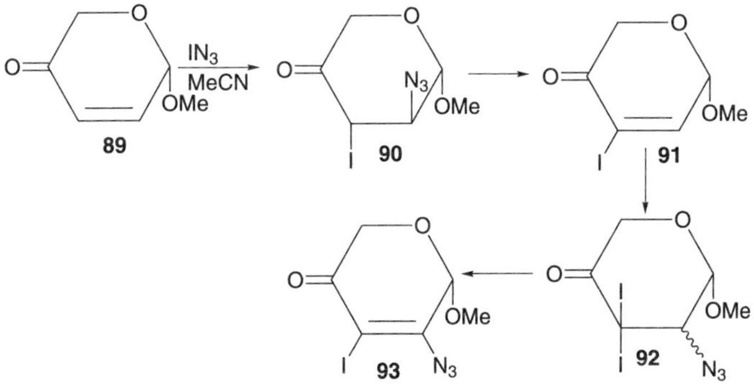

SCHEME 23

eliminated after the first addition to afford **91,** another equivalent of IN_3 added across the newly formed double bonds, followed by elimination of one of the two iodines to afford **93.**

V. HYPERVALENT IODINE

Iodine, found in Nature as I^I, has found wide-ranging applications as salts of hypervalent I^{III} and I^V. Of late, the use of hypervalent iodine has become common in organic synthesis and in carbohydrate synthesis.[128,129] An excellent review devoted solely to applications of hypervalent iodine reagents in carbohydrate chemistry has appeared.[130] Hypervalent iodine reagents have been termed *nonorganometallic reagents* for the manner in which the central iodine acts as a powerful electrophile permitting both ligand exchange and reductive elimination, in simile with organometallic reagents.

Direct oxidation of fully protected glycals, without deprotection, has been reported by Kirschning and co-workers.[131] Using [hydroxy(tosyloxy)iodo]benzene, the Koser reagent,[132] oxidation of either 3,4,6-tri-*O*-acetyl-D-glucal (**94,** Scheme 24) or 3,4,6-tri-*O*-acetyl-D-allal (**95**) afforded the same 3-keto product, **96,** in yields ranging from 32 to 72%. Oxidation of the protected allylic hydroxyl proceeds with either acetic ester, benzyl, or *tert*-butyldimethylsilyl (TBDMS) ethereal protecting groups.[133] Attempting the same reaction with other hypervalent iodine compounds—specifically, $PhI(OCH_3)OTs$, $PhI(OCOCH_3)_2$, or $PhI(OCOCF_3)_2$— did not result in oxidation.

Further investigation of the foregoing reaction involving the Koser reagent, this time with 3,4-di-*O*-*tert*-butyldimethylsilyl-6-*O*-tosyl-D-glucal (**97,** Scheme 25), afforded some insight into the mechanism together with an explanation for the disappointing yields noted with some substrates. Treatment of **97** with a two-fold excess of the Koser reagent resulted in the formation of three separate products.[134] The common first step involves attack of the alkene on the hypervalent iodine to generate the oxonium ion **98.** At this point, the course of reaction may follow one of two routes: elimination of the iodine species followed by cleavage of the Si-O bond to afford **100,** or attack of a hydroxyl group (from the Koser reagent) to give **101,** which may then decompose through one of two different paths to afford

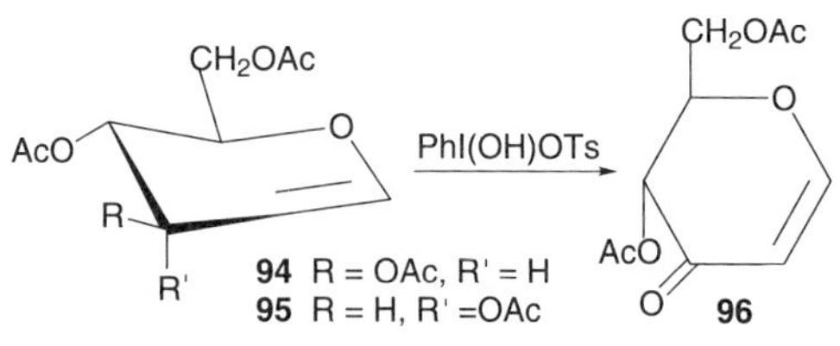

SCHEME 24

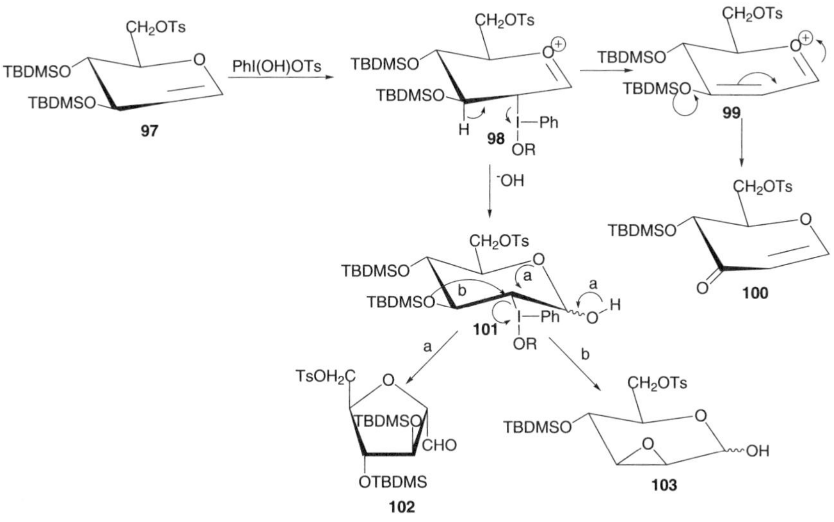

SCHEME 25

either **102** or **103**. In practice, the reaction is found to generate **100, 102,** and **103** in a 1:2:1.1 ratio. The occurrence of these alternative paths was not discussed in conjunction with protecting groups other than TBDMS ethers.

The reaction between 3-deoxyglycals and the hypervalent iodine reagent iododiazidobenzene [PhI(N$_3$)$_2$], generated *in situ* from iodosylbenzene and trimethylsilyl azide,[135] resulted in the addition of an azido group at C-3.[136] Treatment of 4,6-di-*O*-acctyl-3-deoxy-D-glucal (**104**, Scheme 26) with PhI(N$_3$)$_2$ resulted in a 1:8 mixture of 4,6-di-*O*-acetyl-3-azido-3-deoxy-D-lyxal (**105**) and 4,6-di-*O*-acetyl-3-azido-3-deoxy-D-xylal (**106**). Similar results were obtained with either a benzoic ester or TBDMS ether-protecting groups in yields ranging from 30 to 50%. Treatment of 4,6-di-*O*-acetyl-3-deoxy-3-*C*-methyl-D-gulal to the same conditions afforded a 1:1 anomeric mixture of 4,6-di-*O*-acetyl-3-deoxy-3-*C*-methyl-D-*threo*-hex-2-enopyranside, albeit in a low yield. The mechanism is similar to the introduction of an azido group to a silyl enol ether.[137,138]

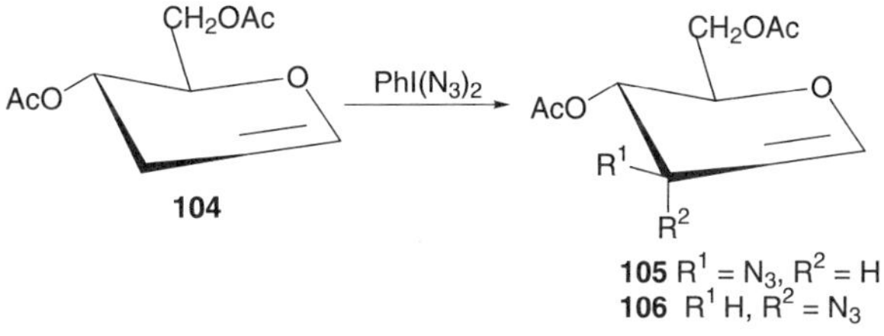

SCHEME 26

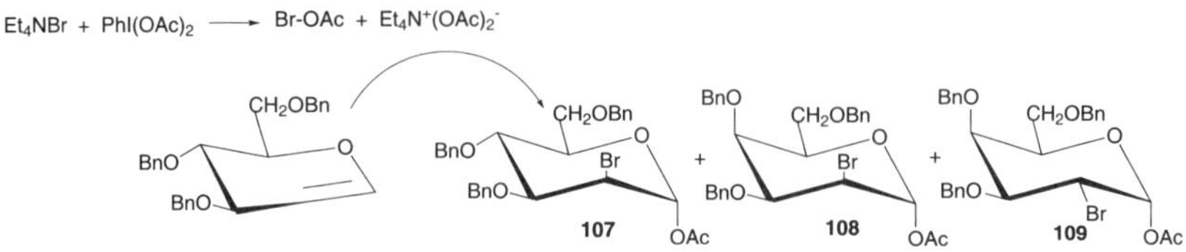

Et₄NBr + PhI(OAc)₂ ⟶ Br-OAc + Et₄N⁺(OAc)₂⁻

SCHEME 27

Treatment of glycals with PhI(OAc)₂ and Et₄NBr was reported to effect acetoxybromination of the glycal unsaturation.[139] The combination of reagents afforded the transient BrOAc species (Scheme 27). In all cases mixtures of products having the α-D-*galacto,* β-D-*talo,* α-D-*gluco,* and β-D-*manno* configurations were obtained. The ratio could be varied slightly depending on the reaction conditions, but in each case the α-D-*galacto* and β-D-*talo* products were found to occur in nearly equal proportions, with lesser amounts of the α-D-*gluco* product, and negligible amounts of the β-D-*manno* product. Treatment of silylated glycals with the hypervalent iodine reagent PhI(OAc)₂–MeSiN₃ was found to be an effective way in which to concomitantly deblock the C-3 position and to oxidize it to the enone.[140] In the case of 3-deoxy sugar, an azido group was introduced at C-3.

Fukase and co-workers[141–143] reported the activation of thioglycosides using a catalytic amount of a hypervalent iodine compound prepared from iodosobenzene (PhIO) and trifluoromethanesulfonic anhydride (Tf₂O), namely, diphenyliodonium triflate (Ph₂IOTf), either with[141,142] or without[143] NBS as an activator. Similarly, glycosylation has been effected by use of (Ph₂IOTf) in combination with various triflate salts, for example: Tf₂O, trimethylsilyl triflate, Yb(OTf)₃, Sc(OTf)₃, to afford a preponderance of β glucosides. The introduction of perchlorate salts, in conjunction with either a triflate salt or a silver salt, led to a greater amount of the α glucoside being formed. The formation of the β glucosides was accomplished both with and without anchimeric assistance in the form of an intermediate acetoxonium species. In the case of a nonparticipating group at C-2, the reactions were performed in acetonitrile—proceeding through the intermediacy of an α nitrilium species, known to afford β-glycosides.[17,19] The formation of the α anomer occurred in all cases in the absence of a participating group at C-2, and is rationalized in terms of anomeric stabilization.[144] Excellent α selectivity was obtained, using nonparticipating protecting groups at C-2, with bis(trifluoroacetoxy)iodobenzene,[145] although the only compounds reported where simple ethyl glycosides. In all cases, however, the mechanism is the same: activation of the thioglycoside to form a cationic intermediate (**105**, Scheme 28), followed by generation of an oxacarbenium intermediate by expulsion of the activated thioglycoside (**106**), and nucleophilic attack of the new aglycone as nucleophile. Similar activation of thioglycosides have been reported using IDCP, or NIS/TfOH.[146]

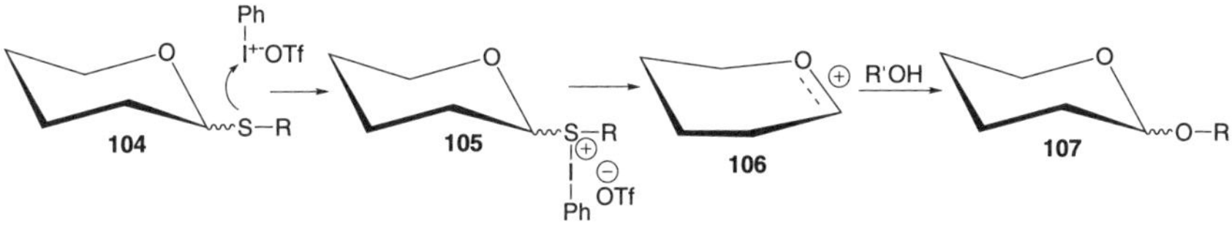

SCHEME 28

The use of the hypervalent iodine reagent diacetoxyiodobenzene (DIB) together with iodine in cyclohexane has been found to be effective in the cleavage of C-1–C-2 bonds of carbohydrates having all hydroxyl groups, save the anomeric hydroxyl group, protected. The course of reaction proceeds through a radical alkoxy species, which decomposes to afford the acyclic product, leaving a formic ester protecting the former ring-oxygen atom. In the case of 2-deoxy sugars, introduction of iodine at C-2 was noted.[147] The reaction proceeded poorly when only a catalytic amount of iodine was employed, and not at all in the absence of iodine. The mechanism is thought to involve formation of the alkoxy radical, followed by β fragmentation and introduction of an acetic ester from the reagent itself. In the case of **108** (Scheme 29), a 1:1 diastereomeric mixture of **110** was obtained. The greater constraint of a cyclic 2,3-acetal afforded a 19:1 mixture of the *erythro* product. In the absence of an electron-withdrawing group at C-2, the formation of an iodo species is obtained, presumably because of the greater reactivity of the carbon radical.

Dorta and co-workers[148] reported the use of organoselenium(IV) compounds in conjunction with iodine, to effect selective anomeric oxidation, or radical β fragmentation followed by recyclization (Scheme 30).[148] Reaction with carbohydrates having unprotected primary hydroxyl groups led to cyclization. Protection of the hydroxyl group at C-5 resulted in formation of the corresponding lactone. Performing the reaction with C-5 protected, but C-3 unprotected, merely resulted in formation of the corresponding lactone. The use of iodosylbenzene in the presence of elemental iodine gave only the fragmentation/cyclization products, with no report of lactone formation.[149,150]

An extension of the foregoing to include acid-catalyzed transacetalation resulted in the formation of furanose sugars. By treating the protected xyloside **115**

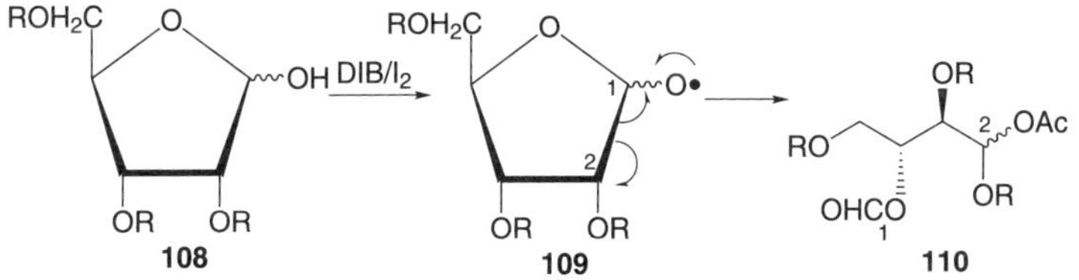

SCHEME 29

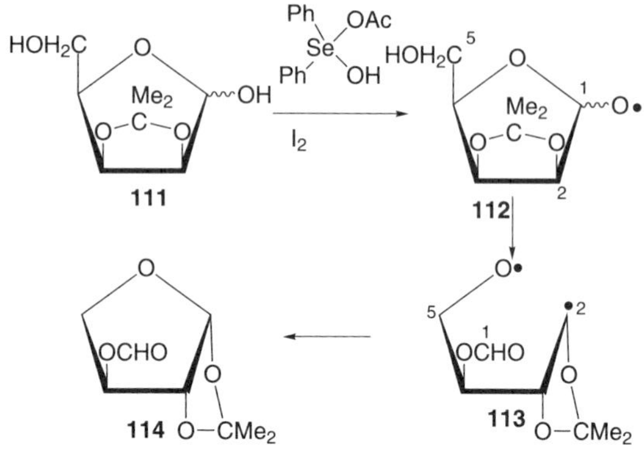

SCHEME 30

(Scheme 31) with DIB/I$_2$, fragmentation to the formic ester **117** occurred. An acetoxy group originating from DIB was appended to the radical at C-1 to give **118**. Treatment of the resulting acyclic compound with acid affords the cyclized products **119** and **120** in high yield.[151] The yields are high, although the reaction times are somewhat long. Anomeric control is nonexistent. The amount of the benzyl glycoside obtained was maximized by limiting the amount of acid in the reaction mixture.

The formation of lactones has also been reported using DIB/I$_2$.[152] By choosing uronic acids as substrates, essentially the same combination of radical fragmentation/cyclization occurs as just described; with the difference that upon cyclization, the carboxylate moiety now closes the ring, forming a lactone. Treatment of uronic and glyculosonic acids with DIB in the presence of iodine was found to generate acyloxy radicals.[153] Subsequent treatment of the mixture with the weakly nucleophilic acetate from DIB results in introduction of an acetoxy group. Using iodosylbenzene, which lacks the competing acetoxy nucleophile of DIB,

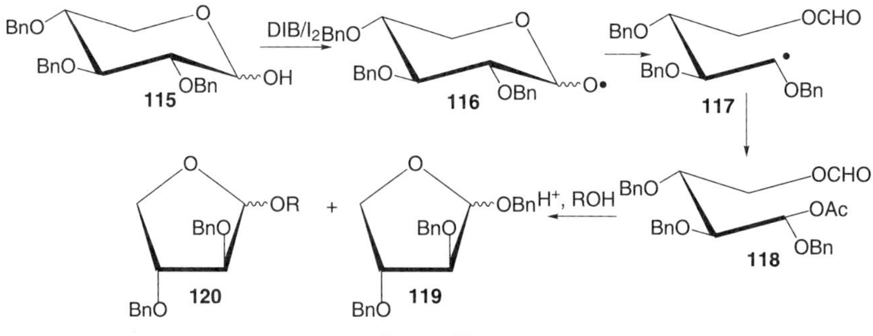

SCHEME 31

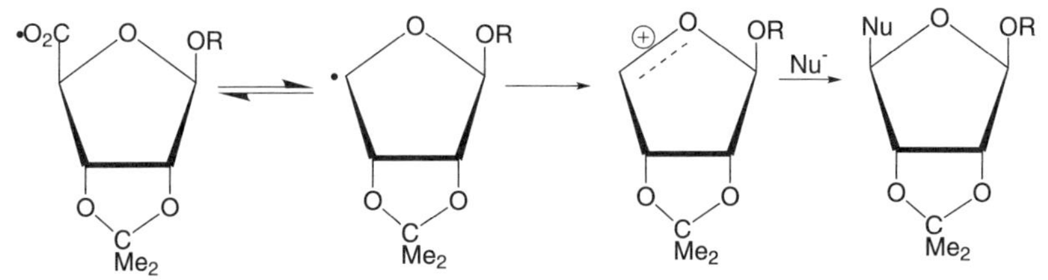

SCHEME 32

introduction of an acetamide from a Ritter-type process can occur with the solvent acetonitrile acting as a nitrogen source (Scheme 32). In the case of pyranoses, no stereoselectivity is observed for the introduction of the nucleophile.

The use of the hypervalent iodine reagent [bis(trifluoroacetoxy)iodo]benzene has been reported to be effective in the synthesis of C-nucleoside-like compounds. Radical decarboxylation of a suitably protected uronic acid, initiated photochemically, followed by addition of a heterocyclic base provided the C-nucleoside in high yield.[154] The mode of action involves initial radical formation of **122** (Scheme 33), followed by introduction of the base and radical coupling.[155] The anomeric selectivity was high in some examples, and low in others—lepidine gave the highest proportion of the β anomer. Isolated yields were poor to moderate.

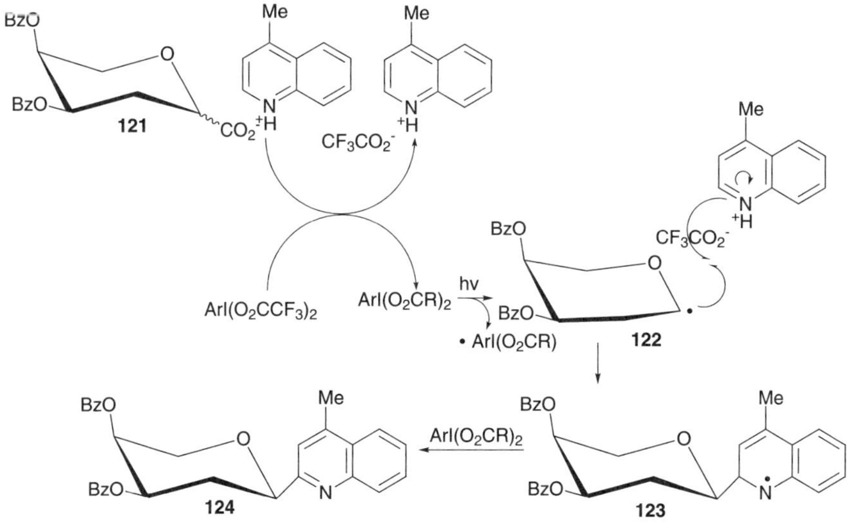

SCHEME 33

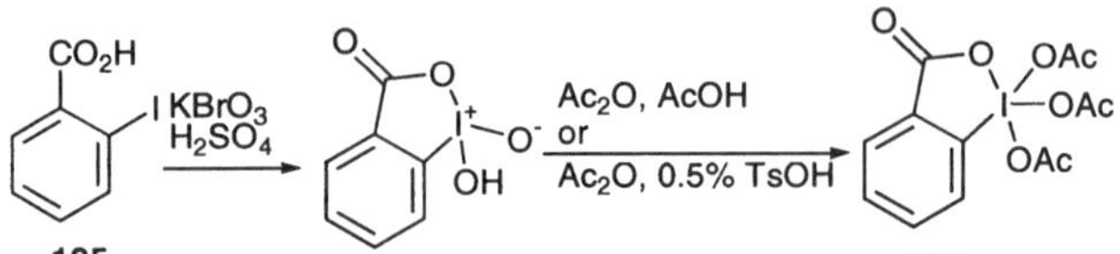

[Scheme 34: compound 125 (2-iodobenzoic acid) → 126 (1-hydroxy-1,2-benziodoxol-3(1H)-one 1-oxide) via KBrO₃, H₂SO₄; → 127 (Dess–Martin reagent) via Ac₂O, AcOH or Ac₂O, 0.5% TsOH]

SCHEME 34

A similar approach has been reported using diacetoxyiodobenzene[156]; isopropylidene protecting groups were used exclusively with DIB, yields were somewhat higher than those reported for the previous method, and retention of chirality was observed with a diastereomeric excess of 95% (Table I).

1. Dess–Martin Reagent

The hypervalent I^{III} compound 1,1,1-triacetoxy-1,1-dihydro-1,2-benziodoxol-3(1H)-one (**127,** Scheme 34), the Dess–Martin reagent, has found widespread application as a mild and effective means for the oxidation of primary or secondary alcohols to the corresponding carbonyl—for primary hydroxyl groups there is no risk of over-oxidation to the carboxylic acid.[157,158] The original synthesis of the agent, in which the cyclic tautomer of 2-iodoxybenzoic acid (**125**), 1-hydroxy-1,2-benodoxl-3(1H)-one-1-oxide (**126**), was acetylated with Ac₂O in AcOH to form **127,** has been supplanted in the literature by a milder, more efficient means of acetylating **126.** By using a trace of TsOH in place of AcOH, the reagent is reported to form the Dess–Martin reagent in better yield, and with a higher degree of purity than the original preparation.[159] Care must be taken in the preparation of the Dess–Martin reagent, as it has been reported to explode upon severe heating or impact.[160]

The observation that an old bottle of the Dess–Martin reagent proved more reliable than freshly prepared reagent prompted further investigation. It was reported that the rate of oxidation is accelerated by the addition of small amounts of water to the reaction, typically perfomed in CH_2Cl_2. Homogeneity of the reaction mixture is critical.[161] This observation conflicts with an earlier report on Dess–Martin oxidation, which claimed that only freshly prepared Dess–Martin reagent was effective in oxidation.[162] The oxidation proceeds cleanly with high yields, for example, in the oxidation of 2′,5′-di-*O-tert*-butyldimethylsilyladenosine to the corresponding 3′-ketonucleoside.[163] The use of the Dess–Martin reagent has been applied to oxidation of protected carbohydrates. Rao and co-workers[164] reported oxidation of the free hydroxyl groups of 1,2:5,6-di-*O*-isopropylidene-α-D-glucofuranose, methyl 4,6-*O*-benzylidene-α-D-glucopyranose, or 1,2:4,5-di-*O*-isopropylidene-β-D-frucopyranose to the corresponding ketones. In each case high yields were obtained.

TABLE I

Cleavage of Carbohydrates by Hypervalent Iodine Reagents

Starting Material	Product	Yield	Ref.
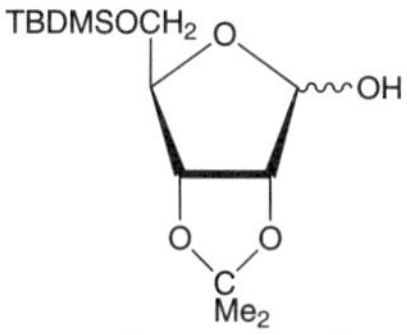	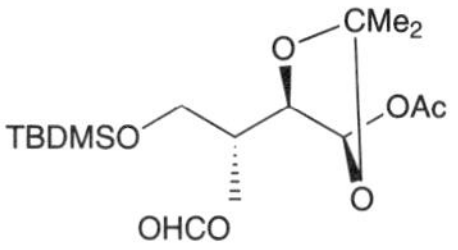	84%	147
	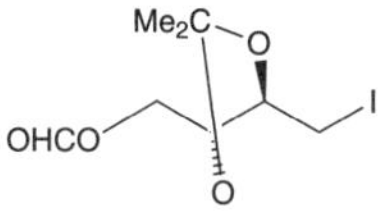	58%	147
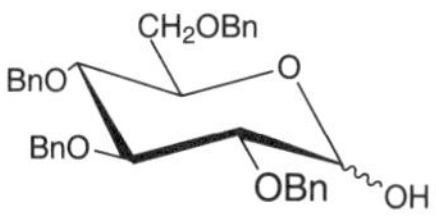	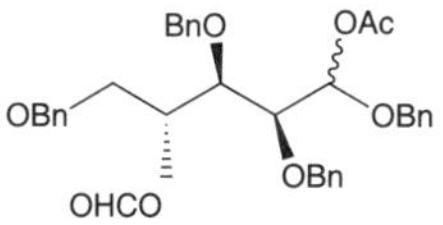	81%	147
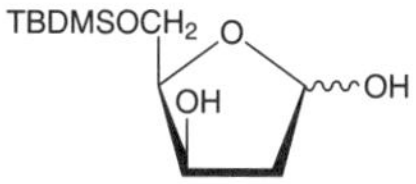	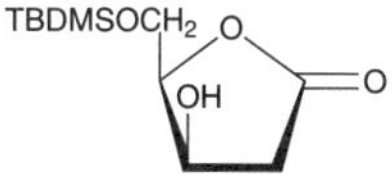	77%	148
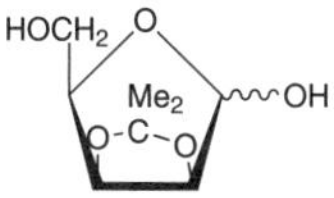	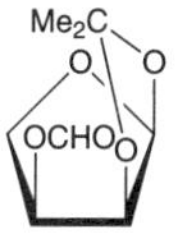	25%	148
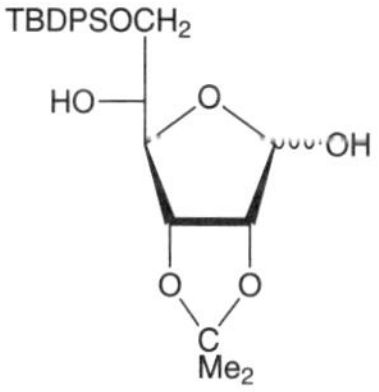	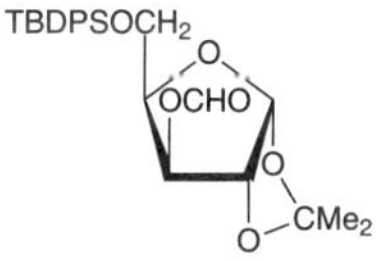	73%	148
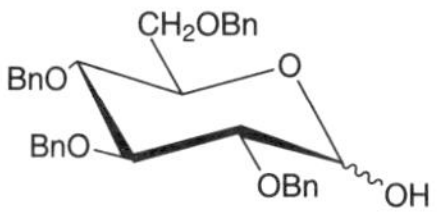	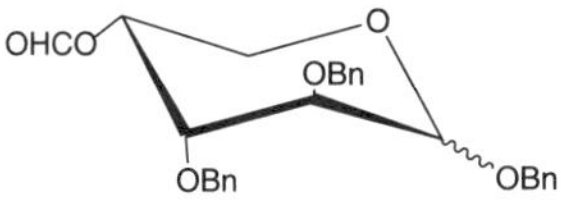	86%	149
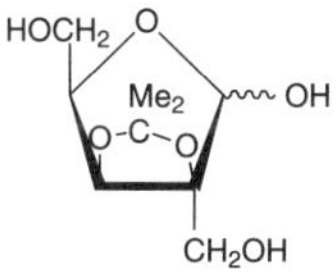	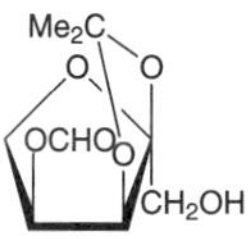	63%	150

VI. Reactions Involving Phosphorus

1. Rydon Reagents

The introduction of the Rydon reagents[165–167] in the 1950s marked the beginning of a long string of phosphorus-based iodinating agents. The Rydon reagents, generated from nucleophilic attack of phosphorus onto a iodine containing electrophile, methyl iodine or iodine itself have both been used, acts to form a highly reactive alkoxyphosphonium salt (Scheme 35). Reaction of a hydroxyl with this highly electrophilic species, followed by displacement with the newly liberated iodide is driven by the highly favorable creation of a stable phosphorus(V) oxide. This route is a particularly attractive technique in carbohydrate synthesis in that the steric bulk of the alkoxyphosphonium species permits a degree of selectivity in the introduction of iodine to less sterically hindered sites. The selective iodination depicted in Scheme 35 demonstrates this selectivity, in which iodination at C-6 was obtained in 60% yield, with no iodination at C-2 reported.[168]

The applicability of the Rydon reagent in carbohydrate chemistry, and in particular nucleoside chemistry, was quickly recognized. Treatment of 2′,3′-O-isopropylideneuridine with the Rydon reagent resulted in a 77% yield of the product having a 5′-iodo group.[169] Reaction with 5′-O-acetyluridine, however, failed to afford any iodinated species. Rather, a mixture of methyl phosphonates was obtained. A possible rationalization for this lies in the adjacent hydroxyl group reacting with the cationic phosphorus center, leading to expulsion of phenol. Reaction with 5′-O-p-nitrobenzoylthymidine (**134**, Scheme 36), lacking a hydroxyl group at C-2′, gave the desired iodinated product, **137,** with net retention of configuration at C-3′ due to the intermediate formation of an anhydro bridge from C-2′ to C-3′ (**136**).

Further applications of the reagent with nuclesosides have been examined.[170] An interesting observation in the iodination of 5′-O-tritylthymidine was that a small amount of the 3′,5′-dideoxy-3′,5′-diiodo species was obtained in addition to the expected proponderance of the C-3′ iodinated product. Presumably, HI produced during the course of the reaction effected the removal of the acid-sensitive trityl ether.

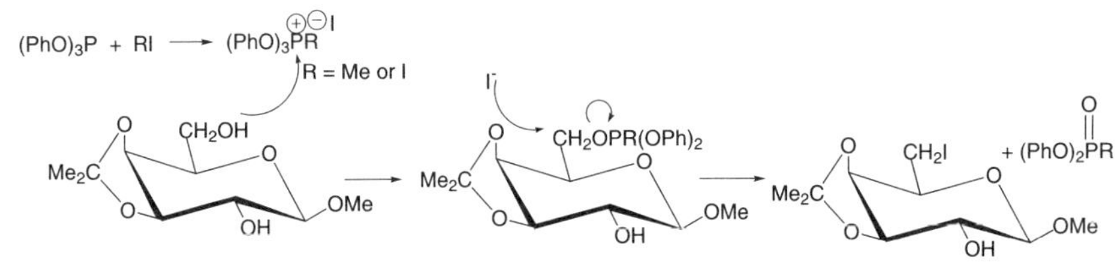

Scheme 35

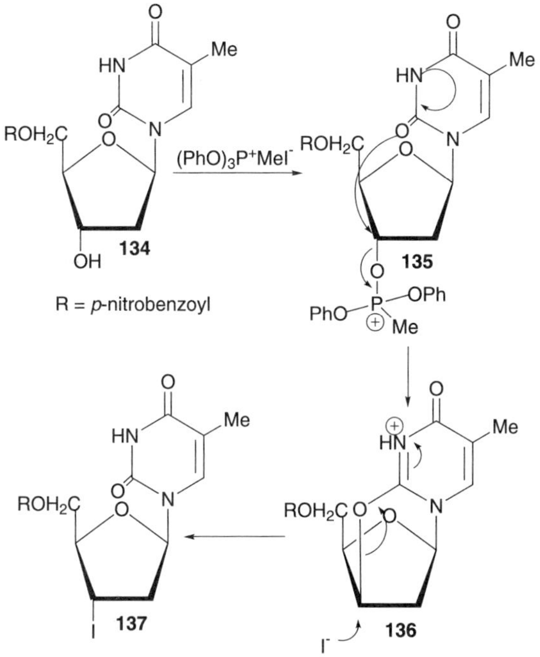

SCHEME 36

2. The Garegg–Samuelsson Reagent

The Garegg–Samuelsson reagent (GSR)—an ensemble of iodine, triphenylphosphine, and imidazole—is an extremely effective means of converting a hydroxyl group to the corresponding alkyl iodide, and constitutes a second generation of the Rydon reagent. This method was initially applied to the formation of alkenes from vicinal *trans*-diols.[171] An improvement, wherein 2,4,5-triiodoimidazole (prepared by treatment of imidazole with an excess of iodine[172]) was used in conjunction with imidazole, was found to afford greater yields of the alkene in some cases.[173]

The course of iodination with the GSR is thought to proceed through formation of an initial complex with the phosphine (**139,** Scheme 37) followed by addition

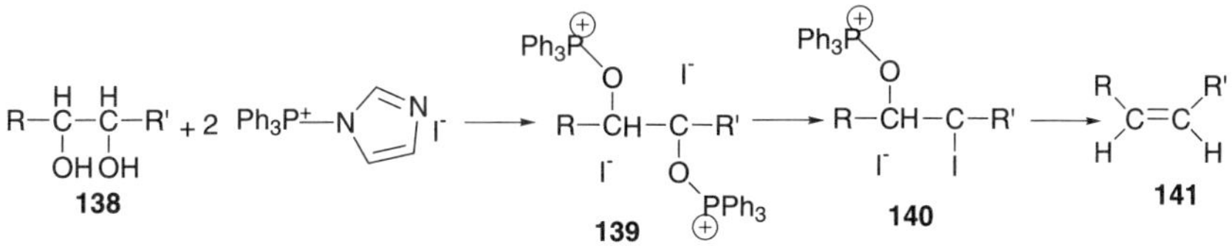

SCHEME 37

of iodine in a manner reminiscent of an Arbusow reaction[174] to generate **140**. Elimination of a second triphenylphosphine oxide, facilitated by the high stability of the phosphorus–oxygen double bond, follows to afford the alkene **141**. Imidazole likely forms a complex with both the triphenylphosphine and iodine, in addition to acting as a base. In the case of vicinal *cis*-diols, the system forms alkenes only if tetrabutylammonium iodide and a large excess of potassium iodide are added, presumably generating the *trans*-iodinated intermediate *in situ*. This method has found widespread applications in carbohydrate synthesis; for example, in the syntheses of gentamine C_{1a},[175] or (-)-frontalin[176] from D-glucose. The use of GSR is effective in the creation of precursors to $2',3'$-dideoxynucleosides, currently of much interest as anti-HIV agents. The formation of an alkene from the vicinal diol—for example, in the synthesis of d4T[177]—leads to an alkene that is readily hydrogenated to the corresponding $2',3'$-dideoxynucleoside.[178–180] Some examples of the use of the GSR are presented in Table II.

The initial use of the GSR focused solely on its application to alkene formation. When applied to carbohydrates, protected at all positions save for one, replacement of the free hydroxyl by iodine with inversion of configuration was observed.[181,182] The reaction itself is heterogeneous; the carbohydrate–phosphorus adduct is very polar, insoluble in the usual reaction medium, toluene. The iodinated product, in general, is less polar and dissolves in the reaction medium. This method has proved successful in effecting iodination at C-2 of various protected pyranose sugars, a reaction that is normally difficult.[93] A rationale for the ease of this substitution at C-2 lies in the positive character of the intermediate, which serves to obviate the unfavorable dipolar interactions responsible for the generally poor reactivity observed for negatively charged nucleophiles toward substitution at C-2 of pyranose rings.

Treatment of methyl α-D-glucopyranoside with GSR in toluene did afford some of the expected methyl 6-deoxy-6-iodo-α-D-glucopyranoside,[183] but was accompanied by formation of methyl 3,6-anhydro-α-D-galactopyranoside. By using a more-polar solvent system, namely, 2:1 toluene:acetonitrile—increasing the solubility of the intermediate, the displacement proceeded to give the desired methyl 6-deoxy-6-iodo-α-D-glucopyranoside in good yield. The steric bulk of the triphenylphosphine is responsible for the observed chemoselectivity. Using a more-polar solvent also allowed for a shorter reaction time and at a lower temperature than in toluene alone. An analogous method has proved successful for the displacement of a hydroxyl group with a bromine group.[184] Treatment of methyl α-D-glucopyranoside with just a two-fold excess of the GSR, followed by further treatment with a four-fold excess of the reagent, resulted in iodination at C-6 and C-4 to form methyl 4,6-dideoxy-4,6-diiodo-α-D-galactopyranoside. The selectivity observed was similar to that seen with chlorination achieved through the use of sulfuryl chloride in pyridine,[185] and may be similarly rationalized on the basis of unfavorable dipolar interactions in the developing transition state.

TABLE II
Applications of the Garegg–Samuelsson Reagent

Starting Material	Product	Yield	Ref.
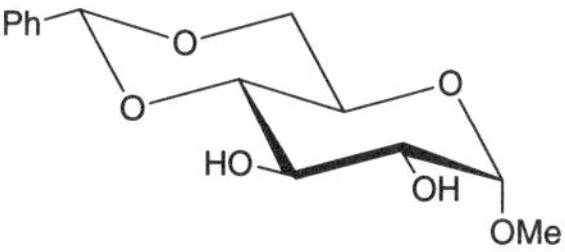	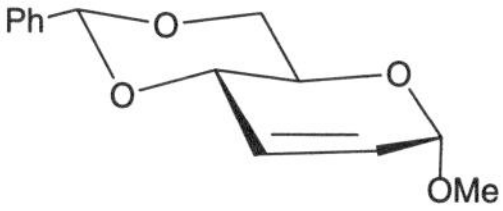	74%	171
	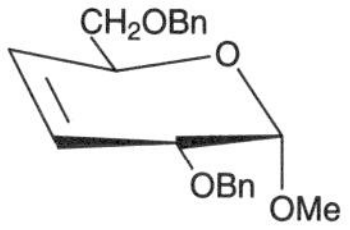	95%	147
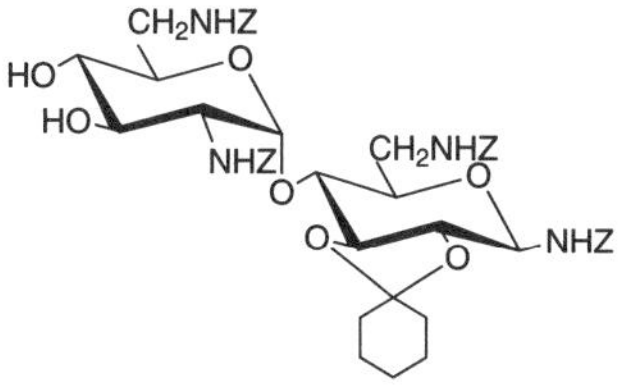	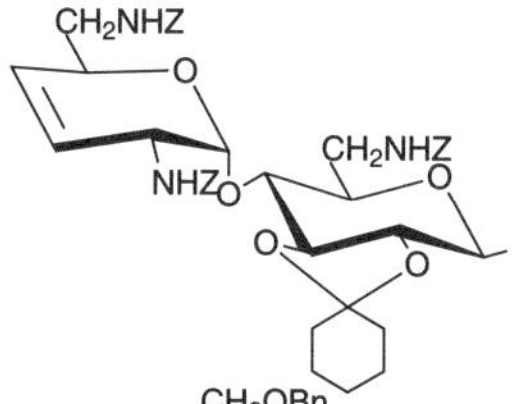	42%	175
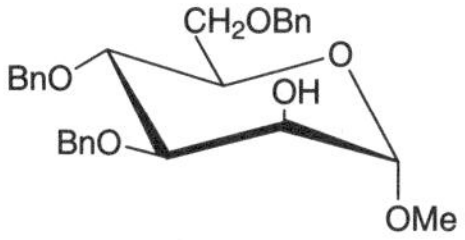	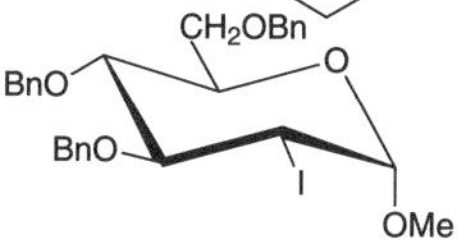	82%	181
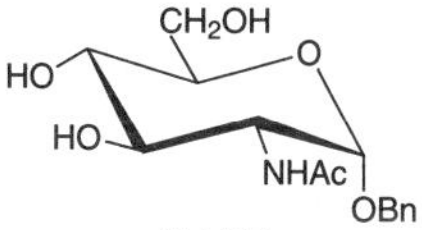	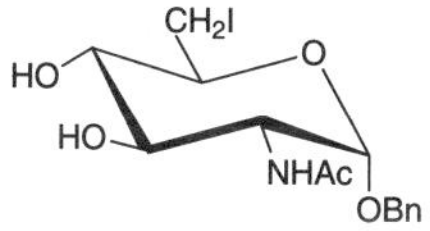	70%	184
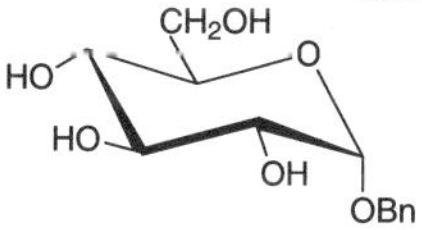	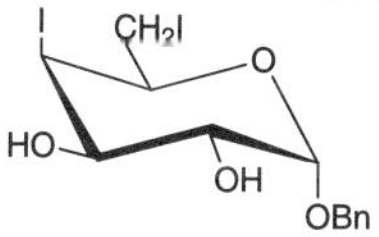	70%	184

The use of the Garegg–Samuelsson method was reported to be effective in the synthesis of benzyl 2-benzyloxycarbonylamino-2,3,4,6-tetradeoxy-6-iodo-α-D-*erythro*-hex-3-enopyranoside (**143,** Scheme 38) from benzyl 2-benzyloxycarbonylamino-2-deoxy-α-D-glucopyranoside (**142**).[186] When the procedure was repeated on a smaller scale, a poor yield of the desired compound was obtained, together with two previously unreported side products.[187] On a 1 mmole scale, a mixture consisting of 39% benzyl 2-benzyloxycarbonylamino-2,3,6-trideoxy-3,6-diiodo-α-D-allopyranoside (**144**), 24% benzyl 2-benzyloxycarbonylamino-2,4,6-trideoxy-4,6-diiodo-α-D-galactopyranoside (**145**), 13.5% of the desired **143,** and 10% of **146**—the position of the alkeneic proton could not be determined unambiguously using a 500 MHz NMR spectrometer. Repeating the reaction

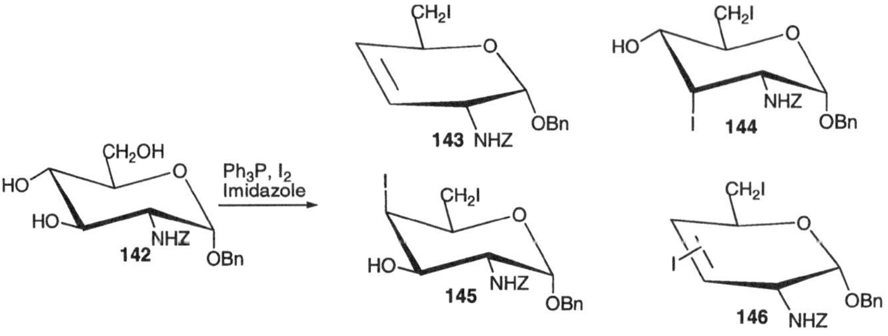

SCHEME 38

exactly as described by Garegg and co-workers[186] on a 10 mmole scale afforded 56% of the desired **143,** 17% of **146,** and 7% of the β anomer of **143.**

The use of GSR has been reported to effect deoxygenation of an α-hydroxy ketone.[188] Treatment of methyl 2,6-di-*O*-pivaloyl-α-D-*ribo*-hexopyranosid-3-ulose (**147,** Scheme 39) with triphenylphosphine, imidazole, and iodine, resulted in deoxygenation to afford methyl 4-deoxy-2,6-di-*O*-pivaloyl-α-D-*erythro*-hexopyranosid-3-ulose (**148**). Presumably the reaction proceeds via initial formation of the iodo compound, followed by loss of iodine through the enol form, a precedent for which has been reported.[189]

A modification of GSR has been reported by Classon and co-workers.[190] The idea remains the same: create a covalently bound phosphorus cation and displace with a nucleophile—in this case, a halogen. Both bromine and iodine have been used.[190] Three different systems were evaluated: (*1*) chlorodiphenylphophine, iodine–bromine, and imidazole; (*2*) *p*-(dimethylamino)phenyldiphenylphosphine, iodine–bromine, and imidazole; or (*3*) polymer-bound triphenylphosphine, iodine–bromine, and imidazole. The last two were found to be very similar to just triphenylphosphine itself, and displayed reactivity inferior to the first system. The polymer-bound reagent does allow for easier removal of triphenylphosphine oxide produced in the course of the reaction. As with the original procedure, and consistent with a S$_N$2 mechanism, inversion of configuration occurred. Again, as with the original method vicinal diols were readily converted into alkenes.[191] This

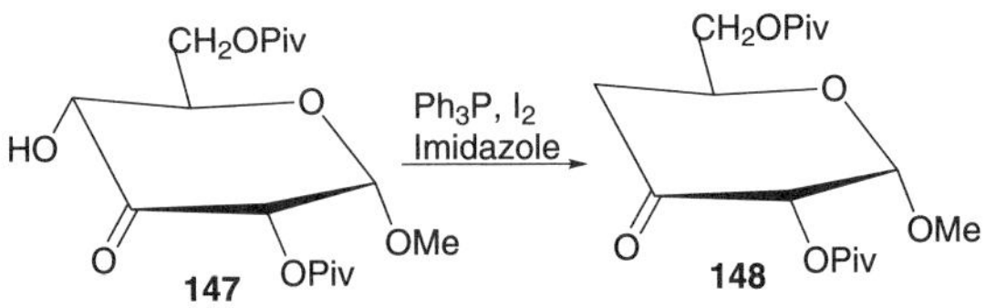

SCHEME 39

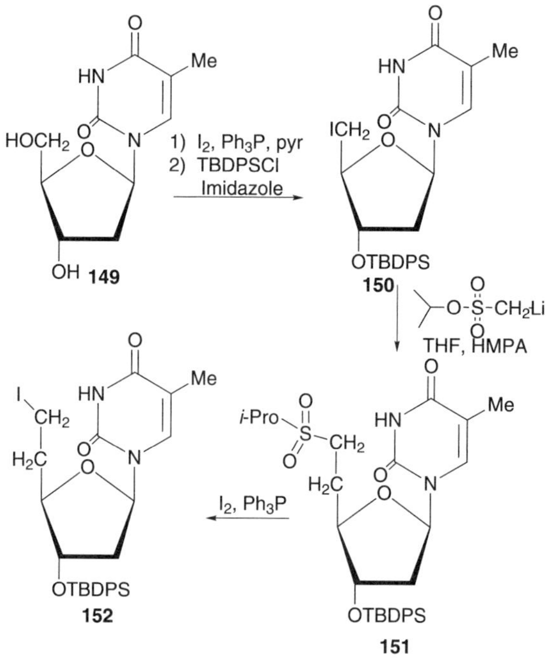

SCHEME 40

technique has been applied to the synthesis of 2′,3′-dideoxy-3′-C-hydroxymethyl-4′-thionucleosides.[192]

In an attempt to form 5′-deoxy-5′-iodothymidine, Huang and co-workers[193] undertook the procedure just described but obtained a disappointing yield. Performing the reaction using only triphenylphosphine and iodine in pyridine, followed by treatment with TBDPSCl, afforded 3′-O-tert-butyldimethylsilyl-5′-deoxy-5′-iodothymidine (150, Scheme 40) in an impressive 88% yield for two steps. The same reagent system was successful in converting a sulfonate (151) to the corresponding iodide (152) in a derivative of thymidine.[194] This type of replacement of a sulfonate with triphenylphosphine–iodine had previously been noted.[195]

A slight variation on the use of the GSR afforded an interesting transfer of an allyl alcohol through a nucleoside. Beginning with the unsaturated uridine derivative 153 (Scheme 41), reaction with chlorodiphenylphosphine afforded the phosphinate 154. Treatment with iodine in acetonitrile–toluene acted to add iodine to the unsaturation, with concomitant elimination between C-2′ and C-3′. The iodine is readily converted to an acetate by treatment with tetrabutylammonium acetate, from which the hydroxyl group is readily obtained after deprotection.[196] Beginning with the "opposite" molecule, namely with a methylene group at C-3′ and the hydroxyl at C-2′, the same sequence was noted. A similar displacement

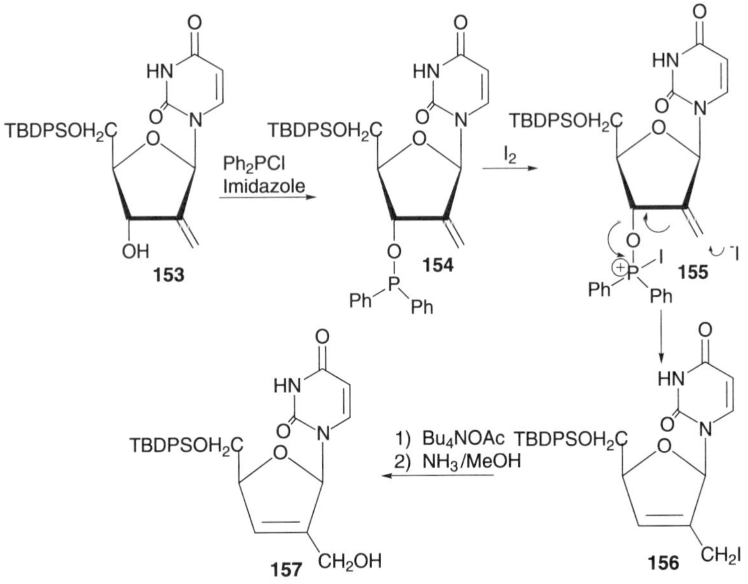

SCHEME 41

of a hydroxyl group by iodine has been achieved using triphenylphosphine and N-iodosuccinimide in DMF.[197] This method, which is an extension of a previously reported bromination,[198] is particularly useful in converting primary hydroxyl groups into the corresponding iodides in the presence of secondary hydroxyl groups.

VII. INSTALLATION AND REMOVAL OF PROTECTING GROUPS USING IODINE

Cyclic acetals permit the simultaneous protection of pairs of hydroxyl groups, as such, they represent an extremely efficient protecting group. Furthermore, either 1,2 or 1,3 diols may be selectively protected rendering the use of cyclic acetals of even greater value to organic chemists in general, and to carbohydrate chemists in particular. The versatility of iodine is demonstrated in that it has been reported to be effective in both the removal and installation of cyclic acetals.

The use of a catalytic amount of iodine in acetone was found to effect the addition of isopropylidene cyclic acetals to a wide variety of mono-[199] and disaccharides.[200] In the case of disaccharides, some cleavage of the glycosidic link was also observed. The proposed rationale for this acetalation is that iodine, acting as a Lewis acid, polarizes the carbon–oxygen bond (Scheme 42), increasing the electrophilicity of the carbon, leading to dehydration. In this instance, the iodine acts in the same manner as a proton in the acid-catalyzed introduction of a cyclic acetal. Hydroiodic

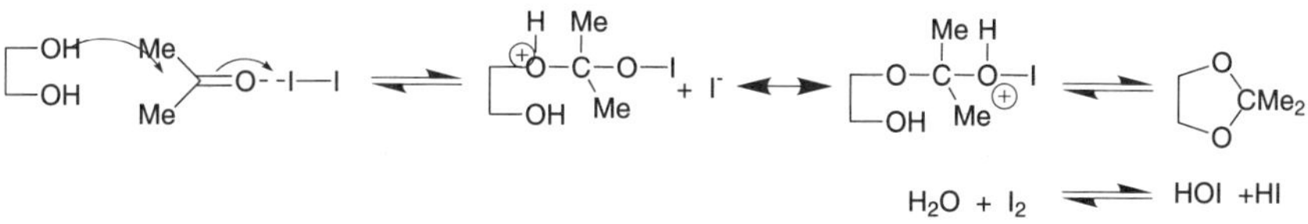

SCHEME 42

acid, present in small amounts in the reaction mixture, is responsible for the observed cleavage of the glycosidic bonds, albeit in modest yields. Acid-catalyzed formation of acetals is an equilibrium process, limiting this otherwise elegant technique to the installation of smaller cyclic acetals arising from volatile progenitors.

The initial discovery of the ability of solutions of iodine in methanol to cleave cyclic acetals was a serendipitous report[123] of unexpected products obtained from an attempt to add the pseudohalogen nityl iodide (prepared by mixing silver nitrate and iodine in methanol) to a 2,3-anhydro sugar. Cleavage of a benzylidene acetal was observed, which served to foment a flurry of reports on the use of iodine in methanol to effect deprotection of acid-sensitive protecting groups. Repeating the foregoing reaction using just iodine in methanol afforded the same deprotection. "Reactions of importance in chemistry are discovered by conception, by misconception and by accident" (Ref. 201, p. 15).

A catalytic amount of iodine in methanol has been reported to remove cyclic acetals and thioacetals.[124] The exact nature of the reaction is unclear. Iodine is known to oxidize methanol to produce formaldehyde and HI.[13] Similarly, chain structures of the type $I_2 \cdot 2CH_3OH$, where iodine is associated with two oxygens, are known to be produced. As well, the formation of charge transfer complexes is also known.[202] A reasonable explanation is that polarization of the ring-oxygen atoms, acting in concert with the trace amount of acid produced, is responsible for the cleavage observed. Some examples of the use of solutions of iodine in methanol to effect cleavage and addition of cyclic acetals are presented in Table III.

The use of iodine in refluxing methanol has also been found to be effective in the cleavage of TBDMS ethers.[203] Presumably, the trace amount of acid known to be produced is responsible for the cleavage of the acid-sensitive silyl ether.[204] The advantage of using iodine in methanol for the removal of protecting groups lies in its simplicity. No overly harsh chemicals are used; indeed, tinctures of iodine have been used to treat wounds since 1828.[205] As well, the implementation of the reaction is simple, no special care need be taken to exclude moisture or oxygen, and the reagents are inexpensive.

The use of a 1% solution of iodine in methanol at 40 °C has been reported to effect the cleavage of trityl and dimethoxytrityl ethers.[206] Again, the ability of solutions of iodine in methanol to release of small amounts of HI into the reaction mixture was credited. In order to further examine the role of acid in the deprotection, a variety of bases were added to the reaction mixture. In the

TABLE III
Some Examples of the Installation and Removal of Cyclic Acetals Using "Iodine-in-methanol"

Starting Material	Product	Reagents	Yield	Ref.
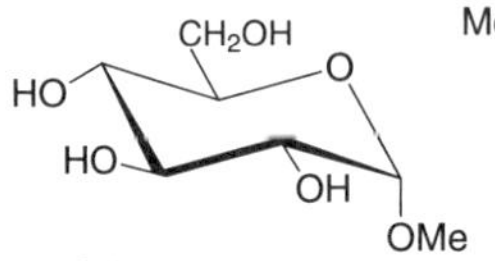	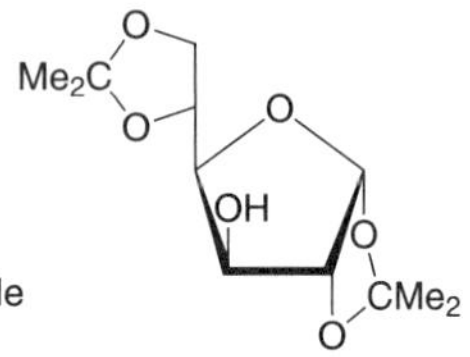	I_2/Me_2CO	70% 80%	199 200
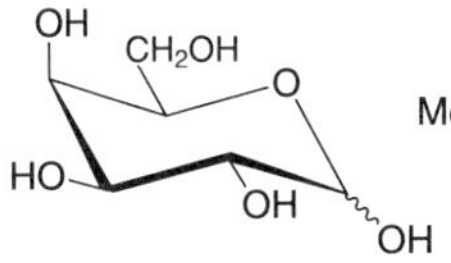	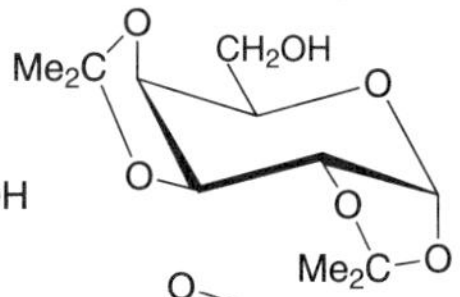	I_2/Me_2CO	70% 84%	199 200
Lactose	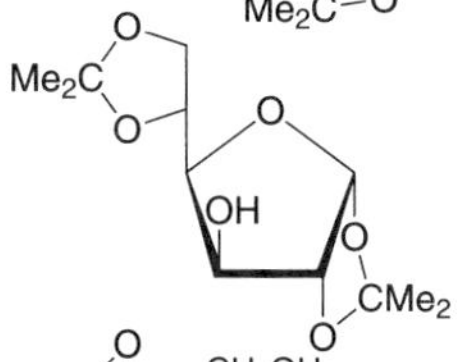	I_2/Me_2CO	15%	200
Maltose	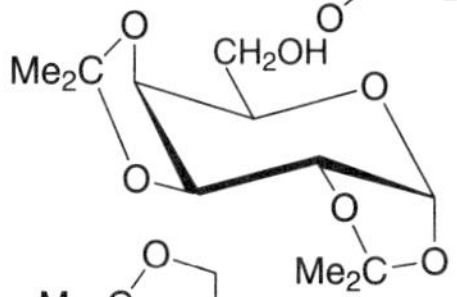		12%	
	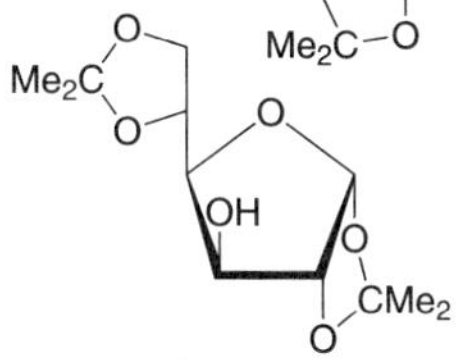	I_2/Me_2CO	16%	200
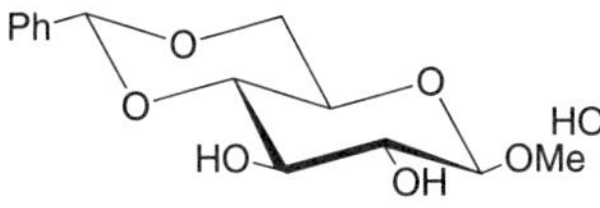	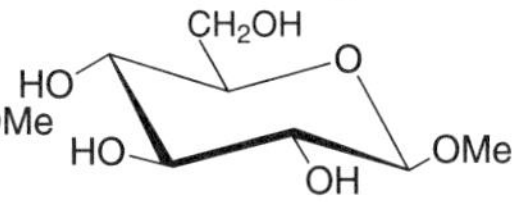	I_2/MeOH	>90%	124
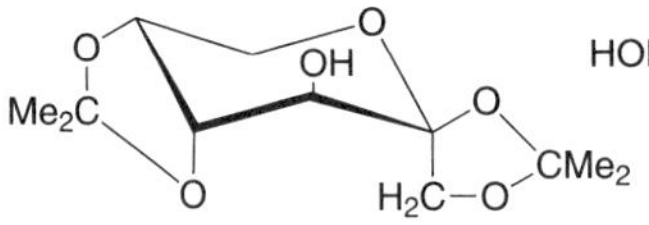	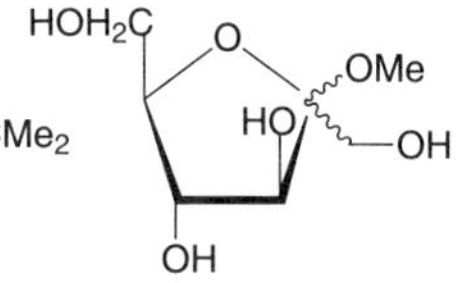	I_2/MeOH	13.5% α 76.5% β	124
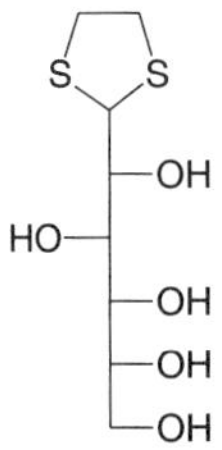	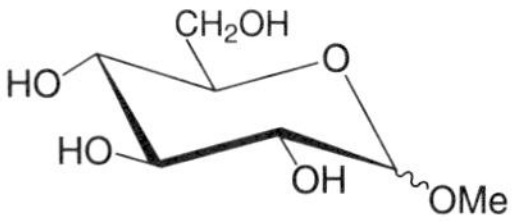	I_2/MeOH	18% β 72% α	124

presence of a proton scavenger yields of the reactions were poor. The reaction was also attempted in refluxing carbon tetrachloride—no deprotection was observed even after 2 weeks. Interestingly, it was found that TBDMS ethers were stable to the reaction conditions at 40 °C; a report, already noted, claimed that TBDMS ethers were readily cleaved in 90 to 120 min by a refluxing solution of iodine in methanol. A further report on the cleavage of TBDMS ethers (with no carbohydrate examples) using refluxing iodine in methanol supports the assertion that they are cleaved under such conditions; this study also noted that the time required for the removal of triisopropylsilyl (TIPS) and *tert*-butyldiphenylsilyl (TBDPS) ethers was substantially slower, so much slower that the selective removal of a TBDMS ether could be achieved in the presence of either a TIPS or a TBDPS ether.[207] The use of iodine monobromide has been reported to be very effective in the cleavage of TBDMS ethers.[208] The deprotection with IBr was complete in as little at 90 s at room temperature. In some cases the reaction with IBr was exothermic enough it had to be performed at −15 to −17 °C, and was still complete in a matter of minutes. Presumably the reaction proceeds through the same path as described for iodine. In the case of the the interhalogen IBr, the greater Lewis acidity imparted by the enhanced dipole hastens the process.

Tetrahydropyranyl (THP) ethers, another species known to be unstable to acid, have similarly been reported to be cleaved by solutions of iodine in methanol.[209] At room temperature, cleavage of the THP ethers was complete in 1.5 to 8 h. As with the previous example using iodine in methanol at lower than reflux temperature, TBDMS ethers were stable to these conditions. The ability to tune the reactivity of the iodine in methanol system by simply controlling the temperature is of value in selective deprotection. This is even more useful when fluorine, known to remove only silyl ethers,[105] is exploited. Given that methoxymethyl ethers, essentially acetals, are known to be cleaved under acidic conditions, it seems likely they too should be subject to removal by solutions of iodine in methanol. Sundry examples of deprotections using iodine in methanol are presented in Table IV.

A variant on iodine deprotections is a report on the cleavage of PMB ethers by a mixture of cerium(III) chloride heptahydrate and sodium iodide in refluxing acetonitrile.[210] A proposed mechanism involves polarization of the ethereal link (**158,** Scheme 43), followed by generation of a quinone-like species, and attack by nucleophilic iodide to afford *p*-methoxybenzyl iodide.

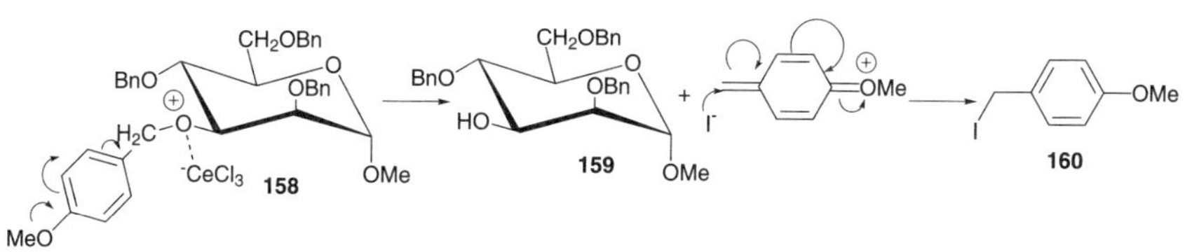

Scheme 43

TABLE IV
Miscellaneous Deprotections Using Iodine in Methanol

Starting Material	Product	Temp (°C)/Time	Yield	Ref.
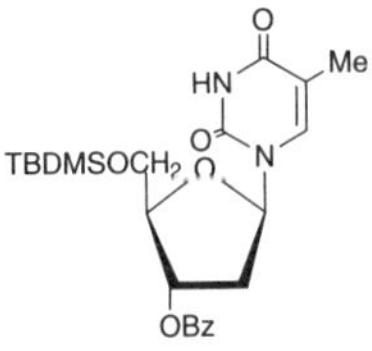	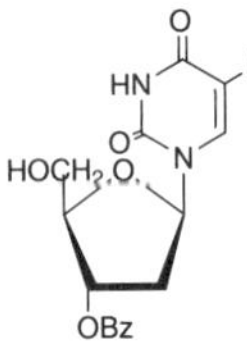	65/2 h	86%	203
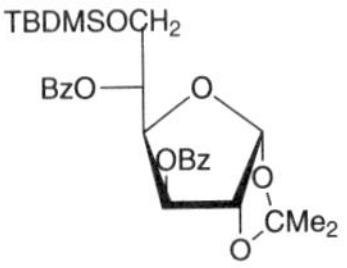	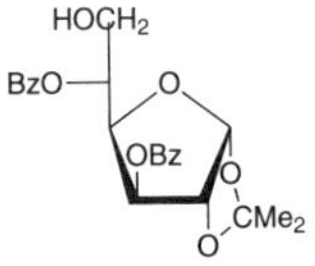	65/90 min	96%	203
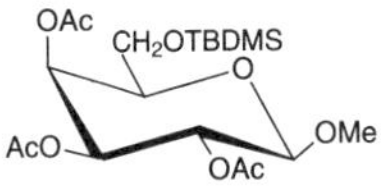	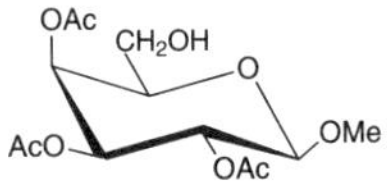	rt/5 min (IBr)	80%	208
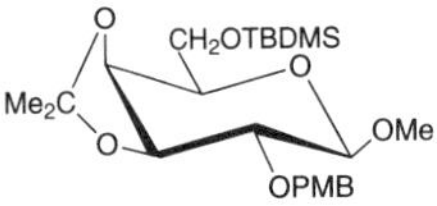	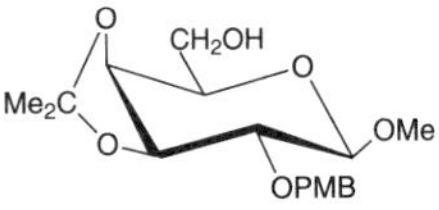	rt/30 min (IBr)	80%	208
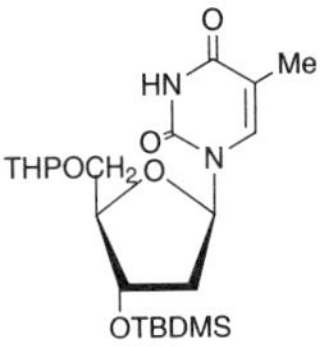	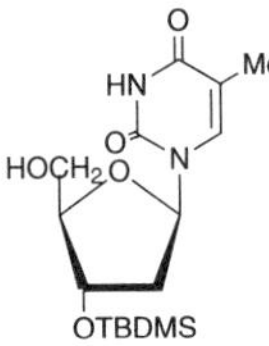	40/3 h	90%	209
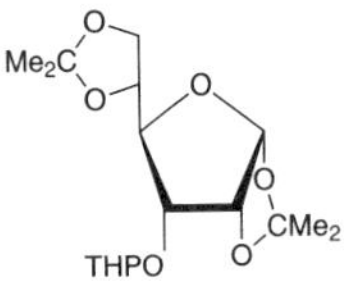	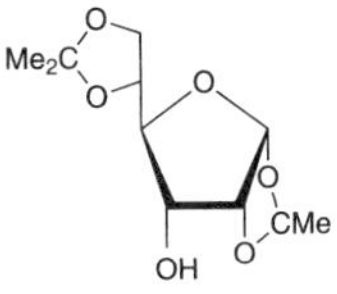	rt/1 h	90%	209
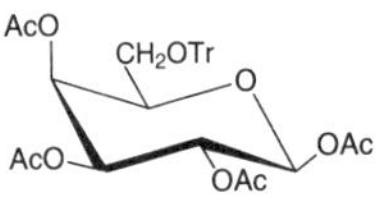	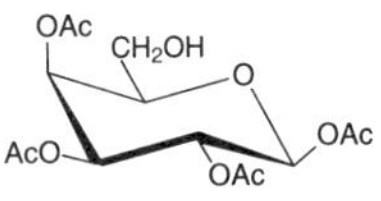	40/not given	84%	206
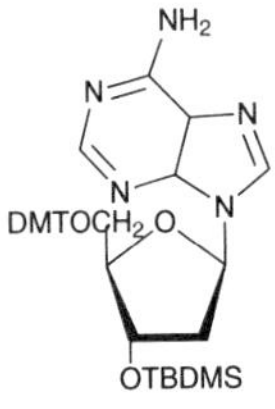	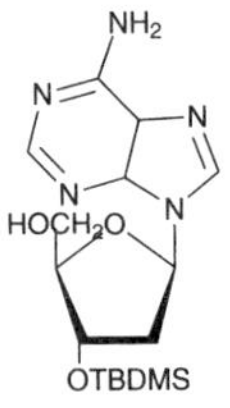	40/not given	87%	206

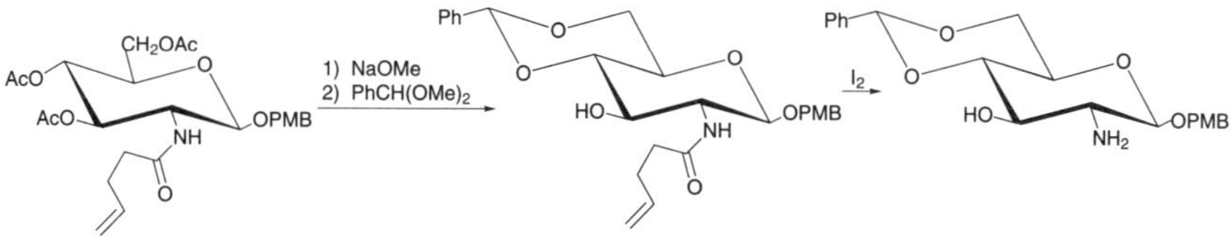

SCHEME 44

The protection and deprotection of the amino functionality is critical to peptide chemists. Given the prominence of glycosylamines in carbohydrate chemistry, the masking of amines is a concern to carbohydrate chemists as well. The use of the *n*-pent-4-enoyl group, installed in high yield through the action of *n*-pent-4-enoic anhydride, is of some value in that the protected compounds so obtained are often readily recrystallized, rendering processing more efficient.[211] Once in place, the *n*-pent-4-enoyl amide is cleaved in a manner similar to the activation of an *n*-pentenoyl glycoside discussed earlier. Reaction with three equivalents iodine in aqueous THF affects cyclization, with subsequent cleavage of the N—C bond. Attempting the reaction with less than three equivalents of iodine resulted in no reaction–consistent with earlier reported instances of iodolactonization involving *n*-pentamides.[212,213] Despite the mildly oxidizing conditions, *p*-methoxybenzyl (PMB) ethers are reported to be stable to these conditions. Presumably, this is due to the lack of formation of HI in solution, as the PMB ether has been reported to be cleaved by the action of iodine in refluxing methanol.[214] Similarly unaffected by iodine in THF are *n*-pentenyl glycosides and benzylidene acetals (Scheme 44).

Iodine has been reported to aid in the addition of acetate protecting groups.[215] Treatment of carbohydrates suspended in acetic anhydride with iodine was found to convert them into the corresponding peracetates very rapidly at room temperature. The reaction is rapid enough that cooling is critical on large scales. The Lewis acidity of molecular iodine was thought to polarize the anhydride carbonyl group, thus permitting reaction (Scheme 45). Further, it was discovered that primary benzyl ethers were cleaved by this method and replaced with acetoxy groups, whereas secondary benzyl ethers remained unaffected. This ability to selectively replace a benzyl ether, stable to hydrolysis but labile to hydrogenolysis, with an acetoxy group having the opposite stability is of great use in orthogonal protecting group strategies.

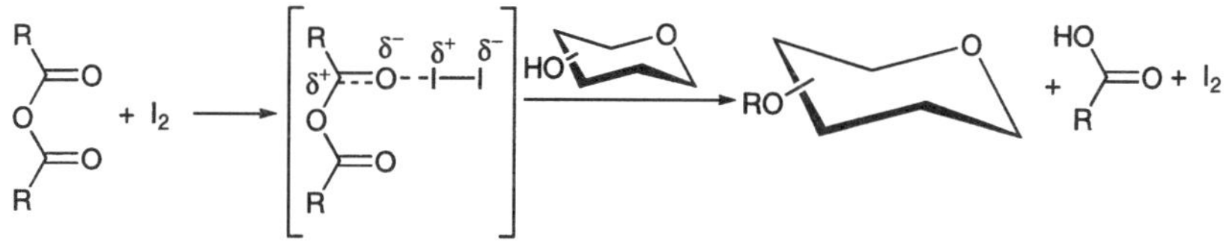

SCHEME 45

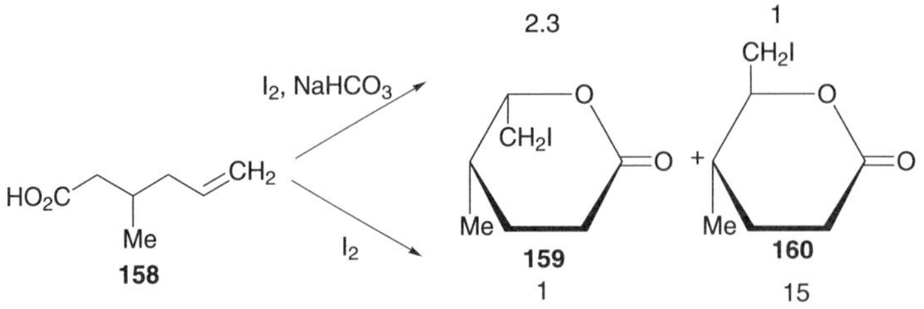

SCHEME 46

VIII. IODOCYCLIZATIONS

Formation of a highly electrophilic iodonium species, transiently formed by treatment of an alkene with iodine, followed by intramolecular quenching with a nucleophile leads to iodocyclization. The use of iodine to form lactones has been elegantly developed. Bartlett and co-workers[216] reported on what they described as thermodynamic versus kinetic control in the formation of lactones. Treatment of the alkenoic acid **158** (Scheme 46) with iodine in the presence of base afforded a preponderance of the kinetic product **159,** whereas the same reaction in the absence of base afforded the thermodynamic product **160.** This approach was used in the synthesis of serricorin. The idea of kinetic versus thermodynamic control of the reaction was first discussed in a paper by Bartlett and Myerson[217] from 1978. It was reasoned that in the absence of base, thermodynamic control could be achieved in that a proton was available to allow equilibration to the most stable ester. In the absence of such a proton, for example by addition of base, this equilibration is not possible, and the kinetic product is favored.

Iodoetherification has been used in the synthesis of *cis* and *trans* linalyl oxides. Mechanistically, the rate of departure of the hydroxyl protecting group in the starting γ,δ unsaturated ether was found to be critical to the stereochemistry of the product.[218] The alkyl group R' (Scheme 47) must be sufficiently bulky as to

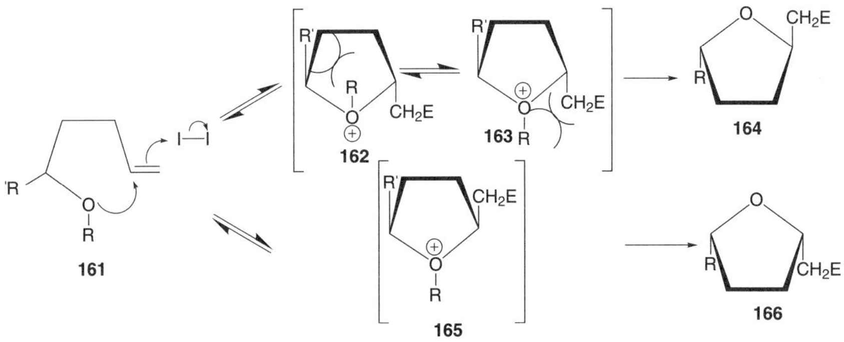

SCHEME 47

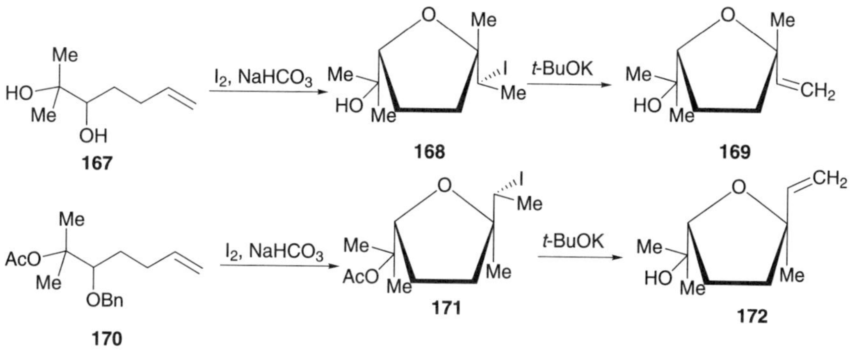

SCHEME 48

permit some steric selectivity, yet be small enough to allow cyclization. Similarly, the rate at which the R group on the oxygen leaves will determine if the lower structure is favored thermodynamically, and not just kinetically. Benzyl ethers were found to just give a 2:1 *cis*:*trans* ratio. Treatment of **167** (Scheme 48) with iodine, followed by *t*-BuOK elimination of iodine afforded the *trans* linalyl oxide, whereas **170** afforded the *cis* isomer.

Tamaru and co-workers[219] have examined electrophile-mediated cyclizations under three different sets of conditions. Treatment of 4-penten-1,3-diols (**173, 176**, Scheme 49) was attempted with 1.5 equivalents of iodine in Et_2O in the presence of $NaHCO_3$; treatment with 1.2 equivalents of NIS in CH_2Cl_2; or treatment with 1.5 equivalents IDCP in $CHCl_3$. In all cases, the yields were high, although the *cis* to *trans* ratio varied markedly. Treatment with iodine alone was found to afford the highest selectivity, giving preponderantly the *cis* product. Interestingly, treatment of 6-phenyl-hex-4,6-diol-2-ene (**176**) to the above conditions afforded mainly the tetrahydropyran **177**.

Glucopyranosides allylated at C-6 were cyclized by the action of IDCP in wet CH_2Cl_2. Reactions were rapid, offering yields of 75–85% after 10 min. A variety

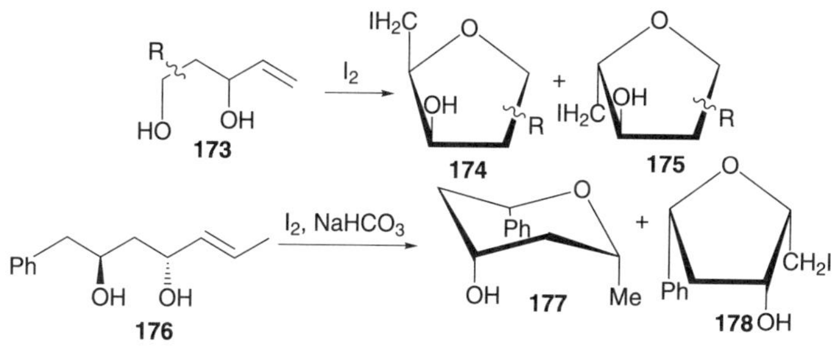

SCHEME 49

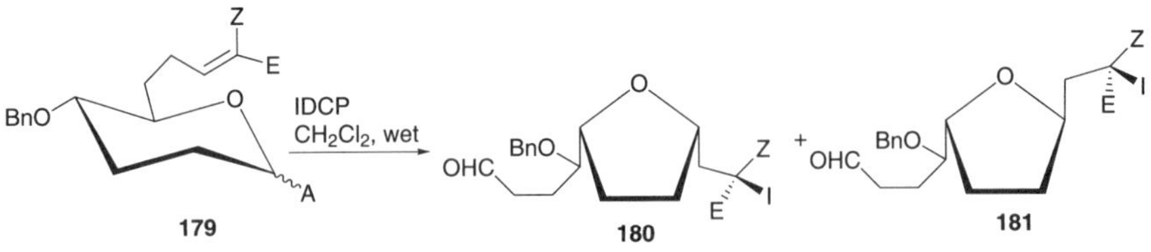

[Scheme 50 diagram]

A	cis:trans ratio	
A = α-OEt, Z = E = H	1.5	1
A = α-OCH$_2$CF$_3$, Z = E = H	1.5	1
A = α/β-OTr, Z = E = Pr	1	0
A = α/β-OTr, Z = Pr, E = H	8	1
A = α/β-OTr, Z = H, E = Pr	20	1

SCHEME 50

of compounds were cyclized in this manner. The ratio of the *cis* to *trans* product obtained was found to vary with the size of the aglycone; very bulky groups, such as trityl ethers, were found to be the only ones that forced the *cis* configuration. The Z alkenes were found to be less likely to give the *cis* product. The only cases were solely one isomer was obtained was with a very bulky (Tr) aglycone, and even then mixtures were generally the rule (Scheme 50)

The use of iodoetherification has been applied to the synthesis of fused 6,6-rings.[220] Treatment of the cyclohexyl dialkenyl compound **182** (Scheme 51) with NIS in CH$_2$Cl$_2$ afforded the tetrahydrofuran **183**, which, upon solvolysis with AgBF$_4$ in DMSO afforded the tetrahydropyran **184**. Further iodocyclization, followed by ring expansion, afforded the tricyclic **185**. This type of methodology was exploited in the synthesis of brevotoxins.

A report on the use of iodolactonization which gave products complementary to those obtained by Sharpless asymmetric epoxidation[221] has been reported.[222]

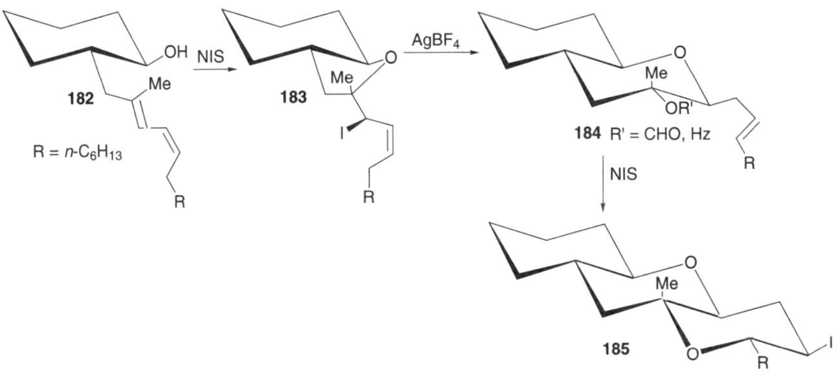

[Scheme 51 diagram]

SCHEME 51

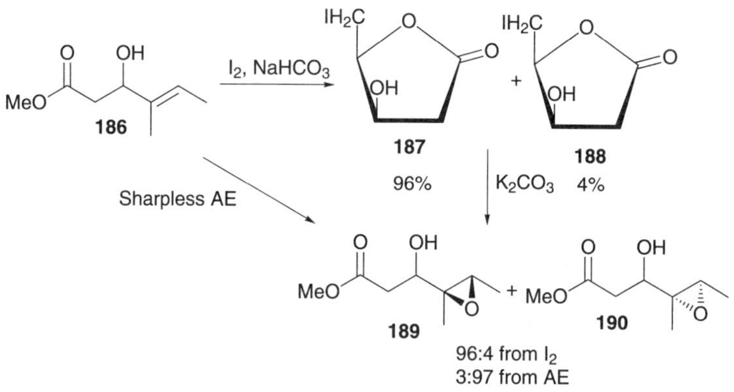

SCHEME 52

Treatment of the allylic alcohol **186** (Scheme 52) to the Sharpless conditions afforded **189** and **190,** in the opposite ratio of that obtained by treatment of with iodine in the presence of base followed by saponification of the methyl ester. The same report noted that with a more substituted alkene, the formation of tetrahydropyran rings was found to preponderate, although small (∼10%) amounts of the tetrahydrofurans were still evident.

The use of iodoetherification has been applied to a total synthesis of sucrose.[223] Beginning with the vinyl glycoside **191** (Scheme 53), removal of the silyl protecting group, followed by treatment with iodine and potassium *tert*-butoxide in THF in the presence of silica gel afforded an anomeric excess for the β anomer of **192** of 96%. Performing the reaction without SiO_2 resulted in a 1:1 mixture of anomers. In a blank experiment, the α anomer was treated with silica gel and was found to be stable, implying the acidic silica gel was actually directing the course of the reaction, rather than converting the α anomer to the β. This type of interaction with a heterogeneous reagent is not unheard of.[224–227]

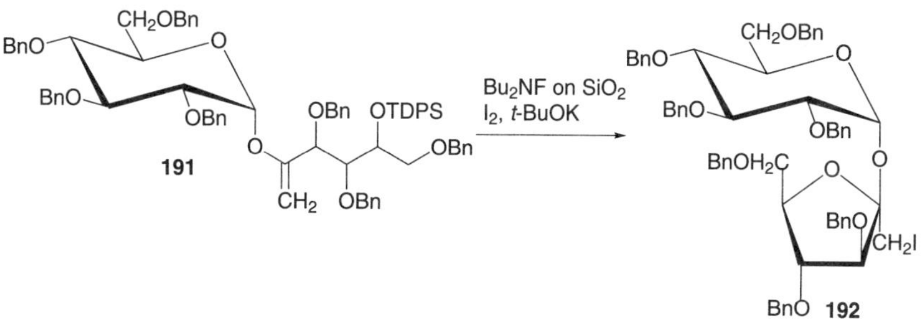

SCHEME 53

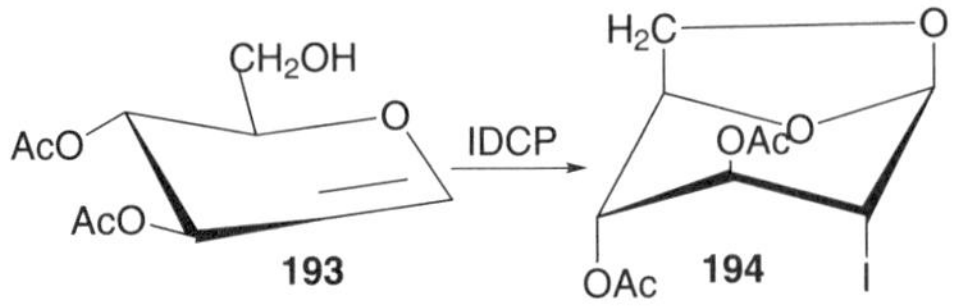

SCHEME 54

In the same manner that iodoetherification is effected by treating an alkene with an electrophilic iodine species in the presence of a nucleophile, glycals participate similarly in this type of reaction. Treatment of 3,4-di-*O*-acetyl-D-glucal with the electrophilic iodine from IDCP, resulted in formation of the cyclized 1,6-anhydro-3,4-di-*O*-acetyl-2-deoxy-2-iodo-β-D-glucopyranose was obtained (Scheme 54).[228] The same reaction has likewise been reported via the use of tributylstannyl ethers, which serve to enhance the nucleophilicity of oxygen.[229]

Formation of 3,6-di-*O*-tributylstannyl-D-glucal was prepared by treating the glycal with 1.5 equivalents of $(Bu_3Sn)_2O$; the enhanced reactivity of the hydroxyl group at C-3 of a glycal accounts for the selectivity observed.[230,231] Treatment of the distannylated glucal with NIS afforded mainly oxidation of the hydroxyl group at C-3. Hydride transfer from C-3 to cationic iodine resulted in the stable enone **196** (Scheme 55) being formed, with only traces of the cyclized product **197**.[232] Further investigation, using the more-polar acetonitrile as a solvent, revealed that only the iodocyclized product was obtained, together with trace amounts of **196**. The use of molecular iodine was found to afford solely the cyclized product, regardless of the solvent used. Presumably, the ability of the halogen to act as a Lewis acid facilitates this process—in the less polar solvent stabilization of the cationic species is more difficult, favoring oxidation. The reaction was found to give the 1,6-anhydro sugar in solely the D-*gluco* configuration. Attack on the

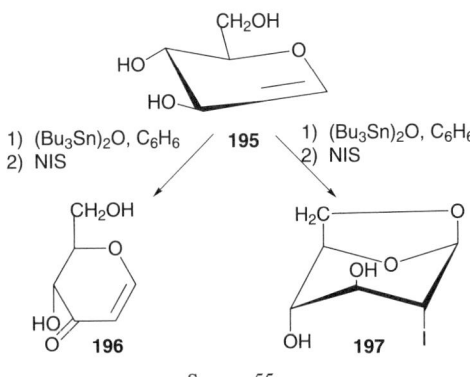

SCHEME 55

electrophilic iodine must have occurred from the bottom of the plane, consistent with previous studies concerning addition of halogens to glycals. Again, the justification for this was ascribed to the purported reverse anomeric effect. The authors report the formation of 10% of the 1,6-anhydro sugar in the D-*manno* configuration.

A modification of the foregoing method has been reported[233] wherein the amount of $(Bu_3Sn)_2O$ was reduced by half (0.75 equiv.), forming solely the C-6 stannylene ether. With the nucleophilicity of only the C-6 oxygen enhanced, none of the D-*manno* product previously reported was obtained. Further elaboration by treatment with NaN_3 afforded the 2-azido compound, in the D-*gluco* configuration. The observed net retention of configuration comes about as a result of the intermediate formation of the 2,3-anhydro species, which opens upon treatment with nucleophile to afford the *trans*-diaxial product, providing ready access to 2-amino sugars.

The use of iodoetherification has been exploited in the synthesis of precursors to mono- and bis-THF acetogenins,[234] potent antitumor and pesticides. Also, has been used in the synthesis of tetrahydrofuran analogs of precursors to pseudomonic acid.[235] As well, NIS has been used as the iodine source, for example in the synthesis of other acetogenins.[236] Other compounds synthesized via iodocyclization include: (±)-(α)-multistriatin,[237] (±)-velbanamine, and (±)-isovalbanamine,[238] and (+)-citreoviral.[239]

IX. MISCELLANEOUS REACTIONS

Oxidation of alcohols to the corresponding carbonyl derivative has been reported using both *N*-bromo and *N*-chlorosuccinimide.[240] The use of *N*-iodosuccinimide in the presence of tetra-*n*-butylammonium iodide has been found to be a convenient method for the same oxidation, and was found to give higher yields than when the corresponding bromides were used.[241] It is particularly convenient in that, much like the Dess–Martin reagent, there is no danger of over-oxidation of primary alcohols, with the advantage that the extra effort of preparing the Dess–Martin periodinane is avoided. The reaction was found not to proceed at all in the absence of tetra-*n*-butylammonium iodide, or if NIS was replaced with iodine itself. The mechanism is thought to be similar to the corresponding oxidation with either *N*-bromo,[242] or *N*-chlorosuccinimide.[243] For example, in the oxidation of 2,3:5,6-di-*O*-isopropylidene-D-mannose (**198**, Scheme 56), transient formation of the intermediate **200** via radical **199**, followed by release of HI to afford the desired lactone. There is a high likelihood that attack of the iodide from the ammonium salt is facilitated by the presence of NIS. Neither molecular iodine nor any intermediate hypoiodite species were detected in the course of the reaction.

Glycosyl fluorides are becoming increasingly common in synthesis.[244] An elegant synthesis of glycosyl fluorides has been reported starting from the thioglycoside, using molecular iodine and fluorine.[245] The reaction proceeds by

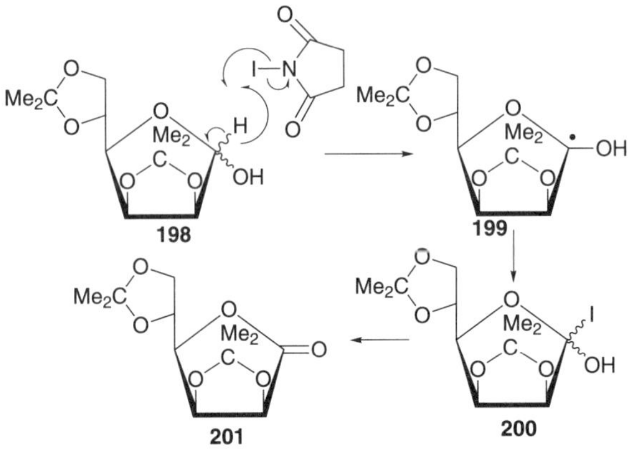

SCHEME 56

activation of the thioglycoside with iodine to afford an oxacarbenium intermediate, which is attacked by nucleophilic fluoride. Many iodine-containing reagents have been discussed to achieve this activation, including the use iodine itself. Yields in the reaction are not spectacular, ranging from 40–57%, and the anomeric selectivity is moderate. In the example of Scheme 57, a preponderance of the β anomer is formed when the reaction is performed in the absence of molecular sieves. With sieves present, the selectivity is the opposite, a 30:1 ratio in favor of the α anomer. Similar examples, all performed in the absence of sieves, resulted in a preponderance of the β anomer. The preference for the β anomer results from the anchimeric assistance of the acetyl group. As no water is produced during the course of reaction and the acetonitrile was rigorously dried beforehand, perhaps the sieves were simply acting as a surface to direct the course of glycosylation. Although this method has a certain elegance in its simplicity, disappointing yields, together with the inherent danger of working with elemental fluorine, render this a less than satisfactory procedure.

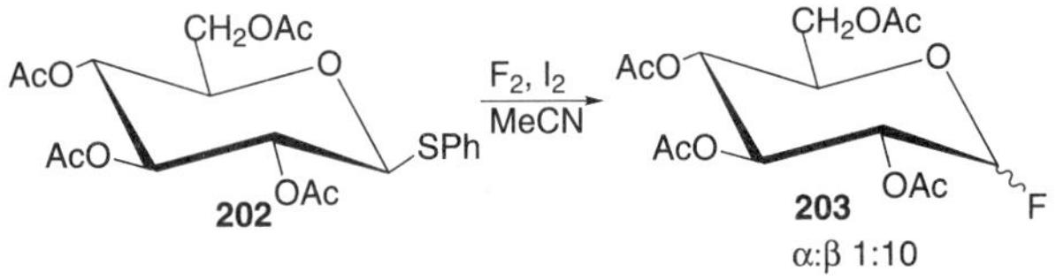

SCHEME 57

References

(1) A. J. Woytek, in K. Othme (Ed.), *Encyclopedia of Chemical Technology*, 3rd ed., Vol. 10, Wiley–Interscience, New York, 1980, pp. 630–654.
(2) A. A. E. Penglis, *Adv. Carbohydr. Chem. Biochem.*, 38 (1981) 195–285.
(3) T. Tsuchiya, *Adv. Carbohydr. Chem. Biochem.*, 48 (1990) 91–277.
(4) P. J. Card, *J. Carbohydr. Chem.*, 4 (1985) 451–487.
(5) J. T. Welch, *Tetrahedron*, 43 (1987) 3123–3197.
(6) W. A. Szarek, *Adv. Carbohydr. Chem. Biochem.*, 28 (1973) 225–306.
(7) P. Tomasik and C. H. Schilling, *Adv. Carbohydr. Chem. Biochem.*, 53 (1998) 263–343.
(8) C. McCloskey, R. E. Pyle, and G. H. Coleman, *J. Am. Chem. Soc.*, 66 (1944) 349–350.
(9) E. A. Talley, D. E. Reynolds, and W. L. Evans, *J. Am. Chem. Soc.*, 65 (1943) 575–582.
(10) H. R Goldschmid and A. S. Perlin, *Can. J. Chem.*, 39 (1961) 2025–2034.
(11) D. E. Reynolds and W. L. Evans, *J. Am. Chem. Soc.*, 60 (1938) 2559–2561.
(12) C. G. J. Verhart, C. T. M. Fransen, B. Zwanenburg, and G. J. F. Chittenden, *Recl. Trav. Chim. Pays-Bas*, 115 (1996) 133–139.
(13) F. R. Cruickshank and S. W. Benson, *J. Phys. Chem.*, 73 (1969) 733–737.
(14) B. Helferich and R. Gootz, *Ber.*, 62, (1929), 2791–2796.
(15) F. J. Kronzer and C. Schuerch, *Carbohydr. Res.*, 34 (1974) 71–78.
(16) J.-P. Praly and R. U. Lemieux, *Can. J. Chem.*, 65 (1987) 213–223.
(17) A. Marra, J.-M. Malletn C. Amatore, and P. Sinaÿ, *Synlett*, (1990), 572–574.
(18) I. Braccini, C. Derouet, J. Esnault, C. Hervé du Penhoat, J.-M. Mallet, V. Michon, and P. Sinaÿ, *Carbohydr. Res.*, 246, (1993), 23–41.
(19) A. J. Ratcliffe and B. Fraser-Reid, *J. Chem. Soc. Perkin Trans. I*, (1989) 1805–1810.
(20) R. K. Ness, H. G. Fletcher, Jr., and C. S. Hudson, *J. Am. Chem. Soc.*, 72 (1950) 2200–2205.
(21) R. W. Jeanloz and P. J. Stoffyn, *Methods Carbohydr. Chem.*, 1 (1962) 221–227.
(22) J. Thiem and B. Meyer, *Chem. Ber.*, 113 (1980) 3075–3085.
(23) A. Klemer and M. Bieber, *Liebigs Ann. Chem.*, (1984) 1052–1055.
(24) B. Ernst and T. Winkler, *Tetrahedron Lett.*, 30 (1989) 3081–3084.
(25) T. Uchiyama and O. Hindsgaul, *Synlett*, (1996) 499–501.
(26) G. Böhm and H. Waldemann, *Liebigs Ann. Chem.*, (1996) 613–619.
(27) G. Böhm and H. Waldemann, *Liebigs Ann. Chem.*, (1996) 621–625.
(28) U. Schmid and H. Waldmann, *Tetrahedron Lett.*, 37 (1996) 3837–3840.
(29) K. P. R. Kartha and R. A. Field, *Carbohydr. Lett.*, 3 (1998) 179–186.
(30) K. P. R. Kartha, M. Aloui, and R. A. Field, *Tetrahedron Lett.*, 37 (1996) 8807–8810.
(31) D. R. Mootoo, V. Date, and B. Fraser-Reid, *J. Chem. Soc., Chem. Commun.*, (1987) 1462–1464.
(32) D. R. Mootoo, V. Date, and B. Fraser-Reid, *J. Am. Chem. Soc.*, 110 (1988) 2662–2663.
(33) R. R. Schmidt, *Angew Chem., Int. Ed. Engl.*, 25 (1986) 212–235.
(34) H. Paulsen, *Angew Chem., Int. Ed. Engl.*, 21 (1982) 155–173.
(35) R. R. Schmidt, *Adv. Carbohydr. Chem. Biochem.*, 50 (1994) 21–124.
(36) D. R. Mootoo, P. Konradsson, U. Udodong, and B. Fraser-Reid, *J. Am. Chem. Soc.*, 110 (1988) 5583–5584.
(37) P. Konradsson, D. R. Mootoo, R. E. McDevitt, and B. Fraser-Reid, *J. Chem. Soc., Chem. Commun.*, (1990) 270–272.
(38) B. Fraser-Reid, Z. Wu, U. Udodong, and H. Ottosson, *J. Org. Chem.*, 55 (1990) 6068–6070.
(39) J. R. Merritt and B. Fraser-Reid, *J. Am. Chem. Soc.*, 114 (1992) 8334–8336.
(40) B. Fraser-Reid, *Acc. Chem. Res.*, 29 (1996) 57–66.
(41) R. Madsen and B. Fraser-Reid, in S. H. Khan and R. A. O'Neil (Eds.), *Modern Methods in Carbohydrate Synthesis*, Vol. 1, Harwood Academic Publishers, U.S., 1996, pp. 155–170.

(42) J. C. Lopez and B. Fraser-Reid, *J. Chem. Soc., Chem. Commun.,* (1991) 159–161.
(43) G. J. Boons, A. Burton, and P. Wyatt, *Synlett,* (1996) 310–312.
(44) T. Norberg, in S. H. Khan and R. A. O'Neil (Eds.), *Modern Methods in Carbohydrate Synthesis,* Vol. 1, Harwood Academic Publishers, U.S. , 1996, pp. 82–106.
(45) J. Kihlberg, D. A. Leigh, and D. R. Bundle, *J. Org. Chem.,* 55, (1990), 2860–2863.
(46) P. Sinaÿ, *Pure Appl. Chem.,* 63, (1991), 519–528.
(47) J. E. Huheey, *Inorganic Chemistry Principles of Structure and Reactivity,* 2nd ed.; Harper & Row: New York, 1978.
(48) I. Fleming, *Frontier Orbitals and Organic Chemical Reactions,* John Wiley & Sons: New York, 1977.
(49) R. P. R. Kartha, M. Aloui, and R. A. Field, *Tetrahedron Lett.,* 37 (1996) 5175–5178
(50) G. H. Veeneman and J. H. van Boom, *Tetrahedron Lett.,* 31 (1990) 275–278.
(51) G. J. Boons, R. Geurtsen, and D. Holmes, *Tetrahedron Lett.,* 36 (1995) 6325–6328.
(52) L. A. J. M. Sliedregt, K. Zegelaar-Jaarsveld, G. A. van der Marel, and J. H. van Boom, *Synlett,* (1993) 335–337.
(53) G. J. Boons, P. Grice, R. Leslie, S. V. Ley, and L. L. Yeung, *Tetrahedron Lett.,* 34 (1993) 8523–8526.
(54) C. W. Andrews, B. Fraser-Reid, and J. P. Bowen, *J. Am. Chem. Soc.,* 113 (1991) 8293–8298.
(55) K. P. R. Kartha and R. A. Field, *Tetrahedron Lett.,* 38 (1997) 8233–8236.
(56) K. P. R. Kartha and R. A. Field, *J. Carbohydr. Chem.,* 17 (1998) 693–702.
(57) T. Yamanoi, K. Nakamura, S. Sada, M. Goto, Y. Furusawa, M. Takano, A. Fujioka, K. Yanagihara, Y. Satoh, H. Hosokawa, and T. Inazu, *Bull. Chem. Soc. Jpn.,* 66 (1993) 2617–2622.
(58) T. Yamanoi, K. Nakamura, H. Takeyama, K. Yanagihara, and T. Inazu, *Bull. Chem. Soc. Jpn.,* 67 (1994) 1359–1366.
(59) R. U. Lemieux, K. B. Hendriks, R. V. Stick, and K. James, *J. Am. Chem. Soc.,* 97 (1975) 4056–4062.
(60) T. Murase, H. Ishida, M. Kiso, and A. Hasegawa, *Carbohydr. Res.,* 184 (1988) C1–C4.
(61) A. Hasegawa, T. Nagahama, H. Ohki, K. Hotta, H. Ishida, and M. Kiso, *J. Carbohydr. Chem.,* 10 (1991) 493–498.
(62) Y.-L. Li and Y.-L. Wu, *Tetrahedron Lett.,* 37 (1996) 7413–7416.
(63) G.-J. Boons and T. Stauch, *Synlett,* (1996) 906–908.
(64) K. P. R. Kartha, L. Ballell, M. McNeil, and R. A. Field, RSC Perkin Division 2nd Intl. Conference on Biological Challenges for Organic Chemistry, University of St. Andrews, Scotland, July 11–15, 1999; K. P. R. Kartha, L. Ballell, and R. A. Field, RSC Carbohydrates and Biological and Medicinal Chemistry Sector Spring Meeting, York University, March 25–26, 1999, pp. 27.
(65) J. Staněk and V. Schwarz, *Coll. Czech. Chem. Commun.,* 20 (1955) 42–45.
(66) C. Prévost, *Compt. Rend.,* 196 (1933) 1129–1131.
(67) R. U. Lemieux and S. Levine, *Can. J. Chem.,* 40 (1962) 1926–1932.
(68) R. U. Lemieux and B. Fraser-Reid, *Can. J. Chem.,* 43 (1965) 1460–1475.
(69) S. Winstein, E. Grunwald, and L. L. Ingraham, *J. Am. Chem. Soc.,* 70 (1948) 821–828.
(70) P. R. Bradley and E. Buncel, *Can. J. Chem.,* 46 (1968) 3001–3006.
(71) R. U. Lemieux and S. Levine, *Can. J. Chem.,* 42 (1964) 1473–1480.
(72) C. Y. Zheng, H. Slebocka-Tilk, W. Nagorski, L. Alvarado, and R. S. Brown, *J. Org. Chem.,* 58 (1993) 2122–2127.
(73) A. Fürst and P. A. Plattner, *Abstr. 12th Intern. Congr. Pure Appl. Chem.,* New York, (1951) 409.
(74) R. U. Lemieux and A. R. Morgan, *Can. J. Chem.,* 43 (1965) 2190–2198.
(75) J. Thiem, H. Karl, and J. Schwenter, *Synthesis,* (1978) 696–698.
(76) J. Thiem and P. Ossowski, *Liebigs Ann. Chem.,* (1983) 2215–2226.
(77) J. Thiem, S. Köpper, and J. Schwentner, *Liebigs Ann. Chem.,* (1985) 2135–2150.
(78) L. Laupichler, C. E. Sowa, and J. Thiem, *Bioorg. Med. Chem.,* 2 (1994) 1281–1294.

(79) J. Thiem and P. Ossowski, *J. Carbohydr. Chem.,* 3 (1984) 287–313.
(80) G. R. J. Thatcher, ed., *The Anomeric Effect and Associated Stereoelectronic Effects;* ACS Symposium Series 539, ACS, Washington, 1993
(81) C. L. Perrin and K. B. Armstrong, *J. Am. Chem. Soc.,* 115 (1993) 6825–6834.
(82) A. R. Vaino, S. S. C. Chan, W. A. Szarek, and G. R. J. Thatcher, *J. Org. Chem.,* 61 (1996) 4514–4515.
(83) M. A. Fabian, C. L. Perrin, and M. L. Sinnott, *J. Am. Chem. Soc.,* 116 (1994) 8398–8399.
(84) P. G. Jones, A. J. Kirby, I. V. Komarov, and P. D. Wothers, *J. Chem. Soc., Chem. Commun.,* (1998) 1695–1696.
(85) C. L. Perrin, M. A. Fabian, J. Brunckova, and B. K. Ohta, *J. Am. Chem. Soc.,* 121 (1999) 6911–6918.
(86) D. Horton, W. Priebe, and M. Sznaidman, *Carbohydr. Res.,* 205 (1990) 71–86.
(87) D. Horton, W. Priebe, and O. Varela, *Carbohydr. Res.,* 130 (1984) C1–C3.
(88) D. Horton and W. Priebe, *Carbohydr. Res.,* 136 (1985) 391–396.
(89) G. Belucci, M. Ferretti, G. Ingrosso, F. Marioni, A. Marsili, and I. Morelli, *Tetrahedron Lett.,* (1972) 3527–3530.
(90) P. Barici, G. Beluci, F. Marioni, I. Morelli, and V. Scartoni, *J. Org. Chem.,* 37 (1972) 4353–4357.
(91) S. J. Danishefsky and J. Y. Roberge, *Pure Appl. Chem.,* 67 (1995) 1647–1662.
(92) R. W. Friesen and S. J. Danishefsky, *J. Am. Chem. Soc.,* 111 (1989) 6656–6660.
(93) A. C. Richardson, *Carbohydr. Res.,* 10 (1969) 395–402.
(94) R. W. Friesen and S. J. Danishefsky, *Tetrahedron,* 46 (1990) 103–112.
(95) F. E. Mcdonald and S. J. Danishefsky, *J. Org. Chem.,* 57 (1992) 7001–7002.
(96) S. J. Danishefsky, K. Koseki, D. A. Griffith, J. Gervay, J. M. Paterson, F. E. Mcdonald, and T. Oriyama, *J. Am. Chem. Soc.,* 114 (1992) 8331–8333.
(97) D. A. Griffith and S. J. Danishefsky, *J. Am. Chem. Soc.,* 112 (1990) 5811–5819
(98) T. Hamada, A. Nishida, and O. Yonemitsu, *J. Am. Chem. Soc.,* 108 (1986) 140–145.
(99) S. J. Danishefsky, J.Gervay, J. M. Peterson, F. E. Mcdonald, K. Koseki, D. A. Griffith, T. Oriyama, and S. P. Marsden, *J. Am. Chem. Soc.,* 117 (1995) 1940–1953.
(100) C. W. Harwig, D. J. Gravert, and K. D. Janda, *Chemtracts—Organic Chemistry,* 12 (1999) 1–26.
(101) U. Zehavi and A. Patchornik, *J. Am. Chem. Soc.,* 95 (1972) 5673–5677.
(102) J. M. Fréchet and C. Schuerch, *Carbohydr. Res.,* 22 (1972) 399–412.
(103) S. J. Danishefsky, K. F. McClure, J. T. Randolph, and R. B. Ruggeri, *Science,* 260 (1993), 1307–1309.
(104) J. Y. Roberge, X. Beebe, and S. J. Danishefsky, *Science,* 269 (1995) 202–204.
(105) E. J. Corey and B. B. Snider, *J. Am. Chem. Soc.,* 94 (1972) 2549–2550.
(106) R. J. Ferrier and N. Prasad, *J. Chem. Soc. (C),* (1969) 570–575.
(107) B. K. Banik, M. S. Manhas, and A. K. Bose, *J. Org. Chem.,* 59 (1994) 4714–4716.
(108) M. Koreeda, T. A. Houston, B. K. Shull, E. Klemke, and R. J. Tuinman, *Synlett,* (1993) 90–92.
(109) J. C. López and B. Fraser-Reid, *J. Chem. Soc., Chem. Commun.,* (1992) 94–96.
(110) B. Fraser-Reid and D. E. Iley, Can. *J. Chem.,* 57 (1979) 645–648.
(111) I. Szczerek, J. S. Jewell, R. G. S. Ritchie, W. A. Szarek, and J. K. N. Jones, *Carbohydr. Res.,* 22 (1972) 163–172.
(112) R. J. Ferrier and N. Prasad, *J. Chem. Soc. (C),* (1969) 581–586.
(113) A. Hantsch, *Ber.,* 33, (1900), 524–527
(114) L. Birchenbach, J. Goubeau, and E. Berniger, *Ber.,* 65, (1932), 1339–1344.
(115) A. Hassner, J. E. Kropp, and G. J. Kent, *J. Org. Chem.,* 34 (1969) 2628–2632.
(116) W. A. Szarek, D. G. Lance, and R. L. Beach, *J. Chem. Soc., Chem. Commun.,* (1968) 356.
(117) W. A. Szarek, D. G. Lance, and R. L. Beach, *Carbohydr. Res.,* 13 (1970) 75–81.
(118) W. A. Szarek, J. S. Jewell, I. Szczerek, and J. K. N. Jones, *Can. J. Chem.,* 47 (1969) 4473–4476.
(119) R. G. S. Ritchie and W. A. Szarek, *Can. J. Chem.,* 50 (1972) 507–511.

(120) F. Cramer, H. P. Bär, H. J. Rhaese, W. Sänger, K. H. Scheit, G. Schneider, and J. Tennigkeit, *Tetrahedron Lett.,* (1963) 1039–1042.
(121) J. S. Brimacombe, J. G. H. Bryan, T. A. Hamor, and L. C. N. Tucker, *J. Chem. Soc., Chem. Commun.,* (1968) 1401–1402.
(122) J. S. Brimacombe, F. Hundey, and M. Stacey, *Carbohydr. Res.,* 13 (1970) 447–450.
(123) J. S. Jewell and W. A. Szarek, *Carbohydr. Res.,* 16 (1971) 248–250.
(124) W. A. Szarek, A. Zamojski, K. N. Tiwari, and E. R. Ison, *Tetrahedron Lett.,* 27 (1986) 3827–3830.
(125) R. G. S. Ritchie and W. A. Szarek, *Carbohydr. Res.,* 18 (1971) 443–445.
(126) D. Lafont and G. Descotes, *Carbohydr. Res.,* 166 (1987) 195–209.
(127) G. Kuswik and G. Grynkiewicz, *Pol. J. Chem.,* 54 (1980) 1319–1322.
(128) R. M. Moriarty and R. K. Vaid, *Synthesis,* (1990) 431–447.
(129) H. Waldmann, in H. Waldmann (Ed.), *Organic Synthesis Highlights II,* VCH, New York, 1995, pp. 223–230.
(130) A. Kirsching, *Eur. J. Org. Chem.,* (1998) 2267–2274.
(131) A. Kirschning, G. Dräger, and J. Harders, *Synlett,* (1993) 289–290.
(132) G. F. Koser and R. H. Wettach, *J. Org. Chem.,* 41 (1976) 3609–3611.
(133) A. Kirschning, *J. Org. Chem.,* 60 (1995) 1228–1232.
(134) A. Kirschning, *Liebigs Ann.,* (1995) 2053–2056.
(135) M. Arimoto, H. Yamaguchi, E. Fujita, Y. Nagao, and M. Ochiai, *Chem. Pharm. Bull.,* 37 (1989) 3221–3224.
(136) A. Kirsching, S. Domann, G. Dräger, and L. Rose, *Synlett,* (1995) 767–769.
(137) P. Magnus, A. Evans, and J. Lacour, *Tetrahedron Lett.,* 33 (1992) 2933–2036.
(138) P. Magnus and J. Lacour, *J. Am. Chem. Soc.,* 114 (1992) 3993–3994.
(139) M. A. Hashem, A. Jung, M. Ries, and A. Kirsching, *Synlett,* (1998) 195–197; See also erratum: *Synlett,* (1998), 692.
(140) A. Kirschning, U. Hary, C. Hary, M. Ries, and L. Rose, *J. Chem. Soc., Perkin Trans I,* (1999) 519–528.
(141) K. Fukase, A. Hasuoka, I. Kinoshita, and S. Kusumoto, *Tetrahedron Lett.,* 33 (1992) 7165–7168.
(142) K. Fukase, A. Hasuoka, I. Kinoshita, Y. Aoki, and S. Kusumoto, *Tetrahedron,* 51 (1995) 4923–4932.
(143) K. Fukase, A. Hasuoka, and S. Kusumoto, *Tetrahedron Lett.,* 34 (1993) 2187–2190.
(144) A. J. Kirby, *The Anomeric Effect and Related Stereoelectronic Effects at Oxygen;* Springer: New York, 1983.
(145) L. Sun, P. Li, and K. Zhao, *Tetrahedron Lett.,* 35 (1994) 7147–7150.
(146) K. Zegelaar-Jaarsveld, H. I. Duynstee, G. A. van der Marel, and J. H. van Boom, *Tetrahedron,* 52 (1996) 3575–3592.
(147) P. de Armas, C. G. Fransisco, and E. Suárez, *Angew. Chem., Int. Ed. Engl.,* 31 (1992) 772–774.
(148) R. L. Dorta, C. G. Fransisco, and E. Suárez, *Tetrahedron Lett.,* 35 (1994) 2049–2052.
(149) P. de Armas, C. G. Fransisco, and E. Suárez, *J. Am. Chem. Soc.,* 115 (1993) 8865–8866.
(150) R. L. Dorta, C. G. Fransisco, and E. Suárez, *Tetrahedron Lett.,* 34 (1993) 7331–7334.
(151) J. Inanaga, Y. Sugimoto, Y. Yokoyama, and T. Hanamoto, *Tetrahedron Lett.,* 33 (1992) 8109–8112.
(152) C. G. Fransisco, C. G. González, and E. Suárez, *Tetrahedron Lett.,* 37 (1996) 1687–1690.
(153) C. G. Francisco, C. G. González, and E. Suárez, *Tetrahedron Lett.,* 38 (1997) 4141–4144.
(154) H. Togo, M. Aoki, and M. Yokoyama, *Tetrahedron Lett.,* 32 (1991) 6559–6562.
(155) H. Togo, M. Aoki, T. Kuramochi, and M. Yokoyama, *J. Chem. Soc. Perkin Trans. I,* (1993) 2417–2427.
(156) E. Vismara, G. Torri, N. Pastori, and M. Marchiandi, *Tetrahedron Lett.,* 33 (1992) 7575–7578.
(157) D. B. Dess and J. C. Martin, *J. Org. Chem.,* 48 (1983) 4155–4156.

(158) D. B. Dess and J. C. Martin, *J. Am. Chem. Soc.,* 113 (1991) 7277–7287.
(159) R. E. Ireland and L. Liu, *J. Org. Chem.,* 58 (1993) 2899.
(160) J. B. Plumb and D. J. Harper, *Chem. Eng. News,* July 16 (1990) 3.
(161) S. D. Meyer and S. L. Schreiber, *J. Org. Chem.,* 59 (1994) 7549–7522.
(162) A. R. de Lera and W. H. Okamura, *Tetrahedron Lett.,* 28 (1987) 2921–2924.
(163) M. J. Robins, V. Samano, and M. D. Johnson, *J. Org. Chem.,* 55 (1990) 410–412.
(164) H. S. P. Rao, P. Muralidharan, and S. Pria, *Ind. J. Chem.,* 36B (1997) 816–818.
(165) S. R. Landauer and H. N. Rydon, *J. Chem. Soc.,* (1953) 2224–2234.
(166) D. G. Coe, S. R. Landauer, and H. N. Rydon, *J. Chem. Soc.,* (1954) 2281–2288.
(167) H. N. Rydon and B. L. Tonge, *J. Chem. Soc.,* (1956) 3043–3056.
(168) N. K. Kochetkov and A. I. Usov, *Tetrahedron,* 19 (1963) 973–983.
(169) J. P. H. Verheyden and J. G. Moffatt, *J. Am. Chem. Soc.,* 86 (1964) 2093–2095.
(170) J. P. H. Verheyden and J. G. Moffatt, *J. Org. Chem.,* 35 (1970) 2868–2877.
(171) P. J. Garegg and B. Samuelsson, *Synthesis,* (1979) 469–470.
(172) K. J. Brunings, *J. Am. Chem. Soc.,* 69 (1947) 205–208.
(173) P. J. Garegg and B. Samuelsson, *Synthesis,* (1979) 813–814.
(174) B. A. Arbusow, *Pure Appl. Chem.,* 9 (1964) 307–335.
(175) D. H. R. Barton, D. Zheng, and S. D. Géro, *J. Carbohydr. Chem.,* 1 (1982) 105–118.
(176) S. Jarosz, D. R. Hicks, and B. Fraser-Reid, *J. Org. Chem.,* 47 (1982) 935–940.
(177) F. A. Luzzio and M. E. Menes, *J. Org. Chem.,* 59 (1994) 7267–7272.
(178) G. L. Tong, W. W. Lee, and L. Goodman, *J. Org. Chem.,* 30 (1965) 2854–2855.
(179) J. P. Horwitz, J. Chua, M. A. Da Rooge, M. Noel, and I. L. Klundt, *J. Org. Chem.,* 31 (1966) 205–211.
(180) J. R. McCarthy Jr., M. J. Robins, L. B. Townsend, and R. K. Robins, *J. Org. Chem.,* 31 (1966) 1549–1553.
(181) P. J. Garegg and B. Samuelsson, *J. Chem. Soc. Perkin Trans. I,* (1979) 978–980.
(182) P. J. Garegg and B. Samuelsson, *J. Chem. Soc. Perkin Trans. I,* (1980) 2866–2869.
(183) P. J. Garegg, R. Johansson, C. Ortega, and B. Samuelsson, *J. Chem. Soc. Perkin Trans. I,* (1982) 681–683.
(184) B. Classon, P. J. Garegg, and B. Samuelsson, *Can. J. Chem.,* 59 (1981) 339–343.
(185) H. J. Jennings and J. K. N. Jones, *Can. J. Chem.,* 43 (1965) 2372–2386.
(186) P. J. Garegg, R. Johansson, and B. Samuelsson, *J. Carbohydr. Chem.,* 3 (1984) 189–195.
(187) Z. Pakulski and A. Zamojski, *Carbohydr. Res.,* 205 (1990) 410–414.
(188) A. P. Rauter, A. C. Fernandes, S. Czernecki, and J.-M. Valery, *J. Org. Chem.,* 61 (1996) 3594–3598.
(189) K. Dak, A. P. Rauter, A. E. Strütz, and H. Weidmann, *Liebigs Ann. Chem.,* (1981) 1768–1773.
(190) B. Classon, Z. Liu, and B. Samuelsson, *J. Org. Chem.,* 53 (1988) 6126–6130.
(191) Z. Liu, B. Classon, and B. Samuelsson, *J. Org. Chem.,* 55 (1990) 4273–4275.
(192) J. Brånalt, I. Kvarnström, G. Niklasson, S. C. T. Svensson, B. Classon, and B. Samuelsson, *J. Org. Chem.,* 59 (1994) 1783.
(193) J. Huang, E. B. McElroy, and T. S. Widlanski, *J. Org. Chem.,* 59 (1994) 3520–3521.
(194) D. K. Baeschlin, M. Daube, M. O. Blättler, S. A. Benner, and C. Richert, *Tetrahedron Lett.,* 37 (1996) 1591–1592.
(195) S. Oae and H. Togo, *Bull. Chem. Soc. Jpn.,* 56 (1983) 3802–3812.
(196) P. Ioannidis, P. Söderman, B. Samuelsson, and B. Classon, *Tetrahedron Lett.,* 34 (1993) 2993–2994.
(197) S. Hanessian, M. M. Ponpipom, and P. Lavallée, *Carbohydr. Res.,* 24 (1972) 45–56.
(198) M. M. Ponpipom and S. Hanessian, *Carbohydr. Res.,* 18 (1971) 342–344.
(199) K. P. R. Kartha, *Tetrahedron Lett.,* 27 (1986) 3415–3416.

(200) C. G. J. Verhart, B. M. G. Caris, B. Zwanenburg, and G. J. F. Chittenden, *Recl. Trav. Chim. Pays–Bas,* 117 (1992) 348–352.
(201) D. H. R. Barton and W. B. Motherwell, *Pure Appl. Chem.,* 53 (1981) 15–31.
(202) F. A. Cotton and G. Wilkinson, *Advanced Inorganic Chemistry,* 5th ed.; John Wiley & Sons: Toronto, 1988.
(203) A. R. Vaino and W. A. Szarek, *J. Chem. Soc., Chem. Commun.,* (1996) 2351–2352.
(204) H. Wetter and K. Oertle, *Tetrahedron Lett.,* 26 (1985) 5515–5518.
(205) C. J. Mazac, in K. Othme (Ed.), *Encyclopedia of Chemical Technology,* 3rd ed., Vol. 13, Wiley–Interscience, New York, 1981, pp. 649 677.
(206) J. L. Wahlstrom and R. C. Ronald, *J. Org. Chem.,* 63 (1998) 6021–6022.
(207) B. H. Lipshutz and J. Keith, *Tetrahedron Lett.,* 39 (1998) 2495–2498.
(208) K. P. R. Kartha and R. A. Field, *Synlett,* (1999) 311–312.
(209) K. S. Ramasamy, R. Bandaru, and D. Averett, *Synth. Commun.,* 29 (1999) 2881–2894.
(210) A. Cappa, E. Marcantoni, and E. Torregiana, *J. Org. Chem.,* 64 (1999) 5696–5699.
(211) R. Madsen, C. Roberts, and B. Fraser-Reid, *J. Org. Chem.,* 60 (1995) 7920–7926.
(212) S. Nadji, D. Reichlin, and M. J. Kurth, *J. Org. Chem.,* 55 (1990) 6241–6244.
(213) O. Kitagawa, T. Hanano, N. Kikuchi, and T. Taguchi, *Tetrahedron Lett.,* 34 (1993) 2165–2168.
(214) A. R. Vaino and W. A. Szarek, *Synlett,* (1995) 1157–1158.
(215) K. P. R. Kartha and R. A. Field, Tetrahedron, 34 (1997) 11753–11766.
(216) P. A. Bartlett, D. P. Richardson, and J. Myerson, *Tetrahedron,* 46 (1984) 2317–2327.
(217) P. A. Bartlett and J. Myerson, *J. Am. Chem. Soc.,* 100 (1978) 3950–3952.
(218) S. D. Rychnovsky and P. A. Bartlett, *J. Am. Chem. Soc.,* 103 (1981) 3963–3964.
(219) Y. Tomaru, S.-I. Kawamaura, and Z.-I. Yoshida, *Tetrahedron Lett.,* 26 (1986) 2885–2888.
(220) P. A. Bartlett and P. C. Ting, *J. Org. Chem.,* 51 (1986) 2230–2240.
(221) K. B. Sharpless, C. H. Behrens, T. Katsuki, A. W. Lee, V. S. Martin, M. Takatani, S. M. Viti, F. J. Walker, and S. S. Woodward, *Pure Appl. Chem.,* 55, (1983), 589–604.
(222) A. R. Chamberlin, M. Dezube, P. Dussault, and M. C. McMills, *J. Am. Chem. Soc.,* 105 (1983) 5819–5825.
(223) A. G. M. Barrett, B. C. B. Bezuidenhoudt, and L. M. Melcher, *J. Org. Chem.,* 55 (1990) 5196–5197.
(224) P. J. Garegg and P. Ossowski, *Acta Chem. Scand.,* B37 (1983) 249–250.
(225) M. Nishizawa, D. M. Garcia, Y. Noguchi, K. Komatsu, S. Hatakeyama, and H. Yamada, *Chem. Pharm. Bull.,* 42 (1994) 2400–2402.
(226) G. H. Posner and D. S. Bull, *Tetrahedron Lett.,* 37 (1996) 6279–7282.
(227) W. H. Pirkle and J. M. Finn, *J. Org. Chem.,* 46 (1981) 2935–2938.
(228) H. B. Mereyala and K. C. Venkataramanaiah, *J. Chem. Research (S),* (1991) 197.
(229) S. David and S. Hanessian, *Tetrahedron,* 41, (1985) 643–663.
(230) I. D. Blackburne, P. M. Fredericks, and R. D. Guthrie, *Aust. J. Chem.,* 1976 29 381–391.
(231) M. S. Shekhani, K. H. Khan, K. Mahmood, P. M. Shah, and S. Malik, *Tetrahedron Lett.* 31 (1990) 1669–1670.
(232) S. Czernecki, C. Leteux, and A. Veyrières, *Tetrahedron Lett.,* 33 (1992) 221–224.
(233) D. Tailler, J.-C. Jacquinet, A.-M. Noirot and J.-M.Beau, *J. Chem. Soc. Perkin Trans. I,* (1992) 3163–3164.
(234) H. Zhang, M. Seepersaud, S. Seepersaud, and D. R. Mootoo, *J. Org. Chem.,* 63 (1998) 2049–2052.
(235) N. Khan, H. Xiao, B. Zhang, X. Chang, and D. R. Mootoo, *Tetrahedron,* 55 (1999) 8303–8312.
(236) P. Bertrand, H. E. Sukkari, J.-P. Gesson, and R. Renoux , *Synthesis,* (1999) 330–335.
(237) P. A. Bartlett and J. Myerson, *J. Am. Chem. Soc.,* 101 (1979) 1625–1627.
(238) S. Takano, M. Hirama, and K. Ogasawara, *J. Org. Chem.,* 45 (1980) 3729–3730.
(239) D. R. Williams and F. H. White, *Tetrahedron Lett.,* 27 (1986) 2195–2198.

(240) R. Filler, *Chem. Rev.,* 63 (1963) 21–43.
(241) S. Hanessian, D. H.-C. Wong, and M. Therien, *Synthesis,* (1981) 394-396.
(242) N. Venkatasubramanian and V. Thiagarajan, *Can. J. Chem.,* 47 (1969) 694–697.
(243) N. S. Srinivasan and N. Venkatasubramanian, *Tetrahedron,* 30 (1974) 419–425.
(244) T. Benneche, *Synthesis,* (1995) 1–27.
(245) R. D. Chambers, G. Sandford, M. E. Sparrowhawk, and M. J. Atherton, *J. Chem. Soc. Perkin Trans. I,* (1996) 1941–1944.

STRUCTURE OF ANOMERIC GLYCOSYL RADICALS AND THEIR TRANSFORMATIONS UNDER REDUCTIVE CONDITIONS

By Jean-Pierre Praly

Unité Mixte de Recherche CNRS—Université Claude-Bernard n° 5622, ESCPE—Lyon, Bât. 308
43 Boulevard du 11 Novembre 1918, 69622, Villeurbanne, France

I. Introduction	65
II. Structure of Sugar-Derived Radicals	67
1. Structure of Carbon-Centered Acetal Radicals	68
2. ESR Studies of Carbohydrate-Derived Radicals	69
3. Interpretation in the Framework of the Frontier Molecular Orbital (FMO) Theory	85
III. Radical Methods to Anhydroalditols from Glycosyl Derivatives	86
1. Radical Reduction with Tin and Silicon Hydrides	87
2. Synthesis of Anhydroalditols by Single-Electron Transfer	100
3. Synthesis of Ring-Fused and C-Branched Anhydroalditols	102
4. Stereoselectivity of Hydrogen-Atom Transfer at Anomeric Radicals	103
IV. Diastereoselective Reductions of Substituted Anomeric Radicals	106
1. Stereocontrolled Synthesis of O-Glycosides and Analogs	107
2. Stereocontrolled Synthesis of C-Glycosyl Compounds and Analogues	115
3. Radical Routes to Nucleosides	127
4. Reduction Diastereoselectivity of Substituted Anomeric Radicals	129
V. Synthesis of 2-Deoxy Sugars by Radical-Induced Rearrangement	130
1. Synthesis of 2-Deoxy Sugars by 2,1-Acyloxy Group Rearrangements	131
2. Synthesis of 2-Deoxy Sugars by 2,1-Phosphonyloxy Group Rearrangement	135
3. Mechanism of Radical-Induced 2,1-Migrations of Ester Groups	138
VI. Radical-Mediated Eliminations: Formation of Glycals	139
VII. Conclusion	142
References	143

I. Introduction

For quite some time, the chemistry of carbohydrate derivatives has been based essentially on ionic processes. The glycosidation reaction, which constitutes a continuing challenge in the field of oligosaccharide synthesis, essentially exploits the reactivity of nucleophilic species towards glycosyl cations or electrophilic sugar derivatives. The classic strategies aiming at the protection–deprotection of carbohydrate derivatives also involve in most cases ionic intermediates generated under either acid- or base-catalyzed conditions. Similarly, negatively charged species intervene in most of the long-established methods suitable for the manipulation of functionalities derived from hydroxyl groups in such reactions as nucleophilic substitution or elimination. Intensive researches over many years have led to a precise understanding of the reactivity of ionic intermediates and have allowed the development of a wide array of derived synthetic methods. In spite of their broad applicability, it is clear that the disadvantages of ionic reactions stem in general from the rather harsh conditions often required.

The search for milder conditions for the manipulation of complex molecules, with high selectivity control, inspires many modern synthetic developments. In this respect, free-radical chemistry, which primarily concerns modifications of weakened bonds, offers promising solutions to this problem. However, for many years, radical processes were considered to be nonselective. Since the 1970s, this prejudice has waned as deeper insights into the formation, structure, and reactions of radicals have emerged. Radical chemistry has thus achieved enhanced prominence in organic synthesis, with wide applications for deoxygenation of alcohols,[1] carbon–carbon bond formation,[2–5] or both, based on radical chemistry associated with the thiocarbonyl group.[6] The ready availability of tri-n-butyltin hydride (n-Bu$_3$SnH) and its versatile reactivity towards numerous functional groups has certainly played a major role for the development of a rich palette of radical reductive methods and C–C bond-forming processes.[7,8] The contribution of radical chemistry to organic synthesis is testified by the number of well-documented reviews devoted to synthetic aspects[9–11] including applications to natural product synthesis.[12] In addition, the early synthetic developments in radical chemistry have been comprehensively surveyed in a two-volume compilation.[13] A deeper appreciation of the stereochemistry of radical reactions contrasts with the old prejudice about their alleged low selectivity.[14,15]

Carbohydrates as a class of chiral compounds displaying several types of oxygen-based functional groups amenable to varied modifications offer wide opportunities for radical-based transformations. The tri-n-butyltin hydride-mediated deoxygenation of dithiocarbonates (Barton–McCombie reaction), reported in 1975, constituted a new and mild method for the synthesis of deoxy sugars.[16] It was an inspiring finding which launched the long-lasting and seminal contribution of Barton's group to radical-mediated C–H bond forming methodologies,[17] among other radical-based synthetic developments.[18]

In 1977, Ferrier showed that hydrogen atoms at C-1 and C-5 of pyranoid compounds, and at C-4 of some furanoid derivatives, can be abstracted on treatment with N-bromosuccinimide in refluxing carbon tetrachloride, provided that the reaction is carried out under irradiation or in the presence of a radical initiator such as benzoyl peroxide. Selectively brominated compounds, not accessible by other routes, were among the reaction products.[19,20] Following these early observations, this mild and selective halogenation method was extended further to several classes of sugar derivatives for preparing many brominated carbohydrates and analogues.[21]

The photochemical intramolecular cyclization of 3-oxoalkyl glycosides, which results in spiro-bicyclic acetals,[22] and of hypoiodites of the corresponding 2- or 3-hydroxyalkyl glycosides, which gives spiro-orthoester derivatives,[23] were also early examples of substitution processes resulting from the abstraction of the H-1 hydrogen atom. The discovery that protected glycopyranos-1-yl radicals add to acrylonitrile[24,25] to produce the corresponding C-glycosyl derivatives efficiently and stereoselectively was the starting point of detailed investigations exploiting the ability of carbon-centered radicals to add intra- or inter-molecularly to various unsaturated bonds. This field was first explored by Giese's group, in collaboration with Sustmann's group, for structural studies of carbohydrate radicals. The requirements for efficient creation of C–C bond are discussed in detail in Giese's book,[3] which established reliable radical routes to C-glycosyl compounds.[26]

Radical methods have also been developed in specific classes of sugar derivatives or for purposes not strictly synthesis oriented. For example, sugar nitroxyl derivatives constitute a broad class of persistent oxygen-centered radicals[27] useful for spin labeling.[28] A high proportion of carbohydrate free-radical research is focused on investigating the products of radiolysis[29] or of chemically initiating species in solution,[30,31] with applications to such polysaccharides as starch.[32] Such investigations are also oriented toward a better understanding of radical-induced damage to nucleic acids[33] and DNA.[34–36]

Consequently, over the past two decades, a large volume of data concerning radical-mediated transformations of sugar derivatives has become available. The aim of the present article is to survey only those aspects that concern the anomeric center of cyclic carbohydrate derivatives and focus on structural features and synthetic transformations based on radical-mediated reductions. Although radical addition to alkenes involves initial C–C bond formation followed by C–H bond formation by hydrogen-atom transfer, this type of reaction is not considered here, although a plethora of such examples may be found in the literature.[3,13,37–40] For the sake of consistency and clarity, some aspects involving other radical centers is mentioned. The chemistry of carbon-centered anomeric radicals has been featured in a specific review by Descotes,[41] which also discussed photo-induced substitutions at anomeric and proanomeric carbon[42] among other synthetic applications of photochemistry to the sugar series. Based on a literature coverage, which includes the year 1998, the structural features of glycopyranos-1-yl radicals are discussed first, followed by radical-based reductions. The synthetic aspects are ordered according to their respective applications in the field of anhydroalditol synthesis, diastereoselective preparation of glycosides and C-glycosyl compounds, synthesis of 2-deoxy sugars, and preparation of glycals via radicals.

II. STRUCTURE OF SUGAR-DERIVED RADICALS

A high stereoselectivity was found for the free-radical reduction of glycopyranosyl halides[43,44] and related 1-C-nitro-sugars[45] with tri-n-butyltin hydride.

Addition of protected glucopyranos-1-yl radicals to acrylonitrile and methyl acrylate was shown to result in stereoselective formation of the corresponding C-α-D-glucopyranosyl derivatives, with minor amounts of the β anomers.[46,47] Surprisingly enough, the anomeric ratio of the products was highly dependent on the O-acetyl or O-methyl sugar protecting groups, giving[47] an α:β ratio of over 50 and 3.5, respectively. In each case, as well as for related radical-mediated reactions involving the anomeric center, the newly created bond was generated stereoselectively from the α face of the $^4C_1$(D) pyranosyl ring.[3,13,21–23] These early observations showed convincingly that radical-mediated reactions can proceed with high stereocontrol. In the search of a rationale to account for this stereoselectivity, these observations raised questions about the precise structure of the radical intermediates. They were first presumed to be σ radicals,[43–45] by analogy with acetalic carbon-centered radicals. However, electron spin resonance (ESR) spectroscopy, used to investigate the structure of a broad range of variously configured and protected sugar-derived radicals, provided reliable data that highlight the peculiarities of this class of intermediates.

1. Structure of Carbon-Centered Acetal Radicals

Using methyl-substituted 2-alkoxytetrahydropyrans as anchored models, hydrogen-atom abstraction by triplet benzophenone or *tert*-butyloxy radicals was shown to occur faster for axially oriented hydrogen atoms attached to the acetalic carbon than for their equatorial counterparts. The relative rates for abstraction, first estimated[48–50] to be in the range 8–10, were later found[51] to be higher (10–16), thus confirming the preferential abstraction of axial hydrogen atoms in conformationally fixed cyclic acetals. Moreover, independently of the *cis* or *trans* configuration of the anchored 2-alkoxytetrahydropyrans investigated (such as **1** and **2**, respectively), the same carbon-centered acetal radical, having an equatorially oriented alkoxy group (e.g., **1R**) was detected by ESR spectroscopy, even at low temperature.[52] This favored conformation, compatible with stabilization of the radical center by conjugative delocalization to the p-type lone pair on the ring oxygen atom, was later confirmed by ESR measurements using several 2-alkoxytetrahydropyran derivatives labeled with ^{13}C at the acetalic 2-position.[53] From the observed average ^{13}C hyperfine coupling constant of 99 G, the deviation from planarity for these σ radicals was estimated to be ~11°, with sigma bond angles of 116°.

Reductive decyanation of 2-cyanotetrahydropyran[54] derivatives with sodium in ammonia at −78 °C, which is thought to proceed throught a pyramidal, axial radical giving by electron transfer a configurationally stable axial anion subsequently protonated with retention of configuration, also indicated a strong preference for axial protonation, on the basis of an observed 119:1 selectivity. Similarly, substituted tetrahydropyran derivatives of related structure, subjected to radical decarboxylation at −78 °C, in the presence of either *tert*-butyl mercaptan as an H-atom donor or acrylonitrile as a radicophilic alkene, gave nonequilibrium radical intermediates, which may be trapped with high stereoselectivity from the axial direction.[55]

2. ESR Studies of Carbohydrate-Derived Radicals

a. Protected Hexo- and Pento-pyranos-1-yl Radicals.—Whereas free-radical chain reactions are useful in synthesis, it is more appropriate, for spectroscopic studies, to generate carbon-centered radicals by nonchain processes. In the carbohydrate field, this is best achieved by means of the photochemical cleavage of hexamethylditin to generate trimethyltin radicals, which act as halogen abstractors and may trigger homolysis of the C–Se bond in selenoglycosides.[47,56–58] Such oxygen-free solvents as benzene, toluene, fluorobenzene, and oxolane, or (more rarely) deuteriated methanol can be used, as well as hexabutylditin in benzene or oxolane. By this method and others, such as exposure of sugar bromides to ^{60}Co γ-rays at low temperature[59] and the Norrish type I photocleavage of *C*-glycosyl compounds having a pivaloyl group[47,56,60,61] at C-1, protected glycopyranos-1-yl radicals are readily generated from accessible sugar derivatives (Table I).

The ESR spectra obtained at several temperatures and by different methods from 2,3,4,6-tetra-*O*-acetyl-α-D-glucopyranosyl bromide (**3**), phenyl 2,3,4,6-tetra-*O*-acetyl-1-seleno-β-D-glucopyranoside (**4**), or *tert*-butyl (2,3,4,6-tetra-*O*-acetyl-β-D-glucopyranosyl) ketone (**5**) were comparable, indicating formation of the same 2,3,4,6-tetra-*O*-acetyl-D-glucopyranos-1-yl radical (**3R**).[47,56,62,66,67] The ESR data indicate a similar structure for the 2,3,4,6-tetra-*O*-methyl-D-glucopyranos-1-yl radical (**6R**) obtained upon photolysis from *tert*-butyl (2,3,4,6-tetra-*O*-methyl-β-D-glucopyranosyl) ketone (**6**).[56] In radical **7R**, ^{13}C-labeling at C-1 allowed detection of the α-^{13}C hyperfine coupling constant. The measured value (4.73–4.75 mT) led to the conclusion that radical **7R** and its analogues **3R, 6R, 8R,** and **9R** may be regarded as π-type radicals, with an almost planar radical center.[56] The radical structures were more precisely established on the basis of the observed hyperfine splitting due to the geminal and vicinal hydrogen atoms at C-1, C-2, C-3, and C-5 (α, β, and γ positions, respectively). The main features of the ESR spectra of radicals **3R, 6R, 8R,** and **9R** are the relatively small coupling constants of the β-hydrogen atoms at C-2 and the occurrence of two γ-hydrogen couplings of relatively large magnitude.[56] On the basis of the W-like arrangement of bonds in the C-1–C-3 subunit in these pyranos-1-yl radicals, the smaller of the two

TABLE I
Generation, Structure, and ESR Data for Radicals 3R–40R

Substrate	Method[a]	Conformation[b]	T °C[c]	g^d	$a(\alpha\text{-H})$	$a(\beta\text{-H})$	$a(\gamma_1\text{-H})$	$a(\gamma_2\text{-H})$	$a(\text{Other})$	Ref.
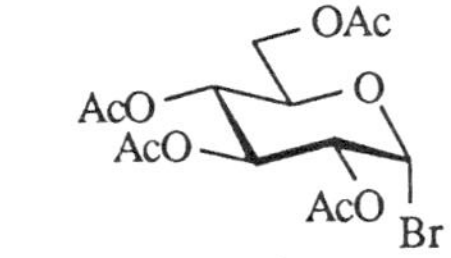 3	A	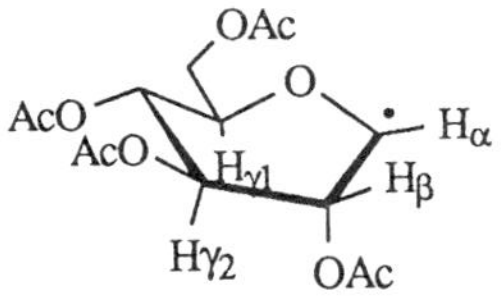 3R	−30	2.0031	1.800	1.364	0.348	0.145		56
	B		+13	2.00311 (3)	1.817	1.245	0.355	0.160		62
	A		+20	2.0031	1.796	1.407	0.345	0.141		56
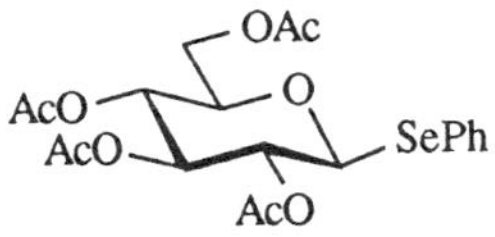 4	C	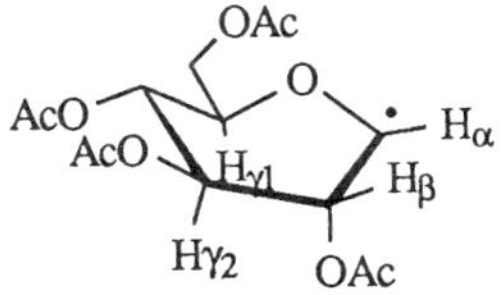 3R	+16	2.0031	1.799	1.392	0.345	0.143		56
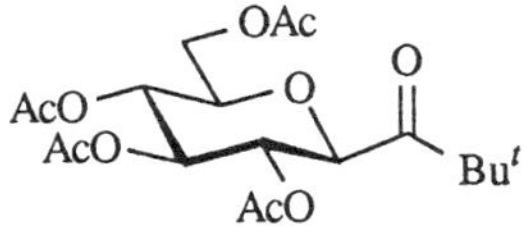 5	D	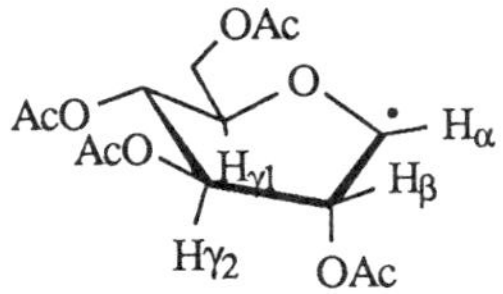 3R	−27	2.0031	1.844	1.179	0.377	0.173		47,56
	D		+40	2.0031	1.820	1.291	0.356	0.145		47,56
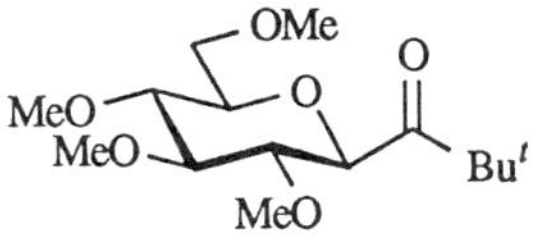 6	D	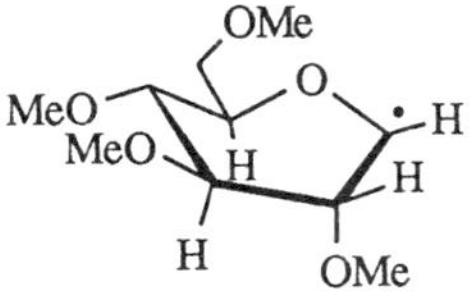 6R	−22	2.0032	1.799	1.058	0.388	0.221		47,56
	D		+3	2.0032	1.799	1.162	0.382	0.205		47,56
	D		+44	2.0032	1.792	1.274	0.359	0.184		47,56

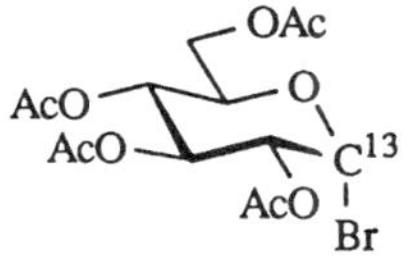

Structure								Ref
7	**7R**	A A	+15 +30	2.00313 2.00313	1.788 1.808	1.285 1.308	0.358 0.360	0.146 0.143	4.730 f 4.751 f	56 56
8	**8R**	A	−6	2.0032	1.815	1.172		0.154	0.047 g	56
9	**9R**	C C C	−24 +16 +33	2.0035 2.0035 2.0033	1.802 1.802 1.802	1.002 1.200 1.199	0.384 0.376 0.376	0.164 0.142 0.142		56 56 56
10	**10R**	A	+30	2.0031	1.706	2.893	0.207	0.040		56

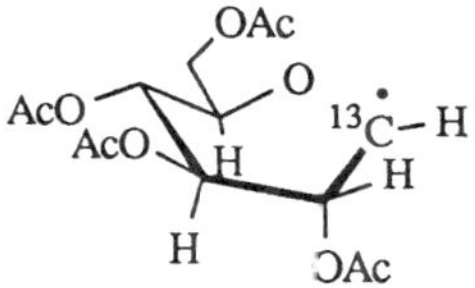

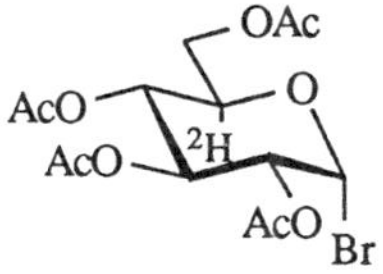

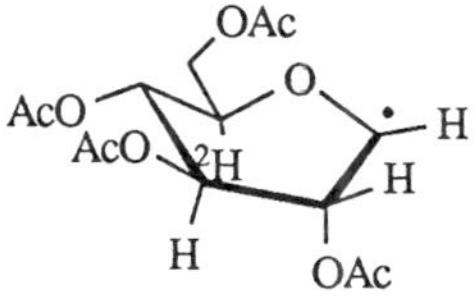

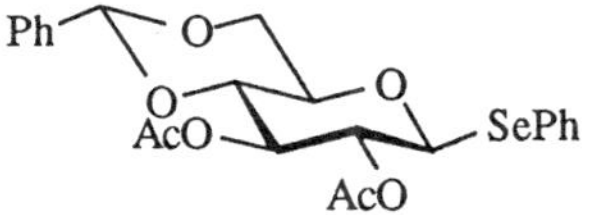

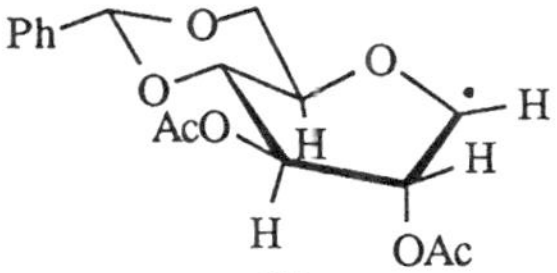

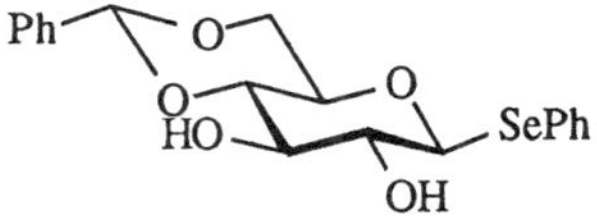

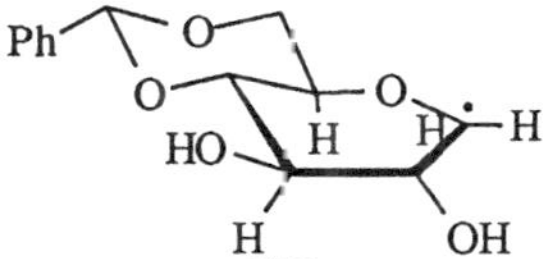

(continued)

TABLE I—*Continued*

Substrate	Method[a]	Conformation[b]	T °C[c]	g^d	Hyperfine Splittings (mT)[e]					Ref.
					$a(\alpha\text{-H})$	$a(\beta\text{-H})$	$a(\gamma_1\text{-H})$	$a(\gamma_2\text{-H})$	$a(\text{Other})$	
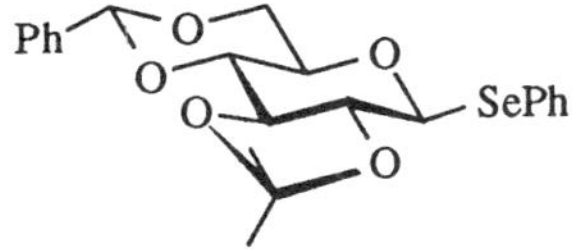 11	E	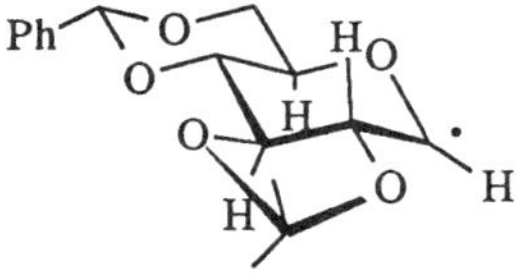 11R	+17	2.00313 (3)	1.393	3.445	0.156	0.084	6.35 *f*	56,57
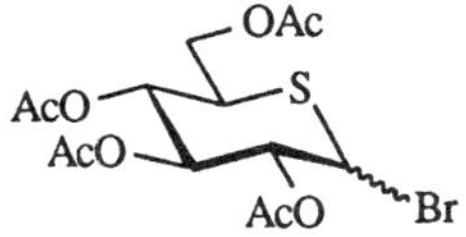 12	B	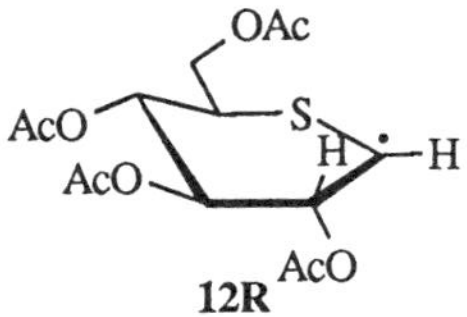 12R	+24	2.00441 (1)	1.732	1.883	0.359			62
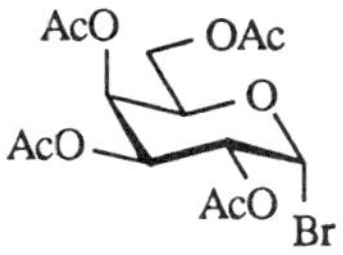 13	B	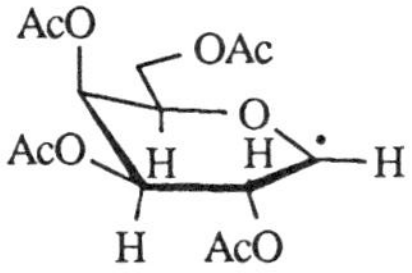 13R	+12	2.00319 (3)	1.648	2.873	0.268			62
	A		+30	2.0031	1.684	2.759	0.253			56
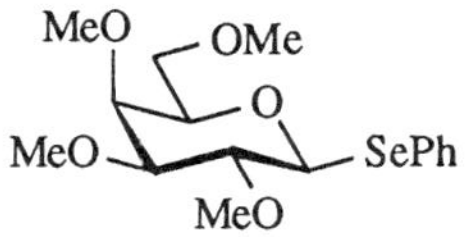 14	A	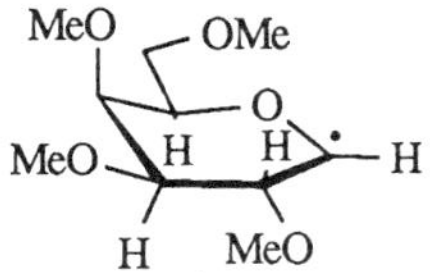 14R	+66	2.0031	1.712	2.597	0.243	<0.03		56
	A		+75	2.0031	1.716	2.578	0.228	<0.03		56
	A		+90	2.0031	1.726	2.540	0.237	<0.03		56

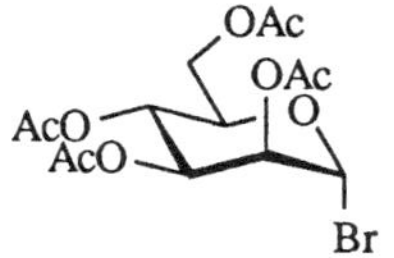 ![15]	A	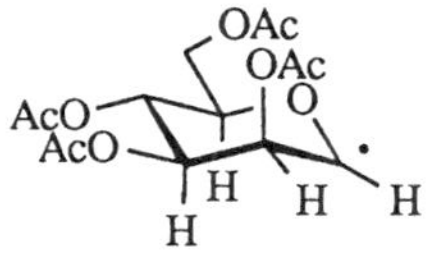 ![15R]	+20	2.00307	1.852	0.353	0.307	<0.03		56
15	A	15R	+95	2.0031	1.834	0.424	0.307	<0.03		56
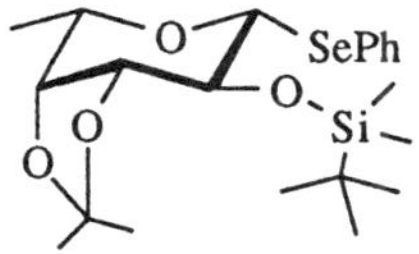 ![16]	E	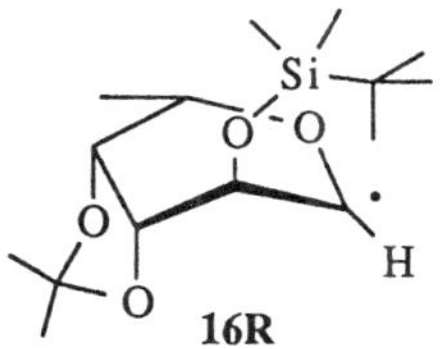 ![16R]	+7	2.00295	1.845	0.140	0.375	0.375		63
16		16R								
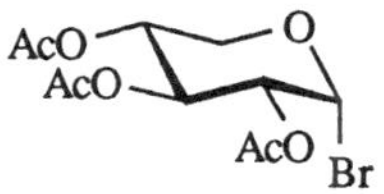 ![17]	A	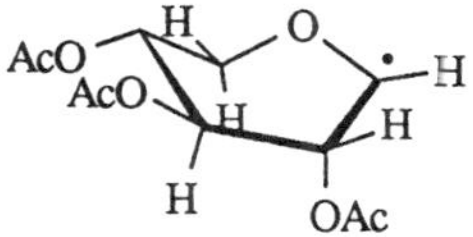 ![17R]	−2	2.0031	1.816	1.189	0.176	0.176	0.048[h]	56
17	A	17R	+5	2.0031	1.829	1.221	0.176	0.176	0.051[h]	56
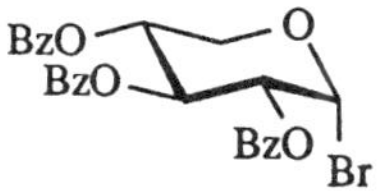 ![18]	F	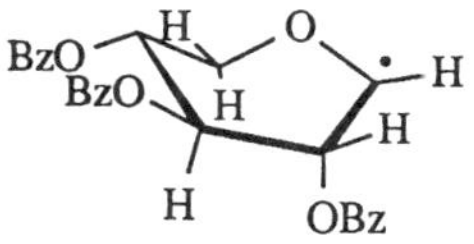 ![18R]	+22	2.0031	1.845	1.250	0.114	0.114		56
18		18R								

(*continued*)

TABLE I—Continued

Substrate	Method[a]	Conformation[b]	T°C[c]	g[d]	Hyperfine Splittings (mT)[e]					Ref.
					$a(\alpha\text{-H})$	$a(\beta\text{-H})$	$a(\gamma_1\text{-H})$	$a(\gamma_2\text{-H})$	$a(\text{Other})$	
19	B	19R	+30	2.00306	1.826	0.473	0.193	0.091		56
	B		+36	2.00306	1.833	0.486	0.209	0.099		56
	B		+45	2.0031	1.835	0.499	0.210	0.106		56
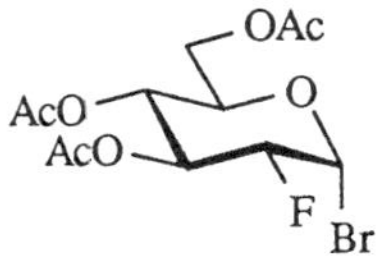 20	B	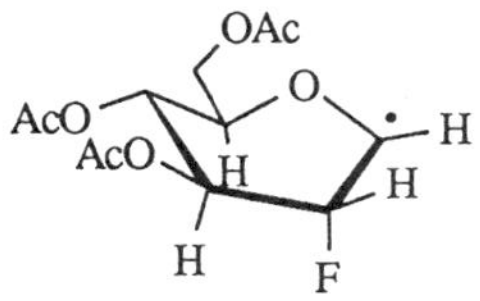 20R	−1	2.0038(2)	1.905	0.435	0.375	0.310	13.62[i]	57
21	B	21R	−9	2.00328 (3)	1.732	4.768 <0.03	0.187	0.031		57
	B		+23	2.00329 (3)	1.734	4.750 0.015	0.180	0.083	≤0.01[j]	57
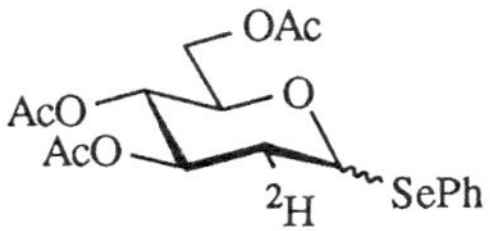 22	B	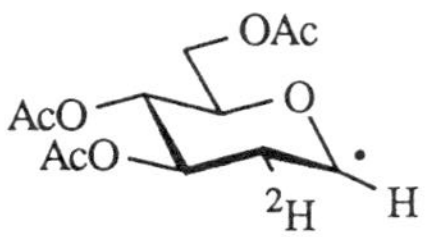 22R	−9	2.00323 (3)	1.728	4.07 (1)	0.184	0.083	<0.01[k]	57

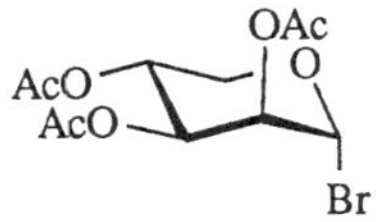

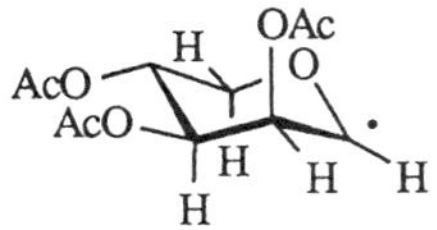

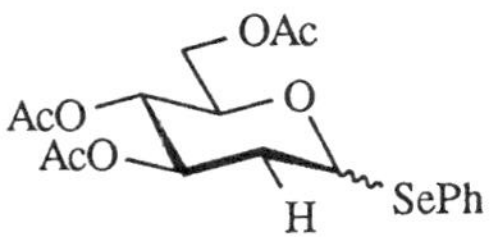

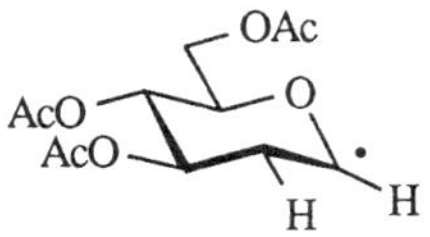

74

Compound	Method	Radical		g						Ref.
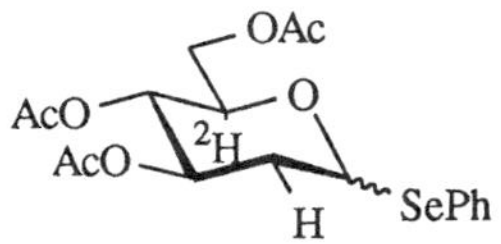 **23**	B	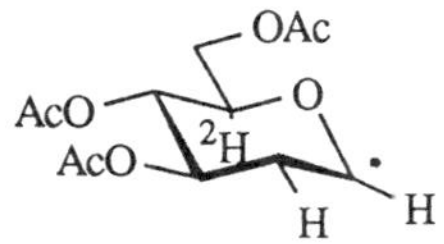 **23R**	−9	2.00320(5)	1.725	4.769 0.014	—	0.079	0.028[g] 0.010[j]	57
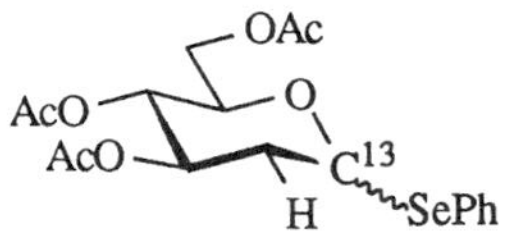 **24**	B	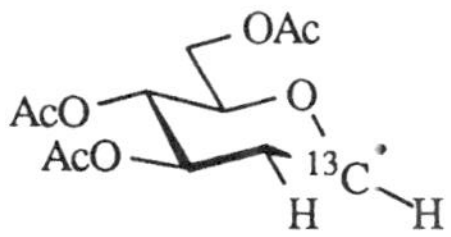 **24R**	−9	2.00347(5)	1.728	4.768 0.010	0.187	0.081	5.774[f]	57
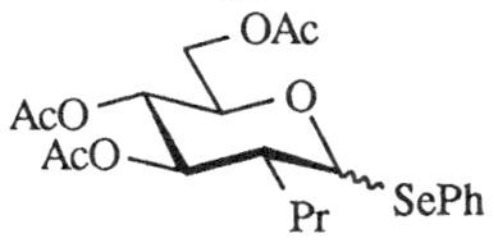 **25**	B	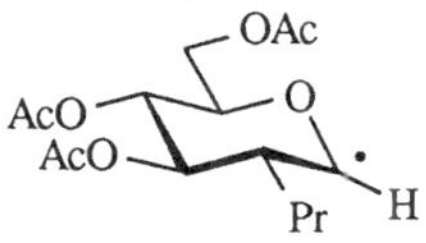 **25R**	+50	2.0032(1)	1.700	3.733	0.190	0.080	0.055[l]	57
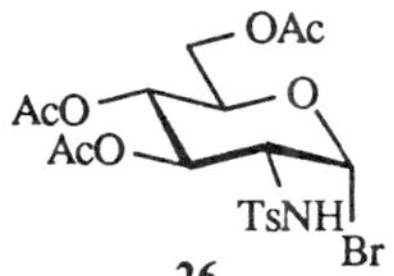 **26**	B	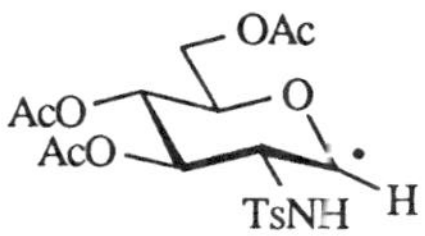 **26R**	−16	2.0031(1)	1.633	3.580	0.190	0.073	0.185[m]	57
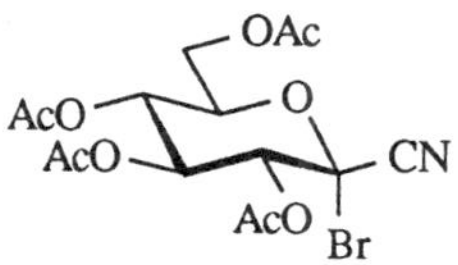 **27**	G	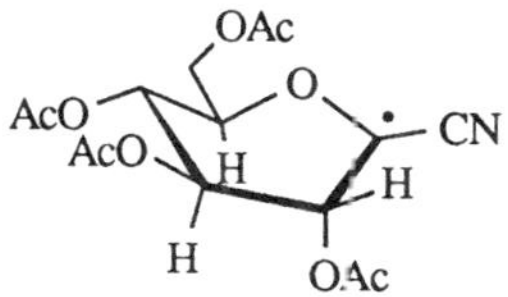 **27R**	+17 +24	2.00360(5)		1.343 1.370	0.397 0.397	0.103 0.103	[n] [n]	58 58

(continued)

TABLE I—Continued

Substrate	Method[a]	Conformation[b]	T°C[c]	g^d	Hyperfine Splittings (mT)[e]				Ref.
					$a(\alpha\text{-H})$	$a(\beta\text{-H})$	$a(\gamma_1\text{-H})$	$a(\gamma_2\text{-H})$	$a(\text{Other})$	
28	G	28R	+7	2.00331		2.646	0.346	0.045	$0.025^{j,n}$	58
			+37	(1)		2.566	0.347	0.046		58
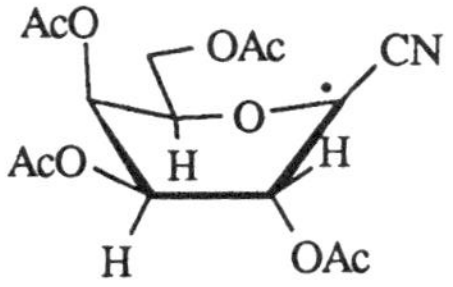 28R'		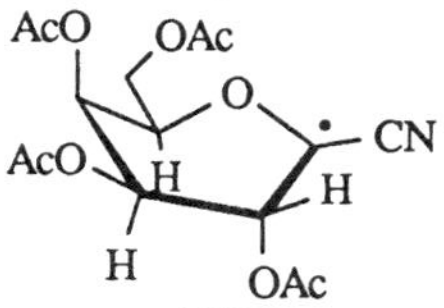 28R''	+37	2.00372 (5)		0.070	0.347	0.046	0.025^n	58
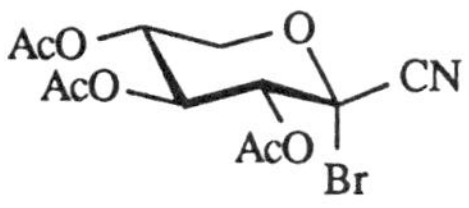 29	G	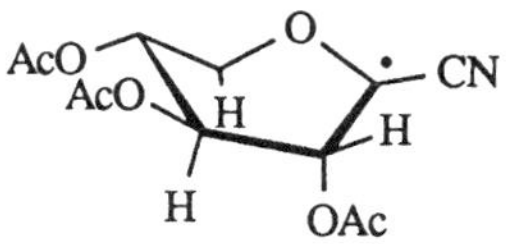 29R	+37	2.00334 (1)		1.630	0.240	0.227	$<0.020^{n,o}$	58
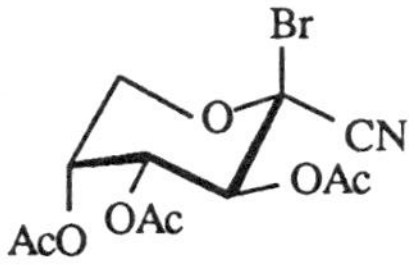 30	G	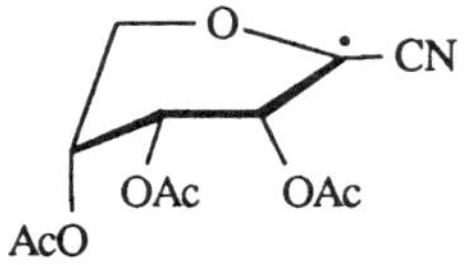 30R	+31	2.00338 (1)		2.266	0.320	0.050	$0.025^{n,o}$	58

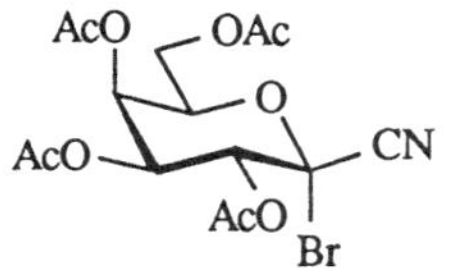

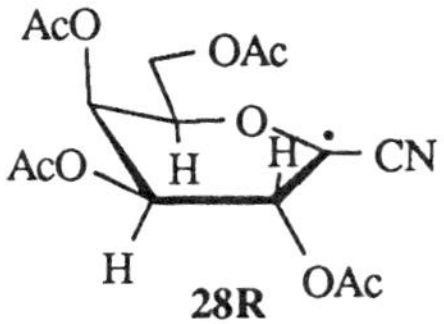

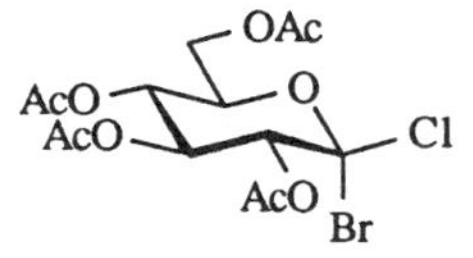

31	A	31R	−16	2.00465 (3)	2.333	0.137	0.533^p 58
			+4		2.149	0.139	0.521^p 58
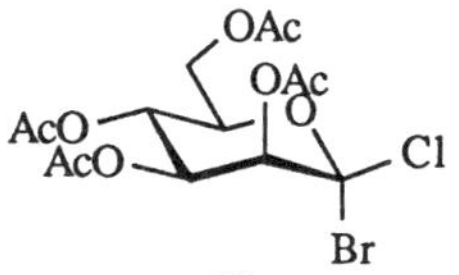 32	B	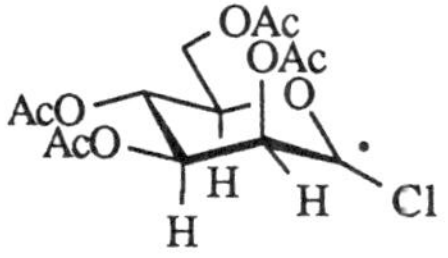 32R	+23		0.275	0.133	0.690^p 58
33	B	33R	+30		0.587	0.163	0.408
							0.103^{q,r} 58
34	B	34R	+2	2.0037 (1)	0.530	0.350	0.132^{q,r} 58
			+2		0.530	0.350	0.102^q 58
			+24		0.547	0.335	0.128^{q,r} 58
			+24		0.547	0.335	0.103^q 58

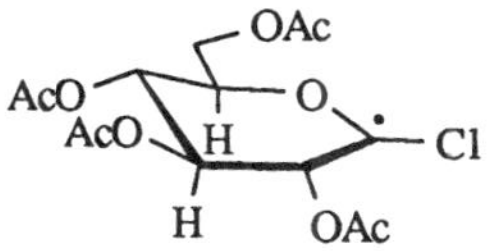

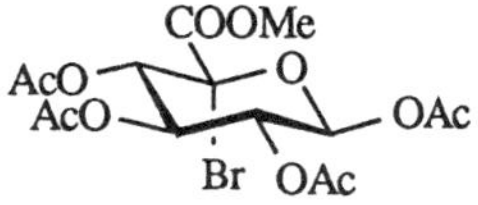

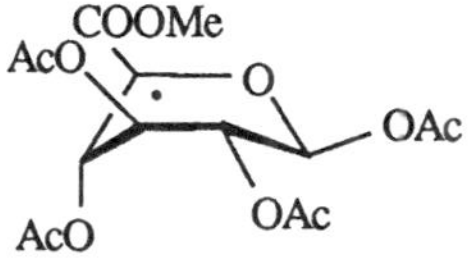

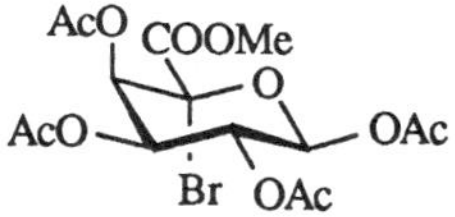

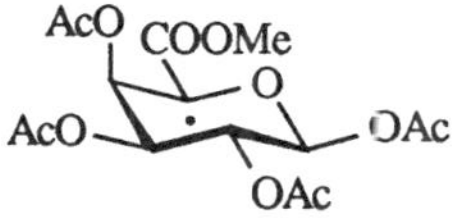

(*continued*)

TABLE I—Continued

Substrate	Method[a]	Conformation[b]	T°C[c]	g[d]	\multicolumn{4}{c}{Hyperfine Splittings (mT)[e]}	Ref.
					$a(\alpha\text{-H})$	$a(\beta\text{-H})$	$a(\gamma_1\text{-H})$	$a(\gamma_2\text{-H})$ a(Other)	
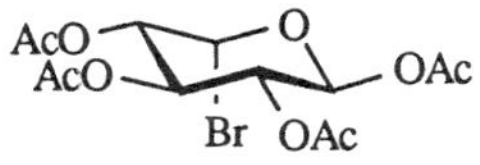 **35**	B	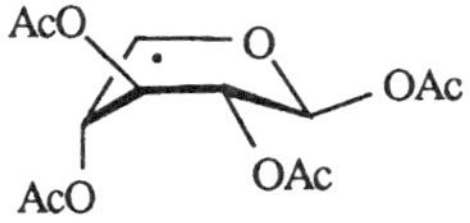 **35R**	−8	2.00307 (1)		1.441	0.186	1.870[s]	58
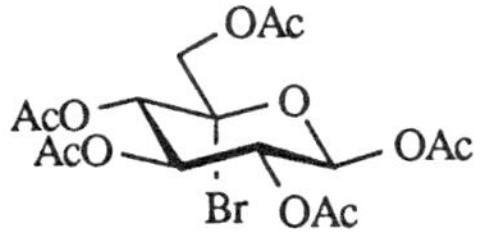 **36**	B	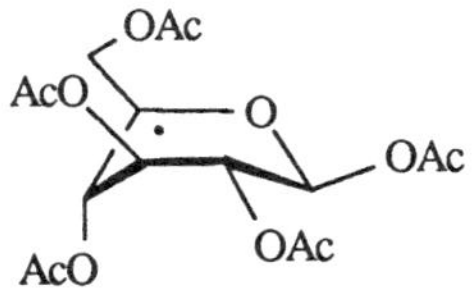 **36R**	+12	2.00302 (5)		1.328	0.200	0.106 1.085[t] 0.831[t]	58
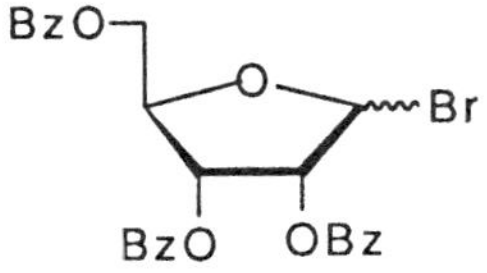 **37**	E	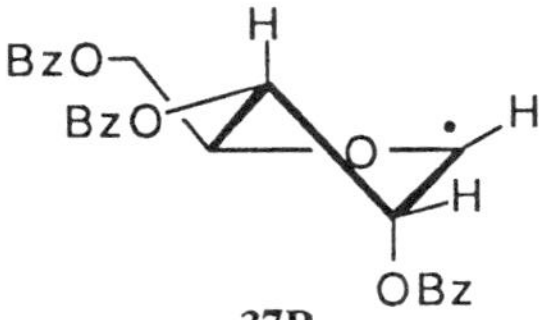 **37R**	+7	2.0031	1.80	0.99	0.084	0.29	64
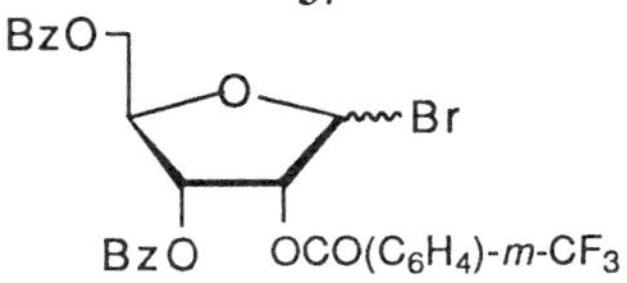 **38**	E	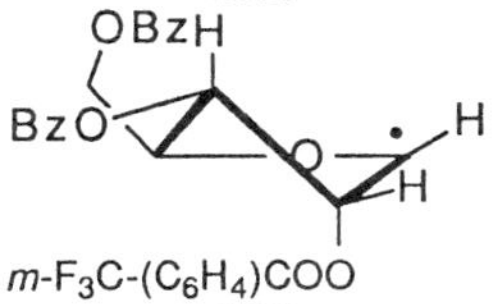 **38R**	+7	2.0031	1.83	1.00	0.084	0.29	64

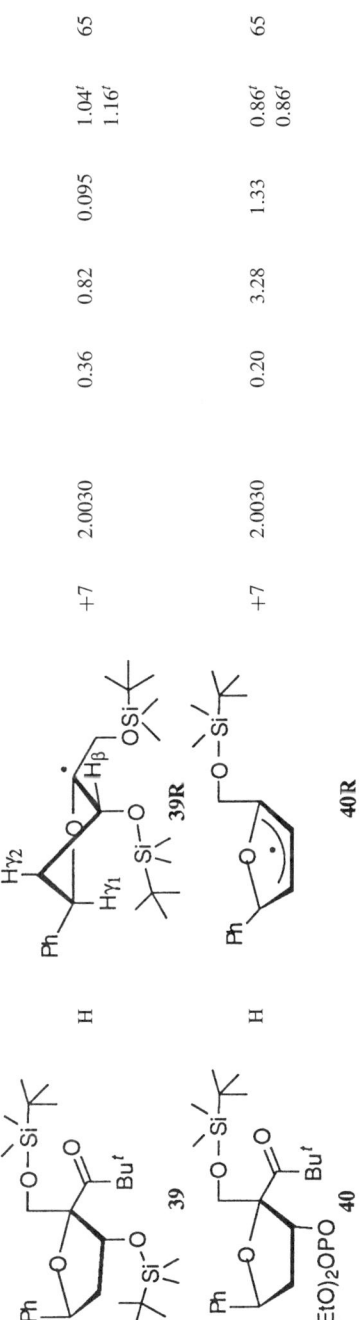

39	H	+7	2.0030	0.36	0.82	0.095	1.04^t 1.16^t	65
40	H	+7	2.0030	0.20	3.28	1.33	0.86^t 0.86^t	65

[a] UV photolysis: A, Me$_6$Sn$_2$, THF; B, Me$_6$Sn$_2$, benzene; C, Me$_6$Sn$_2$ toluene; D, photolysis in 2-propanol; E, Bu$_6$Sn$_2$, benzene; F, Bu$_6$Sn$_2$, THF; G, Me$_6$Sn$_2$, fluorobenzene; H, photolysis in benzene.
[b] α-, β-, γ1-, and γ2-positions are as indicated for **3R** and **39R**.
[c] ±2°C[56].
[d] Estimated errors in the last digit are given in parentheses.
[e] ±0.002–±0.005 mT[56], ±0.003 mT[57,58,62].
[f] α-^{13}C.
[g] γ-^{2}H.
[h] γ-^{1}H.
[i] β-F.
[j] δ-^{1}H.
[k] β-^{2}H.
[l] 2γ-H from the exocyclic CH$_2$ group.
[m] β-N.
[n] a_N in the range 0.300–0.307 mT.
[o] γ$_1$-^{1}H or δ-^{1}H.
[p] a_{Cl}.
[q] 3 δ-H, CH$_3$.
[r] two rotational isomers of the CO$_2$Me group of similar coupling constants.
[s] α-H.
[t] β-H from the exocyclic CH$_2$ group.

γ-hydrogen couplings $a(\gamma_2$-H) was assigned to H-3. This point was confirmed by studying radical **8R**, in which ^{2}H-labeling at C-5 entails the replacement of the larger of the two γ-H couplings by a deuterium triplet coupling.[56] Interestingly, the 2,3-di-O-acetyl-4,6-O-benzylidene-D-glucopyranos-1-yl radical (**9R**), which has a conformationally rigid bicyclic structure, displays hyperfine splittings similar to those observed for the previous species. The data for radicals **3R**, and **6R–9R** were interpreted in terms of a slightly twisted $B_{2,5}$ boat conformation, which corresponds to a dihedral angle as predicted by the β-H coupling constants (the β-C–H bond adopts an orthogonal arrangement with respect to the p-orbital at C-1) and a more pronounced W-arrangement of the singly occupied orbital and the quasiaxial γ-C–H bond than in the $^{1,4}B$ conformation. The 4,6-O-benzylidene-D-glucopyranos-1-yl radical (**10R**) showed a larger $a(\beta$-H) hyperfine splitting (2.9 mT), which is consistent with a half-chair equilibrium conformation or a chair conformation flattened at the C-1 side.[56] The even larger $a(\beta$-H) value found for the 4,6-O-benzylidene-2,3-O-isopropylidene-D-glucopyranos-1-yl radical (**11R**) indicated a chair conformation. The half-chair and chair conformations of **10R** and **11R** showed that their structures were more strongly fixed by additional contraints, due to either hydrogen bonds capable of existing between vicinal free hydroxyl groups in an aprotic organic medium, or to a *trans*-fused tricyclic structure, respectively.[56] The 2,3,4,6-tetra-O-acetyl-5-thio-D-glucopyranos-1-yl radical (**12R**) exhibited a 1.822 mT β-hyperfine coupling constant, which could indicate a somewhat distorted $B_{2,5}$ boat conformation,[62] probably as a result of changes in bond lengths and angles associated with the presence of the bulkier sulfur atom in the ring.

Another study showed that the peracetylated D-glucopyranos-1-yl radicals (**3R**) generated and observed by ESR at 77 K in CD$_3$OD as a matrix retained the $^4C_1$ chair conformation of its precursor **3**.[59] When the sample temperature was allowed to rise, new signals, indicating partial interconversion from the $^4C_1$ to the more usually observed $B_{2,5}$ conformer, were detected. This interconversion occurred only at temperatures close to the softening point of the glasses. The $B_{2,5}$ conformation in **3R** is also supported by synthetic results showing successful intramolecular addition[68] of anomeric radicals onto an allyloxy group attached to C-3, whereas a similar cyclization failed when the allyloxy group was attached to C-6. However, because a skew-boat conformation is structurally flexible, as is demonstrated by the temperature coefficient of the β-H couplings, the ESR data may represent only the most populated equilibrium conformation of the glucopyranos-1-yl radicals under the experimental conditions.[56]

The origin of an interconversion along the path between the D-*gluco* precursors and the $B_{2,5}$ radicals observed was explained in terms of a stabilizing interaction between the p-orbital of the unpaired electron and the σ*-orbital (LUMO) of the β-C–OR bond, favored by a coplanar arrangement. This interaction is increased by the neighboring ring oxygen atom, which raises the SOMO energy of the radical species, so that the steric destabilization of the nonchair conformation is overcompensated by the stabilizing SOMO/LUMO interaction. As inferred from the conformation found for **10R,** the stabilizing interaction between the alkoxyalkyl radical and the β-C–OR bond is strong enough to induce a $^4C_1 \to B_{2,5}$ conformational change, but cannot break additional hydrogen bonds.[56,66,67]

In the D-galactose series, 2,3,4,6-tetra-O-acetyl-α-D-galactopyranosyl bromide (**13**) and phenyl 2,3,4,6-tetra-O-methyl-1-seleno-β-D-galactopyranoside (**14**) led[56,66,67] to the corresponding radicals **13R** and **14R.** They both display a(β-H) coupling constants of ∼2.8 mT and a vanishing a(γ$_2$-H) coupling at C-3 consistent with a half-chair conformation or a flattened $^4C_1$ chair conformation. This behavior can be explained by the existence of steric interactions between the substituents at C-3 and C-4 if the radicals approach the twisted $B_{2,5}$ boat conformation as the glucosyl radicals do.[56] Accordingly, whereas the 3,4,6-tri-O-benzyl-2-O-*tert*-butyldimethylsilyl-D-galactopyranosyl radical favored the $^4C_1$ chair conformation,[69,70] the 2,6-di-O-*tert*-butyldimethylsilyl-3,4-O-isopropylidene-D-galactopyranosyl radical,[69,70] and the 2-O-*tert*-butyldimethylsilyl-3,4-O-isopropylidene-L-fucopyranosyl[63] radical (**16R**) were found to adopt $B_{2,5}$ and $^{2,5}B$ conformations, respectively. The ESR spectra of the 2,3,4,6-tetra-O-acetyl-D-mannopyranos-1-yl radical (**15R**), produced from the corresponding α-bromide[56] indicate the preservation of the $^4C_1$ conformation of the substrate in the radical species. This is shown by the disappearance of one γ-H coupling and observation of a small β-H coupling constant (∼0.3–0.4 mT). This value implies an almost orthogonal arrangement of the p-orbital at C-1 and the β-C–OR bond. For radical **15R,** the stabilizing interaction of the singly occupied p-orbital and the β-C–OR bond is best achieved in the chair conformation, which remained unaltered on the way from the precursor to the radical.

In accord with the conclusion drawn from the study of hexopyranos-1-yl radicals, the 2,3,4-tri-O-acetyl-D-xylopyranos-1-yl radical (**17R**) and its benzoylated analog **18R** achieved stabilization by conversion into the $B_{2,5}$ conformation.[56,67] The flexibility of such pentopyranos-1-yl radicals as **17R** and **18R** accounts for the observation of two other conformers ($^{1,4}B$ and $_4C^1$) in which the acetoxy substituent at C-2 is also axially oriented.[46] Conversely, 2,3,4-tri-O-acetyl-D-lyxopyranos-1-yl radical (**19R**) preserves the $^4C_1$ conformation of the starting material, in accord with the similar behavior of the related mannopyranos-1-yl radical (**15R**).[56]

b. Conformational Effects in 2-Deoxyglucopyranos-1-yl Radicals.—The actions of trimethyltin radicals[57] on various peracetylated glycopyranosyl bromides and phenyl selenoglycopyranosides produced a series of diversely substituted 2-deoxy glycopyranos-1-yl radicals with the following structures: 3,4,6-tri-O-acetyl-2-deoxy-2-fluoro-D-glucopyranos-1-yl (**20R**), 3,4,6-tri-O-acetyl-2-deoxy-D-*arabino*-hexopyranos-1-yl (**21R**) and its ^{13}C-enriched analog (**24R**), 3,4,6-

tri-O-acetyl-2-deoxy-2-deuterio-D-glucopyranos-1-yl (**22R**), 3,4,6-tri-O-acetyl-2-deoxy-D-*arabino*-(5-^{2}H)-hexopyranos-1-yl (**23R**), 3,4,6-tri-O-acetyl-2-deoxy-2-C-propyl-D-glucopyranos-1-yl (**25R**) and 3,4,6-tri-O-acetyl-2-deoxy-2-(p-toluenesulfonamido)-D-glucopyranos-1-yl (**26R**). Among this series, which differs from the D-*gluco* series by the replacement of the 2-acetoxy group by F, H, ^{2}H, propyl, or a p-toluenesulfonamido moiety, the $B_{2,5}$ conformation was found only for radical **20R**. The other radicals (**21R–26R**) having groups at C-2 that are much less or not at all electronegative adopted $^4C_1$ conformations. Although the bulkiness of the p-toluenesulfonamido group might impede a conformational change, the boat conformation observed in **20R** appeared associated with the strongly electronegative fluorine atom at C-2, which adopts an axial disposition. Within the time scale of the ESR experiment, the initial chair conformation is largely converted into a boatlike conformation. This fact, in contrast to the behavior of radicals **21R–26R** showed again that the electronegativity, or σ acceptor strength of the substituent at C-2 (fluorine atom or ester group) plays a key role in determining the structure of glycosyl radicals. Unambiguous proof of the conservation of the $^4C_1$ conformation in **21R** and also the relative assignment of the β- and γ-H hyperfine splittings has been obtained from specifically deuterated precursors leading to radicals **22R** and **23R**, regiospecifically ^{2}H-labeled at either C-2 or C-5, respectively.[57] With a ^{13}C-labeled material having a methylene group at C-2, an out-of-plane bend of ∼6° has been estimated for the radical **24R**. This bent angle corresponds to ∼2.1% of 2s character of the semi-occupied p-orbital, that is, ∼10% on the way from a "pure" sp^2-hybridized π-radical to a "pure" sp^3-hybridized σ-radical.[57] This may be compared with the glucopyranos-1-yl radical (**7R**) as a π-type radical[56,66] with an out-of-plane bend of ∼3.9° (compared with 19.5° in an sp^3-hybridized σ radical and 11° in carbon-centered acetalic radicals[53]) and a 2s character of the semi-occupied orbital estimated to be 1%. A theoretical analysis[54] led to the conclusion that the semiempirical methods AM1 and PM3 provide poor models for anomeric stabilization in radicals and are not useful for predicting radical conformations. *Ab initio* calculations showed that the pyramidalization at the radical center is small.[54]

c. Conformations of C-1 Substituted Pyranos-1-yl Radicals.—Cyclic sugars bearing a cyano group at the anomeric center have, at this position, captodative radical-stabilizing groups that favor their regioselective photobromination at C-1.[21] The axially brominated 2,3,4,6-tetra-O-acetyl-1-bromo-β-D-glucopyranosyl cyanide (**27**) can be obtained in high yield in a crystalline and stable form, and a series of acetylated 1-bromo-glycopyranosyl cyanides having in particular the D-*galacto* (**28**), D-*xylo* (**29**), and D-*arabino* (**30**) configurations has been prepared.[21] Peracetylated β-D-glycopyranosyl chlorides also undergo regio- and stereo-selective radical-mediated bromination at C-1 to afford the corresponding 1-bromo-β-D-gluco- and manno-pyranosyl chlorides **31** and **32**, respectively.[21,71]

From these precursors, acetylated 1-cyano- and 1-chloro-pyranos-1-yl radicals were generated from the corresponding bromides by bromine abstraction with trimethyltin radicals.[58] The 1-cyano-D-glucopyranos-1-yl radical (**27R**) provided ESR data quite similar to those recorded from the C-1-unsubstituted radical (**3R**), and therefore the $B_{2,5}$ conformation was assigned to **27R**. Taking into consideration the resemblance of the ESR data obtained from the D-*galacto*-configured radicals **13R** and **28R,** which both adopt a half-chair conformation, it was concluded that the cyano group does not influence the radical conformation. However, at temperatures above +7 °C, an additional four-line signal appeared. This was clearly assigned to another conformer of **28R**, possibly in the $^{1,4}B$ (**28R′**) or $B_{2,5}$ (**28R″**) conformation, which were not unequivocally identified. The 1-cyano xylosyl radical **29R** exhibited a slight increase in the β-H coupling constant as compared to the C-1-unsubstituted radical (1.630 mT in **29R,** compared to 1.221 mT in **17R**), which was interpreted as a less-pronounced $B_{2,5}$ boat character, maybe due to the influence of the cyano group. In the 1-cyano-D-*arabino*-pyranos-1-yl radical **30R,** the development of a fully evolved boat conformation is prevented, presumably because of unfavorable 2,4-steric interaction of the acetoxy groups favoring the H_4 half-chair conformation. For the 1-chloro-D-*gluco*- and D-*manno*- radicals **31R** and **32R,** which were found to exist in the ^4H half-chair and $^4C_1$ chair conformations, respectively,[58] the ESR data showed some differences from those recorded for the C-1 unsubstituted radicals. The hyperfine coupling variations observed could stem from conformational changes and/or pyramidalization at C-1. The fact that electronegative elements at a radical center should induce pyramidalization, as do two oxygen atoms in acetal radicals, strongly supports the last hypothesis.

d. Non-Anomeric Carbohydrate Radicals.—It is interesting to compare the conformations of glycopyranos-1-yl radicals with those of analogous species having the radical center at other sites on the pyranose ring. In particular, the C-5 bromo derivatives of methyl acetylated-β-D-glycopyranosiduronates (D-*gluco*: **33,** D-*galacto*: **34**), obtained in high yield by radical bromination of precursors with captodatively substituted carbon atoms, and other C-5 bromides (D-*xylo*: **35,** D-*gluco*: **36**) lead[58] to the corresponding C-5 centered radicals **33R–36R** (5-methoxycarbonyl-, 5-unsubstituted-, and 5-acetoxymethyl-pyranos-5-yl radicals). These radicals resemble anomeric radicals because in both cases the radical center bears an alkoxy substituent. It is not surprising, therefore, that conformations favoring a coplanar arrangement between the singly occupied orbital and the β-C–OR bond at C-4 were deduced from analysis of the ESR data. In effect, boat structures compatible with the axial orientation of the acetoxy group at C-4 were proposed for **33R, 35R,** and **36R,** whereas the $^4C_1$ conformation of the bromo precursor was retained in the D-*galacto*-configured radical **34R.**

Finally, it is noteworthy that various deoxypyranos-2-, -3-, and -4-yl radicals[64,66,67,72] generated in benzene by means of trialkyltin radicals, either regiospecifically[66,67,72] or by rearrangement,[62,64] have been investigated. Regardless of the configuration of the precursors (D-*gluco*, D-*manno*, D-*altro*, D-*allo*, D-*galacto*), which all adopt $^4C_1(D)$ chair conformations, the ESR data, interpreted with the help of regiospecific ^2H labeling, showed that they retained to a large extent the conformation of their parent compounds. The structure of 2-deoxypyranos-2-yl radicals was not influenced by their anomeric configuration.[66,72] The $^4C_1$ chair conformation was also found for a deoxypyranos-2-yl radical formed by a 2→1-migration of a diphenylphosphonyloxy group.[64] These radicals showed no obvious tendency to undergo transformation into conformations in which the radical p-orbital could be stabilized by a parallel arrangement with the β-C–OR bond, as has been observed for some D-glycopyranos-1-yl radicals (D-*gluco*, D-*galacto*, D-*xylo*, D-*arabino*). Interestingly, use of either toluene or THF as the solvent was found inappropriate in this series, because hydrogen abstraction from the solvent occurred. The seemingly higher reactivity of carbohydrate radicals with radical centers at C-2, C-3, and C-4 was considered to stem from the absence of significant thermodynamic stabilization,[62,72] unless this is the consequence of the electrophilic character of such carbon-centered radicals induced by acetoxy groups (see Sections III and V). This evidence in terms of conformation and reactivity emphasizes the role played by the ring oxygen atom in stabilizing such α-alkoxy radicals as pyranos-1-yl and pyranos-5-yl radicals.

e. Glycofuranos-1-yl Radicals.—In comparison with the pyranose situation, much less information is available concerning protected furanose-derived radicals.[45,64] ESR studies of the anomeric furanoid radicals generated from 2,3,5-tri-*O*-benzoyl-α-D-ribofuranosyl bromide (**37**) and 3,5-di-*O*-benzoyl-2-[3-(trifluoromethyl)benzoyl]-α-D-ribofuranosyl bromide (**38**) indicated a dihedral angle between the proton at C-2 and the half-filled p-orbital of about 60°. This leads to the conclusion that the 2,3,5-tri-*O*-benzoyl-D-ribofuranos-1-yl radical (**37R**) and its 3,5-di-*O*-benzoyl-2-*O*-[3-(trifluoromethyl)benzoyl] analogue (**38R**) existed in the $_2E$ conformation depicted in Table I. In radical **39R**, for which quantum chemical calculations suggests a pyramidalization at the radical carbon center,[65] the C–O bond attached to the furanose ring, in the ^3E conformation, is in axial orientation, whereas for both **39R** and **40R**, the phenyl group adopts an equatorial disposition. The 2′-deoxyuridin-1′-yl radical, studied by ESR, laser flash photolysis, and theoretical calculations, is considered to exist in a $_3E$ conformation with a slight pyramidalization at the radical center.[61]

f. Radical Species Derived from Free Sugars.—It is of interest to compare these observations with those made earlier when free *myo*-inositol,[73] mono- and disaccharides,[74,75] and neutral polysaccharides[76] where exposed to hydroxyl radicals (OH·) generated with the Ti^{III}-H_2O_2 couple in a flow system appropriate for ESR studies. Abstraction of hydrogen atoms attached to the various carbon centers in the substrates took place rather nonselectively, thus forming several carbohydrate-derived radicals and resulting in complex, superimposed ESR signals. However, analysis of these signals permitted the identification of most of the species initially formed, as well as that of carbonyl-conjugated radicals derived from the former.[73–77] Hydrogen abstraction by OH• was found more selective in the case of D-galacturonic acid,[76] galacturonan,[76] and sugar lactones[78] because the spectra were dominated by signals from radicals formed by abstraction of the hydrogen from the carbon adjacent to the carboxyl group. For furanose compounds derived from D-fructose, hydrogen-atom abstraction from the carbon atom adjacent to the alicyclic oxygen is favored.[79] The general conclusion drawn from these extended studies was that radicals derived from free carbohydrates mostly adopt the structural conformation of the pyranose parent sugar (namely the $^4C_1$ chair conformation).[79] However, the ESR signals found for unprotected hexopyranos-5-yl radicals pointed to predominant conformations, around the C-5–C-6 bond, corresponding to a parallel orientation of the C-6–OH bond with respect to the SOMO at C-5.[74] When hydrogen abstraction occurred at C-6, the SOMO adopted an eclipsed arrangement with the C–O endocyclic bond.[74] This favors a stabilizing SOMO-σ^*(β-C–O) interaction, which decreases the nucleophilic character of the corresponding carbon-centered radicals, as shown by comparing the rates of addition of carbohydrate-derived radicals to methacrylic acid.[80] Hydrogen bonding involving the OH-4 group was invoked to account for structural peculiarities in the pyranos-5-yl radicals derived from uronic acids.[76] These studies were also extended to furanose-derived radicals.[61,75,78,79] Time-resolved ESR spectroscopy has been used for the identification of transient radical intermediates generated photochemically from simple sugars by using triplet photoinitiators.[81]

3. Interpretation in the Framework of the Frontier Molecular Orbital (FMO) Theory

On the basis of frontier molecular orbital (FMO) theory, the initial explanation of the fact that, in sugar-derived α-alkoxyalkyl radicals, β-C–OR or β-C–F bonds at C-2 try to achieve or preserve an eclipsed, that is, axial orientation with respect to the singly occupied p-orbital, invoked a stabilizing interaction between the SOMO and the σ^*-LUMO of the β-C–OR bond. The phenomenon was called a "quasi-anomeric effect" because it resembles in character the normal anomeric effect. The oxygen lone-pair enhances this stabilization because[21] its interaction with the unpaired electron at C-1 raise the energy of the SOMO, rendering it more favorable for the interaction with the σ^* orbital.[67] Introduction of a cyano group at C-1 should result, in terms of FMO theory, in a stabilization of the SOMO by the interaction with the π and π^* orbitals of the cyano group. Consequently, the lowering of the SOMO should lead to a diminished interaction with the σ^*

orbital of the β-C–OR bond. Taking into account the fact that captodatively substituted glycopyranos-1-yl and glycopyranos-5-yl radicals do not differ in terms of structure from their unsubstituted analogs, a refined picture was needed. Thus, the interpretation was completed by considering also the interaction of the π-type lone-pair orbital of the oxygen atom with the σ*-LUMO of the β-C–OR bond.[58,82] The latter contribution could be considered to be the result of a "homo-anomeric" effect between the lone pair at oxygen and the β-C–OR bond at C-2. The combination of both stabilizations, termed a *quasi-homo-anomeric stabilization effect*, supplies the driving force for the change from the $^4C_1$ to the $B_{2,5}$ conformation. This effect, estimated to be larger than 4.2–5.0 kJ/mol (1.0–1.2 kcal/mol)[62] is just strong enough to compensate the repulsive interactions prevailing in boat-like conformations, but cannot break additional hydrogen bonds. Kinetic studies with differently substituted tetrahydropyranyl models, and ESR investigations of the corresponding radicals have been rationalized conclusively in the light of the "quasi-homo-anomeric interaction."[83,84]

Whereas the nature of carbohydrate precursors influences their reactivity, the structure of radical intermediates is of great importance as regard to the stereoselectivity of the possible radical-mediated transformations, as further illustrated in the following articles dealing with radical-mediated reductions.

III. Radical Methods to Anhydroalditols from Glycosyl Derivatives

As already discussed, generation of carbohydrate-derived radicals in the presence of trimethyl- or tributyl-tin radicals produced upon photolysis from hexamethyl- or hexabutyl-ditin in solvents not prone to homolytic C–H cleavage is well suited for ESR studies because chemical reactions are minimized. Under these conditions, however, glycosyl radicals are able to dimerize with low stereoselectivity to yield, in 32% (ref. 85) to 42% (ref. 86) total yield, a mixture of *C*-linked carbohydrate dimers, as also obtained by electrochemical reduction of glycosyl halides on silver.[87] The modest recorded yields are explained on the basis of competing radical reactions leading, among others, to minor proportions of reduced byproducts, including anhydroalditols formed by intermolecular hydrogen abstraction from another sugar molecule,[86,88] and 2-deoxy derivatives[86] produced when an intramolecular 2→1 migration of the group at C-2 (see Section V) precedes the hydrogen-abstraction step. A competing photoreaction between the glucosyl bromide **3,** used as the radical precursor, and hexabutylditin was put forward to account for the formation of 1-(2,3,4,6-tetra-*O*-acetyl-D-glucopyranosyl)butane[86] in a ~1:1 α:β ratio. Whereas the radical-mediated formation of *C*-glycosyl compounds[89] (which is not considered here) results, in general, from addition of glycosyl radicals to electrophilic substituted alkenes, the formation of reduced sugar derivatives requires the presence of a suitable hydrogen source. In this case, and as discussed next, hydrogen quenching of simple anomeric radicals constitutes a mild, simple, and versatile method for the synthesis of anhydroalditols, whereas application of this methodology to complex anomeric radicals may lead stereoselectively to O-glycosides or C-glycosyl compounds (see Section IV).

The preparation of anhydroalditols can be achieved by a variety of ionic or catalytic procedures, including hydrogenation of glycals,[90] hydrogenolysis of 1-thio-glycosides,[91] acid-catalyzed dehydration of acyclic alditols,[92] reduction of glycosyl halides with lithium aluminum hydride,[93,94] treatment of glycosyl azides with hydrazine,[95] reaction of silylated[96] or acetylated[97] alkyl glycosides with triethylsilane in the presence of trimethylsilyltriflate as the catalyst, and electrochemically promoted reductive cleavage of glycosides.[98] Although accessible primary and secondary sugar iodides have been converted into the corresponding deoxy sugars by photolysis in an alcohol as the solvent and the hydrogen source (methanol or, preferably, 2-propanol), this procedure is not applicable to the synthesis of 1,4- and 1,5-anhydroalditols.[99,100] Similarly, photolysis of sugar esters (acetates, propanoates, 2-methylpropanoates, and pivalates) in hexamethylphosphoric triamide (HMPA)-water, which produces the corresponding deoxy sugars in good yield,[101–103] was found inefficient for the preparation of anhydroalditols, as indicated by the preferential photocatalyzed hydrolysis[101] of the acetate group, in 1-O-acetyl-2,3:5,6-di-O-isopropylidene-α-D-mannofuranose. Photochemical reactions of glycosyl phenyl sulfones in benzene or benzene–methanol resulted, presumably via radical pathways, in mixtures of products containing the corresponding anhydroalditols.[104,105] Taking into consideration these limitations, these is obviously a need for mild and more-selective routes to anhydroalditols (Scheme 1).

Being applicable to a large variety of anomerically substituted radical precursors, reductive radical methods have proved particularly suitable for this purpose. As discussed later, carbohydrate precursors can lead to anomeric radicals by three main ways: reaction with metal-centered radicals (mainly triorganotin radicals); transformation by single-electron transfer; or reaction of O-acyl-thiohydroxamates in the presence of a tertiary thiol (Barton methodology). Radical deoxygenation via O-acyl-thiohydroxamate derivatives, prepared from precursors containing a carboxylic group, has found applications for stereocontrolled preparations of O-glycosides and C-glycosyl compounds (see Section IV). The possibility of controlling radical chain-reactions makes their scope even broader. In effect, unsaturated carbohydrate precursors can add either intra- or intermolecularly to a radical species. When this process, which results in a new radical transient due to C–X (X = C, P, S) bond-formation is followed by hydrogen quenching, bicyclic or C-2-branched anhydroalditols (see Section III-3) or C-glycosyl compounds (see Section IV-2) may be obtained, depending upon the conditions applied.

1. Radical Reduction with Tin and Silicon Hydrides

a. General.—All monoradical chain processes may be divided into three discrete mechanistic phases: initiation, propagation, and termination.[106] Although many radical chain processes occur spontaneously at moderate temperatures because of such adventitious initiators as laboratory light or ill-defined redox processes, it is usually desirable to facilitate chain initiation by deliberately adding initiators or by other means. At the initiation stage, generation of free radicals from a neutral species typically occurs via homolytic cleavage of a weak σ bond. Initiators generally possess at least one bond having a homolytic bond-dissociation energy of ~313 kJ.mol^{-1} (~75 kcal.mol^{-1}) or lower. The requisite energy for bond breaking is supplied either thermally or photolytically. For thermal initiation, heating to temperatures below 100 °C is generally sufficient, as shown by the half-lives, plotted against temperature, for various initiators, and, in particular, the frequently used azobisisobutyronitrile (AIBN) and benzoyl peroxide.[107] Benzene and toluene, whose boiling points are in this range and are sufficiently stable so as not to interfere with radical processes, are commonly used solvents. In many ways, AIBN and related azo compounds are ideal initiators. They are safe, easily handled, give good radical yields, and have decomposition rates that are almost solvent independent. The only drawback of AIBN is that the 2-cyano-2-propyl radicals that it produces, in addition to molecular nitrogen, are relatively unreactive. For chain reactions to be initiated, a reactive double bond or a weak, reactive bond (such as S–H, Si–H, or Sn–H) must be present in the system because C–H bonds are not usually attacked. Peroxides yield radicals that are more reactive: for example, methyl radicals from acetyl peroxide, benzoyloxy and phenyl radicals from benzoyl peroxide, and *tert*-butoxy radicals (plus methyl radicals) from di-*tert*-butyl peroxide. Their drawback is that their chemistry is more complicated.[107] It is noteworthy that the triethylborane–oxygen system, which produces ethyl radicals at temperatures as low as −80 °C, has been found suitable for elucidating the mechanism of alcohol deoxygenation with *n*-Bu$_3$SnH,[108] and diphenylsilane,[109,110] and for other synthetic purposes.[109,111] Diethylzinc–air,[112] and 9-borabicyclo[3.3.1]nonane[113] can be used for low-temperature, tin-radical mediated reductions and might find applications in the carbohydrate field.

If reactions must be carried out near or below room temperature, for example because of the thermal lability of a reactant, thermal initiators are not very practical for synthetic work, the triethylborane–oxygen system notwithstanding, and the method of choice is usually photoinitiation. Irradiation of mixtures of glycosyl halides, *n*-Bu$_3$SnH, and AIBN in refluxing diethyl ether with visible light allows radical reactions to proceed under very mild conditions.[114,115] However, irradiation with UV light, delivered by mercury discharge lamps with significant emission at 253.7 nm, is more frequently applied. Use of glassware made of UV-transparent material (quartz) ensures optimum light transmission. In less-favorable situations, or if no reaction occurs, or if in doubt, photoinitiators may be added. AIBN, its analogs, and benzoyl peroxide are particularly convenient because they absorb strongly in the near UV.[107,114,115]

The bond dissociation energy of the Sn–H bond in n-Bu$_3$SnH[116,117] is 309 kJ. mol^{-1} (74 kcal.mol^{-1}), making this bond prone to homolysis, and the consequent generation of triorganotin radicals, facile. In the presence of metal-centered radicals, trialkyltin radicals in particular, various protected glycosyl derivatives form glycosyl radicals, which can abstract an hydrogen atom from n-Bu$_3$SnH to yield anhydroalditols in a chain reaction. Although a glycopyranosyl derivative is shown as a representative of the radical precursor, reduction by a radical-chain reaction is not limited in scope, and is applicable to primary, secondary, and tertiary carbon-centered radicals.[13a] Hydrogen transfer from n-Bu$_3$SnH to alkyl radicals occurs[117,118] with rate coefficients around $\sim 2 \times 10^6$ M^{-1}s^{-1} at 25 °C. These rates, which depend little on the substituents at the radical center, provide a method for estimation for carbohydrate radicals.[3] In typical situations, use of a "high" concentration of the hydrogen donor favors quenching of the initially formed radical, thus avoiding its transformation by other pathways (mainly, termination reactions and rearrangement). However, when reduction of the initially formed radical must be avoided, for example so as to favor addition to a C=C double bond in the first step, the concentration of the hydrogen donor must be lowered. Owing to these favorable features, n-Bu$_3$SnH, also commercially available[8] as the deuteride n-Bu$_3$SnD, is the more popular reducing agent, despite its toxicity, its high molecular weight, its hydrophobic character, and difficulties encountered for the removal of organotin derivatives from the products by column chromatography.[119]

Among the methods used to overcome this problem, one consists in treating the crude reaction mixture with aqueous potassium fluoride[120] to produce insoluble fluorotin derivatives that are readily removed by filtration. When the desired organic compound is soluble in acetonitrile, as it is generally the case for protected sugar derivatives, removal of tin residues can be readily achieved by liquid–liquid phase extraction with hexane–acetonitrile as nonmiscible solvents.[121] The possible elimination of tin byproducts on subsequent treatment with either iodine and 1,8-diazabicyclo[5.4.0]undec-7-ene (DBU),[122] or Me$_3$Al,[123] or aqueous 1M NaOH[123] has also been proposed. Treatment of the reaction mixture with NaBH$_3$CN constitutes another method to facilitate purification of the reaction products because the tin derivatives produced are converted back into n-Bu$_3$SnH, which is readily removed by column chromatography.[124] Although it is known that *in situ* regeneration allows n-Bu$_3$SnH to be used in catalytic amounts in the presence of a reducing agent such as NaBH$_4$ or NaBH$_3$CN, this procedure is applicable only when reduction occurs after either a C–C bond-forming process[125,126] or a radical translocation step.[127] Both n-Bu$_3$SnH and (n-Bu$_3$Sn)$_2$O have been used, in catalytic amounts, for the n-Bu$_3$SnH-catalyzed Barton–McCombie deoxygenation of alcohols, in the presence of Me$_3$SiO–(SiHMeO)$_n$–SiMe$_3$, as a cheap, air- and moisture-stable reductant.[128] Among other procedures[129] for improving and broadening the usefulness of tin hydrides in synthesis, the use of polymer-supported organotin hydrides, which can be separated by filtration from the products, is noteworthy.[130–136] This technique, advantageous in terms of easy processing and multiple regenerations of reagents, has been applied successfully in the reduction of 1,2:5,6-di-O-isopropylidene-3-O-(phenoxythiocarbonyl)-α-D-glucofuranose and of nucleoside derivatives.[133] Dehalogenation of water-soluble organic halides in an aqueous medium has been achieved either in the presence of an added detergent,[137] or by using a base-soluble dialkyltin reagent,[138] but use of an amphiphilic tin hydride carrying three methoxyethoxypropyl groups was shown to reduce alkyl halides in either water or organic solvents.[139] Tin hydrides having either a nitrogen-containing substituent[140,141] or three 2-(perfluorohexyl)ethyl substituents[142] are also readily separated from the reaction products by chromatography or extraction, respectively. However, these approaches, have thus far found only limited applications for reduction of carbohydrate derivatives.[133]

Extensive studies of silyl radicals[143] and substituted silanes,[117,144] which are regarded as less toxic than tin derivatives, have been performed in order to broaden their applications in radical chemistry. As the dissociation energy of the Si–H[144] bond is higher than that of the Sn–H bond in trialkyltin hydrides, only those silanes in which the Si–H bond is somehow weakened have found useful synthetic applications,[117] unless special conditions are applied to overcome limitations arising from the high Si–H bond-strength in simple trialkylsilanes and from unfavorable radical philicity, as discussed next. Silanes bearing phenyl,[109,110,145–147] trimethylsilyl,[116,148–152] or alkylthio[153] substituents have been mainly exploited in such reduction reactions. As compared to the Sn–H bond-dissociation energy in n-Bu$_3$SnH (309 kJ.mol^{-1} or 74 kcal.mol^{-1}),[116,117] the Si–H bond dissociation energies (kJ.mol^{-1}/kcal.mol^{-1}) of substituted silanes are as follows: Et$_3$SiH: 376/90,[116,153] 380/91[146]; PhSiH$_3$: 369/88.3[144]; Ph$_3$SiH: ∼347/∼83[146]; (Me$_3$Si)$_3$SiH: 330/79[116,117,149]; and (MeS)$_3$SiH: 345/82.5.[117,153] Consequently, the absolute rate-constants at 25 °C for hydrogen abstraction by various alkyl radicals from tris(trimethylsilyl)silane (∼2 × 10^5 M^{-1}s^{-1}) and phenylsilanes[143] were found lower as compared with those of n-Bu$_3$SnH (∼2 × 10^6 M^{-1}s^{-1})[117,118] by at least one order of magnitude.

The deoxygenation of sugar derivatives having thiocarbonyl groups at primary and secondary positions has been successfully achieved with phenylsilane,[110,147] diphenylsilane,[109,145] or triphenylsilane,[110] with triethylborane–oxygen as initiator,[109,110] or, in boiling toluene, with AIBN or benzoyl peroxide. The need for repeated additions of the initiator indicates relatively short radical chains,[109] which are synthetically useful even at room temperature.[110] Deoxygenation of primary alcohols might result in low yields of the deoxy products at room temperature, because of formation of the corresponding thioformates,[109] but this is not observed when the reaction is conducted at ∼80 °C. In the presence of di-*tert*-butyl hyponitrite as the initiator, triphenylsilane reduces 1-thio sugars efficiently, at 60 °C in 1,4-dioxane.[154] Competitive experiments showed that, after normalization based on the number of hydrogen atoms on the silicon, the three silanes PhSiH$_3$, Ph$_2$SiH$_2$, and Ph$_3$SiH have very similar reactivity,[110] and that the order of reactivity of silanes and tri-*n*-butyltin hydride is as follows:[110]

$$n\text{-Bu}_3\text{SnH} > (\text{Me}_3\text{Si})_3\text{SiH} > \text{PhSiH}_3 \approx \text{Ph}_2\text{SiH}_2 \approx \text{Ph}_3\text{SiH} > \text{Et}_3\text{SiH}$$

Tris(trimethylsilyl)silane, considered an effective substitute for the toxic tributyltin hydride in free-radical chain reductions, has been extensively studied and the reactivity[152] of (Me$_3$Si)$_3$Si• radicals toward functional groups has been found to be as follows: iodide > xanthate > bromide > selenide > isocyanide ≈ nitro > sulfide ≈ chloride > acid chloride.[109,152] It can be used in catalytic amounts by being regenerated *in situ* in the presence of sodium borohydride.[150] Reduction of (Me$_3$Si)$_3$SiBr with LiAlD$_4$ affords (Me$_3$Si)$_3$SiD in high yield.[148] Because of its diminished reactivity, (Me$_3$Si)$_3$SiH was found to be superior to *n*-Bu$_3$SnH for the synthesis of 2-deoxy sugars in refluxing benzene with added AIBN (see Section V).[116] Triethoxysilane used as the solvent for the benzoyl peroxide-initiated deoxygenation of a xanthate group on a secondary carbon atom is another heterosubstituted silane found efficient for deoxygenation.[110] Although the Si–H bond in triethylsilane[153] is quite strong (377 kJ.mol^{-1} or 90.1 kcal.mol^{-1}), its use in radical deoxygenation processes was found possible, in refluxing toluene with benzoyl peroxide as the initiator, whereas AIBN was found inappropriate.[110] Deoxy sugars were prepared in high yields by deoxygenation, in refluxing triethylsilane, with repeated additions of benzoyl peroxide, which acts not only as an initiator, but also as a trap for triethylsilyl radicals, as shown by the formation of triethylsilyl benzoate among the products.[110] The triethylsilane is the principal hydrogen source, as established by deuterium labeling experiments using Et$_3$SiD, whereas product analysis accounts for the similarity of deoxygenation mechanisms of sugar thiocarbonyl derivatives in the presence of either Et$_3$SiH, Ph$_2$SiH$_2$, or *n*-Bu$_3$SnH.[110,155] A phenyl-substituted silane (9,10-dimethyl-9,10-dihydro-9,10-disilaanthracene) proved an effective reagent for AIBN-initiated dehalogenation of organic halides and deoxygenation of sugars via primary or secondary *O*-thiocarbonyl derivatives.[156]

Among the large array of sugar xanthates and related thiocarbonyl derivatives that have been subjected to radical deoxygenation, those having a (methylthio)thiocarbonyloxy group attached to the anomeric position have not been frequently used as glycos-1-yl radical precursors, probably because of synthetic limitations in terms of multistep preparations and possible isomerization resulting in decreased yields.[157] It has been reported that, under phase-transfer catalysis conditions (*n*-Bu$_4$N$^+$ HSO$_4^-$, C$_6$H$_6$, 50% aq NaOH, CS$_2$, CH$_3$I), 3′-*O*-benzoyl-2,3-*O*-isopropylidene-D-apio-D-furanose gave a 1:0.22 inseparable mixture (65% yield) of the expected xanthate, contaminated by the (methylthio)carbonylthiyl isomer.[158] However, liquid–liquid and solid–liquid phase-transfer methods have been developed for the conversion of anomeric free sugar derivatives, bearing

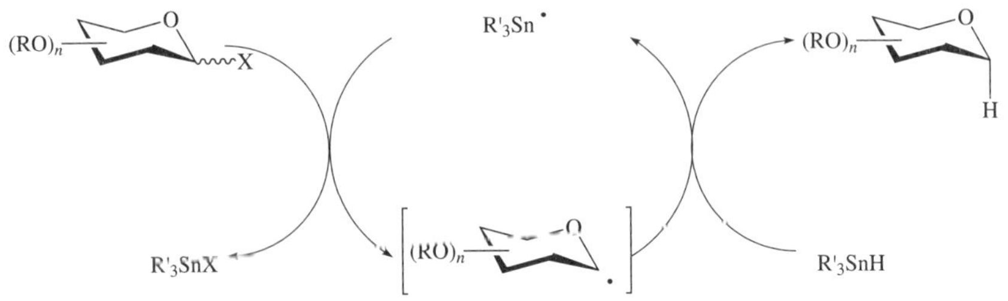

SCHEME 1

acetyl or benzoyl groups, into the corresponding 1-O-(methylthio)thiocarbonyl derivatives.[159] For the generation and transformation of anomeric radicals, the more frequently used precursors (see Table I) are glycosyl halides, isocyanides, isothiocyanates, isoselenocyanates, 1-seleno- and 1-thio-glycosides, because of their ready availability and favorable reactivity in the presence of metal-centered radicals. This last aspect can be understood in terms of a favorable thermodynamic balance based on the bond-dissociation energy of the bonds involved in the synthetic process and the aforementioned stabilization of anomeric radicals. Other readily prepared glycosyl derivatives, such as 1-selenoglycosides[160] and 1-telluroglycosides,[160,161] might find applications in the field. Radical-mediated reductions (Scheme 1) are conducted under an inert atmosphere (argon, nitrogen) to prevent trapping of carbohydrate-derived radicals by molecular oxygen, which may lead to hydroperoxides or hydroxylated compounds.[111,162–164]

b. The Oxygen–Carbon β-Bond Effect.—The oxygen–carbon β-bond effect, also termed the β-*oxygen effect*, was first noted by Barton's group for radical deoxygenation and deamination reactions.[165] For a long time, it has remained a puzzling aspect linked to carbon radical formation. The initial findings were that various thiocarbonyl esters and isonitriles bearing alkoxy and/or acyloxy groups in the β-position underwent deoxygenation and deamination, respectively, on treatment with n-Bu$_3$SnH, at lower temperatures than the corresponding unsubstituted species. The conclusion drawn from these early studies was that "β-bonded oxygen has a marked effect in stabilizing carbon radicals, thus permitting homolytic fission not seen otherwise."[165] Much effort was devoted to this problem because ESR spectroscopy performed on β-alkoxyethyl radicals did not suggest any particular radical stabilization by the β-C–O bond that might have explained these observations.[11,166] Based on the fact that the β-oxygen effect for hydrogen abstraction by the electrophilic (see next) *tert*-butoxyl radical is deactivating, in sharp contrast to the activating β-oxygen effect observed for reduction by stannanes, polar effects developing in transition states of opposite polarities were proposed.[167]

Other investigations with flexible models led to the conclusion that in conformationally unrestricted systems, there is no significant β-oxygen effect[168] in the

Barton–McCombie reaction.[16] In addition, axial xanthates in more conformationally rigid cyclohexyl and dioxanyl derivatives, having in the latter case the xanthate group synclinal to two β-oxygen bonds, were found to react faster as compared to their equatorial counterparts. Molecular mechanics calculations suggested that these results can be satisfactorily explained in terms of different strain energies (torsional and ring) present in the various adduct and fragmented radicals. However, intramolecular competition reactions using a bisxanthate derived from *myo*-inositol 1,3,5-orthoformate showed that radical deoxygenation occurred faster for the equatorially oriented xanthate group, synclinal to two β-oxygen bonds, as compared to the axial group antiperiplanar to two β-oxygen bonds. It is noteworthy that all of the examples described by Barton either involved conformationally rigid systems, in which the thiocarbonyl ester is synclinal to a β-oxygen bond or, for 6-*O*-(*S*-methyl dithiocarbonate)-glycopyranose derivatives, have at least one β-oxygen bond synclinal to the thiocarbonyl ester in all possible staggered conformations. Put together, these data suggest that the relief of unfavorable dipolar interactions, rather than the initially postulated radical stabilization by a β-oxygen bond, may contribute to the acceleration of cleavage when the thiocarbonyl ester is forced into a synclinal relationship with a β-oxygen bond.[168] It is also concluded that polar effects do not contribute significantly to the stabilization of the transition state for fragmentation in the Barton–McCombie reaction.[168]

The conversion, by Norrish type II photocleavages, of various benzylated *O*-phenacyl glycopyranosides, as well as an analogue derived from 2-hydroxytetrahydropyran,[169] into the corresponding lactones, has been studied and interpreted in connection with the generation of anomeric radicals, and the role of the β-oxygen effect. Competition experiments showed that both glucosyl anomers are consumed at the same rate in the course of the photolysis, whereas phenacyl α-glucoside is marginally more reactive than the corresponding α-mannoside. This was taken as an evidence indicating that, at least for this type of facile δ-intramolecular hydrogen-atom abstraction, the difference in activation barriers for removal of an axial or equatorial hydrogen is insignificant. More meaningful was the higher reactivity observed for the tetrahydropyran analog as compared to the carbohydrate models. This effect was interpreted in terms of destabilization of the polarized transition-state for hydrogen abstraction, by the β-oxygen bonds, in the carbohydrate series.[169]

c. Nucleophilic and Electrophilic Properties of Radicals: Thiol-Catalyzed Radical Reductions.—Although the propagation steps in a radical chain-reaction are influenced by various factors such as steric and bond-strength effects, it is worth stressing that the electronic nature of a radical tightly controls its reactivity. Nucleophilic radicals seek electron-poor addition or abstraction sites, whereas electrophilic radicals react at sites of higher electron density.[106] Radicals centered on atoms significantly more electronegative than carbon are considered electrophilic; conversely, atoms less electronegative than carbon form nucleophilic radicals. Hence, alkoxy radicals, chlorine and bromine atoms, and thiyl radicals have electrophilic properties that decrease along the sequence as listed. Triorganostannyl radicals are considered more nucleophilic than triorganosilyl radicals.[106] The philicity of a carbon-centered radical depends on its substitution mode. As a general rule, carbon radicals substituted with electron-releasing groups are nucleophilic, whereas those bearing electron-withdrawing groups are electrophilic. It was found experimentally that one alkoxy substituent increases the nucleophilicity of alkyl radicals, in addition to the reactions to alkenes, more than three methyl groups.[170] In keeping with the behavior of alkyl radicals having an alkoxy substituent, anomeric radicals have a nucleophilic character, as shown in particular by many examples of addition reactions to electron-depleted alkenes.[3,13]

Radical reductions with triethylsilane were found more effective when conducted in the presence of a thiol. This has been rationalized on the basis of favorable polar effects, with thiols acting as polarity-reversal catalysts for hydrogen-atom transfer from organosilanes to alkyl radicals.[171,172] Although the second reaction accounting for the organosilane-mediated reduction of organic halides shown next is exothermic, even for a tertiary alkyl radical [bond dissociation energy: $Me_3C-H = 392$ kJ.mol^{-1}; $Et_3Si-H = 331$ kJ.mol^{-1}], its general sluggishness was interpreted in terms of unfavorable polar factors arising for the abstraction of an electron-rich hydrogen atom from silicon by a nucleophilic alkyl radical.

$$Et_3Si^\bullet + RHal \rightarrow Et_3SiHal + R^\bullet \text{ (fast)} \qquad \text{Eq. } 1$$

$$R^\bullet + Et_3SiH \rightarrow RH + Et_3Si^\bullet \text{ (slow)} \qquad \text{Eq. } 2$$

This reasoning suggested that thiols might help in circumventing such a limitation on the basis of the strength of the S–H bond (384 kJ/mol in MeSH) and of the electrophilic properties of thiyl radicals. The aforementioned slow reaction would be replaced by a catalytic cycle of reactions, both of which will benefit from favorable polar effects:[171]

$$R^\bullet + XSH \rightarrow RH + XS^\bullet \text{ (fast)} \qquad \text{Eq. } 3$$

$$XS^\bullet + Et_3SiH \rightarrow XSH + Et_3Si^\bullet \text{ (fast)} \qquad \text{Eq. } 4$$

The efficient reductions observed for various organic halides when treated with Et_3SiH in two-fold excess in the presence of catalytic amounts of a thiol and di-*tert*-butyl hyponitrite as initiator confirmed the validity of this approach. Silane–thiol couples were also used, in boiling octane with di-*tert*-butyl peroxide as initiator, for the reduction of the methylxanthate of 1,2:5,6-di-*O*-isopropylidene-α-D-glucofuranose, to yield the corresponding deoxy sugar in 60%[173] and 70%[174]

isolated yield, with 4 molar equivalents of silane (Et$_3$SiH,[173] n-Pr$_3$SiH[174]) and tert-dodecanethiol (2 mol%), commercially available as a mixture of isomers. It is noteworthy that the hydrogen-donor ability of thiols was exploited for the enantioselective radical-chain hydrosilylation of prochiral alkenes using optically active thiol catalysts, in particular 2,3,4,6-tetra-O-acetyl-1-thio-β-D-glucopyranose.[154,175] Catalysis methods based on polarity reversal were shown to prevent rearrangements (see Section V) in radical-mediated reductions,[176,177] on account of the fact that the rate constant for hydrogen-atom abstraction from PhSeH by the 2,3,4,6-tetra-O-acetyl-D-glucopyranos-1-yl radical (**3R**) (~3.6×10^6 M^{-1} s^{-1} at 78 °C, a value substantially smaller that that for a primary alkyl radical)[177] was much higher than the rate constant for radical rearrangement. The lower rate of hydrogen abstraction from PhSeH by **3R** was not attributed simply to a stabilizing extended anomeric effect that might retard the quenching with PhSeH. It is possible that the electron-withdrawing β-acyloxy group renders the anomeric radical less nucleophilic, as compared to primary alkyl radicals, and entails a less favorable polarity match for the reaction with PhSeH.[177]

d. Synthesis of Anhydroalditols from Glycosyl Derivatives.—The first reported example of radical-mediated preparation of peracetylated 1,5-anhydro-D-alditols involved treating the corresponding peracetylated α-D-glycopyranosyl chlorides with n-Bu$_3$SnH (4.3 eq) in refluxing benzene containing a catalytic amount of AIBN for ~15 min.[178] Under these conditions, 2,3,4,6-tetra-O-acetyl-1,5-anhydro-D-glucitol (**41**) was obtained in 86% yield by a chain reaction, which also produced tri-n-butyltin chloride, as already exemplified. Comparably high yields were recorded for the synthesis of acetylated 1,5-anhydro-D-mannitol (**42**), -D-galactitol (**43**), -D-iditol, and -D-xylitol (**44**) (70, 75, 84, and 85%, respectively).[178] Similarly, 2-acetamido-3,4,6-tri-O-acetyl-2-deoxy-α-D-glucopyranosyl chloride was reduced with n-Bu$_3$SnH in slight excess (1.2 eq) in refluxing toluene with AIBN. Heating for 1 h 20 min under a nitrogen atmosphere followed by evaporation of the solvent and purification by column chromatography afforded 2-acetamido-3,4,6-tri-O-acetyl-1,5-anhydro-2-deoxy-D-glucitol (**46**) in 80–84% isolated yield on a multigram scale.[179,180] This protocol was also applied to the readily available peracetylated glycopyranosyl bromides,[181] which were found more reactive because they reacted with equimolar amounts of freshly distilled n-Bu$_3$SnH (1.01 eq) in diethyl ether within ~2.5 h.[120] After completion of the visible light-initiated reduction reaction, conversion of the resulting tri-n-butyltin bromide into the insoluble fluoride by stirring the crude reaction mixture with aqueous potassium fluoride provided a simplified purification procedure based on filtration and crystallization, which yielded, on a 100-g scale, the corresponding peracetylated anhydro-D-alditols in high yields[120] (**41**: 86%; **42**: 66%; **43**: 65%).

In the case of **42** and **43**, the reported yields corresponded to two-step transformations, starting from the corresponding peracetates. Similar yields were recorded when the deuteride $n\text{-}Bu_3SnD$ was used in order to investigate the stereoselectivity of the reduction reaction (see Section III, 4).

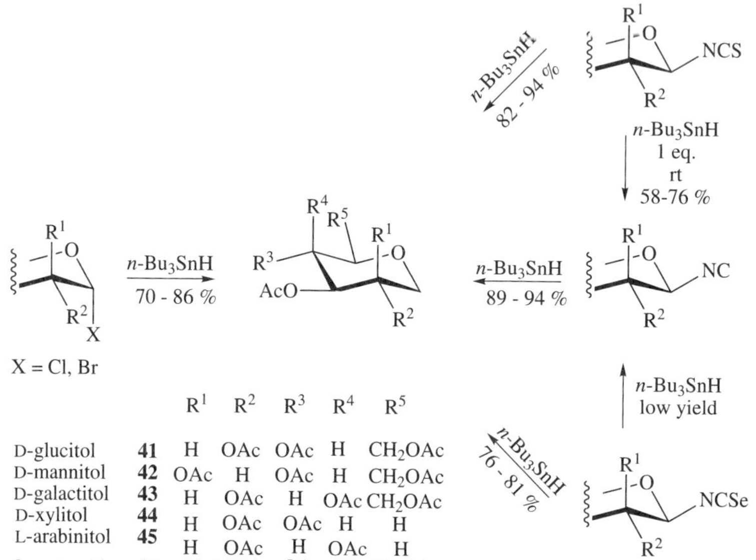

		R^1	R^2	R^3	R^4	R^5
D-glucitol	41	H	OAc	OAc	H	CH_2OAc
D-mannitol	42	OAc	H	OAc	H	CH_2OAc
D-galactitol	43	H	OAc	H	OAc	CH_2OAc
D-xylitol	44	H	OAc	OAc	H	H
L-arabinitol	45	H	OAc	H	OAc	H
2-acetamido -2-deoxy-D- glucitol	46	H	NHAc	OAc	H	CH_2OAc

1,5-Anhydrohexitols **41, 43–46** have also been prepared in 82–94% yields from peracetylated β-D-glycopyranosyl isothiocyanates by treatment with $n\text{-}Bu_3SnH$ and AIBN in refluxing toluene for several hours (6–10 h), the reaction proceeding via intermediate isocyanides.[182] In effect, the peracetylated glycopyranosyl isocyanides, prepared in 58–76% yields from the isothiocyanate precursors by treatment with $n\text{-}Bu_3SnH$ (1 eq) in diethyl ether or dimethoxyethane at 25 °C, were also converted under forcing conditions into the corresponding 1,5-anhydroalditols in somewhat higher yields (89–94%).[182] This observation indicates that the conversion of glycopyranosyl isothiocyanates into 1,5-anhydroalditols involves a desulfurization step that proceeds at room temperature, followed by a reductive deamination reaction that requires heating in the presence of AIBN. Under these conditions, the tin radical attacks the isonitrile group at the carbon atom, whereupon cleavage of the N-glycosyl bond generates a carbohydrate radical and tri-n-butyltin cyanide.[13b] The reactivity of peracetylated glycosyl isothiocyanates might be affected by subtle changes in the reaction conditions because they have been converted in isolated yields as high as 87%, into the corresponding glycosyl thioformamides on treatment with n-Bu$_3$SnH (2 eq) in diethyl ether or toluene for 6 h at 25 °C.[183] Lowering the amount of n-Bu$_3$SnH (to 1 eq) resulted in decreased yields of the glycosyl thioformamides (45–54%), which were formed without incorporation of deuterium when the reaction was conducted in diethyl ether moistened with D$_2$O.[183] In another study, a series of peracetylated 2-deoxy-2-iodo-glycopyranosyl isothiocyanates, prepared from the corresponding glycals by treatment with potassium thiocyanate and iodine, were reduced with n-Bu$_3$SnH (1.2 eq) and AIBN in refluxing diethyl ether.[184] Under these conditions the isothiocyanate group remained unchanged. Reduction proceeded chemoselectively, removing the iodine atom at C-2 to afford 2-deoxypyranosyl isothiocyanates that, on treatment with ammonia, afforded the corresponding N-(2-deoxyglycosyl) thioureas in 31–70% yield.[184] Hence, the isothiocyanate group at the anomeric position can withstand radical reduction, provided mild conditions are employed. Peracetylated glycopyranosyl isoselenocyanates have been shown to react with n-Bu$_3$SnH (1 eq) in refluxing benzene in the presence of AIBN. This method afforded, after 16 h of heating under nitrogen, the 1,5-anhydroalditols **41** and **46** in 76 and 81% yields, respectively.[185] When the reaction was conducted at room temperature, formation of the intermediate isocyanides was also observed, albeit in low yield.[185]

Other 1,5-anhydroalditols were obtained from protected glycopyranosyl bromides by radical reduction. Interestingly, 2,3,4,6-tetra-O-acetyl-1,5-anhydro-5-thio-D-glucitol (**47**)[62] has been prepared in 92% yield from the corresponding bromides (**12**, α, β mixture) on treatment with n-Bu$_3$SnH (1.1 eq) added over 20 h, in refluxing benzene (AIBN). Despite adding n-Bu$_3$SnH slowly, no radical rearrangement was observed. Under a variety of conditions (n-Bu$_3$SnH, hν, or AIBN), 1,5-anhydro-3,4,6-tri-O-benzoyl-D-fructose (**48**) is obtained quantitatively from the corresponding 3,4,6-tri-O-benzoyl-α-D-*arabino*-hexo pyranosyl-2-ulose bromide,[186] even in acrylonitrile as the solvent. In this case, the favored abstraction of hydrogen from the tin hydride by the carbohydrate-derived radical may be explained by the presence of the carbonyl group at C-2, which should decrease the nucleophilic character of the 2-oxo radical and, therefore, its aptitude to add to an electrophilic alkene. 3,4,6-Tri-O-acetyl-2-O-(diphenylphosphonyl)-α-D-mannopyranosyl bromide (**49**) afforded 3,4,6-tri-O-acetyl-1,5-anhydro-2-O-(diphenylphosphonyl)-D-mannitol (**50**) in 77% yield on brief treatment with n-Bu$_3$SnH (concentration: 1.4 M, 19 eq) in dry diethyl ether under irradiation by UV light.[187] Under comparable conditions, 3,4,6-tri-O-acetyl-2-O-(diphenylphosphonyl)-α-D-glucopyranosyl bromide (**51**) afforded a small amount of 3,4,6-tri-O-acetyl-1,5-anhydro-2-O-(diphenylphosphonyl)-D-glucitol (**52**)[64] in admixture with the 2-deoxy derivative resulting from the 2→1 migration of the phosphate ester group, which occurred faster in the D-*gluco* series (see Section V).

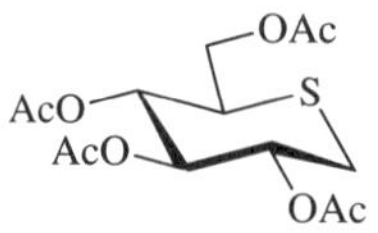

[Structures 47, 48]

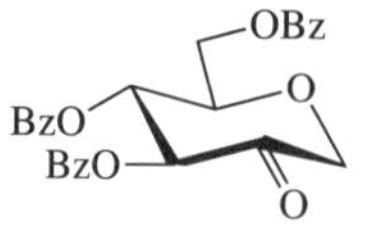

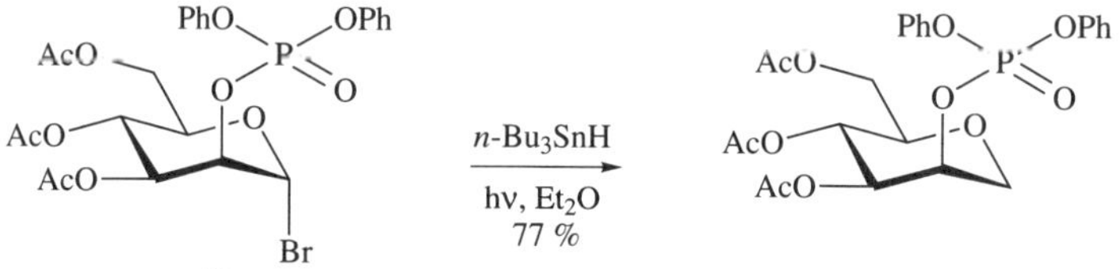

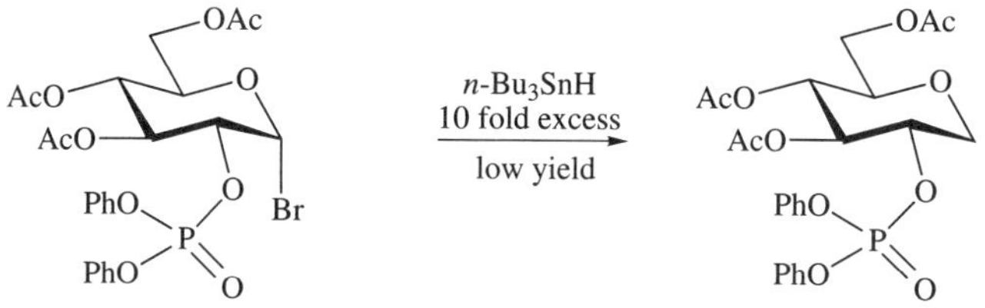

[Structures 49, 50 with reaction: n-Bu₃SnH, hv, Et₂O, 77%]

[Structures 51, 52 with reaction: n-Bu₃SnH, 10 fold excess, low yield]

Access to 1,4-anhydro-D-alditols by radical reduction has received only limited attention, but the most detailed work concerns the reduction of protected glycofuranosyl isocyanides on heating for 2 h at 80 °C in toluene in the presence of n-Bu₃SnH (2 eq) and AIBN.[188] In particular, 2,3,5-tri-*O*-benzoyl-β-D-ribofuranosyl isocyanide (**53**) was converted in 78% yield into 1,4-anhydro-2,3,5-tri-*O*-benzoyl-D-ribitol (**54**) also accessible from phenyl 2,3,5-tri-*O*-benzoyl-1-thio-D-ribofuranoside (**55**).[189]

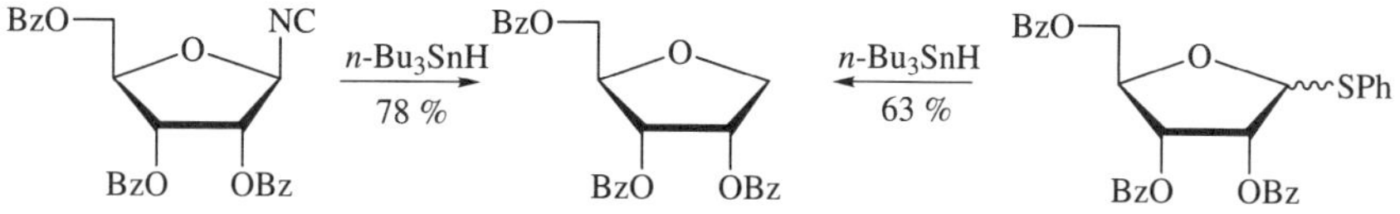

[Structures 53, 54, 55 with reactions: 53 → 54 via n-Bu₃SnH, 78%; 55 → 54 via n-Bu₃SnH, 63%]

The isocyanide method of reduction has been applied successfully for preparing a series of diversely protected 1,4-anhydroalditols:[188] 1,4-anhydro-5-*O*-*tert*-butyldimethylsilyl-2,3-*O*-isopropylidene-D-ribitol (**56**, 81%), 1,4-anhydro-5-*O*-*tert*-butyldimethylsilyl-6-deoxy-2,3-*O*-isopropylidene-D-allitol (**57**, 87%) and -L-talitol (**58**, 87%), 1,4-anhydro-2-*O*-*tert*-butyldimethylsilyl-3,5-*O*-isopropylidene-

D-xylitol (**59,** 81%), and 2-*O*-acetyl-1,4-anhydro-6-deoxy-3,5-*O*-(1,1,3,3-tetraisopropyldisiloxan-1,3-diyl)-D-allitol (**60,** 77%). The reported yields appeared to be independent of the anomeric configuration of the substrates, prepared in two steps from the corresponding glycofuranosyl azides, via the formamides.

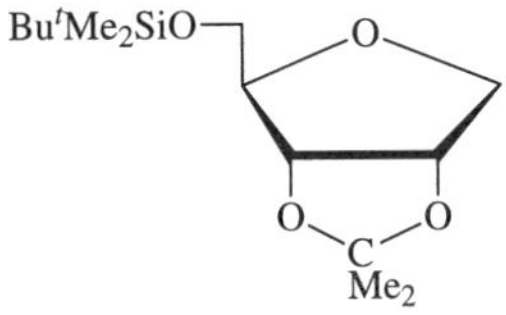

In addition to the conversion of phenyl 2,3,5-tri-*O*-benzoyl-1-thio-D-ribofuranoside (**55**) into 2,3,5-tri-*O*-benzoyl-1,4-anhydro-D-ribitol (**54**) in 63% yield,[189] other examples of the affinity of tin radicals towards the sulfur atom are available. They concern the radical desulfurization of thionolactones with triphenyltin hydride as a possible route to 1,4- and 1,5-anhydroalditols,[190] and the Barton–McCombie deoxygenation of sugar thiocarbonates.[128] This reaction can proceed in the presence of AIBN and *n*-Bu$_3$SnH, used in either stoichiometric (2 eq) or catalytic amounts. In the latter case, *n*-Bu$_3$SnH was produced from (*n*-Bu$_3$Sn)$_2$O (7.5–12 mol%) as a precatalyst, and a polymeric silane Me$_3$SiO–(SiHMeO)$_n$–SiMe$_3$ (5 eq) acting as a reductant, by heating (80–110 °C) in toluene containing *n*-BuOH (5.5 eq). Use of *n*-BuOH makes more efficient both utilization of (*n*-Bu$_3$Sn)$_2$O and regeneration of *n*-Bu$_3$SnH from *n*-Bu$_3$SnOPh,[128] produced by cleavage of the phenylthiocarbonate group. 1,4-Anhydro-2,3:5,6-di-*O*-isopropylidene-D-mannitol was obtained from the corresponding phenylthiocarbonate, in 63 and 72% yields, under either catalyzed or stoichiometric conditions, respectively. As mentioned next, 1-thio sugar derivatives undergo radical reduction in the presence of silanes.

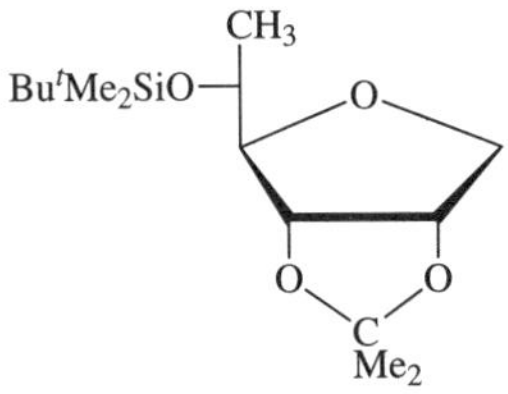

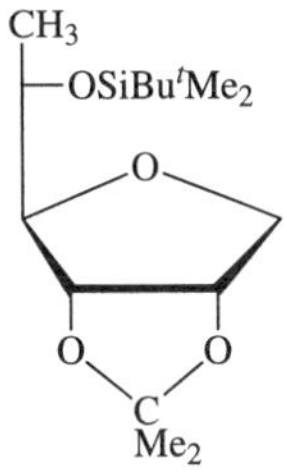

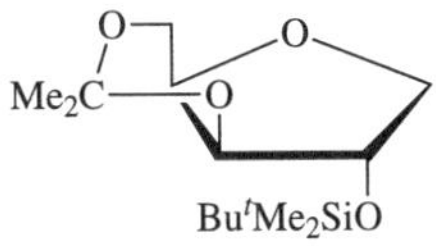

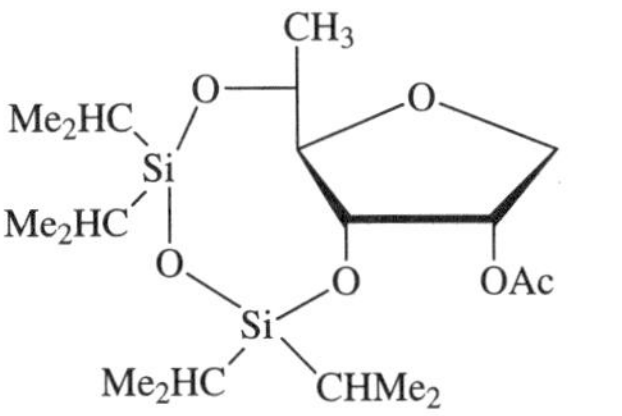

A few syntheses employ silanes as the hydrogen source. Thus, 2,3,4,6-tetra-*O*-acetyl-α-D-glucopyranosyl bromide (**3**) has been converted into **41** in 90% isolated yield on heating in neat PhSiH$_3$ in the presence of AIBN.[191] Compound **41** can also be obtained in comparably high yield (91%) from 2,3,4,6-tetra-O-acetyl-1-thio-β-D-glucopyranose, by treatment with Ph$_3$SiH (1.2 eq) in 1,4-dioxane at 60 °C, in the presence of di-*tert*-butyl hyponitrite.[154] No reaction occurred in the absence of hyponitrite as the initiator (5 mol% based on the thiol). Deoxygenation of 1-O-acetyl-2,3:4,6-di-O-isopropylidene-α-D-mannopyranose (**61**) by triphenylsilane in the presence of *tert*-butyl peroxide was achieved by prolonged heating at 140 °C in a sealed tube to produce 2,3:4,6-di-O-isopropylidene-1,5-anhydro-D-mannitol (**62**) in 73% yield.[192] This deoxygenation method has been successfully applied to a variety of isopropylidene-protected sugars having an acetoxy group attached to either a primary or a secondary carbon atom, even though the reported yields are somewhat lower. In the course of the synthesis of C-apiosyl compounds,[158] solutions containing 3'-O-benzoyl-2,3-O-isopropylidene-D-apio-β-D-furanosyl bromide (**63**), an alkene, and AIBN in refluxing benzene were treated with tris(trimethylsilyl)silane added dropwise. With alkenes made electrophilic by bearing electron-withdrawing groups, C-apiosyl compounds were the major products formed,[3] sometimes in admixture with 1,4-anhydro-3'-O-benzoyl-2,3-O-isopropylidene-D-apio-D-itol (**64**, ~20%). However, when cyclohexene was used as the alkene, compound **64** was isolated in 87% yield, most probably because addition of a nucleophilic carbohydrate radical to cyclohexene, an electron-rich alkene that may act as an hydrogen donor, appears disfavored.

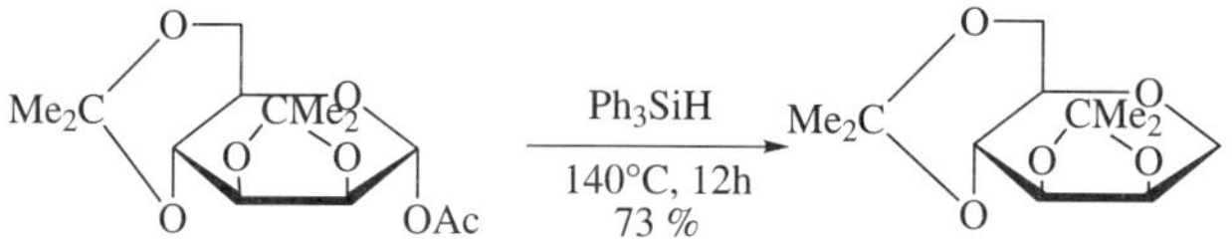

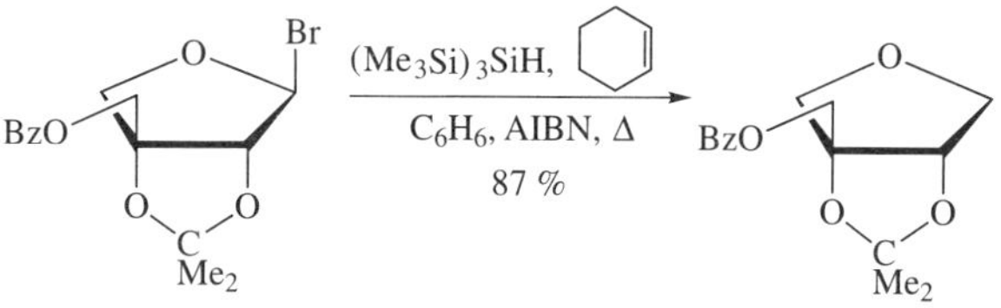

2. Synthesis of Anhydroalditols by Single-Electron Transfer

Peracetylated 1,5-anhydro-D-glucitol (**41**), -D-mannitol (**42**), -D-galactitol (**43**), -D-xylitol (**44**), -D-ribitol (**65**), and 2,3,5-tri-O-acetyl-1,4-anhydro-D-ribitol (**66**) were prepared in high yield from the corresponding glycosyl bromides on treatment with titanocene borohydride (Cp$_2$TiBH$_4$).[193] The reaction also proceeded with catalytic amounts of titanocene borohydride and sodium borohydride in excess, but with a lower selectivity. Titanocene borohydride can be prepared in two steps from commercial titanocene dichloride $Cp_2Ti(IV)Cl_2$ which, on reduction with aluminum in oxolane (THF) under an inert atmosphere, affords bis(titanocene chloride) $(Cp_2Ti(III)Cl)_2$ quantitatively.[194] A subsequent reaction with sodium borohydride in dimethoxyethane solution gave, at room temperature under an inert atmosphere, Cp_2TiBH_4 as crystals in 86% yield.[193] The mechanism of glycosyl halide reduction has been postulated to involve abstraction of a halogen atom[193] by Ti(III) to give a glycosyl radical, although its formation via a radical-anion is conceivable. In the final step, the intermediate radical abstracts a hydrogen atom (^1H or ^2H) from borohydride. Substantial deuterium incorporation at the anomeric carbon, when Cp_2TiBD_4 was used as the reducing agent,[193] and the effectiveness of $(Cp_2Ti(III)Cl)_2$ in promoting stereoselective synthesis of C-glycosyl compounds from acetylated glycopyranosyl bromides in the presence of substituted alkenes, are evidence in support of the proposed radical reaction pathway.[195]

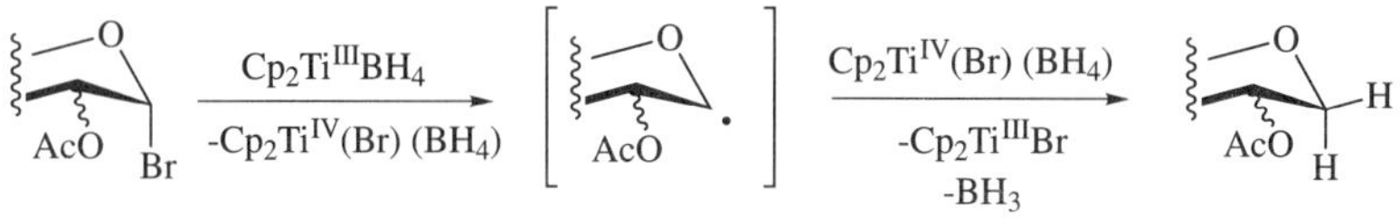

	yield, %
D-glucitol **41**	100
D-mannitol **42**	97
D-galactitol **43**	99
D-xylitol **44**	92

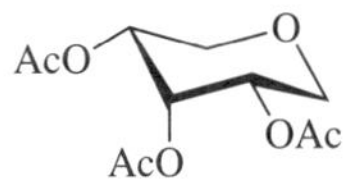

65 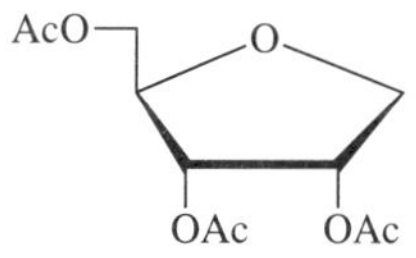   **66**

It has been reported that 2,3,4,6-tetra-O-acetyl-α-D-galactopyranosyl bromide (**13**), which undergoes zinc-induced 1,2-elimination to afford 3,4,6-tri-O-acetyl-1,5-anhydro-2-deoxy-D-*lyxo*-hex-1-enitol (87%) on treatment with zinc-silver/graphite in THF,[196] reacts quite differently in the presence of an added azido sugar (1 eq).[197] Reductive dehalogenation gave predominantly the 2-acetoxy anhydroalditol **43** (75–77% yield)[197] along with minor amounts of the 1,2-elimination product (4–8%). It is assumed that the azido sugar participates first in a single electron-transfer (SET) mechanism from the activated metal to the azide group, with subsequent transfer of the unpaired electron onto the carbon–bromine bond of the glycosyl bromide to produce a radical-anion intermediate.[197] Its collapse forms a bromide anion and the galactopyranosyl radical **13R,** able to abstract a hydrogen atom from oxolane (THF). Hydrogen abstraction from the solvent was proved by using deuterated THF. In keeping with the proposed reaction path, deuterium was incorporated at the anomeric carbon stereoselectively (see Section III,4), whereas recovery of the added azido sugar proved to be almost complete.[197] Based on inhibitions in the presence of such radical traps as carbon tetrachloride or elemental sulfur, mechanistic studies of the zinc–N-base-mediated synthesis of pyranoid glycals[198] pointed to the occurrence of glycosyl radicals along the reaction pathways. When the 1,2-elimination reaction was applied to **3** with an excess of methyl acrylate, a *C*-glucosyl compound, resulting from the addition of glucosyl radical to the electron-deficient double bond, appeared in 29% yield, along with the glucal (71%). Accordingly, adding such a hydrogen donor as *tert*-dodecanethiol (4 eq) to the reaction mixture led to 1,5-anhydroglucitol **41** (25%) and the corresponding acetylated glucal (75%) as the major product.[198]

3. Synthesis of Ring-Fused and C-Branched Anhydroalditols

Each reduction reaction heretofore exemplified utilized glycosyl derivatives that on reaction with a metal-centered radical or a reducing species led regiospecifically to anomeric radicals. It is conceivable to generate anomeric radicals, being stabilized species, from more-reactive radicals prone to intramolecular addition to an endocyclic C=C double bond in appropriately elaborated glycal derivatives.[125] Although intermolecular addition of a nucleophilic, carbon-centered radical to an electron-rich vinyl ether is not favored,[199] it has been shown that intramolecular versions are synthetically useful[200,201] because they proceed at rates[202,203] comparable to those of the corresponding reaction of alkenes. This approach has been exploited with substrates bearing a halogen atom on a pendant group attached to C-3. The intermediate anomeric radicals resulting from the initial cyclization may undergo subsequently hydrogen quenching to produce the corresponding alditols (Scheme 2), unless the reaction conditions are optimized so as to favor a intermolecular C–C bond formation in serial radical reactions.[125]

Attachment of a suitable branch to the C-3 position[125] was achieved by applying known methods resorting to 1,2-dibromoethyl ethyl ether[204] and (bromomethyl)-chlorodimethylsilane.[205,206] In this way, 1,5-anhydro-3-*O*-(2-bromo-1-ethoxyethyl)-2-deoxy-4,6-*O*-isopropylidene-D-*arabino*-hex-1-enitol (**67**), 1,5-anhydro-

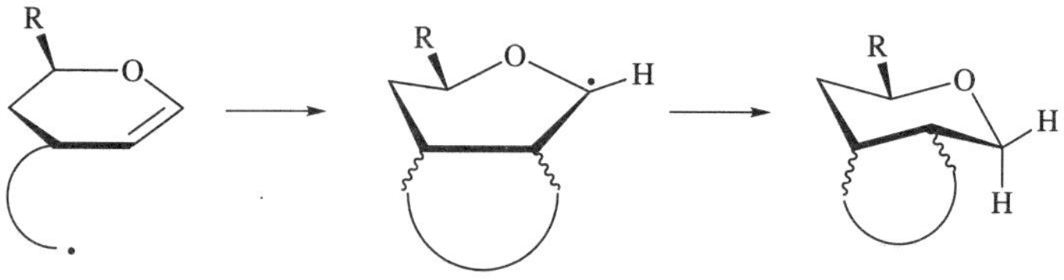

SCHEME 2

4,6-O-benzylidene-3-O-(2-bromo-1-ethoxyethyl)-2-deoxy-D-*ribo*-hex-1-enitol (**68**), their 3-O-bromomethyldimethylsilyl ether analogs (**69**) and (**70**), and 3-O-bromomethyldimethylsilyl-6-deoxy-L-*arabino*-hex-1-enitol (**71**) were prepared in good yields.[125,207] Compounds **67** and **68** were obtained as mixtures of diastereomers which, in the latter case only, could be separated by column chromatography. Such mixed bromoacetals or silicon-tethered bromides lead, in the presence of tri-*n*-butyltin radicals to nucleophilic carbon-centered radicals which undergo 5-*exo-trig* radical cyclization (Fig. 1) provided that a low concentration of n-Bu$_3$SnH is maintained in the medium by either slow addition (conditions A) or by use in catalytic amounts in conjunction with sodium cyanoborohydride (conditions B). The last method led to yields lowered by ~5%. Depending on whether the appendage is preserved or cleaved by subsequent treatments, this methodology results in either 2,3-*cis*-fused-bicyclic alditols or C-2 branched alditols. Cleavage of the siloxane ring was achieved by Tamao oxidation[208] with KHCO$_3$ (1 eq) and KF (4 eq) in 1:1 THF–MeOH, and heating to reflux for 10 h in the presence of 30% H$_2$O$_2$ (20 eq). In the resulting products, the stereochemistry at C-2 is dependent on the stereochemistry of the 3–OH group, in accordance with the expected formation of a *cis*-fused bicyclic system upon cyclization. Although intermolecular addition of a nucleophilic, carbon-centered radical to an electron-rich vinyl ether is not favored,[125] compounds **72** to **75** were obtained by intramolecular addition in good overall yields. However, the bromomethyldimethylsilyl ether **70**, with an axially oriented branch, did not yield any cyclized product, whereas equatorially disposed silyl ethers **69** and **71** did cyclize, albeit to different extents, as shown by the incomplete transformation of the latter. In addition to the 1,5-anhydroalditols presented in Fig. 1, this strategy has been developed further, by reacting the generated anomeric radicals with various radical traps, resulting in stereoselective C–C bond formation. The stereoselectivity of the reduction has not been examined in detail and can only be estimated on the basis of that observed for the radical C–C bond-forming reactions studied. Not surprisingly, a higher α-selectivity was observed for compounds **67** and **69**, which led to cyclized intermediate radicals not hindered on this side. In all of the experiments conducted, the failure to detect products arising from reduction of the initially formed radicals pointed to a fast 5-*exo-trig* cyclization,[3] which proceeded even faster when **67** was converted first into the corresponding iodide. An enhanced propagation at the alkyl iodide–stannyl radical step was put forward to account for the rate acceleration.[125]

4. Stereoselectivity of Hydrogen-Atom Transfer at Anomeric Radicals

2,3,4,6-Tetra-O-acetyl-α-D-glucopyranosyl bromide (**3**) and both anomers of 2,3,4,6-tetra-O-acetyl-D-glucopyranosyl chloride (**73** and **74**, respectively) react with deuterated n-Bu$_3$SnD under thermal or photolytic conditions to give the deuterated 1,5-anhydro-D-glucitols **78** and **79**, with a higher proportion of the 1-α-deuterio isomer **78**,[43,44] as shown by ^{1}H NMR. When the reduction of **3** was

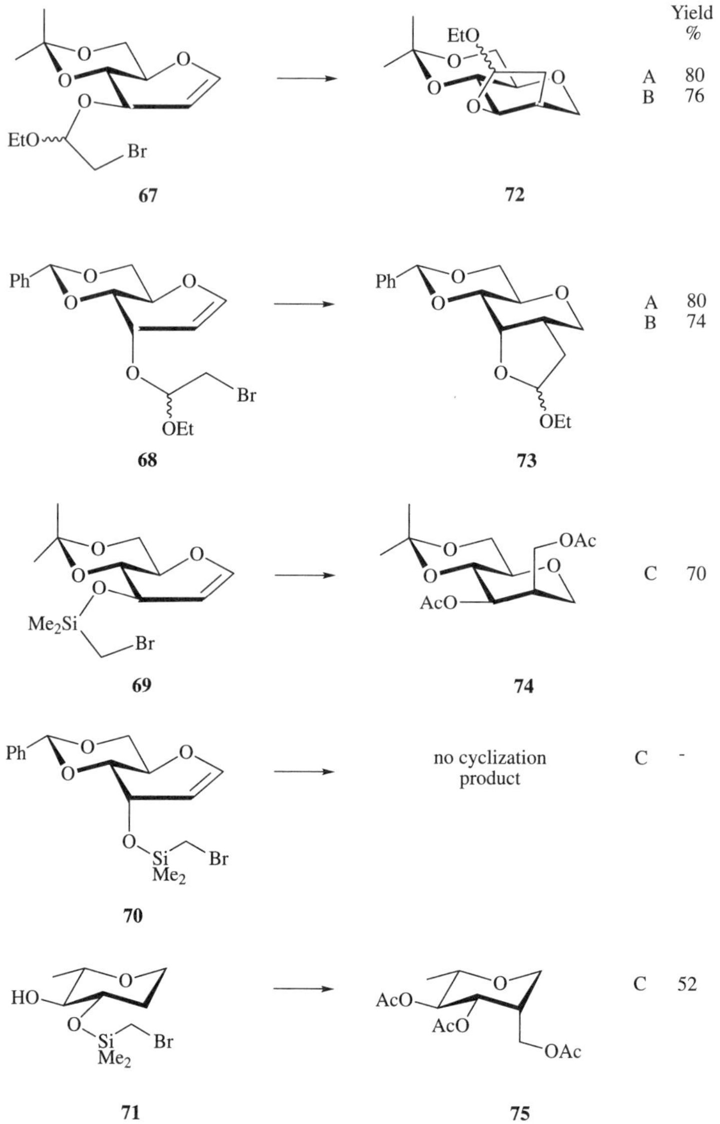

FIG. 1. Synthesis of ring-fused and C-2 branched 1,5-anhydroalditols.[125] Conditions: A, addition of n-Bu$_3$SnH during 12 h; B, n-Bu$_3$SnCl (0.1 eq) and NaBH$_3$CN (2 eq); C, radical cyclization followed by treatment with KF, oxidation, and acetylation.

carried out in THF under irradiation, the ratio of **78** to **79** varied according to the temperature as follows: 98:2, −20 °C; 96:4, 10 °C; 95:5, 30 °C; and 90:10, 60 °C. For reductions performed at 90 °C in toluene with AIBN, this ratio was 82:18 with **3** as the substrate and 83:17 from the β chloride[44] **77**, whereas the isolated yields were[43,44] in the range 65–90%. It was concluded that axial attack of the intermediate glucosyl radical occurs with 16 kJ.mol^{-1} (3.9 kcal.mol^{-1}) lower activation enthalpy than attack from the equatorial direction.[44] When 2,3,4,6-tetra-O-acetyl-α-D-mannopyranosyl bromide (**15**) was subjected to the same treatment, compound **80** was formed in more than 95% relative yield.[44]

The first interpretation of this stereoselectivity, which invoked anomeric σ-radicals, turned out to be inconsistent with the conclusions from ESR studies of glycopyranos-1-yl radicals, shown to adopt conformations having an almost planar π-radical center (see Section II). Because of the structural diversity found for the radicals studied, there is no evident, clear-cut, and general rationale to account for the predominant formation of a new bond from the α-face of the radical, whatever reduction, halogenation,[21] or C–C bond-formation reactions are concerned.[26,46] Furthermore, the stereoselectivity is sometimes largely influenced by the nature of the sugar protecting groups.[47] The 2-butanoyloxycyclohex-1-yl radical and its analogue derived from tetrahydropyran (the 3-butanoyloxytetrahydropyran-2-yl radical) adopt conformations that are compatible with hydrogen transfer from n-Bu$_3$SnH occurring from the axial direction (the less encumbered face), as deduced from the structure of the products.[166] These observations show the influence of steric factors and the *trans*-attack of the tetrahydropyran-2-yl radicals with respect to the non-bonding electron pair of the ring oxygen atom. It is also accepted that the direction of attack at anomeric radicals is under stereoelectronic control from the ring oxygen atom.[46,209] Axial attack on the D-*manno* radical **15R** preserves the overlap between the nonbonding electron-pair of the ring oxygen atom and the unpaired electron of the radical center (or the newly formed bond) on the way to the transition state.[46] For radical **15R**, having in the chair conformation an axially oriented acetoxy group at C-2, steric and stereoelectronic factors cooperate toward attack from the less-encumbered α face. For the D-glucopyranosyl radical **3R**, the $B_{2,5}$ boat conformation places the acetoxy group at C-3 in an almost axial orientation, which hinders the β face, whereas the other face is occupied by the C-2 acetoxy group. Thus, stereoelectronic effects best explain the preferred quenching from the *exo* side of **3R** resulting in an α-stereoselectivity. The lower selectivity observed for reduction reactions and C–C bond-forming processes in **3R,** as compared to **15R,** has been interpreted on the basis of the higher flexibility of boat conformers as compared to chair conformers.[46] There is only a small energy difference between the $B_{2,5}$ and $^{1,4}B$ conformations. According to the stereoelectronic effect, $B_{2,5}$ conformers afford α products, whereas β products are formed from $^{1,4}B$ conformers.[46] This is in keeping with theoretical investigations,[54] which supported the idea that, during the bonding process, sterically and stereoelectronically favored chair transition states are likely to be adopted even by a boat-shaped radical during hydrogen quenching.[26] An alternative explanation has been advanced to explain the preponderance of axial products in these radical reactions, based on the principle of least motion and steric effects.[210]

The favored formation of acetylated 1-α-deuterio-1,5-anhydro-D-glucitol **78** in labeling experiments has been put forward to support the occurrence of carbohydrate radicals as intermediates in the photolytic decomposition of glucopyranosyl phenyl sulfone acetates[104] as well as the Cp_2TiBD_4[193] and zinc–silver/graphite-[197] mediated reduction of glycosyl halides. Cationic intermediates would have favored the opposite stereochemistry.[26]

The favored formation of 1,4-anhydro-1-deuterio-2,3:5,6-di-O-isopropylidene-D-mannitol (**83**) from 2,3:5,6-di-O-isopropylidene-α-D-mannofuranosyl chloride (**82**) with n-Bu_3SnD indicates radical quenching from the less-hindered face.[43] Related examples of halo sugar radical-mediated reductions involving other ether carbon atoms similar in nature to the anomeric center,[21] for instance in hexopyranuronic,[211] 1,5-anhydropentofuranose, and 1,6-anhydrohexopyranose[212] derivatives (at the C-5, C-5, and C-6 positions, respectively) are known to occur with high stereoselectivity.[21] The next section, which deals with stereocontrolled syntheses of O-glycosides and C-glycosyl derivatives, further exemplifies the stereoselectivity and synthetic potential of radical reductions.

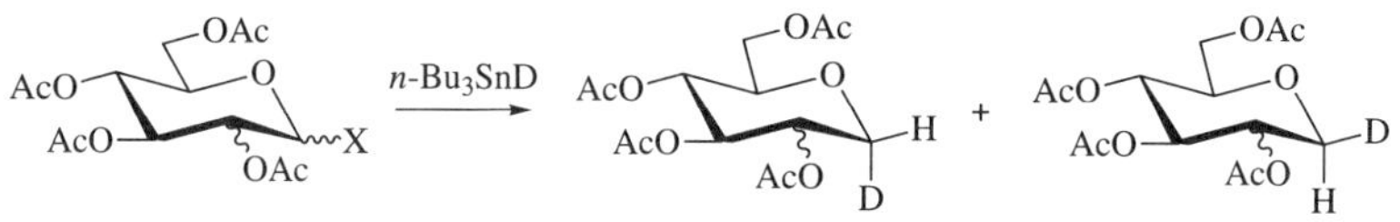

	X	Configuration	*major*	
3	α-Br	D-*gluco*	**78**	**79**
15	α-Br	D-*manno*	**80**	**81**
76	α-Cl	D-*gluco*		
77	β-Cl	D-*gluco*		

82   **83**

IV. DIASTEREOSELECTIVE REDUCTIONS OF SUBSTITUTED ANOMERIC RADICALS

Simple radicals derived from aldoses have been considered exclusively in the preceding sections, from the point of view of structure and transformation. Application of radical reduction to tertiary anomeric radicals, produced from ketoses, higher sugars and other substituted analogues, and having generally alkoxy and alkyl substituents, is discussed next. Radical reductions of such intermediates prove to have high synthetic value because, as discussed next, this process occurs, as in the case of radicals derived from aldoses, with high stereoselectivity. Consequently, new methodologies have been designed for radical-mediated preparations of O-glycosides, C-glycosyl compounds, and other anomerically substituted sugar derivatives.

1. Stereocontrolled Synthesis of O-Glycosides and Analogs

The radical routes leading to O-glycosides may differ, depending on whether they involve a single radical intermediate, regioselectively generated at the anomeric center, or whether a reactive radical can lead by intramolecular atom abstraction to an anomeric radical formed by radical translocation. In both cases, hydrogen quenching of the anomeric radical is the key step in terms of stereoselectivity.

a. Diastereoselective Hydrogen Quenching of Anomeric Radicals.—Diastereoselective quenching of alkoxy-substituted anomeric radicals for the construction of β-glycosides has been first applied to sugar-derived hemithioorthoesters,[213] produced from benzylated sugar thionolactones **84–86**, (D-*gluco*, D-*manno*, and 2-deoxy-D-*arabino* configurations, respectively) by heating, under basic conditions, with methyl iodide and MeOH (10 eq). In hemithioorthoesters **87–89**, obtained as single anomers, the methylthio group was assumed to be equatorial on the basis of stereoelectronic arguments. On photolytic treatment with n-Bu$_3$SnH (5 eq), cleavage of the C–S bond in **87–89** occurred under very mild conditions (toluene, AIBN, 30 °C). Hydrogen quenching from n-Bu$_3$SnH of the resulting carbon-centered anomeric radicals led in high yields to the methyl O-glycosides **90–92**. Applied to thionolactones **84–86**, this methodology was shown to produce β anomers preferentially (β:α ≤ 18:1), with a diastereoselectivity decreasing along the sequence: D-*manno* > D-*gluco* > 2-deoxy-*arabino,* and also in case of thermal initiation of the reaction (80 °C). This methodology is also applicable to construct β-linked disaccharides. Heating 2,3,4,6-tetra-O-benzyl-D-glucothionolactone (**84**) with methyl iodide in the presence of 1,2,3,4-tetra-O-acetyl-β-D-glucopyranose led to the corresponding hemithioorthoester, which, after reduction under photolytic conditions, afforded preferentially O-(2,3,4,6-tetra-O-benzyl-β-D-glucopyranosyl)-(1→6)-(1,2,3,4-tetra-O-acetyl-β-D-glucopyranose) (75% yield, β:α > 10:1).

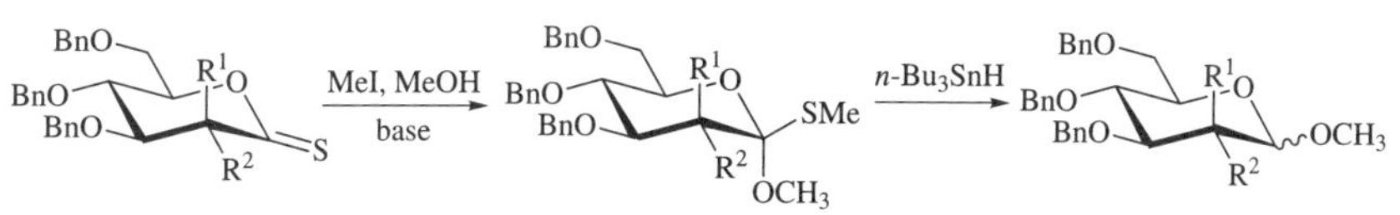

	R¹	R²			%	β:α
84	H	OBn	87	90	87	12:1
85	OBn	H	88	91	85	18:1
86	H	H	89	92	81	6:1

However, the Barton decarboxylation method,[214–216] introduced almost simultaneously in the carbohydrate field, proved to be more versatile. As exemplified next, the preparation of O-glycosides by the Barton O-acyl thiohydroxamate chemistry requires the formation of O-acyl thiohydroxamates of glyculosonic acid glycosides, first by condensation with such a thiohydroxamic acid as N-hydroxy pyridine-2-thione, followed by photochemical or thermal decarboxylation in the presence of a hydrogen donor (tertiary thiol, n-Bu$_3$SnH). O-Acyl thiohydroxamates can be prepared by three closely related protocols. They involve reaction of a carboxylic acid-containing substrate, either with N-hydroxypyridine-2-thione (**93**), in the presence of N,N'-dicyclocarbodiimide (DCC) and N,N-dimethylaminopyridine (DMAP), or with the 3-oxa-2-oxo-1-thiaindolizinium chloride **94**. Conversion of the carboxylic group into the corresponding acyl chloride, followed by condensation with the sodium salt of N-hydroxypyridine-2-thione (**95**) represents the third variant.

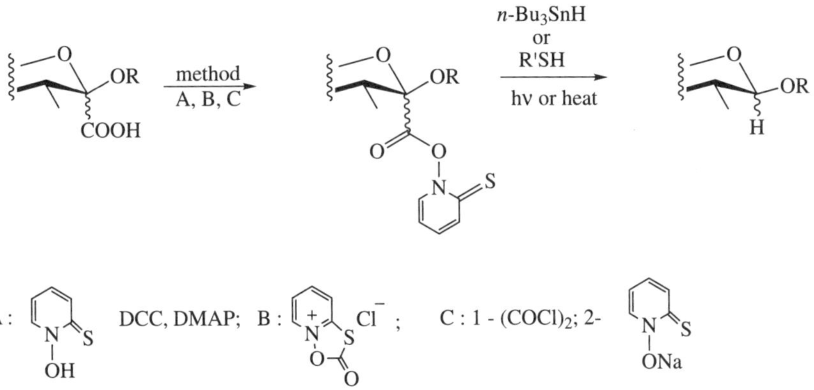

Diastereofacial hydrogen-atom transfer might constitute an elegant solution for the synthesis of O-β-D-mannopyranosides.[217] To this end, methyl and isopropyl 3,4,5,7-tetra-O-methyl-α-D-manno-2-heptulopyranosidonic acid glycosides (**96** and **97**) were converted by reaction with **94** in the presence of triethylamine (NEt$_3$) into the corresponding O-acyl thiohydroxamates. Subsequent white-light photolysis in the presence of *tert*-butylthiol afforded stereoselectively the methylated methyl and isopropyl β-D-mannopyranosides (**98, 99**, β:α > 25:1) in 75 and 48% yields, respectively, from the glyculosonic acids. Isolation of **97** as an impure product (∼65% purity) accounts for the lowered yield recorded for **99**, which was obtained with the same selectivity but in improved yield (68%) by radical reductive decarbonylation,[217] with n-Bu$_3$SnH and AIBN, in refluxing benzene. Acetyl protecting groups can withstand the Barton decarboxylation conditions, so that methyl 2,3,4,6-tetra-O-acetyl-β-D-mannopyranoside (β:α > 10:1) has been prepared in 80% yield (two steps). Good yields and even higher selectivities (β:α > 25:1) were recorded when this strategy was applied to the synthesis of 1,2:3,4-di-O-isopropylidene-6-O-(2,3,4,6-tetra-O-acetyl-β-D-mannopyranosyl)-α-D-galactopyranose (35–56%) and methyl 2,3,4-tri-O-acetyl-6-O-(2,3,4,6-tetra-O-acetyl-β-D-mannopyranosyl)-α-D-glucopyranoside (67%). Other experiments with disaccharides involving methyl 2,3-O-isopropylidene-α-L-rhamnopyranoside as a secondary glycosyl acceptor led to the conclusion that this methodology is not suitable when steric hindrance may impede both the glycosidation and radical steps.[217]

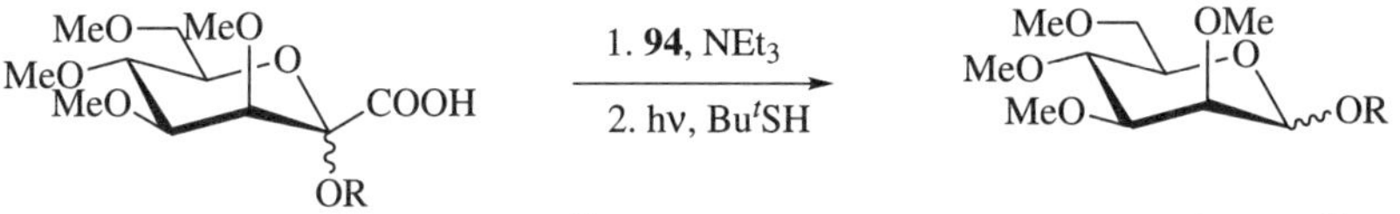

	R		%	α : β
96	Me	98	75	> 25 : 1
97	Pr^i	99	48	> 25 : 1

Other glyculosonic acid derivatives studied, having benzyl, *tert*-butyldimethylsilyl, and isopropylidene protective groups, have been found suitable for stereocontrolled syntheses of the corresponding O-glycosides.[218,219] Methyl 4,5,7-tri-O-benzyl-3-O-*tert*-butyldimethylsilyl-α,β-D-*gluco*-hept-2-ulopyranosidonic acid (**100**) undergoes reductive decarboxylation to afford methyl 3,4,6-tri-O-benzyl-2-O-*tert*-butyldimethylsilyl-β-D-glucopyranoside (**101**) in 65% overall yield (2 steps: O-acyl thiohydroxamate formation, photoinduced reduction) and high stereoselectivity (β:α > 25:1). In contrast, application of the same procedure to the bicyclic, ring-fused 4,5,7-tri-O-benzyl-2,3-O-isopropylidene-α-D-*gluco*-hept-2-ulopyranosonic acid revealed the first example of hydrogen quenching of D-glucopyranos-1-yl radicals from the β-face, as 3,4,6-tri-O-benzyl-1,2-O-isopropylidene-α-D-glucopyranose (**103**) was the sole product obtained, with no evidence found to indicate formation of the alternative diastereoisomer. Accordingly, the α:β ratio was estimated to be >25:1. The highly selective formation of the *cis*-fused isopropylidene acetal **103** constituted, apparently, the sole example of a reaction of a glucopyranosyl radical in which the normal preference for α-facial quenching is reversed. It was proposed that the glucopyranosyl radical leading to **103** is of the σ-type, with a $^OS_2$ conformation identical to that observed for **103**, so that hydrogen quenching can take place from the outside face of the bicyclic system.

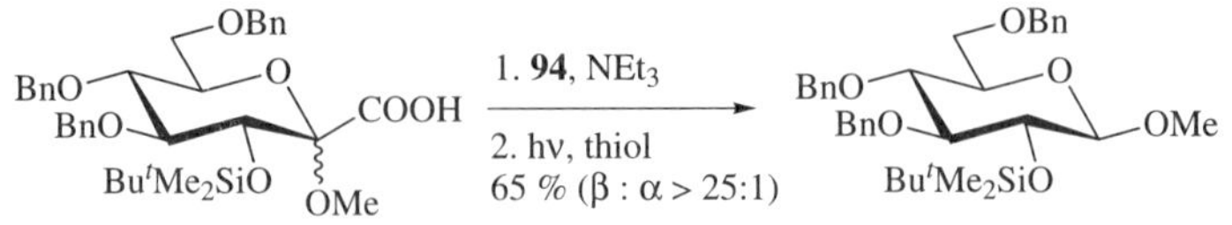

Structures

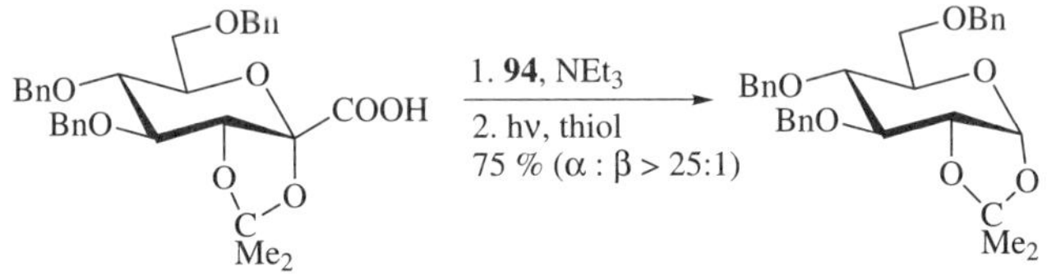

100: BnO, BnO, Bu^tMe_2SiO, OBn, O, COOH, OMe

→ 1. **94**, NEt_3; 2. hν, thiol; 65 % (β : α > 25:1) →

101: BnO, BnO, Bu^tMe_2SiO, OBn, O, OMe

102: BnO, BnO, OBn, O, COOH, O–C(Me_2)–O

→ 1. **94**, NEt_3; 2. hν, thiol; 75 % (α : β > 25:1) →

103: BnO, BnO, OBn, O, O–C(Me_2)–O

Attempts to apply the reductive Barton methodology as a new route toward other O-glycosidic linkages, difficult to prepare otherwise, have met success. Thus, conversion of a series of benzyl- and silyl-protected O- and S-glycosides (anomeric mixtures) of methyl 3-deoxyglyculosonates[220–222] was performed by a three-step sequence, comprising saponification, O-acyl thiohydroxymate preparation by reaction with either **93** or **94**, followed by photolysis in the presence of *tert*-dodecanethiol, giving access to the corresponding 2-deoxy glycopyranosides and their 1-thio analogs. It is noteworthy that C–S bonds are stable under the conditions of Barton decarboxylation,[222] as shown by preparations of phenyl 3,4,6-tri-O-benzyl-2-deoxy-1-thio-β-D-arabinopyranoside (**107**) and phenyl 3,4,6-tri-O-*tert*-butyldimethylsilyl-2-deoxy-1-thio-β-D-lyxopyranoside (**111**), in 48–50% yields. Except for the case of *p*-tolyl 3,4,6-tri-O-benzyl-2-deoxy-β-D-arabinopyranoside (**105**), the overall yields recorded (three steps) were in the range 34–55%, and the selectivity was in each case in favor of the β anomers. Although these are formed predominantly, the presence of α anomers, shown by ^{1}H NMR, pointed to a decreased selectivity for radical hydrogen quenching in the 2-deoxy series.[222] Incomplete stereoselectivity of the hydrogen-atom transfer to alkoxyglycosyl radicals was also observed when a mixture of four diastereoisomeric disaccharide bisthiohydroxamates was subjected to the Barton reductive decarboxylation protocol.[223] The isolated product (76%) was shown by NMR spectroscopy to be a mixture of three disaccharides, with methyl 4-O-benzyl-3-O-(3,4-di-O-benzyl-2,6-dideoxy-β-D-*arabino*-hexopyranosyl)-2,6-dideoxy-β-D-*arabino*-hexopyranoside as the major product, contaminated by the β-(1→3)-α-OMe, and α-(1→3)-β-OMe isomers (80:15:5 ratio, respectively).

As an example in the furanose series, the sodium salt of commercial 2,3:4,6-di-O-isopropylidene-2-keto-α-L-gulonic acid[222] was treated with oxalyl chloride, and then with **95**. Heating the resultant O-acyl thiohydroxamate to reflux in

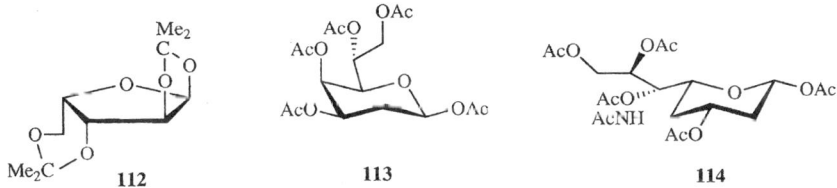

	%	β : α	R		%	β : α
104	40	10 : 1	OMe	108	55	10 : 1
105	12	> 95 : 5	O-C$_6$H$_4$-4-Me	109	34	16 : 1
106	50	11 : 1	O-3β-cholestanyl	110	44	8 : 1
107	50	8 : 1	SPh	111	48	18 : 1

benzene in the presence of *tert*-dodecanethiol led to 1,2:3,5-di-*O*-isopropylidene-α-L-xylofuranose (**112**) in 43% yield. Hydrogen quenching occurred preferentially from the less-hindered β side of the *cis*-fused tricyclic L-xylofuranos-1-yl radical, as discussed for **103**. Reductive decarboxylation methodology applied to 3-deoxy-D-*manno*-2-octulosonic acid and *N*-acetylneuraminic acid derivatives allowed preparations of 2-deoxy derivatives of higher sugars,[224] such as 1,3,4,6,7-penta-*O*-acetyl-2-deoxy-β-D-*manno*-heptose (**113**) and 4-acetamido-1,3,6,7,8-penta-*O*-acetyl-2,4-dideoxy-α-D-*glycero*-D-*galacto*-octose (**114**), obtained in 68 and 27% overall yields (two steps).

The diastereofacial selectivity of radical hydrogen quenching applied to the synthesis of *O*-glycosides is discussed later, in connection with data from related methods concerning radical-mediated preparations of *C*-glycosyl compounds.

b. Inversion of α- to β-*O*-Mannopyranosides by Radical Translocation.—

Although the α-mannopyranoside bond is known as an anomeric linkage formed particularly readily, the β-mannopyranoside linkage is considered to be quite difficultly accessible. Therefore, as an alternative to the Barton decarboxylation method, which suffers from limitations in terms of substrate accessibility and inefficient radical quenching in hindered structures, a method applicable for inverting α-mannopyranosides to β anomers would be of great synthetic potential. This objective might be attainable by a sequence of radical reactions involving first an intra-[225] then an intermolecular hydrogen transfer.[127,226] Based

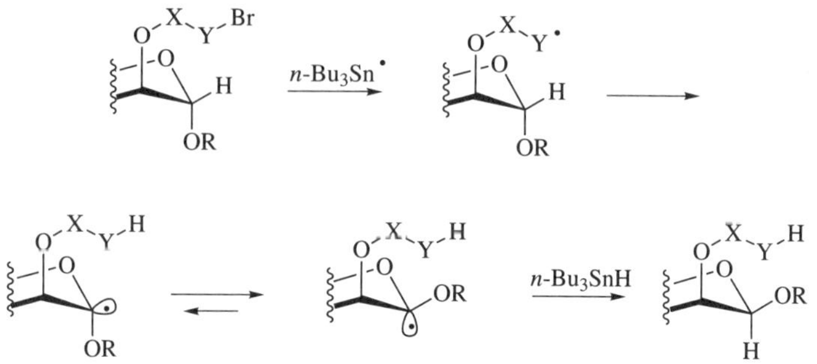

SCHEME 3

on the fact that Norrish type II photocleavage of O-phenacyl 2,3,4,6-tetra-O-benzyl-α-D-mannopyranoside led cleanly to 2,3,4,6-tetra-O-benzyl-D-mannono-1,5-lactone,[169] there was *a priori* no obstacle to the abstraction of the equatorial anomeric hydrogen atom in α-mannopyranosides. Moreover, the initially formed radical can be generated in the vicinity of the anomeric center by attachment to a 2-OH group of a variety of pendant groups, suitable for radical translocation by intramolecular 1,5- or 1,6-hydrogen abstraction. Although it is known that, provided attainment of the required 6-membered transition state, 1,5-hydrogen abstraction is a radical process generally favored over 1,4-[227] and 1,6-hydrogen[23] abstractions, such hydrogen transfers remain influenced by subtle conformational effects.[228]

Turned to practice, different systems (Scheme 3), in which Y• is either an sp^2-hydridized carbon atom or a heteroatom-centered radical, and X various tethers, were found inappropriate,[127,226] in particular because the simple reduction product was formed predominantly, as observed with the bromomethyldimethylsilyl group.[127,228] Carbohydrate derivatives having a 2-O-(1-bromo-2-methoxy-2-propyl) branch, obtained as ~1:1 diastereoisomeric mixtures by treating the precursor alcohol with either 2-methoxypropene and N-bromosuccinimide (NBS) or with 1,2-dibromo-2-methoxypropane and N,N-dimethylaniline, have been investigated with more success. However, dropwise addition of *n*-Bu₃SnH and AIBN to methyl 3-O-benzoyl-4,6-O-benzylidene-2-O-(1-bromo-2-methoxy-2-propyl)-α-D-glucopyranoside in refluxing benzene led to methyl 3-O-benzoyl-4,6-O-benzylidene-α-D-*arabino*-hexopyranosid-2-ulose (**115**) in 70% yield, through almost exclusive 1,4-hydrogen abstraction in such a D-*gluco* model.[127,228] Turning to models having the D-*manno* configuration, it was found that methyl 2-O-(bromomethyldimethylsilyl)-3-O-benzoyl-4,6-O-benzylidene-α-D-mannopyranoside led, on treatment with *n*-Bu₃SnH, to methyl 3-O-benzoyl-4,6-O-benzylidene-2-O-(trimethylsilyl)-α-D-mannopyranoside in 95% yield.[127] Thus, in connection with the geometry of the tether,[127] direct reduction of the initially formed radical was favored over radical translocation. In actual fact, 2-O-

(1-bromo-2-methoxy-2-propyl)-3-*O*-benzoyl-4,6-*O*-benzylidene-α-D-mannopyranoside (**116**), obtained as a diastereoisomeric mixture, was converted, on heating in benzene with *n*-Bu$_3$SnH and AIBN (added dropwise over 5 h), into a complex mixture. The products, after cleavage of the tether by stirring with moist silica gel, were identified as **115**, methyl 3-*O*-benzoyl-4,6-*O*-benzylidene-α-D-mannopyranoside (**117**), methyl 3-*O*-benzoyl-4,6-*O*-benzylidene-β-D-mannopyranoside (**118**), and methyl 3-*O*-benzoyl-4,6-*O*-benzylidene-α-D-glucopyranoside (**119**). The yields were 22, 34, 30, and 8%, respectively, for the ketone, α-mannoside, β-mannoside, and α-glucoside.[127] Formation of these products is indicative of several reaction pathways, including 1,4-hydrogen abstraction leading either to ketone **115** by subsequent radical fragmentation, or to α-mannoside **117** (also formed by direct reduction of the initial radical), and to α-glucoside **119** by hydrogen quenching of the carbon-centered radical at C-2, as well as the desired 1,5-translocation leading to **118**. This view was supported by experiments conducted under varied conditions (different concentrations of *n*-Bu$_3$SnH, use of *n*-Bu$_3$SnD, and tris(trimethylsilyl)silane, a compound known for its decreased reactivity towards the quenching[118] of simple nucleophilic alkyl radicals). The outcome of the reaction was significantly different only when *n*-Bu$_3$SnH was added rapidly over 5 min at 78 °C or when photochemical initiation at room temperature was applied. Formation of the ketone did not occur, and the only isolated products were **117** and **118**, formed in 57–59 and 30–34% yields, respectively.[228]

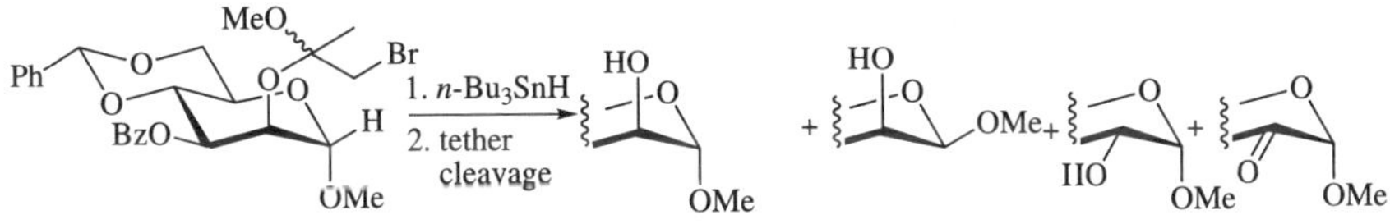

Treatment of methyl 3,4-di-*O*-benzoyl-2-*O*-(1-bromo-2-methoxy-2-propyl)-α-L-rhamnopyranoside with *n*-Bu$_3$SnH and AIBN in refluxing benzene gave a mixture of compounds arising mainly from direct reduction and ring opening.[228] The significant yields of acyclic products formed under a variety of conditions suggested that the 1-alkoxy-glycosy-1-yl radical in the rhamnopyranose series undergoes fragmentation more readily than in the mannopyranose series. Comparable results, showing formation of the desired β-L-rhamnopyranosides in only minor amounts (<8%) or not at all when the tin hydride was added slowly, were obtained when using analogous α-L-rhamnopyranosides with either benzyl or dispiroketal protecting groups at the 2- and 3-positions, with evidence pointing to 1,4-hydrogen abstraction.[228] Although photolytic initiation, at room temperature, gave cleaner reactions, direct reduction (75–89%) was favored over 1,5-hydrogen transfer (11–25%).[228] The same reactivity was observed for the disaccharide derivative methyl [3-O-benzyl-4,6-O-benzylidene-2-O-(1-bromo-2-methoxy-2-propyl)-α-D-mannopyranosyl]-(1 → 6)-2,3,4-tri-O-acetyl-α-D-glucopyrano-side, which, on photolysis at room temperature with dropwise addition of n-Bu$_3$SnH, followed by hydrolysis of the tether, led to methyl (3-O-benzyl-4,6-O-benzylidene-α-D-mannopyranosyl)-(1 → 6)-2,3,4-tri-O-acetyl-α-D-glucopyrano-side and to methyl (3-O-benzyl-4,6-O-benzylidene-β-D-mannopyranosyl)-(1→6)-2,3,4-tri-O-acetyl-α-D-glucopyranoside, in 3:1 ratio.

The possibility of α- to β-mannopyranoside inversion through 1,6-hydrogen transfer was also examined,[226] using methyl 3-O-benzyl-4,6-O-benzylidene-α-D-mannopyranoside having different brominated moieties (o-bromobenzyl or o-bromoaryldimethylsilyl groups) attached to the 2-hydroxyl group. Comparative experiments were carried out by heating 0.05 M solution of **120** and **121** for 14 h at 80 °C with n-Bu$_3$SnH (1.3 eq) and AIBN, followed by DBU–I$_2$ workup.[122] Reduction of the o-bromobenzyl derivative **120** provided the directly reduced α-mannopyranoside **122**, plus two inverted products, the desired methyl 2,3-di-O-benzyl-4,6-O-benzylidene-β-D-mannopyranoside (**124**) and methyl 2,3-di-O-benzyl-4,6-O-benzylidene-α-D-glucopyranoside (**126**) in the ratio of 65:21:13 and a 41% combined yield. The o-bromophenyldimethylsilyl ether **121** led to the analogous three products **123, 125,** and **127**, in the ratio of 33:26:41 and a 58% combined yield. These structures may be rationalized on the basis of bimolecular reduction with n-Bu$_3$SnH by either a direct process (**122, 123**) or after radical translocation (**124, 125**: 1,6-hydrogen transfer; **126, 127**: 1,5-hydrogen transfer). Whereas direct reduction could be minimized by decreasing the concentration of tin hydride, the competition between 1,5-versus 1,6-hydrogen transfer required further investigation using derivatives having acyl groups at the 3 position and, at C-2, o-bromophenyldimethylsilyl or (2-bromo-4,5-difluorophenyl)dimethylsilyl pendant groups, each being more-favorable to radical translocation.[226]

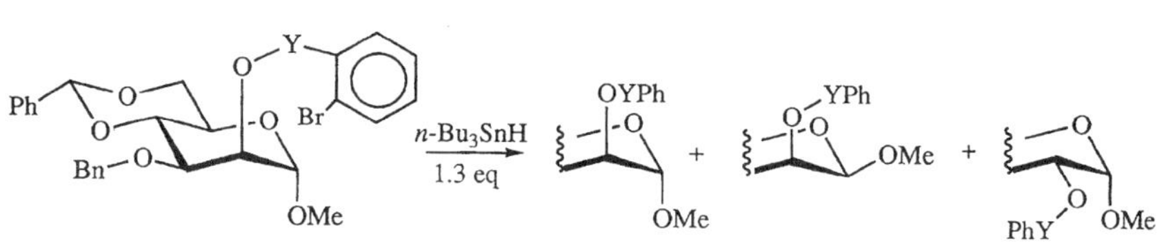

	Y	Yield %	α-manno		β-manno		α-gluco	
120	CH$_2$	41	122	65	124	21	126	13
121	SiMe$_2$	58	123	33	125	26	127	41

Compared to the previous models, the 3-acyl derivatives tested (benzoyl, 2-fluorobenzoyl, 1-naphthoyl, 2-naphthoyl, and pivaloyl) provided increased total amounts of radical-translocation products (β-*manno*/α-*gluco*). For methyl 3-*O*-benzoyl-4,6-*O*-benzylidene-2-*O*-(2-bromo-4,5-difluorophenyl)dimethylsilyl-α-D-mannopyranoside (**128**) and its 3-*O*-(2-naphthoyl) analog **129**, which constitute the more-favorable combination of translocating and protecting groups, all of the products resulted from radical translocation, the 1,6- over the 1,5-hydrogen transfer being surprisingly favored. This may be seen from the 58:42 ratio found for **131** and **133**, having respectively the β-D-*manno*- and α-D-*gluco* configurations.[226] These data confirm the importance of substrate geometry in hydrogen-transfer reactions of aryl radicals, and they demonstrate the validity of the α- to β-mannopyranoside inversion approach, with ~40% isolated yields of the desired β-D-mannopyranosides under such selected conditions.

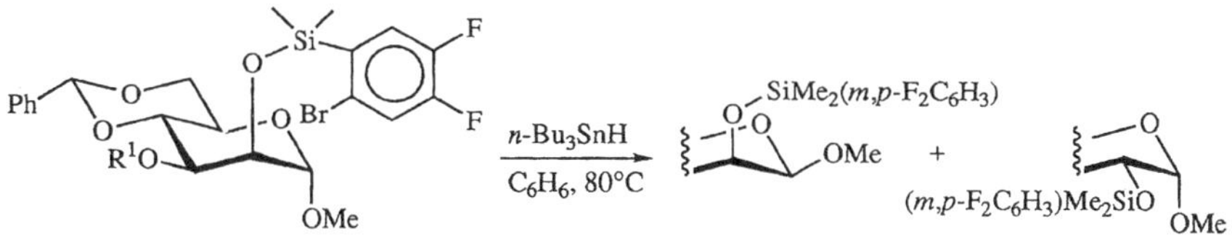

	R^1	Yield %				
128	benzoyl	82	**130**	50	**132**	50
129	2-naphthoyl	68	**131**	58	**133**	42

In conclusion, although conditions have been optimized for the inversion of α- to β-mannosides without interference by detrimental side reactions, the β:α ratio is governed by the intramolecular hydrogen-abstraction step, which remains relatively inefficient. The necessary multistep preparation of suitable substrates constitutes another limitation of the inversion method.

2. Stereocontrolled Synthesis of *C*-Glycosyl Compounds and Analogues

Among glyculosyl derivatives, those having halogen atoms or nitro groups attached to the anomeric center and thio- and seleno-glyculosides are able to undergo radical-mediated reduction in the presence of tin or silicon hydrides by cleavage of weak bonds in chain-reaction processes, as described before for the synthesis of anhydroalditols and *O*-glycosides. Radical reduction, by means of the Barton decarboxylation method, is also applicable to 2-glyculosonic acid derivatives having *C*-substituents at the anomeric carbon atom next to the carboxylic group. Both categories may lead to the corresponding deoxy sugars, which can be considered as *C*-glycosyl derivatives, a class of glycomimetics that has been investigated in detail.[37–40,89]

a. From Glyculosyl Halides and Analogues.—Introduction of a halogen atom at the anomeric center is usually a straightforward task and a common strategy for preparing varied activated sugars, even for glyculoses. Attempts to reduce glyculosyl halides (in which the halide atom adopts generally an axial orientation) under radical conditions proved satisfactory, even though tertiary carbon-centered radicals are involved. This method was exploited for the synthesis of 3-(2,3,4,6-tetra-O-acetyl-β-D-glucopyranosyl)-1-propene (**134**), and its D-*manno*-, and D-*galacto*-analogues (**135, 136**, respectively),[229] using the corresponding α chlorides as precursors. It was found more convenient to prepare them by a one-pot sequence, starting from 2,3,4,6-tetra-O-acetyl-1-bromo-β-D-glycopyranosyl chlorides,[71] which were subjected first to radical-mediated allylation with allyltributyltin, followed by reduction with n-Bu$_3$SnH. The overall yields for the two photoinitiated radical steps were in the range 51–57%, and the yield of the reduction can be estimated to be ∼65%. Only the β-D-C-glycosyl compounds were obtained, with no traces of the α anomers, indicating that the stereoselectivity of hydrogen quenching from the α side is >95:5. It is noteworthy that known methods leading to acetylated 3-(D-glycopyranosyl)-1-propene by either ionic[37,38] or radical[230–234] processes favor formation of α anomers, except when the bulky phthalimido group is present at the 2 position.[232,234] It is to be stressed that the ready availability of glycopyranosylidene dihalides as substrates[71] make this method more practical than the radical decarboxylation procedure applied, as discussed next, to a related stereocontrolled synthesis of benzylated 3-(2-deoxy-β-D-*arabino*-hexopyranosyl)-1-propene.

Methyl 4,5,7-tri-O-acetyl-2,6-anhydro-3-deoxy-D-*gluco*-heptonate (**137**) was obtained as the only product (89% yield) from the corresponding α-chloride,[235] on treatment with n-Bu$_3$SnH (with AIBN) in benzene at 75 °C. Similarly, reduction of the acetylated β-chloride derived from methyl N-acetylneuraminate occurred stereoselectively under comparable conditions, with formation of **138** as a single product (50% yield), by stereoselective quenching from the β side ($^1C_4$ chair conformation).[236,237] Compound **138** is also formed as the minor component of an anomeric mixture, by a samarium diiodide-based deoxygenation method, which might involve anomeric radicals as transients along the reaction path.[238,239]

Reduction of 2,3,4,6-tetra-O-acetyl-1-bromo-β-D-galactopyranosyl cyanide with n-Bu$_3$SnH (with AIBN) in refluxing benzene gave an anomeric mixture of 2,3,4,6-tetra-O-acetyl-D-galactopyranosyl cyanides (**139**, 77% yield) in which the β anomer produced by quenching from the axial direction was found predominant (β:α = 6:4).[240] Similarly, radical reduction of 2,3,4-tri-O-acetyl-1-bromo-D-arabinopyranosyl cyanide afforded 2,3,4-tri-O-acetyl-D-arabinopyranosyl cyanide as an anomeric mixture (**140**, 67% yield, β:α = 1:4). Owing to the favored $^1C_4$ chair conformation of the D-arabinopyranosyl ring, the major α anomer was formed, as in the previous case, by axial quenching. These decreased stereoselectivities might be attributable to the anomeric effect exerted by the cyano group (see Section IV-4).[241–243]

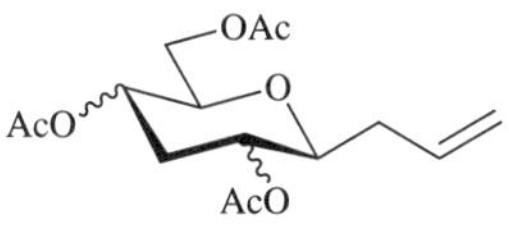

D-gluco   **134**
D-manno   **135**
D-galacto **136**

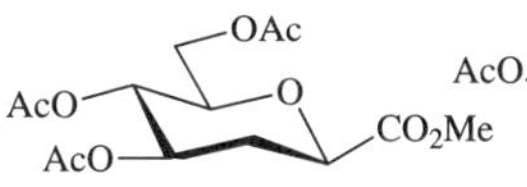

137

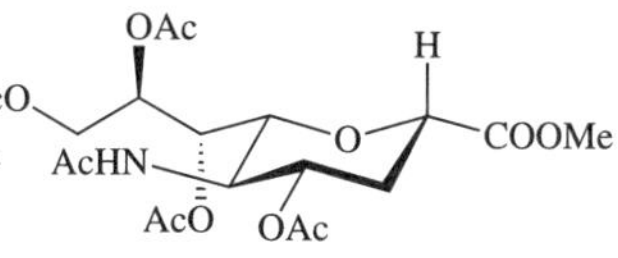

138

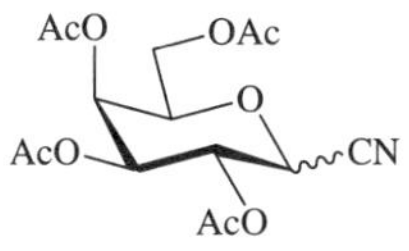

139 α / 139 β

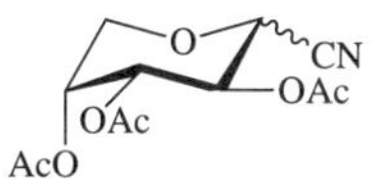

140 α / 140 β

b. From Thio- and Seleno-Glyculosides.—There are only a few examples of radical reduction applied to thio- and seleno-glyculosides, although treatment of phenyl 3,4,5,7-tetra-O-benzyl-2-thio-α-D-*gluco*-heptulopyranoside (**141**) with *n*-Bu$_3$SnD (plus AIBN) in refluxing toluene was reported to afford (2S)-2,6-anhydro-3,4,5,7-tetra-O-benzyl-2-deuterio-D-*gluco*-heptitol (**142**) in 97% yield.[244] This provided a simple means for introducing an asymmmetric substitution (α-deuterium atom at C-2) into **142** without modifying its reactivity as an acceptor in enzymatic galactosylation. In contrast, phenyl 1,3,4,6-tetra-O-benzoyl-2-seleno-α,β-D-fructofuranose (**143**) was partially converted[245] on treatment with tris(trimethylsilyl)silane (plus AIBN) into a 3:2 mixture of 1,3,4,6- tetra-O-benzoyl-2,5-anhydro-D-glucitol (**144**) and 1,3,4,6-tetra-O-benzoyl-2,5-anhydro-D-mannitol (**145**). Rearranged products, expected under the conditions used, were not detected, thus showing that the tertiary fructofuranosyl radical behaves differently from the analogous secondary ribosyl radical, which is prone to benzoyl-group migration (see Section V).

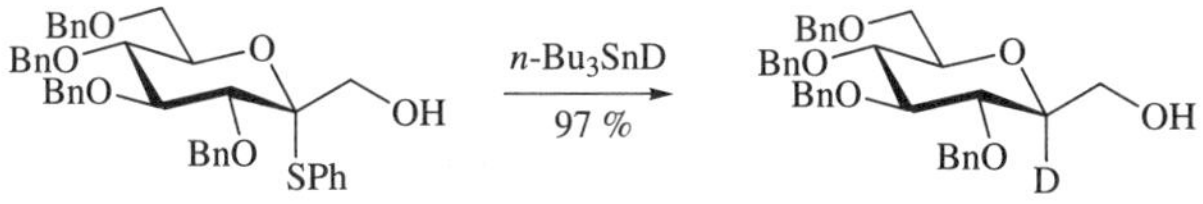

141   **142**

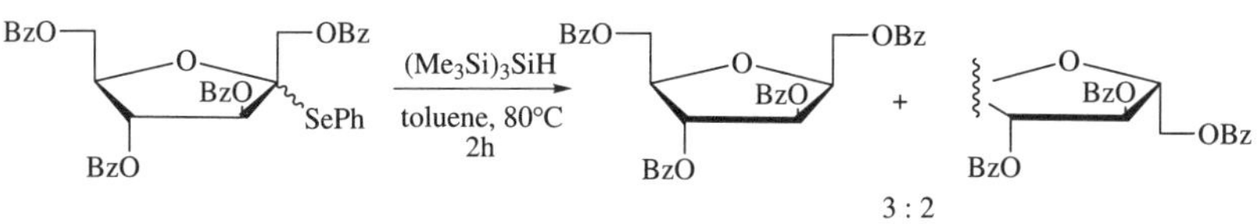

143   **144**   **145**

3 : 2

c. From Glyculosonic Acid Derivatives.—The Barton reductive decarboxylation method has been applied for the stereocontrolled synthesis of 2-deoxy-β-C-glycosyl derivatives.[246,247] To this end, C-substituted derivatives of methyl 4,5,7-tri-O-benzyl-3-deoxy-D-*arabino*-heptulosonate have been prepared from the corresponding 2,3-dideoxy-2-phenylsulfonyl precursors by reductive desulfonylation with lithium naphthalenide and quenching of the enolate thus formed with appropriate alkyl halides (MeI, Me$_3$SiCH$_2$CH$_2$OCH$_2$Cl, allyl and benzyl bromides). After saponification and treatment of the triethylammonium salts with **94** to produce the corresponding O-acyl thiohydroxamates, these were immediately subjected to photolysis (tungsten lamp) in the presence of *tert*-dodecanethiol. Diastereoselective hydrogen-atom transfer to intermediate deoxyglycosyl radicals[246,247] allowed the preparation of 2,6-anhydro-4,5,7-tri-O-benzyl-1, 3-dideoxy-D-*gluco*-heptitol (**146**), 2,6-anhydro-4,5,7-tri-O-benzyl-3-deoxy-1-O-[2-(trimethylsilyl)-ethyl]-D-*gluco*-heptitol (**147**), 4,8-anhydro-6,7,9-tri-O-benzyl-1,2,3,5-tetradeoxy-D-*gluco*-non-1-enitol (**148**), and 2,6-anhydro-4,5,7-tri-O-benzyl-1,3-dideoxy-1-phenyl-D-*gluco*-heptitol (**149**). The "β anomers" were obtained as the sole products in moderate-to-good yields (56, 58, 73, and 92%, respectively for two steps) and in excellent (>95:5) diastereoselectivity in favor of quenching from the axial direction.

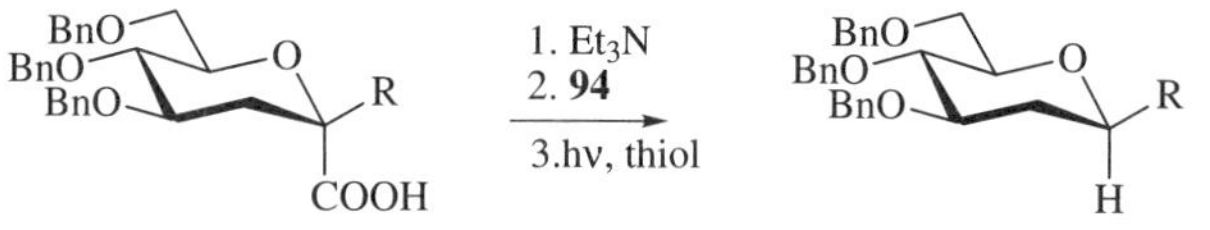

R		%	β : α
Me	**146**	56	>95 : 5
CH$_2$OCH$_2$CH$_2$SiMe$_3$	**147**	58	>95 : 5
CH$_2$CH= CH$_2$	**148**	73	>95 : 5
CH$_2$Ph	**149**	92	>95 : 5

d. From Anomeric Nitro Sugars.—Although anomeric nitro sugars are not frequently used because they require multistep preparation, radical-mediated denitration[248,249] was among the first examples proving the stereoselectivity of radical reduction in the carbohydrate field.[45] Radical denitration of both 2-epimers of 1-O-acetyl-2,5-anhydro-3,4,5,7-tetra-O-benzyl-2-C-nitro-D-*gluco*-heptitol (**150α/150β**) with *n*-Bu$_3$SnH (plus AIBN) in refluxing benzene and under argon, gave exclusively 1-O-acetyl-2,6-anhydro-3,4,5,7-tetra-O-benzyl-D-*glycero*-D-*gulo*-heptitol (**151**) in good yields. Similarly, the *manno* analogues (**152α/152β**) led to 1-O-acetyl-2,6-anhydro-3,4,5,7-tetra-O-benzyl-D-*glycero*-D-*galacto*-heptitol (**153**). The structure of **151** was unambiguously established by deacetylation and further transformation into the corresponding *meso*-pentabenzyl ether.

Reductive denitration had been found suitable for reducing more-complex structures, although the yields were found to be variable.[250] Thus, 5-acetamido-4,7-di-*O*-acetyl-2,6-anhydro-1,3-*O*-benzylidene-8,9-*O*-cyclohexyli- dene-5-deoxy-D-*arabino*-L-*gulo*-nonitol (**156**) was obtained in 97% yield from (4*R*)-5-acetamido-3,6-di-*O*-acetyl-4,8-anhydro-7,9-*O*-benzylidene-1,2-*O*-cyclohexylidene-5-deoxy-4-*C*-nitro-D-*gluco*-L-*erythro*-nonitol (**154**), on treatment with *n*-Bu_3SnH (plus AIBN) in refluxing benzene under nitrogen, whereas, under comparable conditions (toluene, 100 °C), 5-acetamido-2,6-anhydro-1,3-*O*-benzylidene-8,9-*O*-cyclohexylidene-5-deoxy-D-*arabino*-L-*gulo*-nonitol (**157**) was obtained in 31% yield, from the nitrodiol **155.** Generation of *n*-Bu_3SnD *in situ* with *n*-Bu_3SnCl and $LiAlD_4$ led to deuterated **156.** Structural assignments proved, in each case, the α-stereoselectivity of the hydrogen-atom abstraction by the intermediate pyranosyl radicals,[45,250] as also observed when *cis*-annelated nitropyranobenzopyrans were subjected to reductive denitration.[251]

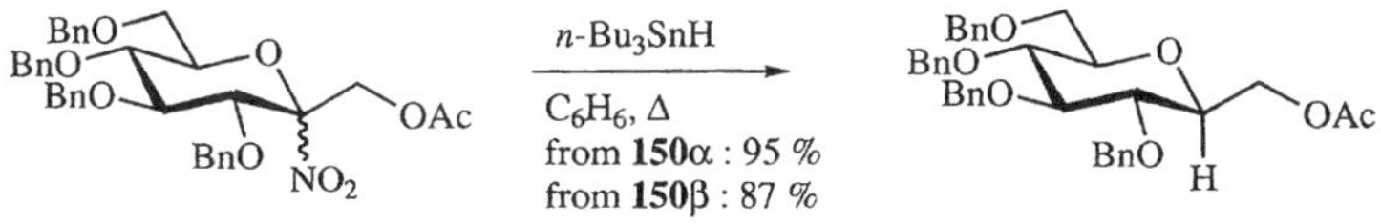

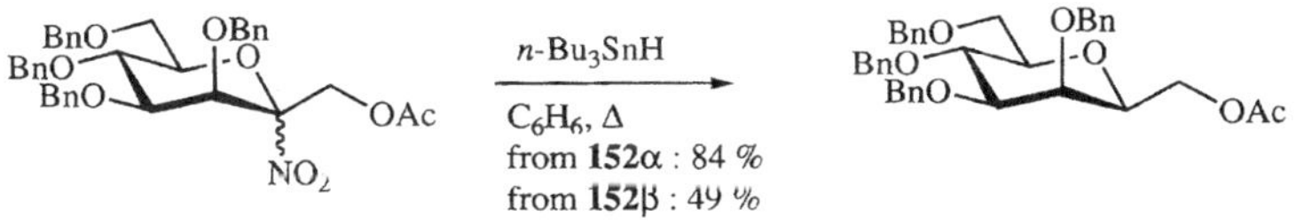

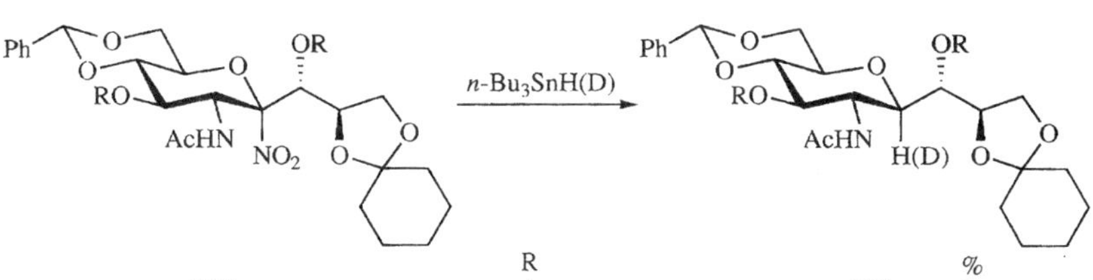

150 α/β **151**

152 α/β **153**

	R		%
154	Ac	**156**	97
155	H	**157**	30

Reductive denitration was also investigated in the furanose series,[45,252] using diethyl (2,3:5,6-di-*O*-isopropylidene-1-nitro-α-D-mannofuranosyl) phosphonate (**158**), an anomeric mixture of the analogous phenyl sulfone (**159**), (4*R*)-4,7-anhydro-2, 3, 4-trideoxy-5, 6:8, 9-di-*O*-isopropylidene-4-*C*-nitro-D-*manno*-nononitrile (**160**), and (7*S*) 6-*O*-acetyl-7-deoxy-1,2:3,4:8,9:11,12-tetra-*O*-isopropylidene-7-nitro-D-*manno*-α-D-*galacto*-dodecose-1,5-pyranose-7,10-furanose (**161**). On treatment with *n*-Bu$_3$SnH (plus AIBN) in refluxing benzene or toluene (nitrogen or argon atmosphere), they were converted into diethyl (2,3:5,6-di-*O*-isopropylidene-β-D-mannofuranosyl) phosphonate (**162**), phenyl (2,3:5,6-di-*O*-isopropylidene-β-D-mannofuranosyl)sulfone (**163**), 4,7-anhydro-2,3-dideoxy-5,6:8,9-di-*O*-isopropylidene-D-*glycero*-D-*galacto*-nonononitrile (**164**), and 6-*O*-acetyl-7,10-anhydro-1,2:3 4:8,9:11,12-tetra-*O*-isopropylidene-D-*erythro*-L-*manno*-α-D-*galacto*-dodeco-1,5-pyranose (**165**) in 78, 61, 58, and 89% yields, respectively. All of the D-mannofuranose-derived reduction products[45,252] were isomerically homogeneous, whereas the D-ribose-derived nitroethers,[45] namely, 1-*O*-acetyl-3,4-*O*-isopropylidene-6-*O*-trityl-α- and β-D-psicofuranosyl nitrite (**166α**, **166β**) gave a ~1:1 diastereoisomeric mixture of 1-*O*-acetyl-2,5-anhydro-3,4-*O*-isopropylidene-6-*O*-trityl-D-allitol (**167**) and -D-altritol (**168**), in 87–90% yield.

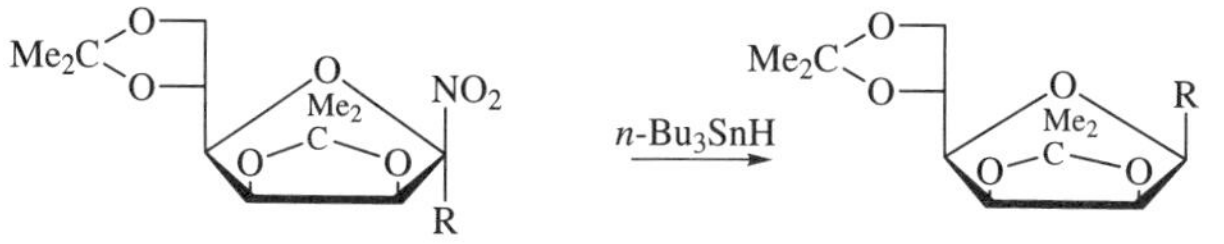

	R		%
158	PO(OEt)$_2$	**162**	78
159 α/β	SO$_2$Ph	**163**	61
160	CH$_2$CH$_2$CN	**164**	58
161		**165**	89

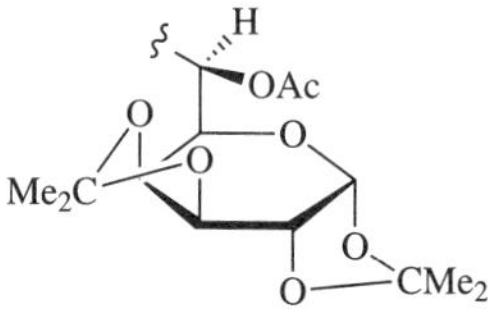

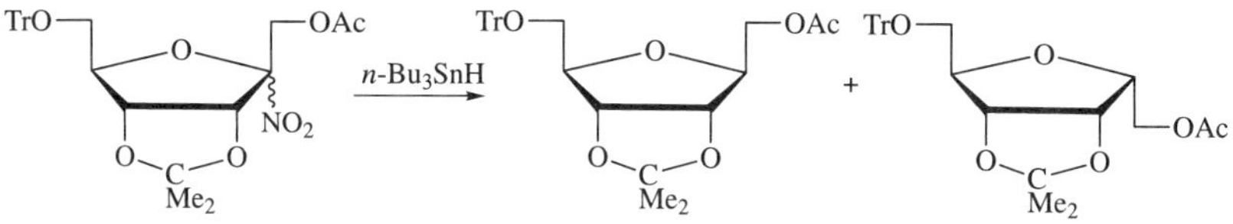

		ratio		%
166α	**167**	46 : 54	**168**	90
166β		48 : 52		87

e. Radical Addition to Exo-Glycals as a Route to *C*-Glycosyl Compounds.—

Intermolecular addition of *C*-, *S*-, or *P*-centered radicals to the exocyclic unsaturated C–C bond of 2,6-anhydro-hept-1-enitols leads to anomeric radicals that, after subsequent hydrogen quenching, can produce *C*-glycosyl compounds or

C-disaccharides. Along this line, the intermolecular addition of malonyl radical to such enitols[253,254] has been explored as a route to new glycomimetics. Malonyl radicals, produced from diethyl chloromalonate in the presence of n-Bu$_3$SnH, are considered as electrophilic radicals and, consequently, they were expected to add to the electron-rich double bond[255] present in such enitols. Hydrogen quenching of the intermediate anomeric radicals constitutes the last step of the chain reaction leading to products. Typical experiments were performed at room temperature by irradiation (UV light, 2 h) of a solution of n-Bu$_3$SnH, diethyl chloromalonate, and a sugar enol ether, in anhydrous THF under a nitrogen atmosphere. In this way, 2,6-anhydro-3,4,5,7-tetra-O-benzyl-1-deoxy-D-*gluco*-1-heptenitol (**169**) and its D-*galacto* analogue (**170**) were converted into diethyl [(2,3,4,6-tetra-O-benzyl-β-D-glucopyranosyl)methyl]malonate (**171**) and the β-D-galactosyl analogue (**172**) obtained in 35 and 32% yields. Although the yields are not very high, the quantitative recovery of the unreacted substrates shows the absence of any side reaction. Slow addition (3 h) of n-Bu$_3$SnH in THF increased the yield of **171** to 50%, with quantitative recovery of unchanged substrate. With (Me$_3$Si)$_3$SiH as the reducing agent, compound **171** was obtained in 30% yield, but no substrate could be recovered. Changing to other solvent such as benzene or acetonitrile lead to decreased yields of addition products. A high degree of stereoselection was observed for hydrogen-atom transfer to the glycopyranosyl radicals (≥95 diastereoisomeric excess, based on the isolated materials) with formation of products of the β configuration. Application of this method to 2,5-anhydro-3,4,6-tri-O-benzyl-1-deoxy-D-*arabino*-1-hexenitol (**173**) led to diethyl [(2,3,5-tri-O-benzyl-α,β-D-arabinofuranosyl)methyl]malonate (**174**) in 30% yield, and with a low stereoselection, as shown by the 10:4 ratio in favor of the β anomer.[253] This method has been developed further for preparing stable analogs of glyceroglycolipids.[254]

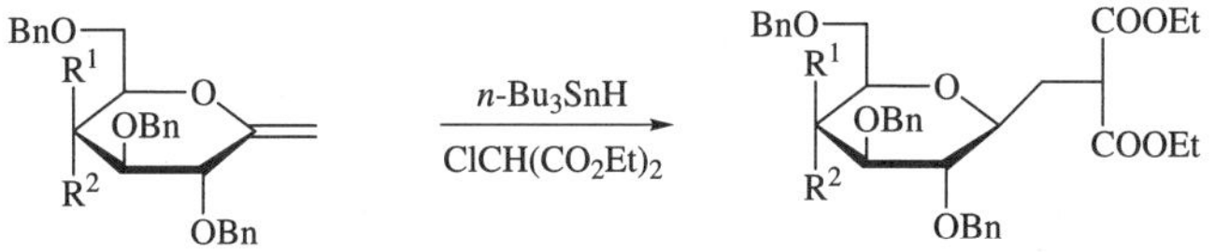

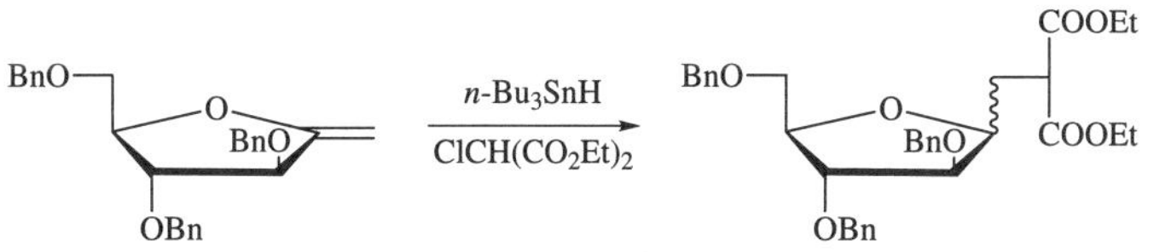

An intramolecular version of this approach has been developed to prepare *C*-linked disaccharides.[70] To this end, 2,6-anhydro-4,5,7-tri-*O*-benzyl-1-deoxy-D-*gluco*-1-heptenitol (**175**) and methyl 4,6-*O*-benzylidene-3-deoxy-3-iodo-β-D-allopyranoside (**176**) were linked by reaction with dichlorodimethylsilane in the presence of butyllithium. Treatment of the tethered iododisaccharide **177** so obtained with *n*-Bu$_3$SnH (plus AIBN) in refluxing toluene produced a pyranos-3-yl radical prone to equatorial addition[46] onto the exocyclic unsaturated C–C bond, whereupon the resulting anomeric radical underwent hydrogen quenching from the α side. Cleavage of the tether with aqueous tetra-*n*-butylammonium fluoride led to the C-β-D-linked disaccharide **178** in 40% yield. Radical addition onto the tri-*O*-benzyl derivative of "2-keto-1-*C*-methylene-D-glucopyranose" (2,6-anhydro-4,5,7-tri-*O*-benzyl-1-deoxy-D-*arabino*-hept-1-en-3-ulose) could not be attempted, because of spontaneous dimerization of such an enone-containing substrate.[256]

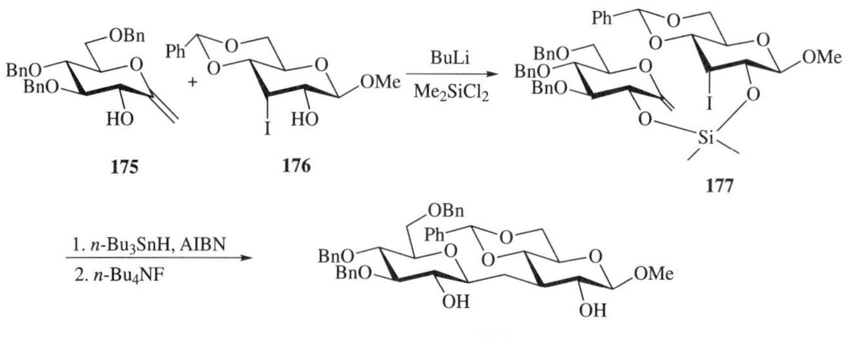

Carbohydrate *gem*-difluoro enol ethers,[257,258] which are readily available from the corresponding glycono-1,4 and -1,5-lactones, are of interest for investigating radical addition to the exocyclic double bond, so as to produce a variety of rarely investigated difluoromethylene-linked *C*-glycosyl compounds (Table II). Molecular orbital calculations (AM1) performed on simple difluoroenol ethers indicated that, whereas the HOMO energies are very similar to those of the corresponding methylene analogs, the LUMO energies are relatively lowered.[259,260] As the important frontier orbital interaction in the radical addition processes under consideration would involve interaction of the radical SOMO with the LUMO of the alkene, such a lowering would lead to an increase in the rate of reaction with nucleophilic radicals. Addition of nucleophilic alkyl radicals, generated from the corresponding halides on slow addition of *n*-Bu$_3$SnH (with AIBN) in degassed benzene at reflux, was shown to occur exclusively at the least-hindered difluoromethylene terminus in 2,5-anhydro-1-deoxy-1,1-difluoro-3,4:6,7-di-*O*-isopropylidene-D-*manno*-hept-1-enitol, with stereospecific hydrogen-atom capture from the unencumbered

TABLE II
Difluoromethylene-linked C-Glycosyl Compounds and C-Disaccharides

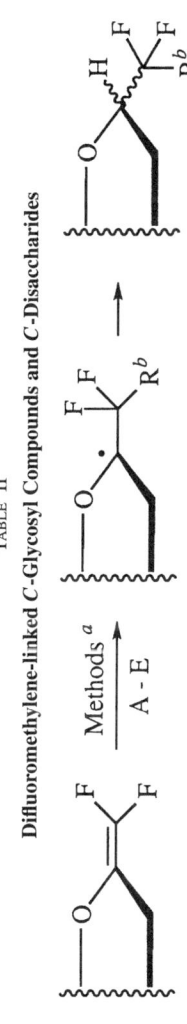

Products	N°	R^b	Method[a]	Yield %	β:α[c]	Refs.
	179	SPh	A	79	β	259
	180	CH$_2$COOEt	B	27	10:3	260
	181		C	14	β	261, 262
	182	SPh	A	92	β	259
	183	(CH$_2$)$_3$CH$_3$	B	33[d]	β	259, 260
	184	(CH$_2$)$_{11}$CH$_3$	B	24[d]	β	259, 260
	185	c-C$_6$H$_{11}$	B	11[d]	β	259, 260

(continued)

TABLE II—Continued

N°	R[b]	Method[a]	Yield %	β:α[c]	Refs.
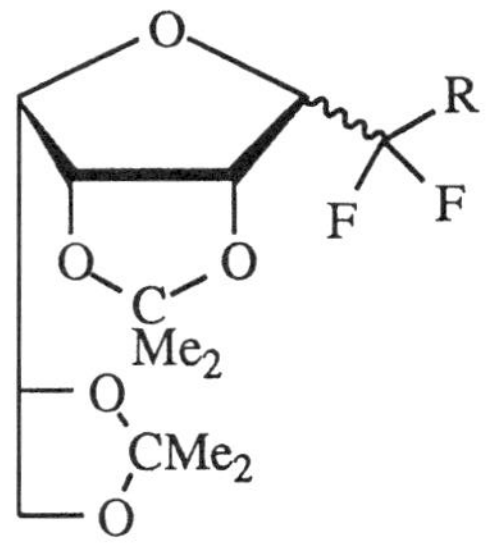 186	CH$_2$COOEt	B	51	α	259, 260
187	(methyl 2,3-di-*O*-acetyl-4-*O*-benzoyl 6-deoxy-α-D-glucopyranosid)-6-yl	B	25[d]	α	259, 260
188	(methyl 2,3,4-tri-*O*-acetyl-6-deoxy-α-D-glucopyranosid)-6-yl	B	40	α	260
189	(1,2:3,4-di-*O*-isopropylidene-6-deoxy-α-D-galactopyranos)-6-yl	B	21	α	260
190	(methyl 2(*R*)-[*tert*-butoxycarbonyl]-amino]-propionate-3-yl	B	14	α	263
191	cyclic bis (CF$_3$)oxazolidinone (see **195**)	B	30	α	263
192	(EtO)$_2$PO	D	23	β	261, 262
		C	29	1 : 0	261, 262
193	(EtO)$_2$PS	E	86	65 : 35	262
194	(BnO)$_2$PS	E	37	65 : 35	262
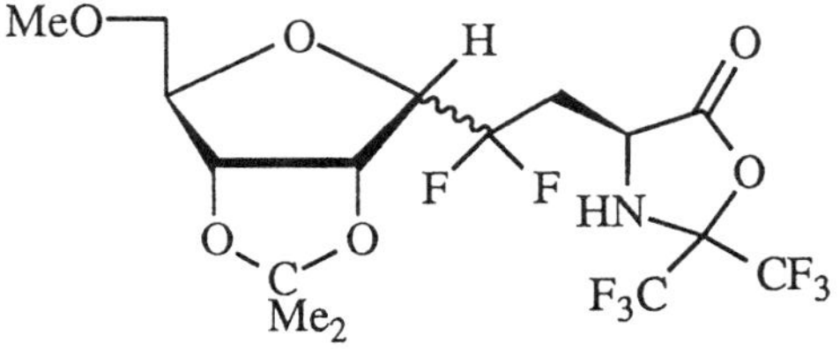 195		B	41	1 : 1	263

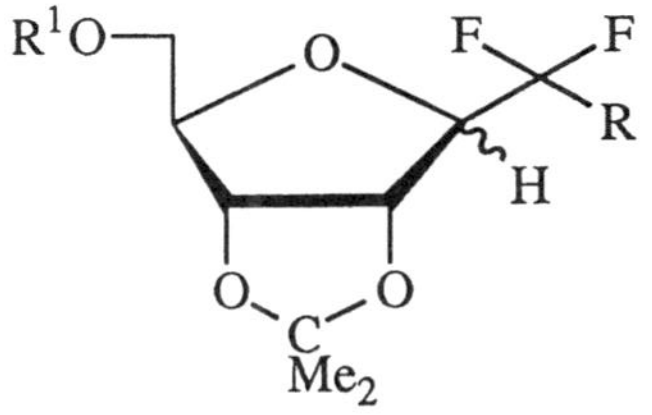

	R group	Method	Yield (%)	α:β	Ref
196	(EtO)$_2$PO, R^1 = H	C	44	β	262
197	(EtO)$_2$PO, R^1 = Bu^{t}Me$_2$Si	D	47	1:0	261
198	(EtO)$_2$PS, R^1 = Bu^{t}Me$_2$Si	C	73	6:1	261
		E	94	9:1	262
199	(B$_n$O)$_2$PS, R^1 = Bu^{t}Me$_2$Si	E	76	9:1	262
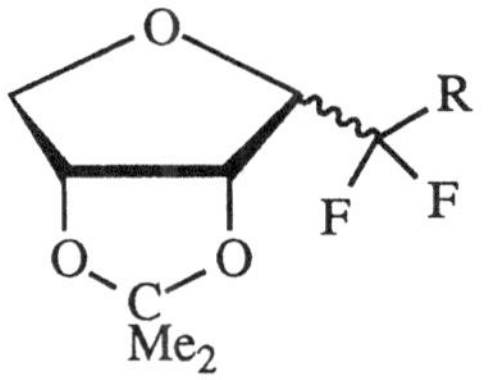 200	(methyl 2,3,4-tri-*O*-acetyl-6-deoxy-α-D-glucopyranosid)-6-yl	B	23	α	260
201	(1,2:3,4-di-*O*-isopropylidene-6-deoxy-α-D-galactopyranos)-6-yl	B	13	α	260
202	cyclic bis(CF$_3$)oxazolidinone (see **195**)	B	20	α	263
203	(EtO)$_2$PO	D	8	1:0	261, 262
		C	36	1:0	261, 262
204	(EtO)$_2$PS	E	66	55:45	262

[a] Methods: A, PhSH, AIBN, C$_6$H$_6$, Δ; C, diethyl(phenylselenyl)phosphonate (3 eq), *n*-Bu$_3$SnH (4 eq), AIBN (0.5 eq) in refluxing benzene; D, diethyl phosphite [(EtO)$_2$P(O)H] (3 eq), di-*tert*-butyl peroxide (0.5 eq.), refluxing octane; E, diethyl or dibenzyl thiophosphite [(EtO)$_2$P(S)H or (BnO)$_2$P(S)H, 3 eq], octane, 145 °C, 13 h, di-*tert*-butyl peroxide (0.5 eq.) added slowly.

[b] RX. *n*-butyl bromide; *n*-dodecanyl iodide; cyclohexyl iodide; ethyl bromo acetate; methyl 2,3-di-*O*-acetyl-4-*O*-benzoyl-6-bromo-6-deoxy-α-D-glucopyranoside; methyl 2,3,4-tri-*O*-acetyl-6-deoxy-6-iodo-α-D-glucopyranoside; 1,2:3,4-di-*O*-isopropylidene-6-deoxy-6-iodo-α-D-galactopyranose; methyl 2(*R*)-[((*tert*-butoxycarbonyl)amino]-3-iodo-propionate; cyclic bis(trifluoromethyl)oxazolidinone bromide.

[c] When the β:α ratio is not indicated, the anomeric configuration of the isolated product is given (α/β).

[d] Unchanged starting material (15/25%) was recovered.

convex face of the bicyclic system. The corresponding C-glycosyl compounds **183–185** were obtained in moderate yield (11–33%) with incomplete transformation of the starting materials, of which 14–25% could be recovered.[259,260] Addition of carbohydrate-derived radicals to a series of carbohydrate *gem*-difluoroenol ethers led, in moderate yields (14–40%), to the corresponding difluoromethylene-linked C-disaccharides **187–189,** and **200, 201,** isolated together with unreacted difluoro enol ether and reduced halide.[259,260] Carbon-centered radicals derived from benzyl 2(*R*)-[(*tert*-butoxycarbonyl)amino]-3-iodopropionate (protected iodoalanine) and cyclic bis(trifluoromethyl)oxazolidinone bromide[263] were also shown to add regio- and stereo-selectively to carbohydrate *gem*-difluoroenol ethers, affording compounds **190, 191,** and **195** in variable yields (14–41%). As deduced from the structure of the products obtained, hydrogen quenching occurred exclusively from the less-hindered convex face in the case of precursors having D-*gulo*- and D-*erythro*- configurations. However, reaction of a 2,5-anhydro-D-*ribo*-difluoroenol ether produced **195** as an equimolar mixture of both anomers,[263] whereas on replacement of the methyl ether at C-6 in the substrate by the bulky *tert*-butyldimethylsilyl ether group, no addition products were observed.[263]

In accordance with frontier molecular orbital theory, it was speculated that carbohydrate difluoroenol ethers would be compatible with the addition of electrophilic radicals, based on a favorable interaction between the HOMO and the SOMO orbitals.[259,260] It was found that addition to 2,5-anhydro-1-deoxy-1,1-difluoro-3,4:6,7-di-*O*-isopropylidene-D-*gulo*-hept-1-enitol of the electrophilic radical derived from ethyl bromoacetate, under the conditions described for nucleophilic radicals, proceeded more efficiently and with similar regio- and stereospecificity, to give the CF_2-glycosyl compound **186** (51%). Only a small quantity of starting material (6%) was recovered, in contrast to the additions of simple alkyl radicals. The enhanced reactivity of difluoroenol ethers toward electrophilic radicals is also indicated by the preparation of ethyl 4,8-anhydro-2,3-dideoxy-3,3-difluoro-5,6,7,9-tetra-*O*-(trimethylsilyl)-D-*glycero*-D-*gulo*-nononate (**180**), and its D-*glycero*-D-*ido* analogues (10:3 ratio) in 27% total yield, in contrast to unsuccessful attempts to achieve comparable syntheses with nucleophilic radicals. By comparison, much higher yields were observed (79–92%), in the AIBN-initiated radical addition of thiophenol to such carbohydrate difluoroenol ethers as 2,6-anhydro-1-deoxy-1,1-difluoro-3,4,5,7-tetra-*O*-(trimethylsilyl)-D-*gluco*-hept-1-enitol and 2,5-anhydro-1-deoxy-1,1-difluoro-3,4:6,7-di-*O*-isopropylidene-D-*manno*-hept-1-enitol[259] to produce **179** and **182.**

In another series of experiments, addition of phosphonyl radicals to carbohydrate *gem*-difluoroenol ethers was investigated as a route to new anomeric carbohydrate difluoromethylene phosphonates.[261,262] Phosphonyl radicals could be produced from either diethyl phosphite in the presence of di-*tert*-butyl peroxide in refluxing octane, or diethyl(phenylselenyl)phosphonate, on treatment with *n*-Bu_3SnH (plus AIBN) added slowly to a benzene solution under reflux. With the first method, the yields were lower, presumably because of the harsh reaction conditions. The latter method, which corresponds to a new free-radical chain reaction involving diethyl(phenylselenyl)phosphonate under mild conditions, led to the products in improved yields, except when the presence of such radical-sensitive protecting groups as benzyl groups in the substrate leads to side reactions. From the results gathered in Table II, it is apparent that the yields of difluoromethylenephosphonothioates are much higher than in the corresponding additions of diethyl phosphites. It is presumed that the weakening of the phosphorus–hydrogen bond in the thiophosphites increases the efficiency in the hydrogen-abstraction step of the propagation sequence.[262] Although the phosphonyl radical adds to the double bond of the difluoroenol ether at the difluoromethylene terminus, with a regioselectivity noted previously for carbon- and sulfur-centered radicals, the stereochemical features of the reaction appeared, as seen from the observed β:α ratio in the furanose series (Table II), entirely contrary to common expectations and previously mentioned observations. Intermediate anomeric radicals resulting from the addition of phosphorus-centered radicals to sugar difluoroenol ethers appear to undergo hydrogen quenching from the opposite direction, as compared to radical intermediates produced by addition of carbon- and sulfur-centered radicals.[261,262] The interpretation proposed to account for this observation is given next.

3. Radical Routes to Nucleosides

Although radical-induced migrations are discussed in more detail in the following section, it is appropriate to mention at this point a route toward nucleosides that involves hydrogen quenching of an anomeric carbon-centered radical produced by 1,2-migration of an acyloxy group.[264–266] To this end, C-2' brominated precursors having a pivaloyloxy group attached to C-1', such as **205,** and analogues differing in the configuration at C-1' and C-2', were treated with either $(Me_3Si)_3SiH$ or *n*-Bu_3SnH, in boiling benzene (with AIBN). As discussed later, the initially formed radical at C-2' can undergo direct reduction to afford **206,** or, when hydrogen quenching in disfavored (use of the less-effective hydrogen donor $(Me_3Si)_3SiH$, low concentration of *n*-Bu_3SnH), pivaloyloxy group migration can occur, to produce ultimately 3',5'-di-*O*-*tert*-butyldimethylsilyl-2'-pivaloyl-α-uridine **207** and its β anomer **208**. It was reported[264] that use of $(Me_3Si)_3SiH$ (1.5 eq) in refluxing benzene leads, in 86% yield, to the α nucleoside **207,** contaminated by traces (<3%) of the epimeric nucleoside **208.** Following the general rule, migration of the pivaloyloxy group took place along the same face of the furanose ring, so that the nucleosides formed differed only in the configuration at C-1'. As expected, hydrogen quenching from the less-hindered β side of the intermediate radical led to **207,** obtained as the major product. Use of $(Me_3Si)_3SiH$ in excess (4 eq) led to a mixture[264] of products (93% yield, reduced/rearranged products in 1:12 ratio).

Not unexpectedly, the directly reduced compound was formed in higher proportions (**206:207:208** = 65:27:5%) in the presence of n-Bu$_3$SnH, although used in slight excess (1.1 eq), as also observed[265] when other brominated precursors related to **205** were treated with n-Bu$_3$SnH. Direct reduction[264] was the only process observed with 5 eq of n-Bu$_3$SnH (**206**: 95%). The tin hydride has another detrimental effect, as the stereoselectivity[264] was lowered (**207:208** = 6:1), when compared with that observed with the more bulky (Me$_3$Si)$_3$SiH (**207:208** = 40:1). These results clearly show that the desired pivaloyloxy group migration occurs when hydrogen quenching is disfavored, a condition that is simply secured by using (Me$_3$Si)$_3$SiH.[264,265] The rate constant of this 1,2-acyloxy rearrangement has been estimated to be ~7.0×10^4 s^{-1}.[264,265] Its driving force may stem from release of steric hindrance, and stabilization of the C-1' radical thus formed, by the uridine moiety.[264] These synthetic data are at variance with a report claiming that use of n-Bu$_3$SnH (amount used not indicated) led only to the rearranged epimeric nucleosides **207** and **208**, obtained in 68% combined yield (10:1.3 mixture).[266] However, this study[266] indicated that treatment of other brominated nucleosides, having a pivaloyloxy group at C-1' with n-Bu$_3$SnH (1.5 eq) led to mixtures in which the major nucleoside formed was contaminated by its epimer, or by 1',2'-didehydro nucleoside resulting from 1,2- elimination, or by the unrearranged reduced product, the latter amounting in one case to 60%.[266] Interestingly, the successful generation of radical species at C-1' in the nucleoside series might have applications in the study of the mechanisms associated with the radical-induced DNA damage.

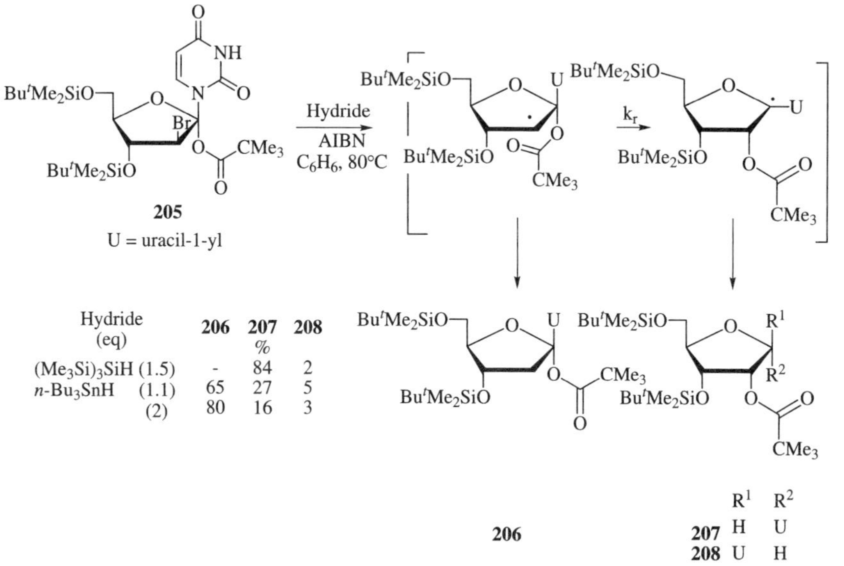

Hydride (eq)	206	207	208
		%	
(Me$_3$Si)$_3$SiH (1.5)	-	84	2
n-Bu$_3$SnH (1.1)	65	27	5
(2)	80	16	3

	R^1	R^2
207	H	U
208	U	H

4. Reduction Diastereoselectivity of Substituted Anomeric Radicals

The aforementioned data show that, in general, 1-substituted-glycopyranos-1-yl radicals undergo hydrogen quenching with high stereoselectivity, comparable to that observed for simple glycopyranos-1-yl radicals, in which hydrogen quenching takes place from the α side of the pyranose ring in the $^4C_1$ chair conformation, or more generally from the axial direction. However, the influence of the structure of the transient radical on the β:α ratio in the products is noteworthy. For example, the β-mannopyranosides **91, 98,** and **99** were obtained with a higher selectivity (18–25:1), as compared to the case of **90** (D-*gluco,* β:α = 12:1), and **92** (2-deoxy-D-*arabino,* β:α = 6:1),[213,217] because of the more pronounced steric hindrance of the β face of the mannopyranose ring. The inclusion of an equatorial silyl ether function at the 2-position results in improved α selectivity[219] for the hydrogen quenching in the Barton decarboxylation method, as shown by the preparation of **101** with a β:α ratio >25:1. The intermediate radical **101R** was assumed[219] to be a σ radical,[48–53] adopting a boatlike conformation (Scheme 4), in which the SOMO is periplanar with the C-2–oxygen bond, so that selective quenching occurred from the *exo*-face.[219] Radical **103R** was also considered as being of the σ-type[219] and adopting a $^0S_2$ conformation like that observed for **103,** with quenching taking place from the outside face of the bicyclic system. For ring-fused furanose radicals,[222] as involved in the preparation of **112,** hydrogen quenching took place from the more accessible side. Low selectivity was observed when steric hindrance of each face was somewhat comparable.

Radical-mediated preparations of *O*-glycosides and *C*-glycosyl compounds in the 2-deoxy series were shown to differ markedly in their stereoselectivities which were generally found much higher for the second class of compounds. This can be seen from the selectivities reported for the synthesis of *O*-glycosides **104–111**, with β:α ratios around 10–20:1, as compared to those found for **146–149**, (β:α > 95:5). Similar high stereoselectivities were observed for other *C*-glycosyl compounds in which the aglycon was an apolar moiety, as for **134–136,** whereas radical-mediated reduction leading to **139** and **140** afforded mixtures of cyanides, indicating a poor selectivity.

These observations were rationalized by invoking, for the *O*-substituted radicals, pyramidalized σ-radicals in the $^4C_1$ chair conformation, with a greater bias for the axial disposition of the single electron[247] in **146R** than in **104R,** taken as examples (Scheme 4), owing to the decreased preference of aliphatic carbon over oxygen substituents for the axial site at the anomeric position (anomeric effect). The anomeric effect exerted by such electron-withdrawing substituents as the nitrile group,[241,243] may account for a more equalized balance between possible forms, and consequently the decreased selectivity observed.

As may be seen from Table II, the stereochemistry found for the products resulting from the addition of a phosphonyl radical to a difluoroenol ether double bond in the furanose series was opposite to that found when addition of carbon- or sulfur-centered radicals were considered. Such an unusual stereochemistry proved completely contradictory to current expectations[265,266] for V-shaped furanoid intermediates and earlier conclusions based on the observed structure of products formed by related processes.[261,262] In general terms, for the furanoside derivatives, the β stereochemistry of the major isomer at the anomeric center was apparently indicative of hydrogen capture from the more-hindered face of the molecule. Tetrahydrofuran-derived difluoroenol ethers, having no functional group at position 2, led to essentially equimolar mixtures of both possible anomeric derivatives.[261–266] The proposed explanation for these stereoselective trends invoked hyperconjugative interaction of the unpaired electron with the C–P bonding electrons, which requires the eclipsed conformations shown for **196R**.[262] Of the two possible intermediates, A or B, steric factors would favor B in which the phosphorus substituent effectively impedes hydrogen-atom abstraction from the β-face. Within the pyranose series, the stereochemical outcome is consistent with operation of the "radical anomeric effect."[261]

V. Synthesis of 2-Deoxy Sugars by Radical-Induced Rearrangement

In the course of ESR studies of the tetra-*O*-acetyl-D-galactopyranos-1-yl radical (**13R**) generated in benzene solution by abstraction of a selenophenyl group or a bromine atom[56,62] with photolytically generated trialkyltin radicals (see Section II), signals corresponding to a second radical species[62] were detected at temperatures above 0 °C. The new species, whose concentration increased with increasing temperatures, was later identified as being the 1,3,4,6-tetra-*O*-acetyl-2-deoxy-α-D-*lyxo*-pyranos-2-yl radical. This observation corresponded to the first 2,1-acetyloxy migration involving sugar-derived radicals, although this rearrangement (the so-called Surzur–Tanner rearrangement) was discovered three decades ago for β-(acyloxy)alkyl radicals.[267–269] This was the starting point of extensive researches, which showed that radical-induced migrations can occur not only with acyloxy groups, but also, as far as carbohydrates are concerned, with diphenylphosphonyloxy groups. These radical ester migrations, which are paralleled by the rearrangement of β-(nitroxy)alkyl and β-(sulfonatoxy)alkyl groups,[270] have been discussed comprehensively.[271] The aforementioned 1,2-migrations of pivaloyloxy groups, initially attached to the C-1' position of nucleosides (see Section IV-3), is opposite, in its motion, to the more commonly encountered 2,1-migrations of acyloxy- and phosphonyloxy-groups, which involve initial formation of an anomeric radical and its subsequent rearrangement, under poor hydrogen-donating conditions, into a C-2-centered radical. From the synthetic point of view, conditions were designed so as to take advantage of 2,1-migrations in pyranos/furanos-1-yl radicals for the synthesis of 2-deoxy sugar derivatives. In addition, much effort was devoted to propose a rationale accounting for such rearrangements, which, on the basis of

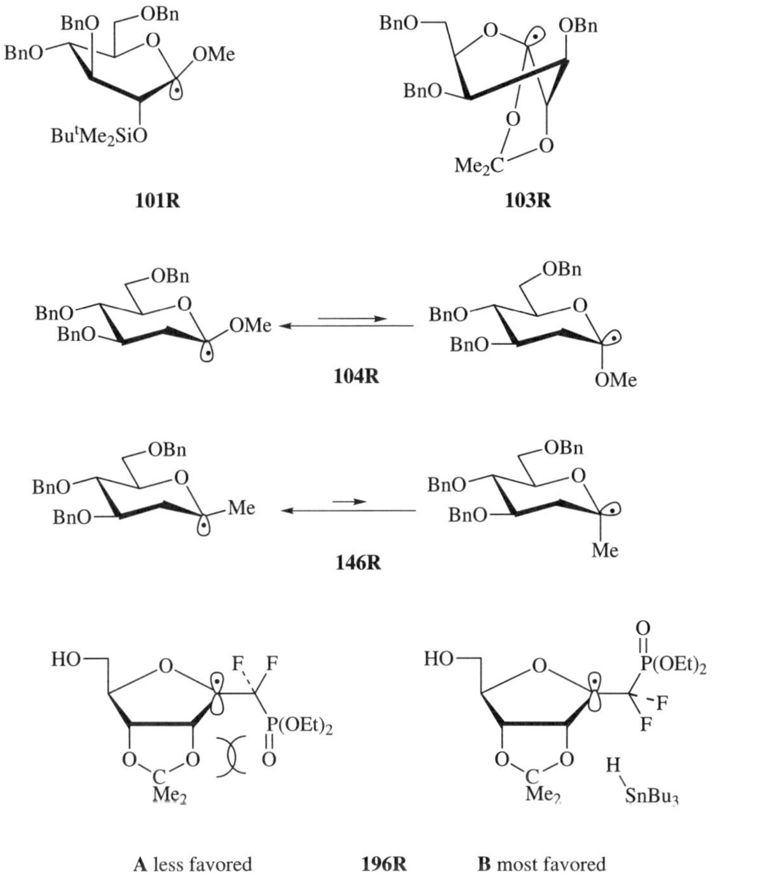

101R

103R

104R

146R

A less favored   **196R**   B most favored

SCHEME 4

the stabilization assumed for anomeric radicals, appeared to occur contrathermodynamically. This also raises questions about the intimate mechanisms and possible transition-state models compatible with the experimental data collected.

1. Synthesis of 2-Deoxy Sugars by 2,1-Acyloxy Group Rearrangements

After it had been recognized that radical rearrangement can occur in a medium having poor hydrogen-donor ability[56,62] so that reduction of the initially produced radicals to anhydroalditols (see Section III) is avoided, reduction of diversely protected sugar derivatives capable of reaction with metal hydrides [such as n-Bu$_3$SnH and (Me$_3$Si)$_3$SiH] was investigated as new routes to 2-deoxy sugar derivatives. The precursors used include the halides or selenides, listed in Table I. With n-Bu$_3$SnH,

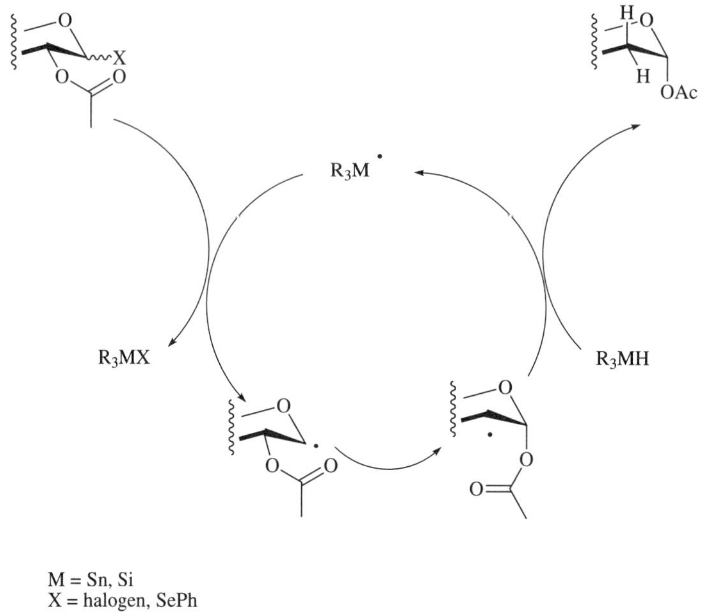

SCHEME 5

M = Sn, Si
X = halogen, SePh

this approach met with success (Scheme 5), provided that the concentration of this effective hydrogen donor (see Section III) is maintained low, generally by slow addition (8–30 h) with a pump-driven syringe, and when interruptions of the radical chain process (rearrangement/reduction) are minimized (inert atmosphere, distilled solvents).

Applied to the bromides **3,** the benzoyl analogue (2,3,4,6-tetra-*O*-benzoyl-α-D-glucopyranosyl bromide), **15, 13,** 2,3,4,6-tetra-*O*-acetyl-α-D-allopyranosyl bromide, 2,3,4-tri-*O*-acetyl-6-deoxy-6-iodo-α-D-glucopyranosyl iodide, 2,3,6-tri-*O*-benzoyl-4-deoxy-4-iodo-α-D-glucopyranosyl iodide, and the bromide **17,** this method leads, respectively,[272–274] to the following 2-deoxy sugars: 1,3,4,6-tetra-*O*-acetyl-2-deoxy-α-D-*arabino*-hexopyranose (**209**), its benzoyl analogue **210,** the β anomer **211,** 1,3,4,6-tetra-*O*-acetyl-2-deoxy-α-D-*lyxo*-hexopyranose (**212**), its benzoyl analogue **213,** 1,3,4,6-tetra-*O*-acetyl-2-deoxy-α-D-*ribo*-hexopyranose (**214**), 1,3,4-tri-*O*-acetyl-2,6-dideoxy-α-D-*arabino*-hexopyranose (**215**), 1,3,6-tri-*O*-benzoyl-2,4-dideoxy-α-D-*threo*-hexopyranose (**216**), and 1,3,4-tri-*O*-acetyl-2-deoxy-α-D-*threo*-pentopyranose (**217**). The isolated yieds were in the range 65–95%, and somewhat lower for the doubly reduced compounds **215** and **216.** The method was found suitable for preparing analogues of **209,** labeled (^{13}C, ^{2}H) at C-1 or C-5.[57] The reverse migration, consisting in the formation of glycopyranos-1-yl radicals from regiospecifically generated glycopyranos-2-yl radicals, has not been observed,[62] and reductions at C-4, C-5, and C-6 occurred without rearrangement.[62,273]

In agreement with ESR studies, the stabilized[275] sulfur-containing radical **12R,** derived from 5-thioglucose, does not undergo rearrangement and produces only the corresponding thioalditol **47**.[62] Inhibition[176] of 2,1-(acyloxyl)alkyl migration has been observed in the presence of added thiophenol or selenophenol, because quenching of anomeric radicals by PhSeH has been established[177] to be 3.6 × 10^6 $M^{-1}s^{-1}$ at 78 °C for the glucosyl radical **3R**, a process much faster than its rearrangement to the glucos-2-yl radical (estimated[62] to be 4.0 × 10^2 s^{-1} at 75 °C). Treatment of 2,3,4,6-tetra-O-acetyl-α-D-glucopyranosyl bromide (**3**) in refluxing benzene with dropwise addition of n-Bu$_3$SnH and AIBN in the same solvent over 3 h led ultimately to the rearranged 2-deoxy derivative **209** and the 1,5-anhydroglucitol **41** in 89:11 ratio. Similar experiments, conducted in the presence of either PhSeH (10 mol%), PhSeSePh (10 mol%) or PhSH (10 mol%) added to the initial benzene solution of the sugar bromide, resulted in the predominant reduction of the initially formed radical, as shown by the measured product ratios: < 5:95, < 5:95, and 10:90, respectively.[176] These results show that the presence of only 10 mol% of PhSeH, PhSeSePh (reduced *in situ* to PhSeH) and, to a lesser extent, PhSH has a dramatic effect on the efficiency of the stannane-mediated 2,1-rearrangement in β-(acyloxy)alkyl radicals.

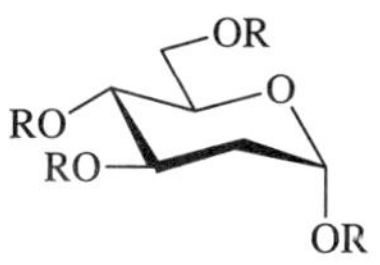

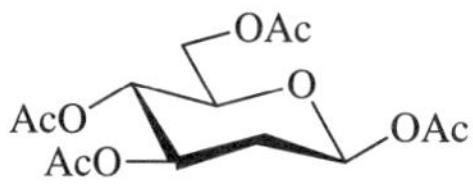

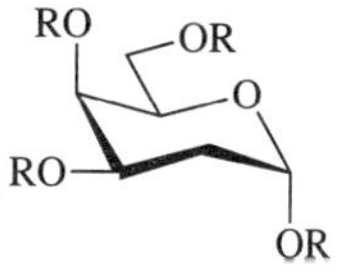

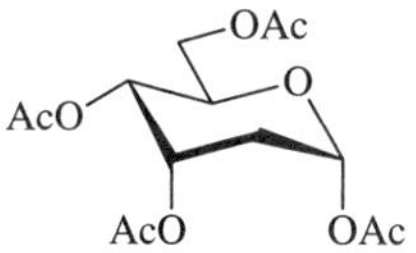

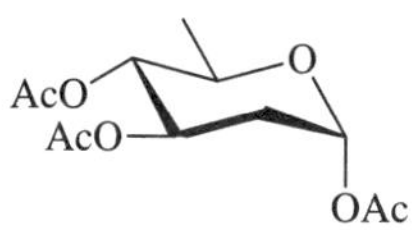

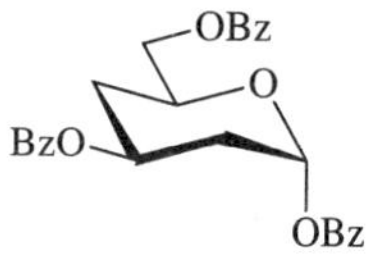

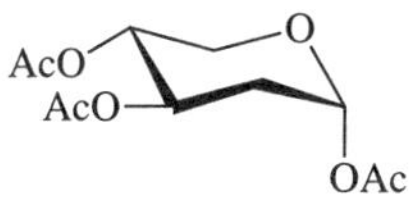

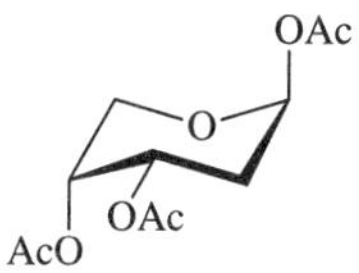

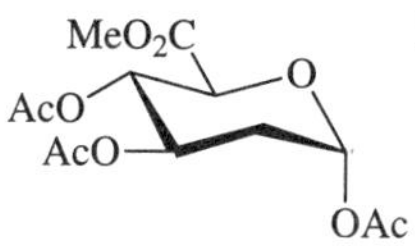

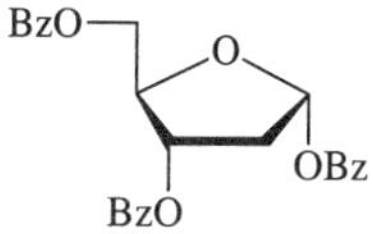

Use of less-efficient hydrogen atom donors, as compared to n-Bu$_3$SnH, constitutes another approach to the synthesis of 2-deoxy sugars. Indeed, (Me$_3$Si)$_3$SiH, with a lower hydrogen-donor ability (see Section III), can be added all at once, and slow addition is not necessary to achieve high yields of rearranged 2-deoxy sugars.[116] Hence, with heating to 80 °C for 1 h, such peracetylated glycopyranosyl bromides as **3** (D-*gluco*), and **13** (D-*galacto*) in the presence of $(Me_3Si)_3SiH$ and AIBN led readily to the rearranged 2-deoxy sugar tetraacetates **209** and **212** in 70–71% yields.[116]

It was found accidentally that various sugar xanthates (mainly 1-*S*-glycopyranosyl-*O*-neopentyl xanthates), readily prepared from glycosyl halides by displacement with sodium *O*-neopentyl xanthate, in refluxing cyclohexane undergo conversion into the corresponding 2-deoxy sugars upon addition of small amounts of dodecanoyl peroxide as the radical initiator.[276] Typically, a solution of the sugar xanthate in degassed cyclohexane, or a 9:1 mixture of cyclohexane and 1,2-dichloromethane is boiled under reflux and dodecanoyl peroxide is added portionwise (2% every 2 h) until the starting material is consumed (6–15% is required, depending on the substrate). The 2-deoxy sugars **209, 213, 217, 218,** and methyl (1,3,4-tri-*O*-acetyl-2-deoxy-α-D-*arabino*-hexopyranos)uronate (**219**) were obtained in good yields from 1-*S*-glycopyranosyl-*O*-neopentyl xanthates (90–96% yield) via 2,1-acyloxy migration in the initially formed radical intermediates. *O*-Ethyl xanthates led also to 2-deoxy sugars, albeit in decreased yields because of the formation of a small amount (∼17%) of the corresponding ethyl 1-thioglycoside, probably by ionic origin. It was found possible to use glycosyl bromides as precursors, but the reaction was relatively inefficient and a much greater amount of peroxide was needed to drive the reaction to completion (Scheme 6).

The success of this radical reduction with cyclohexane as the hydrogen source (which is not restricted to glycos-2-yl radicals)[276,277] rests on the presence of electron-attracting ester groups (acetate, benzoate) flanking the radical center, enhancing its electrophilic character[88] and rendering it favorable to hydrogen atom abstraction from the electron-rich cyclohexane.[276] Use of non-carbohydrate models also pointed to the fact that the electrophilic nature of carbon-centered radicals is decisive for accelerating the rate-determining abstraction step from cyclohexane.[276] The nucleophilic character of anomeric radicals accounts for the absence of 1,5-anhydroalditols among the reaction products. This clean and selective hydrogen-atom transfer from cyclohexane to a saturated carbon radical, which appeared to be unprecedented, was later extended to the radical reduction of iodo sugars.[277]

There are only rare examples of 2,1-(acyloxy)alkyl radical migration in furanoid radicals, even though 1,2-pivaloyloxy group migration is known for nucleoside derivatives (see Section IV-3). ESR Studies of the radicals **37R** and **38R,** generated respectively from 2,3,5-tri-*O*-benzoyl-α-D-ribofuranosyl bromide and 3,5-di-*O*-benzoyl-2-*O*-[3-(trifluoromethyl)benzoyl]-α-D-ribofuranosyl bromide upon irradiation of benzene solutions containing hexabutylditin, showed only signals for the nonrearranged radicals even on heating up to 70 °C.[64] However,

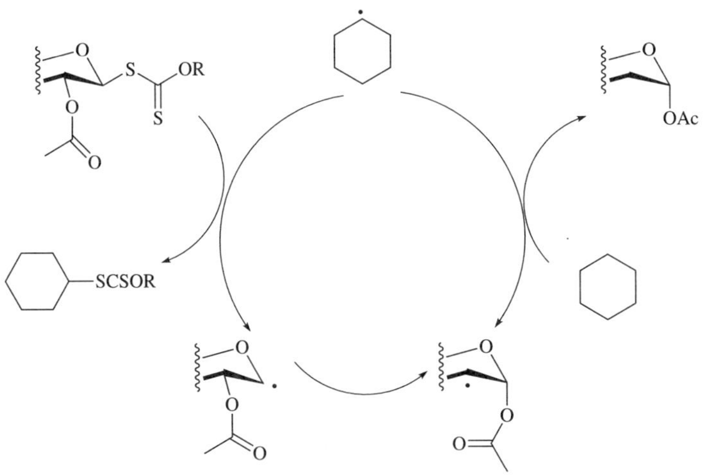

R = Et, CH$_2$CMe$_3$

SCHEME 6

radical rearrangement was observed with phenyl 2,3,5-tri-O-benzoyl-1-seleno-α-D-ribofuranoside, which is more stable under the applied conditions (n-Bu$_3$SnH and AIBN in refluxing toluene for 12 h) than a brominated radical precursor.[273] This procedure led to 1,3,5-tri-O-benzoyl-2-deoxy-α-D-*erythro*-pentofuranose (**220**) in 72% yield.[273]

The rate constant for the 2,1-rearrangement of an acetoxy group has been determined to be 5.3×10^1 s^{-1} at 27 °C,[62,64] so that it is considered to be slow.

2. Synthesis of 2-Deoxy Sugars by 2,1-Phosphonyloxy Group Rearrangement

Among the other known radical rearrangements,[270] only the 2,1-migration of a phosphonyloxy group has been investigated in the carbohydrate field.[64,187,278] To this end, protected 2-O-(diphenylphosphonyl)-D-glycosyl bromides were prepared, from partially acyl-protected carbohydrate derivatives having a free hydroxyl group at the C-2 position, by a straitghtforward two-step procedure.[64,187] The 2-O-(diphenylphosphonyl) group was installed first by treatment with diphenylphosphorochloridate in the presence of 1-methylimidazole. Exchange of the acetoxy/benzoyloxy group at C-1 by a bromine atom was then effected with hydrogen bromide in acetic acid. Irradiation of such glycosyl bromides as **49** and **51** in the presence of n-Bu$_3$SnH (1.1 eq), resulted in a radical chain-reaction, producing quantitatively the 1-O-phosphonylated hexopyranoses: 3,4,6-tri-O-acetyl-2-deoxy-1-O-(diphenylphosphonyl)-β-D-*arabino*-hexopyranose (**221**), and 3,4,6-tri-O-acetyl-2-deoxy-1-O-(diphenylphosphonyl)-α-D-*arabino*-hexopyranose (**222**).

Similarly, 3,4,6-tri-*O*-benzoyl-2-deoxy-1-*O*-(diphenoxyphosphonyl)-α-D-*arabino*-hexopyranose (**223**), 3,4,6-tri-*O*-acetyl-2-deoxy-1-*O*-(diphenylphosphonyl)-α-D-*lyxo*-hexopyranose (**224**), 3,4-di-*O*-acetyl-2,6-dideoxy-1-*O*-(diphenylphosphonyl)-α-D-*arabino*-hexopyranose (**225**) were prepared in quantitative yield. Compounds **221–225** are sensitive to elimination of diphenyl hydrogenphosphate on heating, and to hydrolysis of the anomeric phosphate group; in addition, compound **221** is prone to anomerization.[64,187] Neither the corresponding anhydroalditol nor the 2-deoxy sugar were observed upon treating phenyl 3,4,6-tri-*O*-acetyl-2-*O*-(diphenylphosphonyl)-1-thio-β-D-glucopyranoside with *n*-Bu$_3$SnH and AIBN in refluxing benzene,[278] because of 1,2-elimination leading to the corresponding glucal (see Section VI).

Migration of the phosphonyloxy group occurs along the same side of the pyranose ring,[64,187] as observed for 2,1-migration of the acetoxy group and 1,2-migration of the pivaloyloxy group. Use of *n*-Bu$_3$SnD elucidated the favored direction for quenching (axial/equatorial) of the glycos-2-yl radicals, and was shown to be as follows from the proportion of the 2-deoxy isotopomers: **221**: 40:60; **222, 223**: 90:10; and **225**: 95:5. Hence, incorporation of ^{2}H at C-2 from the β-face (>90%) was favored when the diphenylphosphate group at C-1 is axially oriented, but labeling at C-2 was found unspecific in the rearranged radical leading to **221** in connection with the equatorial disposition of the phosphate group at the anomeric center. Extension of this approach to the furanose series was not developed extensively, probably because of the sensitivity of the products obtained: complete decomposition of 3,5-di-*O*-benzoyl-2-deoxy-1-*O*-(diphenylphosphonyl)-α-D-*erythro*-pentofuranose, prepared from 3,5-di-*O*-benzoyl-2-*O*-(diphenylphosphonyl)-α, β-D-ribofuranosyl bromide, occurred within 5–10 min.[64] Interestingly, the 2-deoxy-D-glycopyranosyl phosphates obtained from the aforementioned rearrangement are glycosyl donors suitable for the construction of 2-deoxy disaccharides and 2-deoxy ribonucleosides.[64]

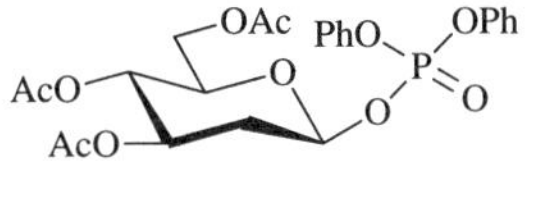

221

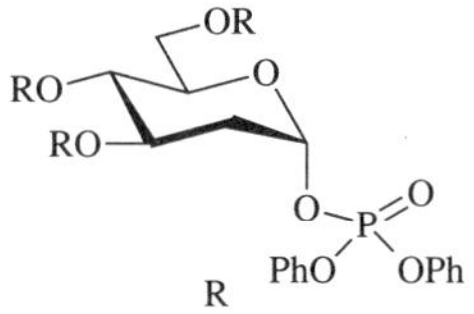

222 Ac
223 Bz

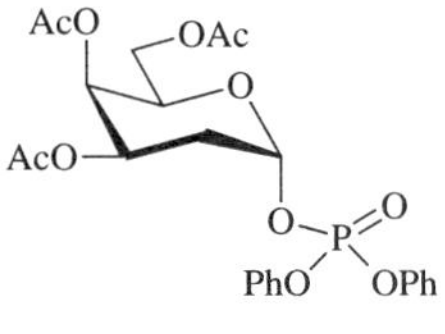

224

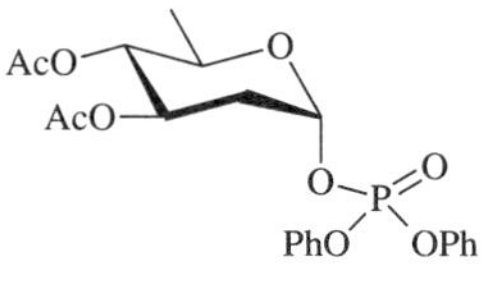

225

2-O-(Diphenylphosphonyl)-D-mannitol (**50**) was obtained from 3,4,6-tri-O-acetyl-2-O-(diphenylphosphonyl)-α-D-mannopyranosyl bromide (**49**) at elevated concentrations (1.4 M) of n-Bu$_3$SnH under UV-light irradiation,[187] in comparatively low (77%) yield (see Section III). At lower concentrations (0.19 M), **50** and **221** were formed as a 1:1 mixture. Similar treatment of 3,4,6-tri-O-acetyl-2-O-(diphenylphosphonyl)-α-D-glucopyranosyl bromide (**51**) with n-Bu$_3$SnH in large excess (10 eq) afforded a small amount of the direct reduction product **52** along with the predominant rearranged[64] product **222**. Hence, rearrangement of the diphenylphosphonyloxy group appears to be a very fast reaction so that, on a preparative scale, the rearrangement products were formed quantitatively without the need for slow addition of n-Bu$_3$SnH. Although the rate constant for the rearrangement of an acetoxy group[62,64] had been determined to be 5.3×10^1 s^{-1} at 27 °C, that for an equatorially oriented diphenylphosphonyloxy group is at least 8×10^6 s^{-1} at 27 °C.[64] However, changing from the D-*gluco* series to the mannopyranosyl bromide **49** (axial phosphate) causes a retardation[187] of the rearrangement by nearly two orders of magnitude ($\sim 9 \times 10^4$ to 1×10^5 s^{-1}). The lower reaction rate reflects the influence of a better stabilized primary anomeric radical of D-*manno* configuration and a less stabilized secondary radical (β-orientation of the diphenylphosphate group at C-1) on the total gain of stabilization energy. In the D-*gluco* series, as compared to the rearrangement rate measured[187] for 3,4,6-tri-O-acetyl-2-O-(diphenylphosphonyl)-α-D-glucopyranosyl bromide (**51**), the rearrangement rate was found, respectively, slightly lower, and significantly faster (25-fold acceleration)[187] for the process leading to benzoylated **223**, and the 2,6-dideoxy sugar **225**. On the basis of the proposed charge-separated transition state assumed for phosphonyloxy group rearrangement, the bulkier (and more electron-withdrawing) benzoyl groups, as compared to acetyl groups, probably hinder, for both steric and electronic reasons,[187] formation of the positively charged pyranose intermediate leading to **223**. Conversely, for the reaction leading to **225**, the increased rate observed is best explained[187] by the absence of acetyl group at C-6, on account of steric (more flexibility to accomodate the $B_{2,5}$ to $^4C_1$ conformational change between the initial and rearranged radicals) and electronic reasons.[187]

Because the migration of (PhO)$_2$P(O)O group is at least 100 times faster than that of the CH$_3$C(O)O group, it is not surprising that selenophenol was found poorly effective at suppressing the 2,1-(phosphonyloxy)alkyl migration.[176]

The seemingly contrathermodynamic rearrangement involving the formation of a secondary glycos-2-yl radical at the expense of an anomeric radical is best understood in terms of the formation of an anomeric C–O bond in place of a simple secondary alkyl C–O bond for both the acetoxy-[62] and the phosphonyloxy group migration.[279] It has been proposed that a stabilizing effect of at least 14.6 kJ.mol^{-1} (3.5 kcal. mol^{-1}) is present in the 1,3,4,6-tetra-O-acetyl-2-deoxy-α-D-arabinopyranos-2-yl radical, which cannot be caused by the radical center (stabilization of the unpaired electron).[62] Formation of C-1–O bonds (particularly axial ones) is assumed to contribute markedly to the stabilization of the rearranged radical.[62] The relief of unfavorable dipolar interactions[168] between the groups attached to C-2 and C-3 (acetoxy, and the bulkier diphenylphosphate) could also favor the 2,1-rearrangement.

3. Mechanism of Radical-Induced 2,1-Migrations of Ester Groups

The fact that precursors of the D-*gluco*- and D-*manno*- configurations lead to rearranged products having, respectively, α- and β-D-*arabino* configurations shows conclusively that β-(acyloxy)alkyl- and β-(phosphonyloxy)alkyl radical rearrangements are *cis*-migrations along the same side of the sugar ring. However, an extended compilation[271] of experimental data points to the fact that the intimate mechanism of 2,1-rearrangements is far from simple. Because this topic has been reviewed[271] in great detail, only the main data and conclusions involving sugar derivatives are mentioned here. In a simplified way, such rearrangements of ester groups can be envisaged as 2,3- and 1,2-shifts (Scheme 7), also termed *five-electron–five-center shifts* and *three-electron–three-center shifts*. The five-electron–five-center shift is believed to be less polarized than the three-electron–three-center shift. It is to be emphasized that these proposed transition states correspond to ideal models for ill-defined processes which might occur through several modes.[271]

2,1-Acetyloxy migration in pyranosyl radicals is believed to occur primarily by the first pathway (2,3-shift), in agreement with labeling experiments that conclusively showed that 3,4,6-tri-*O*-acetyl-2-*O*-benzoyl-α-D-glucopyranosyl bromide enriched with 45% ^{18}O in the benzoyl group (**226**) rearranged to 3,4,6-tri-*O*-acetyl-1-^{18}O-benzoyl-2-deoxy-α-D-*arabino*-hexopyranose (**227**) without scrambling of the ^{18}O atom: the carbonyl oxygen of the benzoyl group became exclusively the alkoxy oxygen in the rearranged product.[62] The presence, in the precursors studied, of such electron-withdrawing groups as acetoxy and benzoyloxy groups prevents the development of a positive charge on the pyranose ring during the rearrangement and, therefore, precludes a polar transition state. This idea is supported by the fact that migration, in an ^{18}O-enriched tetrahydropyran model devoid of electron-withdrawing groups, occurred with ^{18}O-scrambling, as a result of a dual rearrangement mode (2,3- and 1,2-shifts).[280] This raises the question as to whether migration of the chloroacetoxy group, which has been shown to be faster than that of the acetoxy group,[62] occurs mainly by the 2,3-shift mode or with a substantial contribution

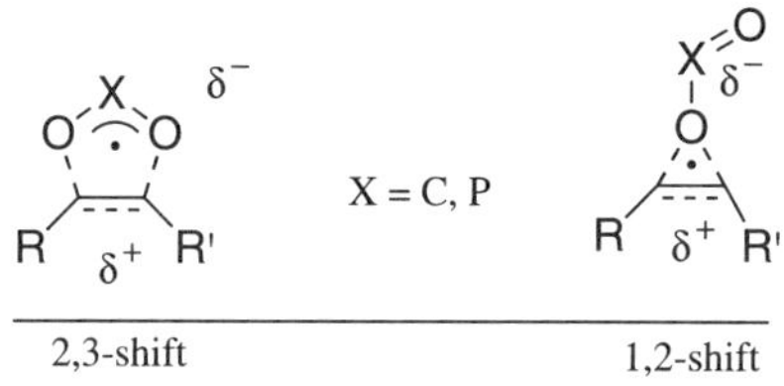

| 2,3-shift | 1,2-shift |

X = C, P

Scheme 7

of the more-polarized 1,2-shift pathway. It has been suggested that, in general, faster ester migration occur to a greater extent through the 1,2-shift pathway.[281]

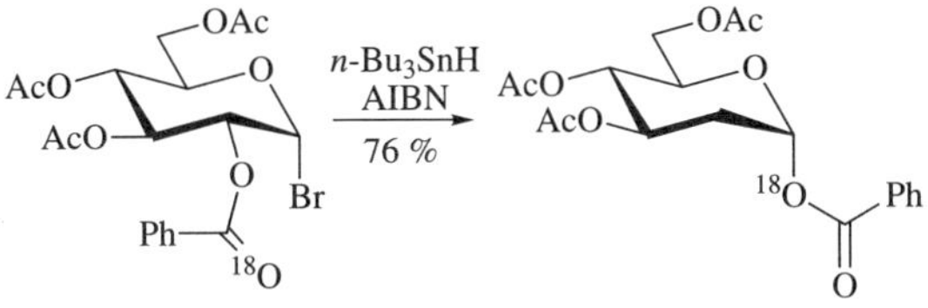

[Scheme showing compound 226 → 227 with n-Bu₃SnH, AIBN, 76%]

226   **227**

Among the possibilities that were initially proposed to explain the 2,1-(phosphonyloxy)alkyl radical rearrangement,[278] only those corresponding to intramolecular processes[279] occurring through a highly polarized transition-state resembling an alkene radical-cation loosely bound to a phosphate anion[282] are in agreement with experimental evidence.[271] Phosphoranyl radicals are not intermediates in this rearrangement,[279] and diffusively free cation-radicals are excluded as the first formed intermediates.[283] However, it appears that the reactions of β-(phosphonyloxy)alkyl radicals, which are accelerated by an increase in solvent dielectric constant,[283] are poorly characterized, and questions remain as whether they are heterolytic or concerted.[283] The changing proportion of 1,2- and 2,3-shifts in going from the (phosphonyloxy)alkyl to the (acyloxy)alkyl migration have been discussed in terms of the conformational equilibria of the two different esters and the Curtin–Hammett principle.[279] Radical-induced S_N1 substitution reactions of phosphate groups, not prone to migration toward hindered carbon-centered anomeric radicals, have been shown to occur via radical cations attacked by the methanol used as the solvent.[284]

VI. Radical-Mediated Eliminations: Formation of Glycals

It is generally accepted that carbon-centered radicals are not prone to β-elimination. This process can compete with radical chain steps only if the β bonds are very weak: for example C–S, C–Br, or C–metal.[3] This may be seen from the data collected in the preceding sections, showing in particular that, under poorly hydrogen-donating conditions, migration of vicinal ester and phosphonyloxy groups occurs to afford mainly rearranged 2-deoxy sugar derivatives, and sometimes minor amounts (<27%) of unsaturated derivatives.[266,278,279] It is not certain whether these compounds result from free-radical pathways, and it has been suggested that 3,4,6-tri-O-acetyl-1,5-anhydro-2-deoxy-D-*arabino*-hex-1-enitol (**228**) was produced from phenyl 3,4,6-tri-O-acetyl-2-O-(diphenylphosphonyl)-1-thio-β-D-glucopyranoside via the thermally labile[64] rearranged 2-deoxy sugar **222**, because of prolonged heating (29 h, 80 °C). Anhydroalditol **52** was shown to be stable under the reaction conditions.[278,279] Based on labeling experiments, it was concluded that thermal conversion of **222** into **228** occurs by *syn*-elimination of diphenyl hydrogenphosphate.[64]

There are only rare examples of elimination as the main reaction leading cleanly to 1,5-anhydro-2-deoxy-1-enitols (glucals) under conditions typical for radical-mediated reductions. However, treatment with n-Bu$_3$SnH of sugar derivatives having, at both C-1 and C-2, atoms or groups attacked by trialkyltin radicals (Br, SePh, or the thiocarbonyl group), was found to be a general[109] method useful for preparing the corresponding glycals. As a matter of fact, compound **228** could be obtained, in 92 and 83% yields, from diastereoisomeric mixtures of peracetylated dibromides and 2-bromo-1-O-(phenoxythiocarbonyl) precursors, respectively.[285]

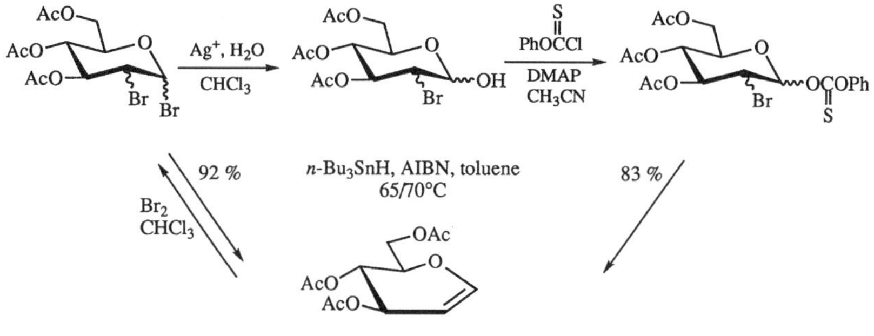

Similarly, phenyl 3,4-O-isopropylidene-1-thio-2-O-(methylthiothiocarbonyl)-β-D-galactopyranoside and phenyl 3-O-benzyl-4,6-O-benzylidene-1-thio-2-O-(methylthiothiocarbonyl)-β-D-galactopyranoside gave, on brief treatment (5–10 min) with n-Bu$_3$SnH (plus AIBN) in refluxing toluene, 1,5-anhydro-2-deoxy-3,4-O-isopropylidene-D-*lyxo*-hex-1-enitol (**229**), and 1,5-anhydro-3-O-benzyl-4,6-O-benzylidene-2-deoxy-D-*lyxo*-hex-1-enitol (**230**) in 93 and 90% yields, respectively.[286]

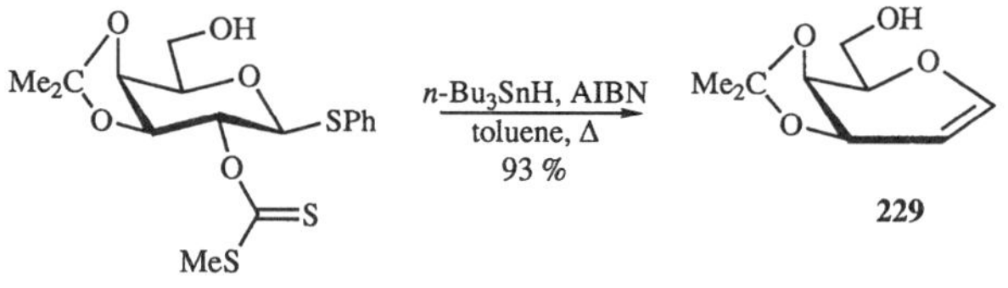

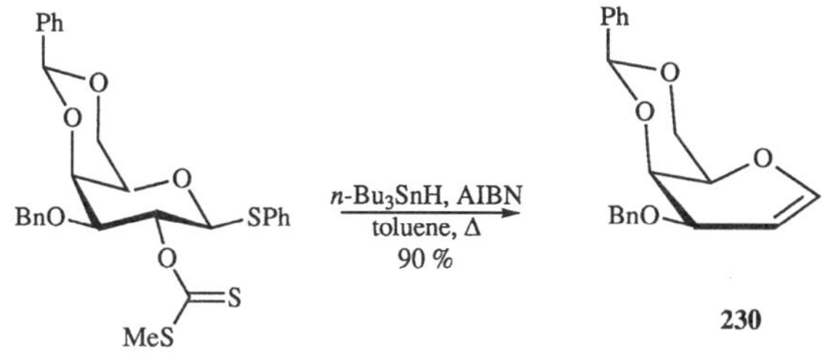

More unexpectedly, in phenyl 1-seleno-α-D-glycopyranosides having an azido group at the C-2 position, radical β-elimination was found to occur readily[287] on heating a benzene solution containing n-Bu$_3$SnH (1.2 eq) and AIBN for 1 h. Under these conditions, 3,4,6-tri-O-acetyl-1,5-anhydro-2-deoxy-D-*lyxo*-hex-1-enitol (**231**), 3-O-acetyl-1,5-anhydro-4,6-O-benzylidene-2-deoxy-D-*arabino*-hex-1-enitol (**232**), and the D-*ribo* analogue (**233**) were obtained in high yields (94–96%). Reduction of the azido group to the amine[288,289] was not observed under the reaction conditions.[287] It has been reported that cyclic thiocarbonates derived from some carbohydrate 1,2-diols can be converted to the corresponding glycals on treatment with n-Bu$_3$SnH (plus AIBN) in boiling toluene for 14 h.[290] Although 1,5-anhydro-4,6-O-benzylidene-2-deoxy-3-O-methyl-D-*arabino*-hex-1-enitol was obtained in 55% yield, this procedure was found to be inappropriate[290] for the synthesis of furanose glycals.[291]

	R^1	R^2	R^3	R^4
D-*gluco*	H	N$_3$	H	OAc
D-*manno*	N$_3$	H	H	OAc
D-*allo*	H	N$_3$	OAc	H

	R^3	R^4
232	H	OAc
233	OAc	H

Competitive formation of glucal **228** (45% yield) has been also reported to occur during the photo-initiated anomerization of bis(dimethylglyoximato)(pyridine) (2,3,4,6-tetra-O-acetyl-α-D-glucopyranosyl)cobalt to the corresponding β-D-glucopyranosyl)cobaloxime, via **3R**.[292] Several processes based on single-electron transfer lead to glucals in variable yields. For example, the electrochemical reduction of **3** affords **228** (20% yield) as a minor product,[87] supporting the idea that **3** undergoes a single electron-transfer reduction to produce a bromide anion and **3R**. Although this species undergoes mainly dimerization, it may be converted, by a further reduction step, into a glucopyranosyl anion, ultimately decomposing into **228** and an acetate anion. Even though these examples appear not particularly useful synthetically for glycal preparation, some related reductive methods were found very effective mainly for the synthesis of pyranoid glycals. They can be considered as metal-mediated eliminations, involving treatment of glycopyranosyl halides with zinc,[196,198,293] or with bis(titanocene chloride),[194,294] whereas phenyl thioglycopyranosides and glycosyl phenyl sulfones, when treated with lithium naphthalenide, undergo reductive lithiation and subsequent elimination of the C-2 substituent.[286] Glycals are produced in high yield from glycosyl phenyl sulfones on treatment with samarium iodide (SmI_2) in the presence of hexamethylphosphoric triamide (HMPA).[295] These methods are outside the scope of the present work, and are not discussed in detail. However, evidence exists showing that glycopyranos-1-yl radicals are produced along the reaction pathway, as proved by the formation, among the reaction products, of 1,5-anhydroalditols[193,197,198] or C-glycosyl compounds,[195,198,295] according to the conditions used.

VII. Conclusion

It is well established from the general literature that radical-mediated reactions have gained a prominent role in organic synthesis, and their contribution to the development of new methodologies applicable in the carbohydrate field is also well recognized. The foregoing discussion, which concerns primarily the structural features of glycos-1-yl radicals and their chemical reactivity under reductive conditions, presents data collected from the literature mainly after 1980. In spite of a rather extensive search, some data might be missing, and apologies are presented for possible omissions. At this time, a deeper insight has been gained in the structure of anomeric radicals. Detailed ESR studies have revealed that glycos-1-yl radicals, and in particular glycopyranos-1-yl radicals, adopt conformations compatible with extended electron delocalization involving the ring oxygen atom, the SOMO, and the β-bond. Glycos-1-yl radicals can abstract hydrogen atom from suitable hydrogen donors, to produce, by radical chain reactions, the corresponding anhydroalditols as a result of C–H bond formation at C-1. When this methodology is applied to more-complex precursors leading to 1-substituted anomeric radicals, hydrogen atom quenching of the intermediates may lead, depending upon the substituent present, to O-glycosides and to C-glycosyl compounds. It is noteworthy that hydrogen quenching generally occurs, with high stereoselectivity and with preferred α selectivity for glycopyranosyl radicals. For this reason, radical approaches have also been explored to achieve inversion of α- to β-D-mannopyranosides, and to develop new routes to C-glycosyl compounds of β-D-configuration that are difficult to prepare otherwise. The mildness of the conditions applied and the high yields achieved under simple conditions make radical-mediated routes more attractive, in many cases, than alternative approaches. The success of radical-mediated reductions rests on the nucleophilic character of glycos-1-yl radicals and the wide palette of reactivity of triorganostannanes, particularly tri-n-butyltin hydride. When glycos-1-yl radicals are generated under poorly hydrogen-donating conditions, they rearrange by ester group migration, leading to 2-deoxy sugar derivatives. Radical-induced 1,2-eliminations are also possible, provided a weak bond is present at C-2 in suitable precursors, and consequently a limited number of radical-induced syntheses of glycals has been proposed.

This chemistry, as well as its applicability with regard to C–C bond-forming processes, owes its development mainly to the availability and unique reactivity of organostannanes, combined with the varied functionalization of sugar derivatives.[296] It is to be emphasized that tin chemistry is very versatile, as it is compatible with ionic processes whose usefulness and applicability to sugar chemistry has been comprehensively surveyed.[297] However, there is now a trend to find new hydrogen-atom donors and substitutes for organostannanes[119] so as to avoid problems arising from their toxicity. In the continuing quest for new synthetic methodologies improved in terms of applicability, selectivity, low cost, and safety, it is highly probable that radical-mediated methodologies will offer new optimized synthetic tools, in addition to the interest that arises from the generation of radical transients in such important biological processes as DNA biosynthesis[298] and cleavage.[299–304]

REFERENCES

(1) W. Hartwig, *Tetrahedron,* 39 (1983) 2609–2645.
(2) D. J. Hart, *Science,* 223 (1984) 883–887.
(3) B. Giese, *Radicals in Organic Synthesis: Formation of Carbon–Carbon Bonds.* Pergamon Press, Oxford, New York, 1986.
(4) D. P. Curran, *Synthesis,* (1988) 417–439.
(5) D. P. Curran, *Synthesis,* (1988) 489–513.
(6) D. Crich and L. Quintero, *Chem. Rev.,* 89 (1989) 1413–1432.
(7) M. Pereyre, J. P. Quintard, and A. Rahm, *Tin in Organic Synthesis,* Butterworths, London, 1986, pp. 35–68.
(8) N. P. Neumann, *Synthesis,* (1987) 665–683.
(9) *Selectivity and Synthetic Applications of Radical Reactions,* Tetrahedron Symposia-in-Print Number 22, B. Giese (Ed.) *Tetrahedron,* 41 (1985) No. 19.
(10) M. Ramaiah, *Tetrahedron,* 43 (1987) 3541–3676.
(11) W. B. Motherwell and D. Crich, in *Free Radical Chain Reactions in Organic Synthesis,* Academic Press, London, 1992.
(12) C. P. Jasperse, D. P. Curran, and T. L. Fevig, *Chem. Rev.,* 91 (1991) 1237–1286.
(13) M. Regitz and B. Giese, *Houben-Weyl, Methoden der Organischen Chemie, C- Radikale,* Band E 19 a / part 1 and 2, Georg Thiem Verlag, Stuttgart, New-York, 1989, a—pp. 150–170, b— pp. 218–219.
(14) N. A. Porter, B. Giese, and D. P. Curran, *Acc. Chem. Res.* 24 (1991) 296–304.
(15) D. P. Curran, N. A. Porter, and B. Giese, *Stereochemistry of Radical Reactions,* VCH, Weinheim, 1996.
(16) D. H. R. Barton and S. W. McCombie, *J. Chem. Soc. Perkin Trans. I,* (1975) 1574–1585.
(17) D. H. R. Barton, J. A. Ferreira, and J. C. Jaszberenyi, in S. Hanessian (Ed.), *Preparative Carbohydrate Chemistry,* Marcel Dekker Inc., New York, 1996, pp.151–172.
(18) D. H. R. Barton, *Tetrahedron,* 48 (1992) 2529–2544.
(19) R. J. Ferrier and R. H. Furneaux, *J. Chem. Soc., Perkin Trans. 1,* (1977) 1993–1996.

(20) R. H. Furneaux, *J. Chem. Soc., Perkin Trans. 1,* (1977) 1996–2000.
(21) L. Somsák and R. J. Ferrier, *Adv. Carbohydr. Chem. Biochem.,* 49 (1991) 37–92.
(22) G. Remy, L. Cottier, and G. Descotes, *Can. J. Chem.,* 61 (1983) 434–438.
(23) J.-P. Praly, G. Descotes, M.-F. Grenier-Loustalot, and F. Metras, *Carbohydr. Res.,* 128 (1984) 21–35.
(24) B. Giese and J. Dupuis, *Angew. Chem. Int. Ed. Engl.,* 22 (1983) 622–623.
(25) R. M. Adlington, J. E. Baldwin, A. Basak, and R. P. Kozyrod, *J. Chem. Soc., Chem. Commun.,* (1983) 944–945.
(26) P. M. Collins and R. J. Ferrier, *Monosaccharides: Their Chemistry and Their Role in Natural Products,* John Wiley & Sons, Chichester, 1995, pp. 179–182.
(27) J. M. J. Tronchet, M. Zsély, and M. Geoffroy, *Carbohydr. Res.,* 275 (1995) 245–258 and references therein.
(28) J. M. J. Tronchet, in R. I. Zhdanov (Ed.), *Bioactive Spin Labels,* Springer Verlag, Berlin, (1992) 355–387, and references therein.
(29) C. von Sonntag, *Adv. Carbohydr. Chem. Biochem.,* 37 (1980) 7–77.
(30) R. van den Berg, J. A. Peters, and H. van Bekkum, *Carbohydr. Res.,* 267 (1995) 65–77 and references therein.
(31) M.A. Shalaby, H. S. Isbell, and H. S. El Khadem, *J. Carbohydr. Chem.,* 14 (1995) 429–437.
(32) J. Raffi, C. Thiéry, C. Battesti, J.-P. Agnel, J. Triolet, and P. Vincent, *J. Chim. Phys.,* 90 (1993) 1009–1019 and references therein.
(33) H. Catterall, M. J. Davies, and B. C. Gilbert, *J. Chem. Soc., Perkin Trans. 2,* (1992) 1379–1385 and references therein.
(34) D. Schulte-Frohlinde and C. von Sonntag, in S. S. Wallace and R. B. Painter (Eds.), *Ionizing Radiation Damage to DNA: Molecular Aspects,* John Wiley & Sons, New York, 1990.
(35) C. von Sonntag, *Int. J. Radiat. Biol.,* 66 (1994) 485–490 and references therein.
(36) B. Giese, A. Ghosez, T. Göbel, J. Hartung, O. Hüter, A. Koch, K. Kroder, and R. Springer, in F. Minisci (Ed.), *Free Radicals in Synthesis and Biology,* Kluwer Academic Publishers, 1989.
(37) D. E. Levy and C. Tang, *The Chemistry of C-Glycosides,* Pergamon, Elsevier Science Ltd, Oxford, 1995.
(38) M. H. D. Postema, *C-Glycoside Synthesis,* CRC Press Inc., Boca Raton, 1995, pp. 161–345.
(39) C. Bertozzi and M. Bednarski, in S. H. Khan and R. A. O'Neill (Eds.), *Modern Methods in Carbohydrate Synthesis,* Harwood Academic Publishers Gmbh, Amsterdam, 1996, pp. 316–351.
(40) B. Giese and H.-G. Zeitz, in S. Hanessian (Ed.), *Preparative Carbohydrate Chemistry,* Marcel Dekker Inc., New York, 1996, pp. 507–525.
(41) G. Descotes, *J. Carbohydr. Chem.,* 7 (1988) 1–20.
(42) G. Descotes, *Top. Current Chem.,* 154 (1990) 39–76.
(43) J.-P. Praly, *Tetrahedron Lett.,* 24 (1983) 3075–3078.
(44) B. Giese and J. Dupuis, *Tetrahedron Lett.,* 25 (1984) 1349–1352.
(45) F. Baumberger and A. Vasella, *Helv. Chim. Acta,* 66 (1983) 2210–2222.
(46) B. Giese, *Angew. Chem. Int. Ed. Engl.,* 28 (1989) 969–980.
(47) J. Dupuis, B. Giese, D. Rüegge, H. Fischer, H.-G. Korth, and R. Sustmann, *Angew. Chem. Int. Ed. Engl.,* 23 (1984) 896–898 and references therein.
(48) C. Bernasconi and G. Descotes, *C. R. Acad. Sci. Ser. C,* 280 (1975) 469–472.
(49) K. Hayday and R. D. McKelvey, *J. Org. Chem.,* 41 (1976) 2222–2223.
(50) B. W. Babcock, R. D. Dimmel, D. P. Jr. Graves, and R. D. McKelvey, *J. Org. Chem.,* 46 (1981) 736–742.
(51) R. D. McKelvey and H. Iwamura, *J. Org. Chem.,* 50 (1985) 402–404.
(52) V. Malatesta, R. D. McKelvey, B. W. Babcock, and K. U. Ingold, *J. Org. Chem.,* 44 (1979) 1872–1873.
(53) R. D. McKelvey, T. Sugawara, and H. Iwamura, *Magn. Reson. Chem.,* 23 (1985) 330–334.

(54) S. D. Rychnovsky, J. P. Powers, and T. J. LePage, *J. Am. Chem. Soc.,* 114 (1992) 8375–8384.
(55) A. J. Buckmelter, J. P. Powers, and S. D. Rychnovsky, *J. Am. Chem. Soc.,* 120 (1998) 5589–5590.
(56) H.-G. Korth, R. Sustmann, J. Dupuis, and B. Giese, *J. Chem. Soc. Perkin Trans. 2,* (1986) 1453–1459.
(57) H.-G. Korth, R. Sustmann, B. Giese, B. Rückert, and K. S. Gröninger, *Chem. Ber.,* 123 (1990) 1891–1898.
(58) H.-G. Korth, J.-P. Praly, L. Somsák, and R. Sustmann, *Chem. Ber.,* 123 (1990) 1155–1160.
(59) H. Chandra, M. C. R. Symons, H.-G. Korth, and R. Sustmann, *Tetrahedron Lett.,* 28 (1987) 1455–1458.
(60) B. K. Goodman and M. M. Greenberg, *J. Org. Chem.,* 61 (1996) 2–3.
(61) C. Chatgilialoglu, T. Gimisis, M. Guerra, C. Ferreri, C. J. Emanuel, J. H. Horner, M. Newcomb, M. Lucarini, and G. F. Pedulli, *Tetrahedron Lett.,* 39 (1998) 3947–3950.
(62) H.-G. Korth, R. Sustmann, K. S. Gröninger, M. Leisung, and B. Giese, *J. Org. Chem.,* 53 (1988) 4364–4369.
(63) E. D. Rekaï, G. Rubinstenn, J.-M. Mallet, P. Sinaÿ, S. N. Müller, and B. Giese, *Synlett,* (1998) 831–834.
(64) A. Koch, C. Lamberth, F. Wetterich, and B. Giese, *J. Org. Chem.,* 58 (1993) 1083–1089.
(65) S. Peukert, R. Batra, and B. Giese, *Tetrahedron Lett.,* 38 (1997) 3507–3510.
(66) R. Sustmann and H.-G. Korth, *J. Chem. Soc. Faraday Trans. I,* 83 (1987) 95–105.
(67) H.-G. Korth, R. Sustmann, J. Dupuis, K. S. Gröninger, T. Witzel, and B. Giese, in H. G. Viehe et al. (Eds.), *Substituent Effects in Radical Chemistry,* D. Reidel Publishing Co., 1986, pp. 297–300.
(68) K. S. Gröninger, K. F. Jäger, and B. Giese, *Liebigs Ann. Chem.,* (1987) 731–732.
(69) B. Giese, unpublished results.
(70) P. Sinaÿ, *Pure Appl. Chem.,* 70 (1998) 407–410.
(71) J.-P. Praly, L. Brard, G. Descotes, and L. Toupet, *Tetrahedron,* 45 (1989) 4141–4152.
(72) H.-G. Korth, R. Sustmann, K. S. Gröninger, T. Witzel, and B. Giese, *J. Chem. Soc. Perkin Trans. 2,* (1986) 1461–1464.
(73) B. C. Gilbert, D. M. King, and C. B. Thomas, *J. Chem. Soc. Perkin Trans. 2,* (1980) 1821–1827.
(74) B. C. Gilbert, D. M. King, and C. B. Thomas, *J. Chem. Soc. Perkin Trans. 2,* (1981) 1186–1199.
(75) B. C. Gilbert, D. M. King, and C. B. Thomas, *J. Chem. Soc. Perkin Trans. 2,* (1983) 675–683.
(76) B. C. Gilbert, D. M. King, and C. B. Thomas, *Carbohydr. Res.,* 125 (1984) 217–235.
(77) B. C. Gilbert, D. M. King, and C. B. Thomas, *J. Chem. Soc. Perkin Trans. 2,* (1982) 169–179.
(78) M. Fitchett and B. C. Gilbert, *J. Chem. Soc. Perkin Trans. 2,* (1986) 1169–1177.
(79) M. Fitchett, B. C. Gilbert, and R. L. Willson, *J. Chem. Soc. Perkin Trans. 2,* (1988) 673–689.
(80) B. C. Gilbert, J. R. Lindsay Smith, S. R. Ward, A. C. Whitwood, and P. Taylor, *J. Chem. Soc. Perkin Trans. 2,* (1998) 1565–1572.
(81) I. A. Shkrob, M. C. Depew, and J. K. S. Wan, *Chem. Phys. Lett.,* 202 (1993) 133–140.
(82) E. Juaristi and G. Cuevas, *Tetrahedron,* 48 (1992) 5019–5087.
(83) A. L. J. Beckwith and P. J. Duggan, *Tetrahedron,* 54 (1998) 4623–4632.
(84) A. L. J. Beckwith and P. J. Duggan, *Tetrahedron,* 54 (1998) 6919–6928.
(85) B. Giese, B. Rückert, K. S. Gröninger, R. Muhn, and H. J. Lindner, *Liebigs Ann. Chem.,* (1988) 997–1000.
(86) A. Alberti, M. A. Della Bona, D. Macciantelli, F. Pelizzoni, G. Sello, G. Torri, and E. Vismara, *Tetrahedron,* 52 (1996) 10241–10248.
(87) M. Guerrini, P. Mussini, S. Rondinini, G. Torri, and E. Vismara, *J. Chem. Soc. Chem. Commun.,* (1998) 1575–1576.
(88) E. Vismara, A. Donna, F. Minisci, A. Naggi, N. Pastori, and G. Torri, *J. Org. Chem.,* 58 (1993) 959–963.
(89) H. Togo, W. He, Y. Waki, and M. Yokoyama, *Synlett,* (1998) 700–717.
(90) L. Zervas, *Ber.,* 63 (1930) 1689–1690.

(91) N. K. Richtmyer, C. J. Carr, and C. S. Hudson, *J. Am. Chem. Soc.,* 65 (1943) 1477–1478.
(92) A. Duclos, C. Fayet, and J. Gelas, *Synthesis,* (1994) 1087–1090.
(93) R. K. Ness, H. G. Fletcher, Jr., and C. S. Hudson, *J. Am. Chem. Soc.,* 72 (1950) 4547–4549.
(94) M. Funabashi and T. Hasegawa, *Bull. Chem. Soc. Jpn.,* 64 (1991) 2528–2531.
(95) H. Paulsen, D. Schnell, and W. Stenzel, *Chem. Ber.,* 110 (1977) 3707–3713.
(96) J. A. Bennek and G. R. Gray, *J. Org. Chem.,* 52 (1987) 892–897.
(97) A. Jeffery and V. Nair, *Tetrahedron Lett.,* 36 (1995) 3627–3630.
(98) J. Zheng, J. L. Gore, and G. R. Gray, *J. Am. Chem. Soc.,* 120 (1998) 2684–2685.
(99) R. W. Binkley, *Adv. Carbohydr. Chem. Biochem.,* 38 (1981) 105 193.
(100) R. C. Roth and R. W. Binkley, *J. Org. Chem.,* 50 (1985) 690–693.
(101) J.-P. Pete, C. Portella, C. Monneret, J.-C. Florent, and Q. Khuong-Huu, *Synthesis,* (1977) 774–776.
(102) P. Jütten and H.-D. Scharf, *J. Carbohydr. Chem.,* 9 (1990) 675–681.
(103) A. Klausener, G. Beyer, H. Leismann, H.-D. Scharf, E. Müller, J. Runsink, and H. Görner, *Tetrahedron,* 45 (1989) 4989–5002.
(104) P. M. Collins and B. R. Whitton, *J. Chem. Soc. Perkin Trans. 1,* (1974) 1069–1075.
(105) P. M. Collins and B. R. Whitton, *Carbohydr. Res.,* 36 (1974) 293–301.
(106) E. L. Laird and W. L. Jorgensen, *J. Org. Chem.,* 55 (1990) 9–27.
(107) C. Walling, *Tetrahedron,* 41 (1985) 3887–3900.
(108) D. H. R. Barton, D. O. Jang, and J. Cs. Jaszberenyi, *Tetrahedron Lett.,* 31 (1990) 3991–3994.
(109) D. H. R. Barton, D. O. Jang, and J. Cs. Jaszberenyi, *Tetrahedron,* 49 (1993) 7193–7214.
(110) D. H. R. Barton, D. O. Jang, and J. Cs. Jaszberenyi, *Tetrahedron,* 49 (1993) 2793–2804.
(111) C. S. Burgey, R. Vollerthun, and B. Fraser-Reid, *J. Org. Chem.,* 61 (1996) 1609–1618.
(112) I. Ryu, F. Araki, S. Minakata, and M. Komatsu, *Tetrahedron Lett.,* 39 (1998) 6335–6336.
(113) V. T. Perchyonok and C. H. Schiesser, *Tetrahedron Lett.,* 39 (1998) 5437–5438.
(114) B. Giese, J. Dupuis, M. Leising, M. Nix, and H. J. Lindner, *Carbohydr. Res.,* 171 (1987) 329–341.
(115) J.-P. Praly, Z. El Kharraf, and G. Descotes, *Carbohydr. Res.,* 232 (1992) 117–123.
(116) B. Giese, B. Kopping, and C. Chatgilialoglu, *Tetrahedron Lett.,* 30 (1989) 681–684.
(117) C. Chatgilialoglu, *Acc. Chem. Res.,* 25 (1992) 188–194.
(118) C. Chatgilialoglu, K. U. Ingold, and J. C. Scaiano, *J. Am. Chem. Soc.,* 103 (1981) 7739–7742.
(119) P. A. Baguley and J. C. Walton, *Angew. Chem. Int. Ed.,* 37 (1998) 3072–3082.
(120) P. Kociensky and C. Pant, *Carbohydr Res.,* 110 (1982) 330–332.
(121) J. R. Rasmussen, C. J. Slinger, R. J. Kordish, and D. D. Newman-Evans, *J. Org. Chem.,* 46 (1981) 4843–4846.
(122) D. P. Curran and C.-T. Chang, *J. Org. Chem.,* 54 (1989) 3140–3157.
(123) P. Renaud, E. Lacôte, and L. Quaranta, *Tetrahedron Lett.,* 39 (1998) 2123–2126.
(124) D. Crich and S. Sun, *J. Org. Chem.,* 61 (1996) 7200–7201.
(125) J. C. López, A. M. Gómez, and B. Fraser-Reid, *J. Org. Chem.,* 60 (1995) 3871–3878.
(126) J. A. Axon and A. L. J. Beckwith, *J. Chem. Soc. Chem. Commun.,* (1995) 549–550.
(127) J. Brunckova, D. Crich, and Q. Yao, *Tetrahedron Lett.,* 35 (1994) 6619–6622.
(128) R. M. Lopez, D. S. Hays, and G. C. Fu, *J. Am. Chem. Soc.,* 119 (1997) 6949–6950.
(129) D. P. Curran, *Angew. Chem. Int. Ed.,* 37 (1998) 1174–1196.
(130) U. Gerigk, M. Gerlach, W. P. Neumann, R. Vieler, and V. Weintritt, *Synthesis,* (1990) 448–452.
(131) J. R. Blanton and J. M. Salley, *J. Org. Chem.,* 56 (1991) 490–491.
(132) C. Bokelmann, W. P. Neumann, and M. Peterseim, *J. Chem. Soc. Perkin Trans. 1,* (1992) 3165–3166.
(133) W. P. Neumann and M. Peterseim, *Synlett,* (1992) 801–802.
(134) H. Kuhn and W. P. Neumann, *Synlett,* (1994) 123–124.
(135) D. P. Dygutsch, W. P. Neumann, and M. Peterseim, *Synlett,* (1994) 363–365.

(136) G. Dumartin, M. Pourcel, B. Delmond, O. Donard, and M. Pereyre, *Tetrahedron Lett.,* 39 (1998) 4663–4666.
(137) U. Maitra and K. D. Sarma, *Tetrahedron Lett.,* 35 (1994) 7861–7862.
(138) R. Rai and D. B. Collum, *Tetrahedron Lett.,* 35 (1994) 6221–6224.
(139) J. Light and R. Breslow, *Tetrahedron Lett.,* 31 (1990) 2957–2958.
(140) D. L. J. Clive and W. Yang, *J. Org. Chem.,* 60 (1995) 2607–2609.
(141) E. Vedejs, S. M. Duncan, and A. R. Haight, *J. Org. Chem.,* 58 (1993) 3046–3050.
(142) D. P. Curran and S. Hadida, *J. Am. Chem. Soc.,* 118 (1996) 2531–2532.
(143) C. Chatgilialoglu, *Chem. Rev.,* 95 (1995) 1229–1251.
(144) Y.-D. Wu and C.-L. Wong, *J. Org. Chem.,* 60 (1995) 821–828.
(145) D. H. R. Barton, D. O. Jang, and J. Cs. Jaszberenyi, *Tetrahedron Lett.,* 31 (1990) 4681–4684.
(146) M. Lesage, J. A. Martinho-Simões, and D. Griller, *J. Org. Chem.,* 55 (1990) 5413–5414.
(147) D. H. R. Barton, D. O. Jang, and J. Cs. Jaszberenyi, *Synlett,* (1991) 435–438.
(148) H. Bürger and W. Kilian, *J. Organomet. Chem.,* 18 (1969) 299–306.
(149) C. Chatgilialoglu, D. Griller, and M. Lesage, *J. Org. Chem.,* 53 (1988) 3641–3642.
(150) M. Lesage, C. Chatgilialoglu, and D. Griller, *Tetrahedron Lett.,* 30 (1989) 2733–2734.
(151) D. Schummer and G. Höfle, *Synlett,* (1990) 705–706.
(152) M. Ballestri, C. Chatgilialoglu, K. B. Clark, D. Griller, B. Giese, and B. Kopping, *J. Org. Chem.,* 56 (1991) 678–683.
(153) C. Chatgilialoglu, M. Guerra, A. Guerrini, G. Seconi, K. B. Clark, D. Griller, J. Kanabus-Kaminska, and J. A. Martinho-Simões, *J. Org. Chem.,* 57 (1992) 2427–2433.
(154) M. B. Haque, B. P. Roberts, and D. A. Tocher, *J. Chem. Soc. Perkin Trans. 1,* (1998) 2881–2889.
(155) D. H. R. Barton, D. O. Jang, and J. Cs. Jaszberenyi, *Tetrahedron Lett.,* 32 (1991) 7187–7190.
(156) M. Oba, Y. Kawahara, R. Yamada, H. Mizuta, and K. Nishiyama, *J. Chem. Soc. Perkin Trans. 2,* (1996) 1843–1848.
(157) P. Di Cesare and B. Gross, *Synthesis,* (1980) 714–715.
(158) M. Drescher and F. Hammerschmidt, *Synthesis,* (1996) 1451–1454.
(159) T. Endo and Y. Araki, *J. Carbohydr. Chem.,* 16 (1997) 1393–1406.
(160) R. V. Stick, D. M. G. Tilbrook, and S. J. Williams, *Aust. J. Chem.,* 50 (1997) 233–235.
(161) S. Yamago, K. Kokubo, S. Masuda, and J.-I. Yoshida, *Synlett,* (1996) 929–930.
(162) S. Moutel and J. Prandi, *Tetrahedron Lett.,* 35 (1994) 8163–8166.
(163) T. Bamhaoud and J. Prandi, *J. Chem. Soc. Chem. Commun.,* (1996) 1229–1230.
(164) S. Mayer and J. Prandi, *Tetrahedron Lett.,* 37 (1996) 3117–3120.
(165) D. H. R. Barton, W. Hartwig, and W. B. Motherwell, *J. Chem. Soc. Chem. Commun.,* (1982) 447–448.
(166) A. L. J. Beckwith, *Chem. Soc. Rev.,* 22 (1993) 143–151.
(167) I. D. Jenkins, *J. Chem. Soc. Chem. Commun.,* (1994) 1227–1228.
(168) D. Crich, A. L. J. Beckwith, C. Chen, Q. Yao, I. G. E. Davison, R. W. Longmore, C. Anaya de Parrodi, L. Quintero-Cortes, and J. Sandoval-Ramirez, *J. Am. Chem. Soc.,* 117 (1995) 8757–8768.
(169) J. Brunckova and D. Crich, *Tetrahedron,* 51 (1995) 11945–11952.
(170) B. Giese, J. Dupuis, T. Hasskerl, and J. Meixner, *Tetrahedron Lett.,* 24 (1983) 703–706.
(171) R. P. Allen, B. P. Roberts, and C. R. Willis, *J. Chem. Soc. Chem. Commun.,* (1989) 1387–1388.
(172) B. P. Roberts, *Chem. Soc. Rev.,* 28 (1999) 25–35.
(173) J. N. Kirwan, B. P. Roberts, and C. R. Willis, *Tetrahedron Lett.,* 31 (1990) 5093–5096.
(174) S. J. Cole, J. N. Kirwan, B. P. Roberts, and C. R. Willis, *J. Chem. Soc. Perkin Trans. 1,* (1991) 103–112.
(175) M. B. Haque and B. P. Roberts, *Tetrahedron Lett.,* 37 (1996) 9123–9126.
(176) D. Crich and Q. Yao, *J. Org. Chem.,* 60 (1995) 84–88.
(177) D. Crich, X.-Y. Jiao, Q. Yao, and J. S. Harwood, *J. Org. Chem.,* 61 (1996) 2368–2373.
(178) J. Augé and S. David, *Carbohydr. Res.,* 59 (1977) 255–257.

(179) M. J. Bamford, J. C. Pichel, W. Husman, B. Patel, R. Storer, and N. G. Weir, *J. Chem. Soc. Perkin Trans. 1,* (1995) 1181–1187.
(180) A Schäfer, G. Klich, M. Schreiber, H. Paulsen, and J. Thiem, *Carbohydr. Res.,* 313 (1998) 107–116.
(181) K. P. R. Kartha and H. J. Jennings, *J. Carbohydr. Chem.,* 9 (1990) 777–781.
(182) Z. J. Witczak, *Tetrahedron Lett.,* 27 (1986) 155–158.
(183) M. Avalos, R. Babiano, C. Garcia-Verdugo, J. L. Jiménez, and J. C. Palacios, *Tetrahedron Lett.,* 31 (1990) 2467–2470.
(184) F. Santoyo González, F. García Calvo Flores, J. Isac-García, F. Hernández-Mateo, P. García-Mendoza, and R. Robles-Díaz, *Tetrahedron,* 50 (1994) 2877–2894.
(185) Z. J. Witczak, *Tetrahedron,* 41 (1985) 4781–4785.
(186) F. W. Lichtenthaler, S. Schwidetzky, and K. Nakamura, *Tetrahedron Lett.,* 31 (1990) 71–74.
(187) A. Koch and B. Giese, *Helv. Chim. Acta,* 76 (1993) 1687–1701.
(188) J. Hiebl and E. Zbiral, *Monatsh. Chem.,* 121 (1990) 683–690.
(189) L. Beigelman, A. Karpeisky, J. Matulic-Adamic, C. Gonzalez, and N. Usman, *Nucleosides Nucleotides,* 14 (1995) 907–910.
(190) K. C. Nicolaou, M. Sato, E. A. Theodorakis, and N. D. Miller, *J. Chem. Soc. Chem. Commun.,* (1995) 1583–1585.
(191) D. O. Jang, *Synth. Commun.,* 27 (1997) 1023–1027.
(192) H. Sano, T. Takeda, and T. Migita, *Synthesis,* (1988) 402–403.
(193) C. L Cavallaro and J. Schwartz, *J. Org. Chem.,* 61 (1996) 3863–3864.
(194) C. L Cavallaro and J. Schwartz, *J. Org. Chem.,* 60 (1995) 7055–7057.
(195) R. P. Spencer and J. Schwartz, *J. Org. Chem.,* 62 (1997) 4204–4205.
(196) R. Csuk, A. Fürstner, B. I. Glänzer, and H. Weidmann, *J. Chem. Soc. Chem. Commun.,* (1986) 1149–1150.
(197) A. Fürstner, J. Baumgartner, and D. N. Jumbam, *J. Chem. Soc. Perkin Trans. 1,* (1993) 131–138.
(198) L. Somsák, J. Madaj, and A. Wisniewski, *J. Carbohydr. Chem.,* 16 (1997) 1075–1087.
(199) B. Giese, *Angew. Chem. Int. Ed. Engl.,* 22 (1983) 753–764.
(200) M. J. Begley, M. Ladlow, and G. Pattenden, *J. Chem. Soc. Perkin Trans. 1,* (1988) 1095–1106.
(201) M. Ladlow and G. Pattenden, *J. Chem. Soc. Perkin Trans. 1,* (1988) 1107–1118.
(202) S.-U. Park, S.-K. Cung, and M. Newcomb, *J. Am. Chem. Soc.,* 108 (1986) 240–244.
(203) A. L. J. Beckwith and D. H. Roberts, *J. Am. Chem. Soc.,* 108 (1986) 5893–5901.
(204) G. Stork, *Bull. Soc. Chim. Fr.,* 127 (1990) 675–680.
(205) H. Nishiyama, T. Katajima, M. Matsumoto, and K. Itoh, *J. Org. Chem.,* 49 (1984) 2298–2300.
(206) G. Stork and M. J. Sofia, *J. Am. Chem. Soc.,* 108 (1986) 6826–6828.
(207) J. C. López and B. Fraser-Reid, *J. Am. Chem. Soc.,* 111 (1989) 3450–3452.
(208) K. Tamao, N. Ishida, and M. Kumada, *J. Org. Chem.,* 48 (1983) 2120–2122.
(209) B. Giese, *Pure Appl. Chem.,* 60 (1988) 1655–1658.
(210) M. L. Sinnott, *Adv. Phys. Org. Chem.,* 24 (1988) 113–204.
(211) D. Medakovic, *Carbohydr. Res.,* 253 (1994) 299–300.
(212) H. Hori, Y. Nishida, H. Ohrui, and H. Meguro, *J. Carbohydr. Chem.,* 9 (1990) 601–618.
(213) D. Kahne, D. Yang, J. J. Lim, R. Miller, and E. Paguaga, *J. Am. Chem. Soc.,* 110 (1988) 8716–8717.
(214) D. H. R. Barton, D. Crich, and W. B. Motherwell, *Tetrahedron,* 41 (1985) 3901–3924.
(215) D. H. R. Barton, C.-Y. Chern, and J. Cs. Jaszberenyi, *Tetrahedron,* 51 (1995) 1867–1886.
(216) D. H. R. Barton, C.-Y. Chern, and J. Cs. Jaszberenyi, *Aust. J. Chem.,* 48 (1995) 407–425.
(217) D. Crich, J.-T. Hwang, and H. Yuan, *J. Org. Chem.,* 61 (1996) 6189–6198.
(218) D. Crich and L. B. L. Lim, *Tetrahedron Lett.,* 32 (1991) 2565–2568.
(219) D. Crich and L. B. L. Lim, *J. Chem. Soc. Perkin Trans. 1,* (1991) 2209–2214.
(220) D. Crich and T. J. Ritchie, *J. Chem. Soc., Chem. Commun.,* (1988) 1461–1463.

(221) D. Crich and T. J. Ritchie, *Carbohydr. Res.,* 190 (1989) C3–C6.
(222) D. Crich and T. J. Ritchie, *J. Chem. Soc. Perkin Trans. 1,* (1990) 945–954.
(223) D. Crich and F. Hermann, *Tetrahedron Lett.,* 34 (1993) 3385–3388.
(224) T. Sugai, G.-J. Shen, Y. Ichikawa, and C.-H. Wong, *J. Am. Chem. Soc.,* 115 (1993) 413–421.
(225) D. P. Curran, D. Kim, H. T. Liu, and W. Shen, *J. Am. Chem. Soc.,* 110 (1988) 5900–5902.
(226) N. Yamazaki, E. Eichenberger, and D. P. Curran, *Tetrahedron,* 35 (1994) 6623–6626.
(227) R. J. Ferrier, D. W. Hall, and P. M. Petersen, *Carbohydr. Res.,* 239 (1993) 143–153.
(228) D. Crich, S. Sun, and J. Brunckova, *J. Org. Chem.,* 61 (1996) 605–615.
(229) J.-P. Praly, G.-R. Chen, J. Gola, G. Hetzer, and C. Raphoz, *Tetrahedron Lett.,* 38 (1997) 8185–8188.
(230) G. E. Keck, E. J. Enholm, J. B. Yates, and M. R. Wiley, *Tetrahedron,* 41 (1985) 4079–4094.
(231) G. E. Keck and J. H. Byers, *J. Org. Chem.,* 50 (1985) 5442–5444.
(232) B. A. Roe, C. G. Boojamra, J. L. Griggs, and C. R. Bertozzi, *J. Org. Chem.,* 61 (1996) 6442–6445.
(233) F. Pontén and G. Magnusson, *J. Org. Chem.,* 61 (1996) 7463–7466.
(234) J. Cui and D. Horton, *Carbohydr. Res.,* 309 (1998) 319–330.
(235) S. Myrvold, L. M. Reimer, D. L. Pompliano, and J. W. Frost, *J. Am. Chem. Soc.,* 111 (1989) 1861–1866.
(236) W. Schmid, R. Christian, and E. Zbiral, *Tetrahedron Lett.,* 29 (1988) 3643–3646.
(237) J. Gervay and T. Q. Gregar, *Tetrahedron Lett.,* 38 (1997) 5921–5924.
(238) S. Hanessian and C. Girard, *Synlett,* (1994) 863–864.
(239) I. R. Vlahov, P. I. Vlahova, and R. J. Linhardt, *J. Am. Chem. Soc.,* 119 (1997) 1480–1481.
(240) L. Somsák, G. Batta, and I. Farkas, *Tetrahedron Lett.,* 27 (1986) 5877–5880.
(241) L. Somsák and M. Szabó, *J. Carbohydr. Chem.,* 9 (1990) 755–759.
(242) F. W. Lichtenthaler and F. Hoyer, *Carbohydr. Res.,* 253 (1994) 141–150.
(243) J.-P. Praly, C. Di Stèfano, L. Somsák, M. Hollósi, Z. Majer, and W. Voelter, *Tetrahedron Asymm.,* 10 (1999) 901–911.
(244) L. Panza, P. L. Chiappini, G. Russo, D. Monti, and S. Riva, *J. Chem. Soc. Perkin Trans. 1,* (1997) 1255–1256.
(245) A. Bouali, G. Descotes, D. F. Ewing, A. Grouiller, J. Lefkidou, A.-D. Lespinasse, and G. Mackenzie, *J. Carbohydr. Chem.,* 11 (1992) 159–169.
(246) D. Crich and L. B. L. Lim, *Tetrahedron Lett.,* 31 (1990) 1897–1900.
(247) D. Crich and L. B. L. Lim, *J. Chem. Soc. Perkin Trans. 1,* (1991) 2205–2208.
(248) N. Ono and A. Kaji, *Synthesis,* (1986) 693–704.
(249) H.-G. Korth, R. Sustmann, J. Dupuis, and B. Giese, *Chem. Ber.,* 120 (1987) 1197–1202.
(250) R. Julina, I. Müller, A. Vasella, and R. Wyler, *Carbohydr. Res.,* 164 (1987) 415–432.
(251) W. Brade and A. Vasella, *Helv. Chim. Acta,* 73 (1990) 1923–1930.
(252) R. Meuwly and A. Vasella, *Helv. Chim. Acta,* 68 (1985) 997–1009.
(253) L. Cipolla, L. Liguori, F. Nicotra, G. Torri, and E. Vismara, *J. Chem. Soc. Chem. Commun.,* (1996) 1253–1254.
(254) L. Cipolla, F. Nicotra, E. Vismara, and M. Guerrini, *Tetrahedron,* 53 (1997) 6163–6170.
(255) T. Linker, T. Sommermann, and F. Kahlenberg, *J. Am. Chem. Soc.,* 119 (1997) 9377–9384.
(256) O. R. Martin and F. Xie, *Carbohydr. Res.,* 264 (1994) 141–146.
(257) W. B. Motherwell, *Aldrichimica Acta,* 25 (1992) 71–80.
(258) M. J. Tozer and T. F. Herpin, *Tetrahedron,* 52 (1996) 8619–8683.
(259) W. B. Motherwell, B. C. Ross, and M. J. Tozer, *Synlett,* (1989) 68–70.
(260) T. F. Herpin, W. B. Motherwell, and M. J. Tozer, *Tetrahedron Asymm.,* 5 (1994) 2269–2282.
(261) T. F. Herpin, J. S. Houlton, W. B. Motherwell, B. P. Roberts, and J.-M. Weibel, *J. Chem. Soc. Chem. Commun.,* (1996) 613–614.
(262) T. F. Herpin, W. B. Motherwell, B. P. Roberts, S. Roland, and J.-M. Weibel, *Tetrahedron,* 53 (1997) 15085–15100.

(263) T. F. Herpin, W. B. Motherwell, and J.-M. Weibel, *J. Chem. Soc. Chem. Commun.*, (1997) 923–924.
(264) T. Gimisis, G. Ialongo, M. Zamboni, and C. Chatgilialoglu, *Tetrahedron Lett.*, 36 (1995) 6781–6784.
(265) T. Gimisis, G. Ialongo, and C. Chatgilialoglu, *Tetrahedron*, 54 (1998) 573–592.
(266) Y. Itoh, K. Haraguchi, H. Tanaka, K. Matsumoto, K. T. Nakamura, and T. Miyasaka, *Tetrahedron Lett.*, 36 (1995) 3867–3870.
(267) J.-M. Surzur and P. Teissier, *C. R. Acad. Sci., Ser. C*, 264 (1967) 1981–1984.
(268) J.-M. Surzur and P. Teissier, *Bull. Soc. Chim. Fr.*, (1970) 3060–3070.
(269) D. D. Tanner and F. C. Law, *J. Am. Chem. Soc.*, 91 (1969) 7535–7537.
(270) D. Crich, A. L. J. Beckwith, G. F. Filzen, and R. W. Longmore, *J. Am. Chem. Soc.*, 118 (1996) 7422–7423.
(271) A. L. J. Beckwith, D. Crich, P. J. Duggan, and Q. Yao, *Chem. Rev.*, 97 (1997) 3273–3312.
(272) B. Giese, K. S. Gröninger, T. Witzel, H.-G. Korth, and R. Sustmann, *Angew. Chem. Int. Ed. Engl.*, 26 (1987) 233–234; *Angew. Chem.*, 99 (1987) 246–247.
(273) B. Giese, S. Gilges, K. S. Gröninger, C. Lamberth, and T. Witzel, *Liebigs Ann. Chem.*, (1988) 615–617.
(274) B. Giese and K. S. Gröninger, *Org. Synth.*, 69 (1990) 66–71.
(275) M. Baudry, M.-N. Bouchu, G. Descotes, J.-P. Praly, and F. Bellamy, *Carbohydr. Res.*, 282 (1996) 237–246.
(276) B. Quiclet-Sire and S. Z. Zard, *J. Am. Chem. Soc.*, 118 (1996) 9190–9191.
(277) J. Boivin, B. Quiclet-Sire, L. Ramos, and S. Z. Zard, *J. Chem. Soc. Chem. Commun.*, (1997) 353–354.
(278) D. Crich and Q. Yao, *J. Am. Chem. Soc.*, 115 (1993) 1165–1166.
(279) D. Crich, Q. Yao, G. F. Filzen, *J. Am. Chem. Soc.*, 117 (1995) 11455–11470.
(280) A. L. J. Beckwith and P. J. Duggan, *J. Chem. Soc. Perkin Trans. 2*, (1993) 1673–1679.
(281) D. Crich and G. F. Filzen, *J. Org. Chem.*, 60 (1995) 4834–4837.
(282) D. Crich and X.-Y. Jiao, *J. Am. Chem. Soc.*, 118 (1996) 6666–6670.
(283) S.-Y. Choi, D. Crich, J. H. Horner, X. Huang, F. N. Martinez, M. Newcomb, D. J. Wink, and Q. Yao, *J. Am. Chem. Soc.*, 120 (1998) 211–212.
(284) S. Peukert and B. Giese, *Tetrahedron Lett.*, 25 (1996) 4365–4368.
(285) T.-S. Lin, J.-H. Yang, M.-C. Liu, and J.-L. Zhu, *Tetrahedron Lett.*, 31 (1990) 3829–3832.
(286) A. Fernandez-Mayoralas, A. Marra, M. Trumtel, A. Veyrières, and P. Sinaÿ, *Carbohydr. Res.*, 188 (1989) 81–95.
(287) F. Santoyo-González, F. G. Calvo-Flores, F. Hernández-Mateo, P. Garcia-Mendoza, J. Isac-Garcia, and M. D. Pérez-Alvarez, *Synlett*, (1994) 454–456.
(288) N. E. Poopeiko, T. I. Pricota, and I. A. Mikhailopulo, *Synlett*, (1991) 342.
(289) M. C. Samano and M. J. Robins, *Tetrahedron Lett.*, 32 (1991) 6293–6296.
(290) J. J. Patroni, R. V. Stick, D. Matthew, G. Tilbrook, B. W. Skelton, and A. H. White, *Aust. J. Chem.*, 42 (1989) 2127–2141.
(291) R. Robles Diaz, C. Rodríguez Melgarejo, I. Izquierdo Cubero, and M. T. Plaza López-Espinosa, *Carbohydr. Res.*, 300 (1997) 375–380.
(292) A. Ghosez, T. Göbel, and B. Giese, *Chem. Ber.*, 121 (1988) 1807–1811.
(293) L. Somsák and I. Németh, *J. Carbohydr. Chem.*, 12 (1993) 679–684.
(294) R. P. Spencer and J. Schwartz, *Tetrahedron Lett.*, 37 (1996) 4357–4360.
(295) P. de Pouilly, A. Chénedé, J.-M. Mallet, and P. Sinaÿ, *Tetrahedron Lett.*, 33 (1992) 8065–8068.
(296) Z. J. Witczak and S. Czernecki, *Adv. Carbohydr. Chem. Biochem.*, 53 (1998) 143–199.
(297) T. B. Grindley, *Adv. Carbohydr. Chem. Biochem.*, 53 (1998) 17–142.
(298) R. Lenz and B. Giese, *J. Am. Chem. Soc.*, 119 (1997) 2784–2794.
(299) A. Gugger, R. Batra, P. Rzadek, G. Rist, and B. Giese, *J. Am. Chem. Soc.*, 119 (1997) 8740–8741.

(300) B. Giese, A. Dussy, E. Meggers, M. Petretta, and U. Schwitter, *J. Am. Chem. Soc.,* 119 (1997) 11130–11131.
(301) D. Crich and Q. Yao, *Tetrahedron,* 54 (1998) 305–318.
(302) D. Crich and X.-S. Mo, *J. Am. Chem. Soc.,* 119 (1997) 249–250.
(303) Z. A. Doddridge, J. L. Warner, P. M. Cullis, and G. D. D. Jones, *Chem. Commun.,* (1998) 1997–1998.
(304) W. K. Poggozelski and T. D. Tullius, *Chem. Rev.,* 98 (1998) 1089–1107.

OLIGOSACCHARIDE–PROTEIN CONJUGATES AS VACCINE CANDIDATES AGAINST BACTERIA

By Vince Pozsgay

National Institute of Child Health and Human Development, National Institutes of Health, 6 Center Dr., MSC 2720, Bethesda, Maryland 20892-2720, U.S.A.

I. Overview of Polysaccharide-Based Antibacterial Vaccines	153
II. Synthesis and Immunological Evaluation of Protein Conjugates of Oligosaccharide Components of Bacterial Polysaccharides	157
1. Studies on Dextran-Related Oligosaccharide–Protein Conjugates	157
2. Protein Conjugates of Oligosaccharide Fragments of Capsular Polysaccharides	161
3. Protein Conjugates of Oligosaccharides Related to Lipopolysaccharides	179
III. Conclusion	193
References	194

I. Overview of Polysaccharide-Based Antibacterial Vaccines*

Complex polymers of mono- and oligo-saccharides are ubiquitous components of the bacterial cell wall.[1–3] The three major classes are the capsular polysaccharides,

---

* Abbreviations in the text.—Abe, abequose; BSA, bovine serum albumin; CRM, cross-reacting mutant; Ds, degree of substitution; DTx, diphtheria toxin; *E., Escherichia;* Gal, galactose; Glc, glucose; *H., Haemophilus;* Hib, *Haemophilus influenzae* type b; HSA, human serum albumin; IgG, immunoglobulin G; IgM, immunoglobulin M; Kdo, 3-deoxy-D-*manno*-octulosonic acid; KLH, keyhole limpet hemocyanin; LD_{50}, 50% lethal dose; LPS, lipopolysaccharide; MALDI-TOF, matrix-assisted laser desorption ionization time of flight; Man, mannose; MW, molecular weight; *N., Neisseria;* O-SP, O-specific polysaccharide; *S., Salmonella; Sh., Shigella; Str., Streptococcus;* TT, tetanus toxoid.

Abbreviations in the formulas and schemes.—Ac, acetyl; All, allyl; BDPS, ^{t}butyldiphenylsilyl; Bn, benzyl; BOM, benzyloxymethyl; Bz, benzoyl; CA, chloroacetyl; CPG, controlled-pore glass; DMT, 4,4′-dimethoxytrityl; Et, ethyl; Gal, D-galactose; Glc, D-glucose; GlcA, D-glucuronic acid; GlcNAc, 2-acetamido-2-deoxy-D-glucose; Hep, heptose; Kdo, 3-deoxy-D-*manno*-octulosonic acid; Man, D-mannose; MBn, 4-methoxybenzyl; Me, methyl; MMT, monomethoxytrityl; PEA, phosphoethanolamine; Ph, phenyl; Phth, phthaloyl; *p*-NO$_2$Bz, 2-nitrobenzoyl; Pr, propyl; Rha, L-rhamnose; Tol, 2-methylbenzoyl; Ts, *p*-tolylsulfonyl; Z, benzyloxycarbonyl.

Single-letter notations for amino acids.—A, alanine; C, cysteine; D, aspartic acid; E, glutamic acid; G, glycine; H, histidine; I, isoleucine; K, lysine; L, leucine; N, asparagine; Q, glutamine; R, arginine; S, serine; T, threonine; V, valine; Y, tyrosine.

the teichoic acids, and the lipooligo- and poly-saccharides. Bacterial cell-surface polysaccharides provide mechanical protection to the cell against the intracellular osmotic pressure, impair phagocytosis, and activate the alternate pathway of complement activation.[4] That pneumococcal capsular polysaccharides are carriers of type-specificity was recognized by the identification of the "soluble specific substances" isolated from pneumococci as polysaccharides devoid of proteinaceous material.[5] This observation was extended to both gram-positive and gram-negative bacteria expressing capsular polysaccharides or teichoic acid-type polymers. Lipopolysaccharides of gram-negative bacteria exhibit the same principle: their O-specific polysaccharide domains alone define serotype specificity. Only few instances are known when different bacteria carry identical capsular polysaccharides (e.g., *E. coli* K12 and K82, and *E. coli* K1; group B *Neisseria meningitidis, Moraxella nonliquefaciens,* and *Pasteurella haemolytica* A2;[6,7] *Enterococcus faecalis,* and *Enterococcus faecium*[8]), or O-specific polysaccharides (e.g., *Shigella sonnei, Plesiomonas aeruginosa*[9]). Fully expressed capsular polysaccharides or teichoic acids of gram-positive bacteria and the O-specific polysaccharides of gram-negative enteric bacteria are essential for virulence of "primary" pathogens.[10] Bacterial cell-surface polysaccharides that have close structural similarity to saccharides found in human tissues may protect the bacteria from response by the host's immune response, and this molecular mimicry may be related to bacterial virulence. Because carbohydrates are secondary gene products and require several enzyme systems even for the construction of a monosaccharide, cell-surface carbohydrate epitopes are highly conserved and, in contrast to proteins, no mutation of bacterial polysaccharides has so far been reported. This permanency of the structures of surface polysaccharides has made them a target for immunoprophylaxis since the discovery by Francis and Tillet[11] that purified capsular polysaccharides are able to elicit anti-polysaccharide antibodies in healthy humans at a single dose of 10 µg. Subsequent work by Heidelberger et al.[12] firmly established that antibodies induced by pneumococcal capsular polysaccharides provide type-specific immunity that may last for up to 8 years. Antibodies to other cellular components of encapsulated bacteria are "of considerably less importance" for protection.[13,14] Recognition of increasing occurrence of bacterial resistance to antibiotics[15] emphasized the need for disease prevention by vaccination. Developments since the early explorations on bacterial polysaccharides as vaccine components led to the licensure of several vaccines as commercial pharmaceuticals consisting of purified cell-surface polysaccharides or teichoic acid-like polymers. These include 23-valent vaccines against pneumococci containing 25 µg of the capsular polysaccharide of each of the 23 most invasive types [16][Pneumovax (Merck) and Pnu-immune (Wyeth Lederle)], a bivalent vaccine against groups A and C *N. meningitidis* [AC Vax (SmithKline Beecham) and Mangivax (Pasteur Merieux)], a tetravalent vaccine against groups A, C, W-135, and Y *N. meningitidis*[16] [Menomune (Pasteur Merieux)], and a vaccine containing the Vi polysaccharide[17] against typhoid fever [Typhim Vi (Pasteur Merieux Connaught)]. Despite their success, capsular polysaccharide vaccines have limitations because of their T-cell dependence (no booster response) and age-related immunogenicity: in contrast to proteins, immune responsiveness to polysaccharides does not usually develop in infants younger than 18 months of age.[17] Although the protective efficacy of the pneumococcal vaccine in infants is far from ideal, it has been suggested that their use in infants should be considered until a better vaccine becomes available.[18] The age-related immune response to plain polysaccharides may also be structure-dependent: in contrast to other polysaccharides, the capsular polysaccharides of group A *N. meningitidis*[19] and type 3 and 18C pneumococci[20] are good immunogens in infants starting at the age of 3 to 6 months and produce protective levels of IgG antibodies. In addition, the molecular size has a crucial role in that larger polysaccharides are generally more immunogenic than smaller ones, although no molecular weight range has been defined for polysaccharide immunogenicity.[21] The utility of capsular polysaccharides as vaccines is further limited by the fact that elderly persons and the those who are immunocompromised react poorly to plain polysaccharides,[22] and some polysaccharides, most notably the α-$(2\rightarrow 8)$-linked polymer of *N*-acetylneuraminic acid of group B *N. meningitidis,* are poor immunogens in humans of all ages.[23]

Based on Landsteiner and Lampl's[24] recognition that nonimmunogenic small molecules (haptens) may be rendered immunogenic by their covalent coupling to proteins, Avery and Goebel[25,26] showed in the 1930s that a conjugate of type-3 pneumococcal capsular polysaccharide with horse-serum globulin elicited antipolysaccharide-specific antibodies in rabbits that conferred both active and passive immunity against the homologous organism. Neither the purified polysaccharide alone nor the intact bacterial cell from which it was obtained elicited polysaccharide-specific antibodies in rabbits. As a pioneering development in antibacterial immunoprophylaxis, a tetanus toxoid conjugate of the capsular material of the gram-negative bacterium *Haemophilus influenzae* type b (Hib) was prepared.[27] This conjugate offers protection against infection by this pathogen, which is a major causative organism of childhood meningitis in young children, with high morbidity and mortality even with antibiotic therapy. Successors of this vaccine may be used in infants at the age of 2 months, who are at greatest risk to infection, to induce protective levels of humoral IgG antibodies against Hib. In countries where these vaccines are routinely used, cases of infant meningitis caused by Hib are virtually eliminated.[28] The spectacular success of the Hib vaccines in controlling a major childhood disease stimulated development of conjugate vaccines consisting of the capsular polysaccharides of numerous other bacterial pathogens. These include *Str. pneumoniae,*[29] *N. meningitidis,*[30,31] group B *Streptococcus,*[32,33] *Salmonella typhi,*[34] and *Staphylococcus aureus.*[35]

The O-specific polysaccharide components of the lipopolysaccharides of gram-negative bacteria fulfil an immunological role similar to that of the capsular polysaccharides in that they are essential for bacterial virulence and express serotype specificity.[36–39] The purified O-specific polysaccharides, when devoid of Lipid A, are nontoxic, have no known pharmacological effects, and because of their low molecular weight (usually less than 30 kDa), they are not immunogenic. Similarly to other haptenic materials, they can be rendered immunogenic by covalent attachment to proteins. It has been proposed that serum antibodies against the O-specific polysaccharides of gram-negative enteric bacteria (e.g., *Salmonella*[40 43] and E. coli[44]) confer serotype-specific immunity, and it has been documented[10,44–49] that protection against infection by enteric bacteria correlates with the level of anti-O-specific polysaccharide-specific antibodies. Based on the hypothesis that a critical level of anti-O-specific polysaccharide-specific serum IgG antibody may protect against infection by the homologous organism by neutralizing the inoculum, a new generation of conjugate vaccines is being developed that contain either detoxified lipopolysaccharides or purified O-specific polysaccharides covalently linked to a bacterial toxoid.[10,48] These experimental vaccines were safe and immunogenic when tested in animals or humans, elicited anti-O-specific polysaccharide-specific IgG antibodies, and provided statistically significant protection against the homologous organism in human trials.[10]

Because of the poor immunogenicity of plain polysaccharides in the most vulnerable populations, carbohydrate vaccine technology in the future will focus on the design and synthesis of conjugate vaccines in which bacterial cell-surface carbohydrates or analogs are covalently linked to immunogenic proteins or their derivatives. In clinical trials, a vaccine is considered successful if it elicits high levels of anti-polysaccharide antibodies with an increased IgG/IgM ratio that can be boosted by a subsequent injection even in young infants, the immunocompromised, and the elderly.[50] Improvement of the current vaccines requires a better understanding of the role of the submolecular properties of glycoconjugates that influence the immune response. These include the length of the saccharide chain, the importance of the non-reducing terminal saccharide unit, the distance between the saccharide and the protein, the coupling chemistry, and the number of saccharide chains per protein molecule. Several studies indicate that, in glycoconjugate vaccines, the length of the saccharide chain may be less important than in plain polysaccharides. For example, albumin conjugates of dextran fragments obtained by depolymerization of the native dextran are immunogenic in animal models.[51] Conjugates of small dextran molecules (MW ≤ 4 kDa) were better immunogens in mice than those of large ones (≤ 40 kDa).[50,52] Interestingly, chicken serum albumin conjugates of dextran-related oligosaccharides containing only 3–5 glucose units were more immunogenic than conjugates of dextran having MW ~ 40 kDa.[52] It should be noted that comparisons of these results should take into account the methods of conjugation: while the small oligosaccharides were coupled through their reducing end to chicken serum albumin, therefore exposing the non-reducing terminus and leaving the entire saccharide chain available for interaction with the B-cell receptors, the dextran conjugates were prepared by multiple-point attachment, probably providing a network of undefined structure in which the available unchanged saccharide length is not known. In a study aimed at defining the saccharide length necessary for polysaccharide-specific immune response, tetanus toxoid conjugates of fragments of the capsular polysaccharide of type III group B *Streptococcus* were prepared. It was found that the conjugate of a fragment representing only $\frac{1}{10}$ of the native polysaccharide elicited anti-polysaccharide antibodies in rabbits, while the unconjugated native polysaccharide was nonimmunogenic.[53] In the Hib polysaccharide system, the extent of saccharide loading (saccharide molecules/protein) was more critical than other structural variables, including the chain length.[54] A similar trend was reported for capsular polysaccharide fragments of several *Str. pneumoniae* serotypes.[55]

The literature on bacterial polysaccharides as vaccine components has been reviewed in this series by Jennings.[56] Since then, the subject has been reviewed by Ala'Aldeen and Cartwright,[57] Dick and Beurret,[58] Dintzis,[59] Egan,[60] Goldblatt,[61] Jennings,[62-65] Lee,[66] Fattom,[35,67] Paradiso and Lindberg,[68] Peeters,[69] and Robbins.[10,45-47]

This chapter covers reports on the evaluation of the immunogenicity of protein conjugates of oligosaccharide components of bacterial cell-surface polysaccharides. Syntheses of representative oligosaccharides and their conjugation are included to illustrate the various chemical approaches used for the preparation of the synthetic immunogens. "Oligosaccharide" mixtures obtained by degradation of native polysaccharides and not purified to homogeneity are not covered. Oligosaccharides corresponding to mycobacterial polysaccharide antigens have recently been reviewed by Aspinall[70] and by Lipták.[71] Synthetic lipid A and its analogs have been reviewed by Rietschel.[38,39,72] Detailed description of the clinical relevance of individual pathogens can be found in the reviews just referred to.

II. Synthesis and Immunological Evaluation of Protein Conjugates of Oligosaccharide Components of Bacterial Polysaccharides

1. Studies on Dextran-Related Oligosaccharide–Protein Conjugates

Although dextrans are not candidates for bacterial immunoprophylaxis, they represent simple, easily available, and useful models to study aspects of carbohydrate immunogenicity that may be relevant to the design of oligosaccharide-based conjugate vaccines. Structurally, native dextrans are homopolymers of α-(1→6)-linked D-glucopyranose residues that have α-(1→2), α-(1→3), and α-(1→4)-linked branches of varying length.[21,73,74] Kabat[75-77] and Richter and Eby[78] studied the minimal molecular requirements for antidextran antibody induction using protein and lipid conjugates of various epitope patterns of linear and branched dextrans. In Kabat's[75] approach, isomaltose (**1**) and isomaltoriose (**2**), prepared by controlled acid hydrolysis of native dextrans, were covalently linked to BSA by the aldonic acid route using the mixed anhydride method. Thus, the disaccharide **1** was oxidized to an equilibrium mixture of the free and lactonized aldonic acid derivative **3** that was activated with isobutyl chloroformate **4** (→**5**) followed by condensation with the protein to afford the conjugate **6**. The BSA conjugates so prepared contained 8 and 14 moles of α-linked glucose and isomaltose, respectively, attached to the ε-amino group of the protein's lysine residues through the polyhydroxy spacer derived from the reducing-end glucose residue of the original di- and trisaccharides.[75] Although the antibodies raised in rabbits against the former conjugate having only one intact D-glucose residue in the hapten part did not precipitate dextran, the antibodies formed toward the disaccharide hapten exhibited extensive precipitation with native dextran. Dextran-precipitating antibodies were likewise elicited in mice and rabbit by protein conjugates of linear dextran fragments up to the hexasaccharide **7**.[76,79,80] An interesting finding in these early studies was that the linear dextran-derived di- to hexa-saccharides were able to inhibit precipitation of the antidextran antibodies by the homologous dextran as a function of their size: in the range studied, the linear hexasaccharide **7** was the best inhibitor.[81] Although dextrans offer a large reservoir of the various

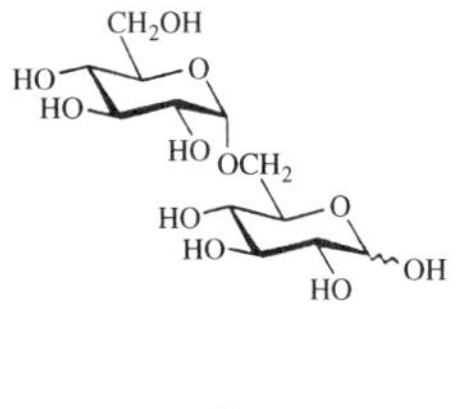

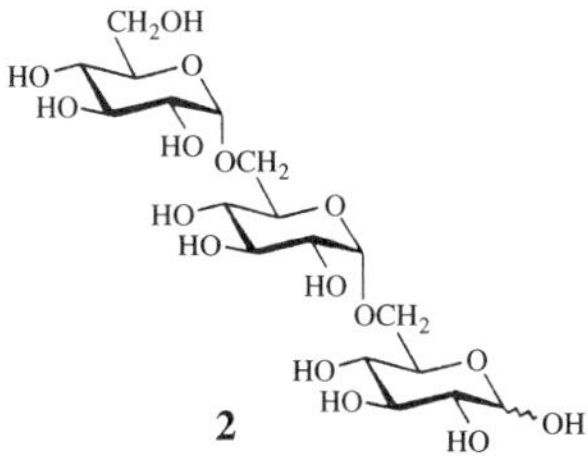

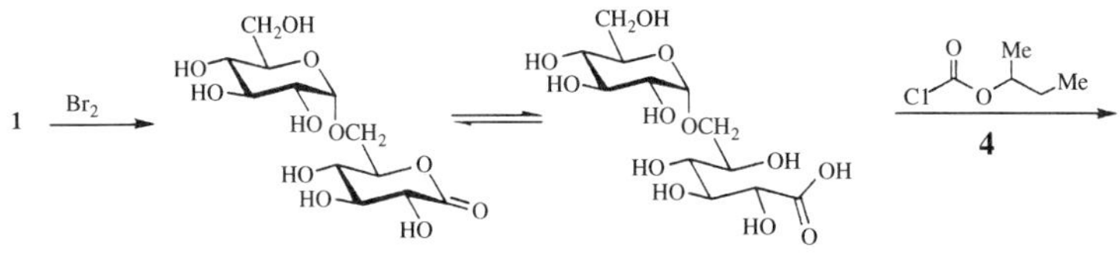

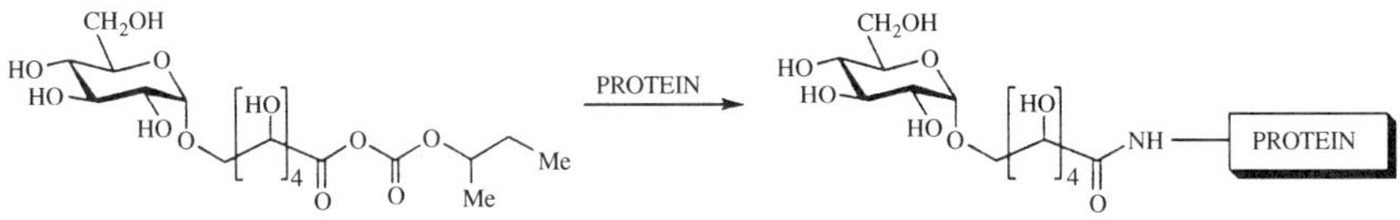

α-D-Glc$p$-(1→6)-[α-D-Glc$p$-(1→6)-]$_4$α-D-Glc

7

oligosaccharide epitopes, difficulties in specific degradation and isolation of the targeted oligosaccharides make chemical synthesis a better approach to sufficient amounts of the required fragments. Eby and Schuerch[82-84] synthesized protein conjugates of a range of dextran fragments that represent various epitope patterns of native dextrans. A typical synthetic sequence[82] started from the glucosyl donor[85] **8** that was condensed with the spacer **9** followed by deprotection. Subsequent glycosylation of compound **10** with the donor **8** afforded **11**, from which removal of the protecting groups provided isomaltose **12** having a terminal free amino group in its aglycon.

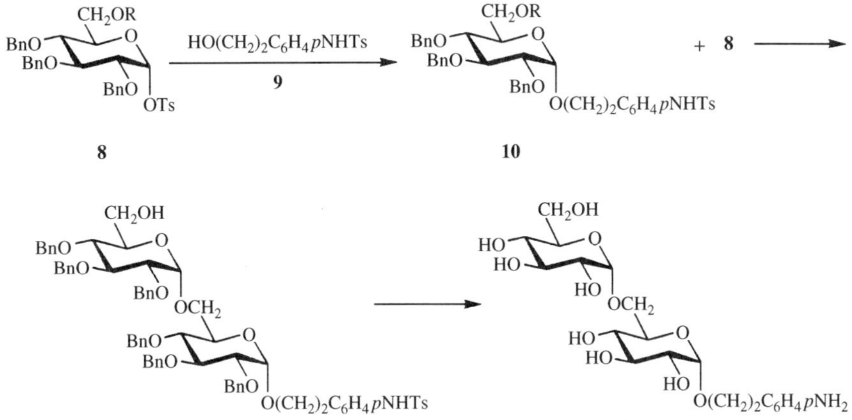

Dextran fragments prepared by chemical synthesis and their protein conjugates obtained either by the diazo coupling[25] or by the isothiocyanate protocol[86] (**13–23**) are listed in Table I. In the diazo-coupling method, a *p*-aminophenyl glycoside **24** is converted into the diazo intermediate **25**, which reacts with the protein to form conjugate **26**. More efficient conjugation was achieved with the isothiocyanate coupling procedure. In this approach the aminophenyl glycoside **24** is treated with thiophosgene to afford the reactive isothiocyanate **27**, which reacts with the ε-amino groups of the lysine residues by forming a stable thiourea linkage in the conjugate **28**. Richter and Eby[78] found that the conjugates listed in Table I induced anti-dextran antibodies in rabbits when injected at weekly intervals up to four times, using 1 mg of the conjugate for each injection. In agreement with Kabat's[75] findings they demonstrated that a single glucosyl residue in not sufficient for inducing dextran-reactive antibodies in the rabbit model.[78] The formation was noted of antibody populations that are directed at distinct glucosyl epitopes having α-(1→2), α-(1→3), and α-(1→6) linkages. The failure of conjugate **21** to induce an antibody population to the α-(1→4) glucosyl epitope may be interpreted as a consequence of the presence of this linkage in glycogen, a self-polysaccharide of

TABLE I
Protein Conjugates of Synthetic Oligosaccharide Fragments of Dextrans

Conjugate Number	Oligosaccharide Structure	Coupling Method[a,b]	Carrier Protein	Saccharide/ Protein Ratio (mol/mol)
13	α-D-Glcp-(1 → 6)-α-D-Glcp-(1 → O-(CH$_2$)$_2$C$_6$H$_4$NH$_2$	Diazo	Edestin	8.4
14	α-D-Glcp-(1 → 6)-α-D-Glcp-(1 → 6)-α-D-Glcp-(1 → O-(CH$_2$)$_2$C$_6$H$_4$NH$_2$	Diazo	Edestin	10
15	α-D-Glcp-(1 → 6)-[α-D-Glcp-(1 → 6)]$_2$-α-D-Glcp-(1 → O-(CH$_2$)$_2$C$_6$H$_4$NH$_2$	Diazo	Edestin	7.5
16	α-D-Glcp-(1 → 6)-α-D-Glcp-(1 → O-(CH$_2$)$_2$C$_6$H$_4$NH$_2$	Diazo	BSA	4.4
17	α-D-Glcp-(1 → 6)-α-D-Glcp-(1 → 6)-α-D-Glcp-(1 → O-(CH$_2$)$_2$C$_6$H$_4$NH$_2$	Diazo	BSA	2
18	α-D-Glcp-(1 → 6)-[α-D-Glcp-(1 → 6)]$_2$-α-D-Glcp-(1 → O-(CH$_2$)$_2$C$_6$H$_4$NH$_2$	Diazo	BSA	1
19	α-D-Glcp-(1 → 6)-α-D-Glcp-(1 → 2)-α-D-Glcp-(1 → O-(CH$_2$)$_2$C$_6$H$_4$NH$_2$	NCS	BSA	7.8
20	α-D-Glcp-(1 → 6)-α-D-Glcp-(1 → 3)-α-D-Glcp-(1 → O-(CH$_2$)$_2$C$_6$H$_4$NH$_2$	NCS	BSA	19
21	α-D-Glcp-(1 → 6)-α-D-Glcp-(1 → 4)-α-D-Glcp-(1 → O-(CH$_2$)$_2$C$_6$H$_4$NH$_2$	NCS	BSA	18
22	α-D-Glcp-(1 → 2)-α-D-Glcp-(1 → O-(CH$_2$)$_2$C$_6$H$_4$NH$_2$	NCS	BSA	32.31
23	α-D-Glcp-(1 → 2)-α-D-Glcp-(1 → O-(CH$_2$)$_2$C$_6$H$_4$NH$_2$	NCS	Hemocyanin	800

[a] Diazo coupling; see ref. 25.
[b] For the NCS (isothiocyanate) coupling, see ref. 89.

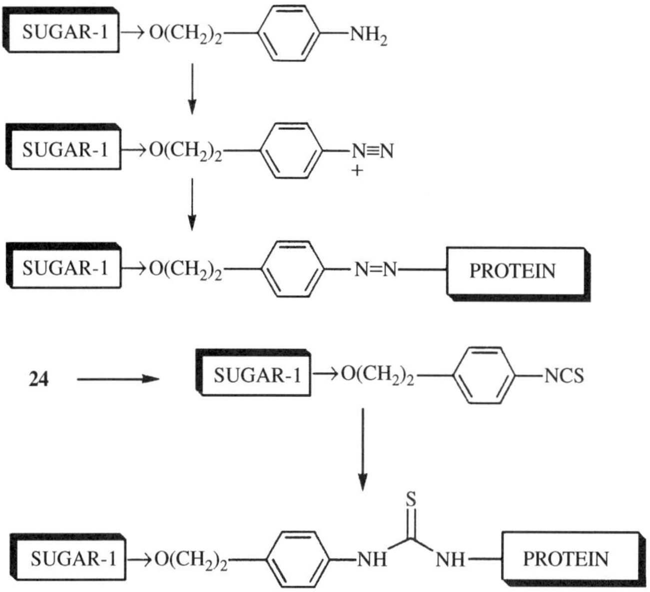

```
SUGAR-1 →O(CH₂)₂—⟨C₆H₄⟩—NH₂                    24
              ↓
SUGAR-1 →O(CH₂)₂—⟨C₆H₄⟩—N≡N⁺                   25
              ↓
SUGAR-1 →O(CH₂)₂—⟨C₆H₄⟩—N=N—[PROTEIN]          26

24  ⟶   SUGAR-1 →O(CH₂)₂—⟨C₆H₄⟩—NCS             27
              ↓
                          S
                          ‖
SUGAR-1 →O(CH₂)₂—⟨C₆H₄⟩—NH—C—NH—[PROTEIN]       28
```

the host. Antiserum induced by conjugates **17** and **18,** which contain an average of two and one saccharide chains per conjugate, respectively, failed to induce immunoprecipitation when mixed with the homologous conjugates. However, conjugate **16,** having only four saccharide chains per BSA, was able to precipitate antidextran antibodies. This finding supports the hypothesis that immunoprecipitation occurs as a consequence of formation of a supramolecular chain between the antibody and the antigen moieties.[78] Interestingly, conjugate **22,** containing an average of 32 saccharide chains per BSA, was only weakly immunogenic. This observation, combined with the low immunogenicity of the conjugates having low saccharide loadings, led to the conclusion that 10–25 oligosaccharide chains per BSA are optimal for eliciting anti-carbohydrate antibodies.[78]

2. Protein Conjugates of Oligosaccharide Fragments of Capsular Polysaccharides

a. *Haemophilus influenzae—H. influenzae* **type b.**—There are six capsulated types of this organism,[87] of which type b (Hib) causes almost all systemic infections, the most dangerous of which is meningitis.[45] The capsular material of Hib is a polymer of the ribofuranose—ribitol—phosphate repeating unit **29** (polyribose—ribitol—phosphate, PRP). Monomers,[88,89] dimers,[90–96] trimers,[90,95–97] a tetramer,[96] pentamers,[94,96–99] hexamers,[94,97,100] a decamer,[101] and an analogue[102] of the pentamer of **29** have been synthesized. In van Boom's solid-phase approach,[92,100] the key starting ribosyl—ribitol moiety **32** was assembled from compounds **30** and **31**. Next, the ribose–ribitol construct **32** was transformed to the succinylated derivative **33** and to the phosphoramidite **34**. Compound **33** was

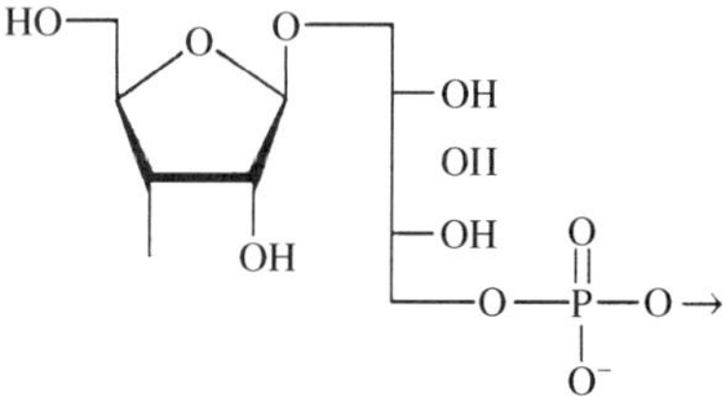

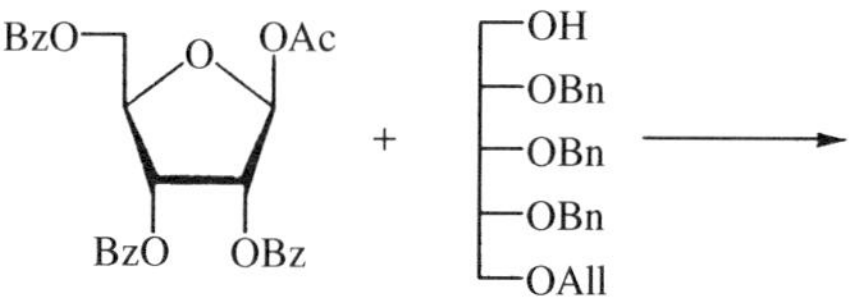

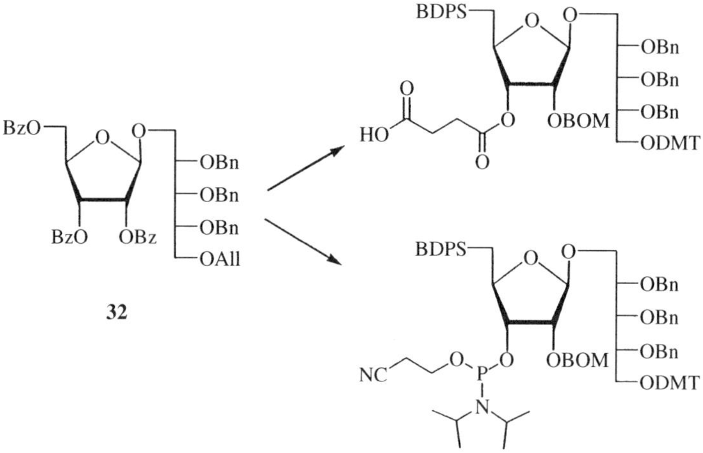

immobilized to amino group-functionalized, controlled-pore glass (**35**), followed by the removal of the protecting group at O-5 of the ribitol moiety to afford **36**. Iterative, multistep chain-extension using the phosphoramidite moiety **34** afforded the hexamer (**37**) of the repeating unit. The synthesis was terminated by the attachment of the aminohexyl linker **38** (→**39**) that served as a spacer between the carbohydrate chain and the protein to minimize steric hindrance. Subsequent release of the oligomers from the solid support and removal of the protecting groups afforded the hexamer of the repeating unit (**40**). Using a similar approach, Kandil[97] prepared oligomers of the Hib capsular material up to a hexamer of the repeating unit **29** and attached them to tetanus toxoid (TT) and to synthetic T-helper cell epitope peptide fragments **41–43** of the outer membrane proteins

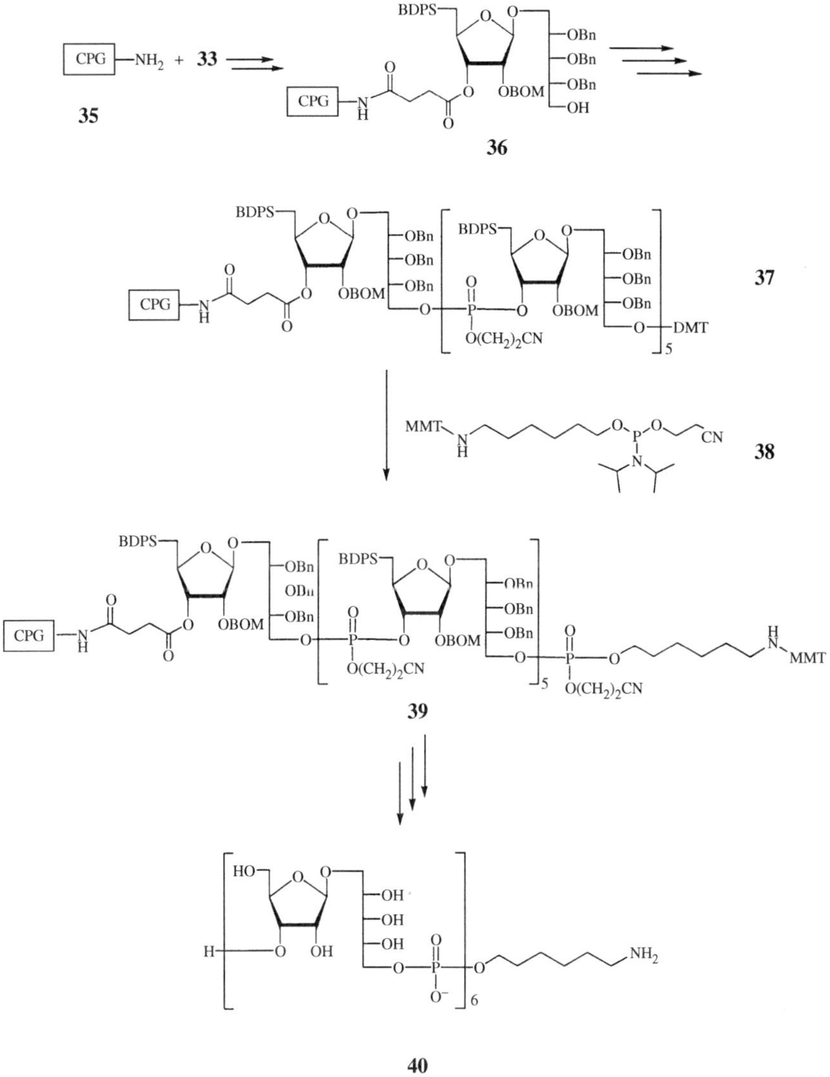

CYAKAQVERNAGLIADSVKDNQITSALSTQ **41**

CDIVAKIAYGRTNYKYNESDEHKQQLNG **42**

VKGYLAGYLAGKGVDAGKLGTVSYGC **43**

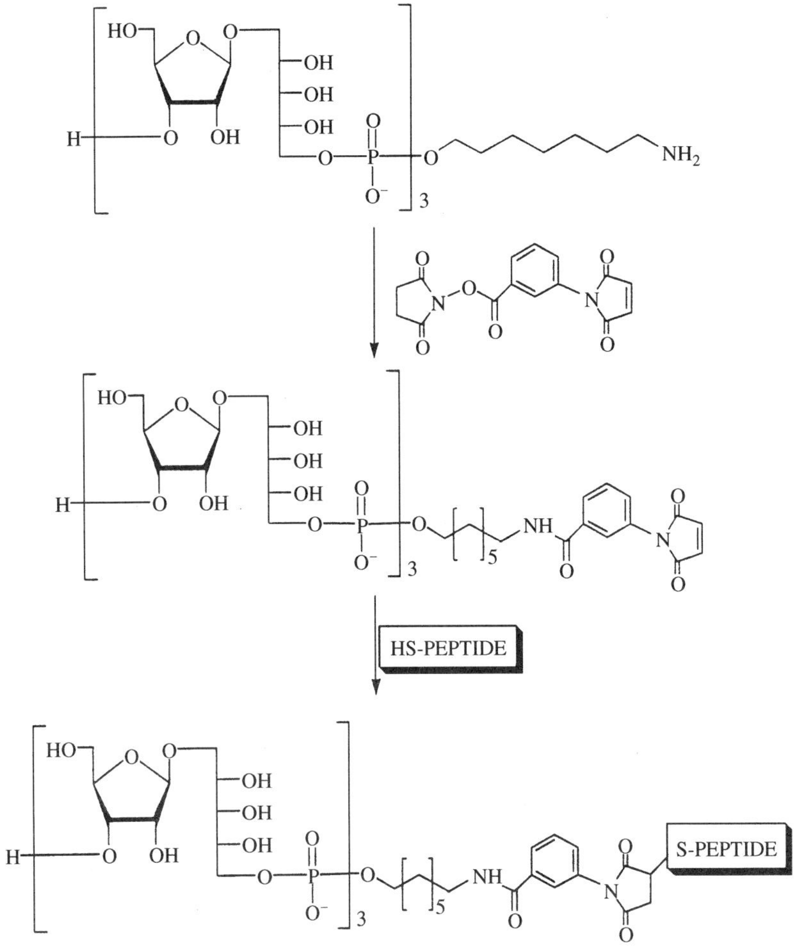

of Hib.[94] In the conjugation process, the saccharide–spacer construct **44** is treated with the heterobifunctional linker **45** to introduce a maleimidoyl group, through which the intermediate **46** so obtained is attached to the sulfhydryl group of immunogenic peptide or protein **47** to afford the conjugate **48**. In the rabbit model, the TT conjugate of the trimer of **29** elicited antibody responses. The conjugate of the trimer with peptide **41** elicited protective antibody responses in rabbits comparable to those obtained with the first-generation commercial conjugate vaccine ProHibit (Connaught) when formulated with alum.[103] The immunogenicity of the saccharide–peptide conjugates was dependent on the site of attachment on the peptide: the anti-saccharide antibody responses were higher for conjugates in which the hapten was linked at the N-terminus (peptides **41** and **42**). This finding was interpreted in terms of the inability of the antigen-presenting cells to cleave the T-cell epitope peptide from the conjugates in which the saccharide components were attached to the C-terminus of the peptide. A similar observation was reported for T-cell epitope peptide conjugates of a pneumococcal polysaccharide.[104] In rabbits, the constructs containing the dimer of **29** attached to T-cell epitope peptides[103] or to proteins were weakly immunogenic.[105] In the range studied the highest antibody responses were observed with peptide conjugates of the synthetic trimer of **29**. Surprisingly, conjugates of the pentamer and the hexamer of **29** with the peptides **41–43** failed to elicited antibody levels higher than those observed with the conjugates of the trimer. This finding has been interpreted as being due to an adverse effect of the hapten moiety on the antigen processing and by its steric hindrance, which may mask the epitope determinants on the peptide.[103] The immunogenicity of the glycopeptide conjugates was enhanced in rabbits when multiple copies of the T-cell epitope peptide—saccharide constructs were presented on a polylysine backbone. The anti-polysaccharide IgG titers obtained with these constructs were comparable to or higher than those obtained with the commercial vaccine ProHibit.[103]

Diphtheria toxin and TT conjugates of synthetic di- to tetra-mers of the repeating unit **29** were prepared using either the thioether chemistry[105] or a random activation approach employing glutaric dialdehyde as the linker.[106] In the first phase of the thioether approach, a sulfhydryl group is introduced in hapten **49** by N-acylation with the active ester **50** (→**51**) followed by hydroxylamine treatment. *In situ* reaction of the thiol **52** so prepared with the N-bromoacetylated carrier protein **53** affords the conjugate **54**, having stable thioether and amide linkages between the sugar and the protein components. The composition of the Hib oligomer–protein conjugates prepared by the thioether approach is shown in Table II. In mice, the TT conjugate of the trimer prepared by the random activation approach induced higher primary and secondary responses than the conjugate of the dimer.[106] Similarly, the TT conjugate of the tetramer (**58**) made by the thioether chemistry elicited consistently higher IgG titers than the trimer conjugates **55** and **56** after both the first and second immunizations. In monkeys, the tetrameric conjugate **58** and the licensed vaccine HibTiter, which contains an average of 20 repeating units, were equally immunogenic, whereas the TT conjugate **56** containing the trimer was less efficient, notwithstanding its higher saccharide/protein ratio.[105] Tetanus toxoid and meningococcal outer-membrane protein conjugates of a synthetic pentamer of **29** elicited anti-Hib polysaccharide antibodies in male mice at a level higher than did the Hib polysaccharide–diphtheria toxoid conjugate.[99] Using conjugates containing 4–12 repeating units of the Hib capsular material and a genetically toxoided mutant (CRM197, a cross-reacting mutant) of DTx, Anderson et al.[54] concluded that in humans the length of the hapten and its exposed terminus are less critical for anti-polysaccharide immunity than the extent of the hapten loading of the conjugate.

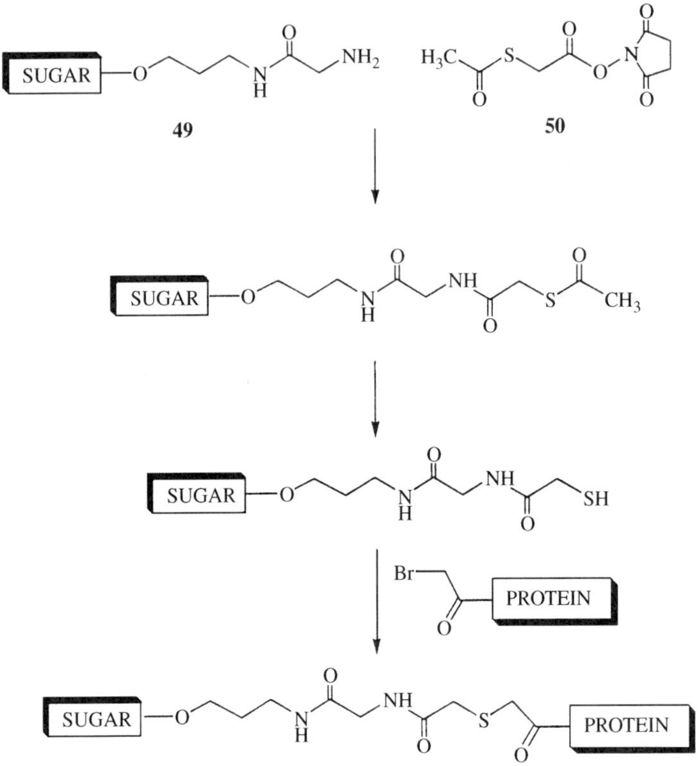

b. *Streptococcus*—Group A.—The cell-wall polysaccharide of the gram-positive group A *Streptococcus* is a polymer of the branched trisaccharide **59**. Pinto and others[107–113] reported chemical synthesis of a number of its di- to hexa-saccharide components with and without a linking moiety, as well as their protein conjugates. Typical features of these syntheses are illustrated in the construction of the pentasaccharide **70**, for which di- and a tri-saccharide blocks were first constructed.[109]

TABLE II
Toxoid Conjugates of Trimers and Tetramers of the Hib Polysaccharide[105] 29

Conjugate Number	Conjugate Composition[a]	Saccharide/Protein Ratio (mol/mol)
55	(RRP)$_3$-DTx	9.9
56	(RRP)$_3$-TT	21
57	(RRP)$_4$-DTx	6.5
58	(RRP)$_4$-TT	5.3

[a] RPR, ribose–ribitol–phosphate; DTx, diphtheria toxin; TT, tetanus toxoid.

Thus, the O-benzylated glucosamine derivative **60** was condensed with the spacer-linked L-rhamnoside **61**, followed by the removal of the base-labile protecting group at the site of the chain extension to afford the disaccharide acceptor **62**. The trisaccharide block **67** was prepared in a similar approach. Thus, condensation of the glucosamine derivative **63** with rhamnoside **64** afforded the disaccharide

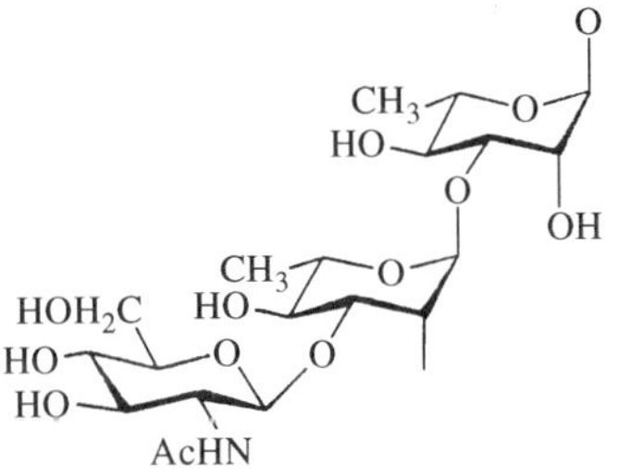

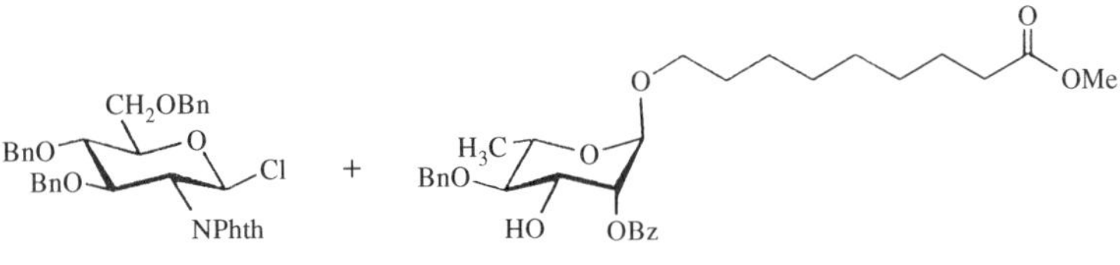

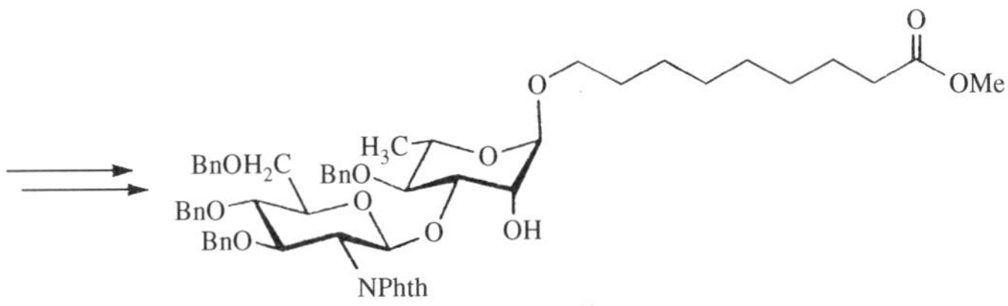

65 in which the allyl group was replaced by chlorine to afford the glycosyl donor **66**. This was condensed with a second rhamnose unit **64** to provide the

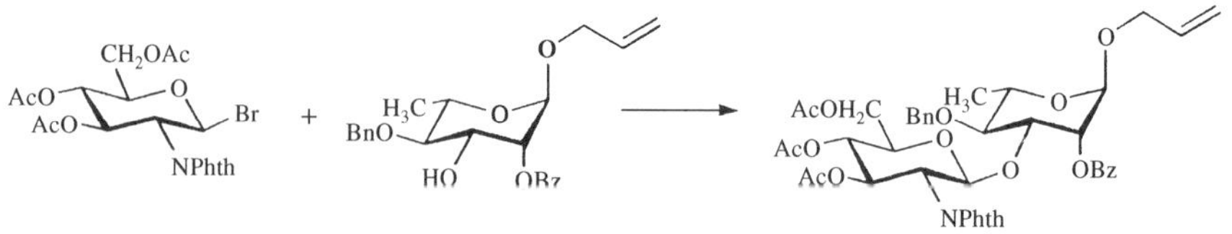

trisaccharide **67**. Next, compound **67** was converted into the glycosyl chloride **68**, which was condensed with the disaccharide block **62** to afford the pentasaccharide **69**, and this was deprotected (→**70**) with subsequent conversion into hydrazide **71**. Covalent attachment of the pentasaccharide **71** to BSA was achieved by the acyl azide method[114] involving treatment of a hydrazide (**72**) with N_2O_4 to form an intermediate acyl azide (**73**), which is allowed to react *in situ* with BSA to afford glycoconjugate **74**. BSA conjugates of the frame-shifted trisaccharide repeating-units **75** and **76** were also prepared, having an average incorporation level of 8–18 mol of saccharides per mol of BSA.[115] Polyclonal antibodies, elicited in rabbits by repeated injections of the conjugates prepared in Freund's complete adjuvant, were highly specific for the homologous immunogen. Antibodies to the linear trisaccharide **75** were equally well recognized by the homologous trisaccharide **75** and by the pentasaccharide **71**, whereas they reacted poorly with the branched trisaccharide **76**. Because of the presence in the pentasaccharide **71** of both epitopes present in the trisaccharides **75** and **76**, polyclonal antibodies raised against the trisaccharide conjugates equally well recognized the branched pentasaccharide antigen. Only the antibodies raised against the pentasaccharide conjugates but not the trisaccharide conjugates recognized the native group A streptococcal polysaccharide, indicating the importance of both the size and the presence of the branching epitope that in this system appears to be essential for antibody recognition.

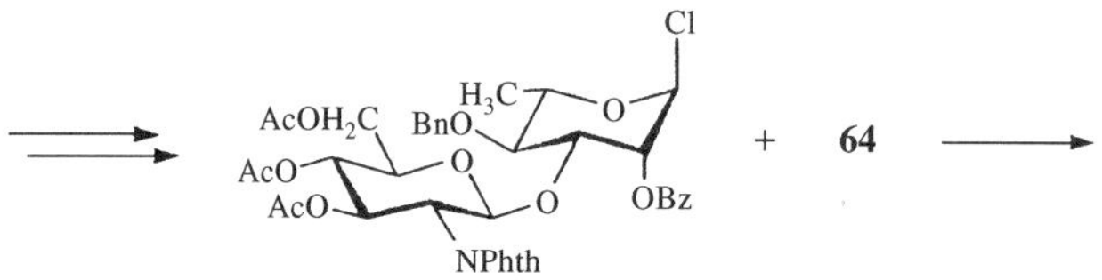

c. *Streptococcus*—*Str. pneumoniae* Type 2.—The capsular polysaccharide (**77**) of this bacterium contains α-linked glucuronic acid as the terminal residue of the side-chain disaccharide.[116] When studying the structural requirements of disaccharides necessary for eliciting anti-type 3 pneumococcal antibodies, Goebel[117] synthesized the *p*-aminobenzyl glycoside of gentiobiuronic acid, a positional isomer of cellobiuronic acid whose presence in the type 3 pneumococcal

67 R = OAll
68 R = Cl

69

70 R = OMe
71 R = $NHNH_2$

72 **73**

74

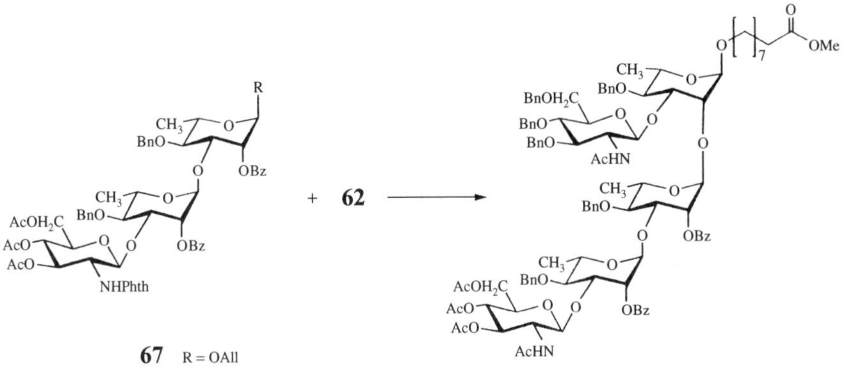

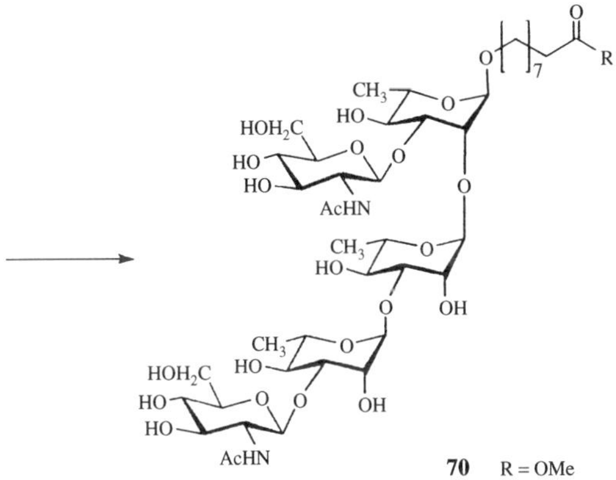

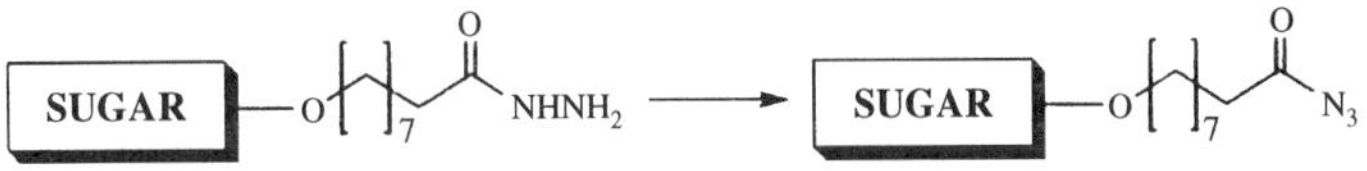

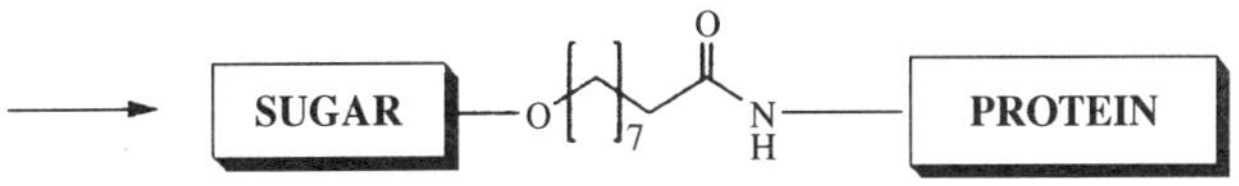

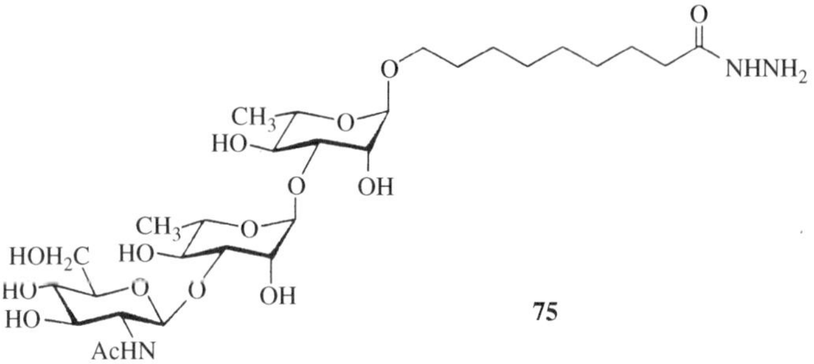

75

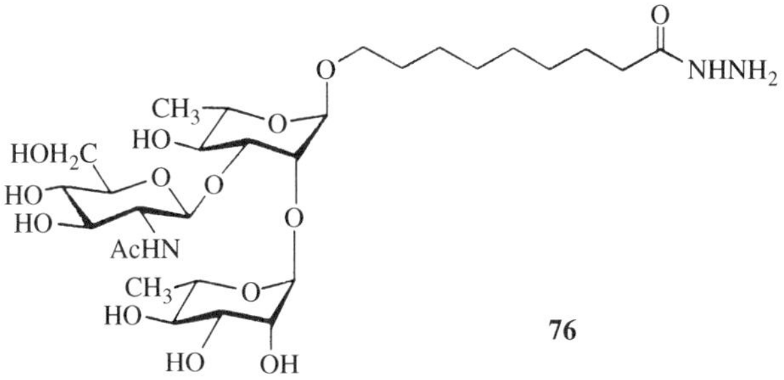

76

3)-α-L-Rha*p*-(1→3)-α-L-Rha*p*-(1→3)-β-L-Rha*p*-(1→4)-α-D-Glc*p*-(1→
         2
         ↑
         1
α-D-Glc*p*A-(1→6)-α-D-Glc*p*

77

polysaccharide was already established at that time. Thus, condensation of the methyl ester **78** and the tetraacetate **79** afforded the disaccharide **80,** which was converted into the glycosyl bromide derivative **81**. Sequential reaction with *p*-nitrobenzyl alcohol followed by removal of the protecting groups yielded the disaccharide glycoside **82,** which was subjected to catalytic reduction to afford the *p*-aminobenzyl derivative **83**. Horse globulin conjugates of gentiobiuronic acid **83,** glucuronic acid **84,** and cellobiuronic acid **85** prepared by the diazo method elicited antibodies in rabbits (without an adjuvant) that precipitated the homologous antigens linked to an unrelated carrier protein.[117] Rabbit anti-gentiobiuronic acid sera conferred passive protection in mice against challenge by type 2 pneumococcus, despite the unnatural anomeric configuration of the uronic acid moiety in the conjugate. This remarkable observation was the first to show that even a monosaccharide can act as a hapten in a protein conjugate to elicit protective antibodies. Interestingly, no agglutinating activity was observed in the sera of rabbits immunized to the gentiobiuronic acid conjugate with either type 2 or type 3 pneumococci. This study concluded that the protection is due to antibodies directed toward the glucuronic acid moiety in the conjugates, and correctly established the presence of a glucuronic acid residue in the type 2 pneumococcal polysaccharide 35 years before the structure of this polysaccharide was elucidated.

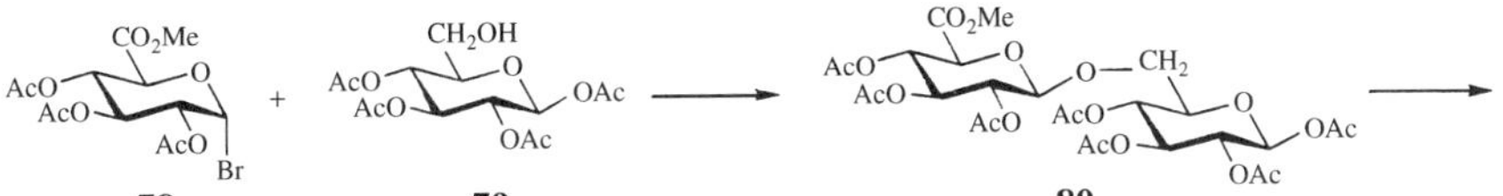

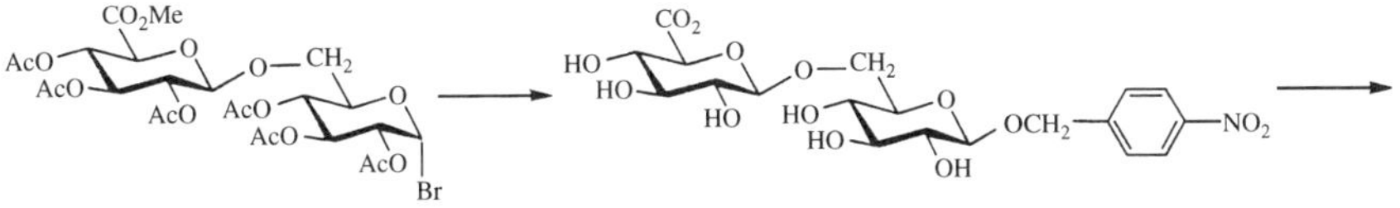

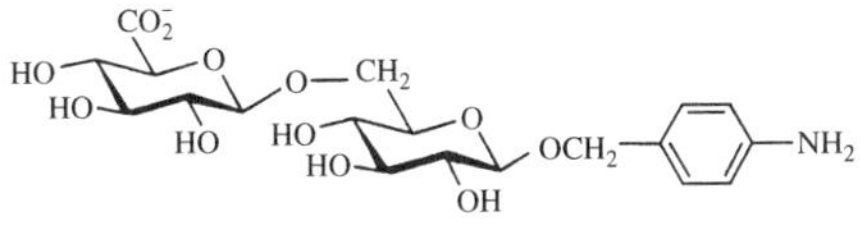

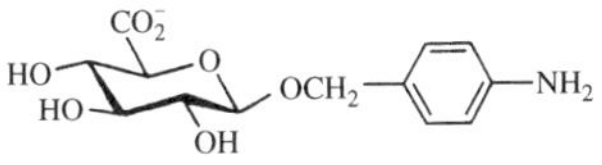

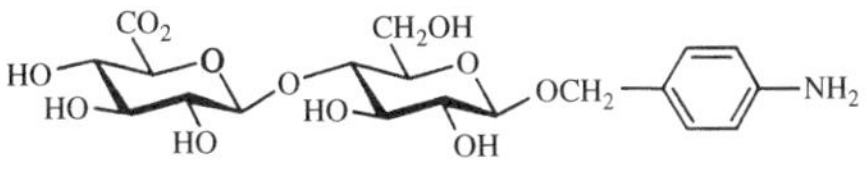

d. *Streptococcus*—*Str. pneumoniae* Type 3.—The first disaccharide conjugated to a protein was derived from the capsular polysaccharide of *Streptococcus pneumoniae* type 3, whose repeating unit is β-(1→3)-linked cellobiuronic acid[117–119] **86**. Controlled acid hydrolysis of the type 3 pneumococcal polysaccharide afforded the disaccharide **87**, which was converted into the methyl ester[120] **88** from which the glycosyl bromide derivative **89** was prepared.[121] Conversion into the *p*-nitrobenzyl glycoside **90** followed by deprotection and catalytic reduction of the nitro group afforded the aminobenzyl glycoside **85**, which was conjugated to horse globulin using the diazo method (p. 8).[118] This material elicited cellobiuronic acid-specific antibodies in rabbits that agglutinated type 3 pneumococcus and precipitated the capsular polysaccharide **86**. The specificity of this reaction was demonstrated by the failure of antibodies raised against horse globulin conjugates of cellobiose, glucose, or glucuronic acid to cross-react with the same polysaccharide.[118] Rabbits immunized with the artificial cellobiuronic acid immunogen developed resistance against challenge by the homologous organism,

3)-β-D-GlcpA-(1→4)-β-D-Glcp-(1→ **86**

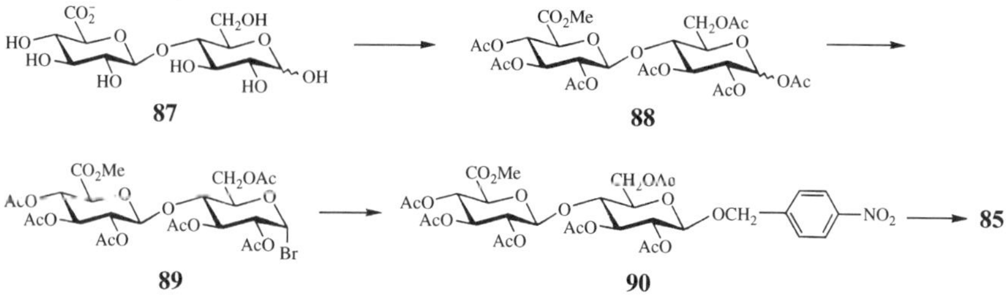

87 → **88**

89 → **90** → **85**

and the rabbit antisera provided passive protection in mice not only against type 3, but also against type 2 and type 8 pneumococci that bear a similar or an identical disaccharide moiety in their respective repeating units. Neither active nor passive immunity could be induced with the corresponding cellobiose conjugate or horse globulin conjugates containing glucuronic acid as the hapten.[119]

In order to study the role of multiple antigenic subunits in oligosaccharide-based conjugate vaccines, Snippe *et al.*[122] isolated a trimer (**91**) of cellobiuronic acid from the type 3 pneumococcal polysaccharide by acid-catalyzed partial depolymerization and condensed it with bovine serum albumin, keyhole limpet hemocyanin, and tetanus toxoid using the method of Svenson and Lindberg.[123] Thus, hexasaccharide **91** was reacted with the amine **92** in the presence of NaCNBH$_3$ to afford the pentasaccharide glycoside **93**, which was treated with thiophosgene. The resulting isothiocyanate **94** was reacted *in situ* with the proteins to yield BSA, KLH, and TT conjugates of the pentasaccharide **93** in which the molar ratio of the saccharide chains to the protein was 9, 6, and 16, respectively. Immunization of mice with the BSA and the KLH conjugates (without an adjuvant) elicited low amounts of

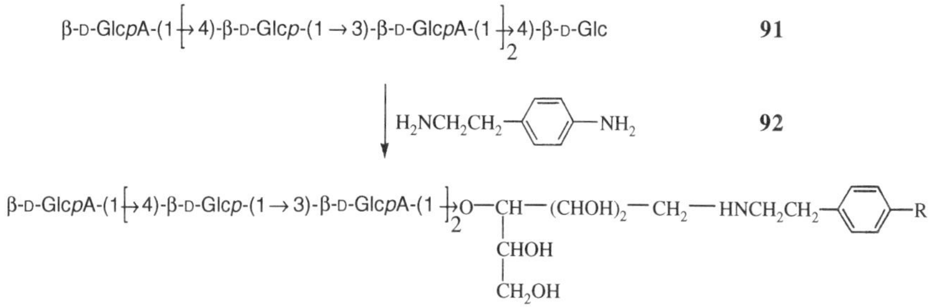

β-D-GlcpA-(1→4)-β-D-Glcp-(1→3)-β-D-GlcpA-(1→4)-β-D-Glc   **91**

↓ H$_2$NCH$_2$CH$_2$—⟨ ⟩—NH$_2$ **92**

β-D-GlcpA-(1→4)-β-D-Glcp-(1→3)-β-D-GlcpA-(1→2 O—CH—(CHOH)$_2$—CH$_2$—HNCH$_2$CH$_2$—⟨ ⟩—R
                                                        |
                                                        CHOH
                                                        |
                                                        CH$_2$OH

93 R = NH$_2$
94 R = NCS

anti-type 3 polysaccharide specific IgM antibodies. The importance of the carrier protein was demonstrated by the observation that although one additional injection only was needed to elicit high amounts of IgG antibodies for the group treated with the KLH conjugate, repeated booster injections were necessary to augment the IgG levels with BSA conjugate. Antisera raised against the native polysaccharide had no detectable levels of IgG antibodies and repeated injections failed to increase the IgM antibody level. After only one injection, the BSA conjugate provided better protection in mice against challenge with live bacteria for 3 weeks after immunization than either the KLH or the TT conjugates. The protection could be enhanced by repeated injections. For example, a single booster injection of the KLH conjugate resulted in complete protection against challenge with 25-fold of mean lethal dose of live bacteria for at least 23 weeks. Passive protection in mice by antisera elicited by the KLH conjugate was superior to that provided by the antisera raised against the unconjugated polysaccharide. Because the antisera raised against the latter contained no or only very low levels of IgG, the passive-protection experiment demonstrated that antisera containing both IgM and IgG antibodies offer better protection than those lacking IgG.

e. *Streptococcus—Str. pneumoniae* **Type 17F.**—To study the immunogenicity of small oligosaccharide fragments of the capsular polysaccharide **95** of type 17F pneumococcus as part of carbohydrate–protein conjugates, Veeneman *et al.*[124] synthesized its overlapping di-, (**96**) tri-, (**97**) and tetra-saccharide fragments (**98**).

In the approach to **98**, condensation of the thiorhamnoside donor **99** with D-arabinitol **100** followed by the removal of the allyl group afforded the alcohol **101** that was phosphitylated with the reagent **102**. Reaction of the resulting phosphoramidite **103** with the disaccharide **104** afforded an intermediate phosphitetriester that was oxidized to afford the phosphodiester **105,** which was deprotected to give **98**. KLH conjugates containing 484, 1017, and 1654 chains of the di-, tri-, and tetrasaccharides, respectively, were prepared by the thioether approach (p. 11).[105] When injected in female mice without an adjuvant, only a low anti-polysaccharide IgG response was obtained. Inclusion of the adjuvant Quil A significantly enhanced the titers for the di- and the trisaccharide, but not for the tetrasaccharide conjugates: the di- and trisaccharide conjugates used with the adjuvant offered protection against challenge with 25-fold mean lethal doses of *Str. pneumoniae* type 17B, whereas no or little protection was provided by the tetrasaccharide conjugate. Surprisingly, the tetrasaccharide conjugate was not antigenic when tested with antibodies raised against the native polysaccharide **95**. This, and the failure of the tetrasaccharide conjugate to induce protecting antibodies was rationalized by a conformation for this unit that is distinctly different from the conformation of this sequence in the native polysaccharide and from the partial conformations in its di- and trisaccharide fragments.

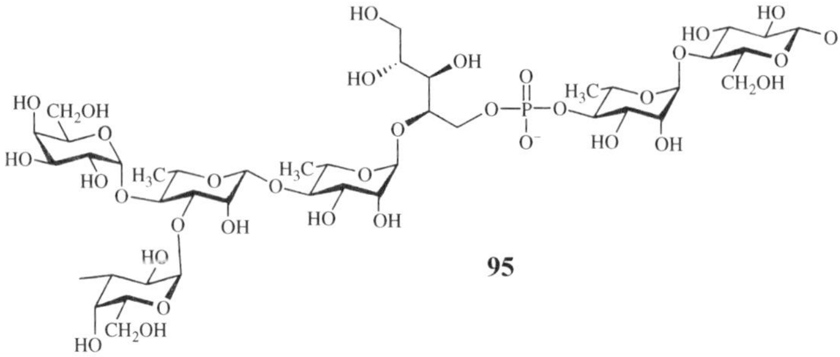

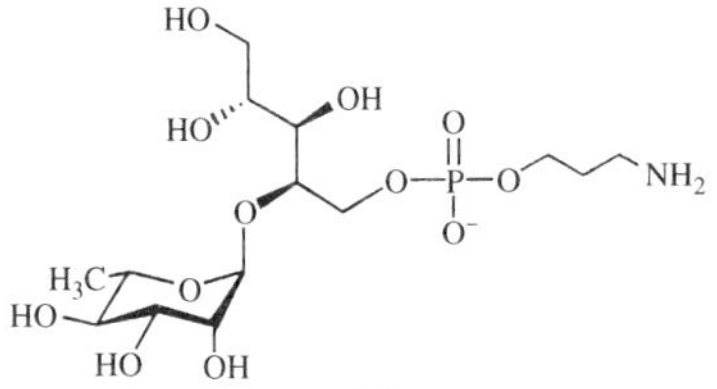

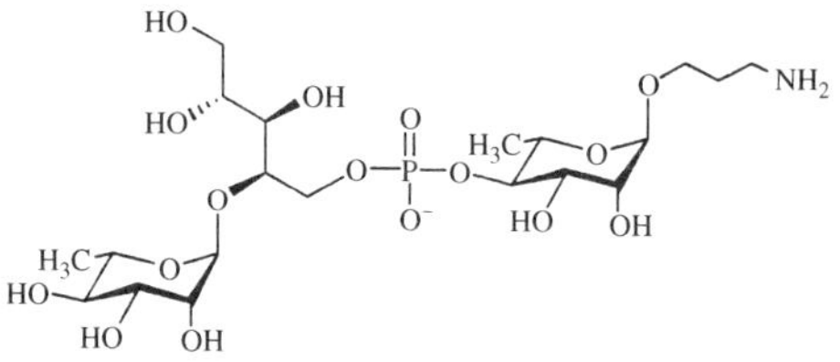

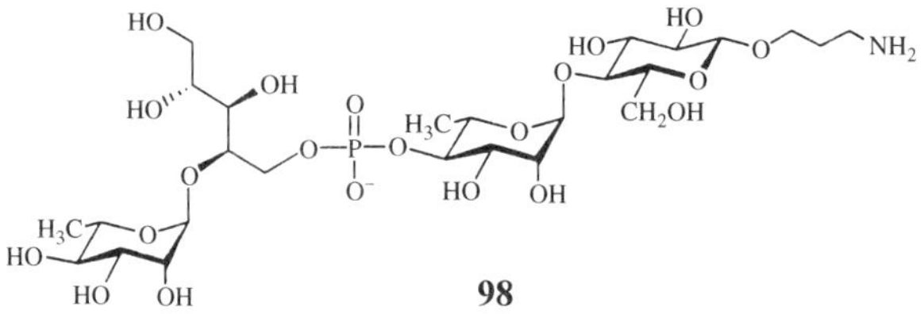

f. *Streptococcus*—*Str. pneumoniae* Type 23F.—In order to probe the epitope specificity of antibodies induced by oligosaccharide fragments of capsular polysaccharides, de Velasco et al.[125] studied the immunogenicity of synthetic tri- and tetra-saccharide fragments[126,127] **107, 108** of the type 23F pneumococcal polysaccharide **106** as part of KLH conjugates. In the synthesis of **108**, the key compound was the tetrasaccharide thioglycoside **113** that was assembled from the monosaccharide precursors **109–112**. Next, this intermediate was coupled with the spacer **114** followed by regioselective deprotection to afford the alcohol **115**, which was coupled with glycerol phosphonate **116**. Oxidation of the resulting intermediate afforded the phosphodiester **117**, which was subjected to deprotection steps to afford the aminopropyl glycoside[126] **108**. Covalent attachment of **107** and **108** to keyhole limpet hemocyanin was achieved by using thioether chemistry to afford conjugates that contained an average of 940 and 1163 chains of the tri- (**107**) and

175

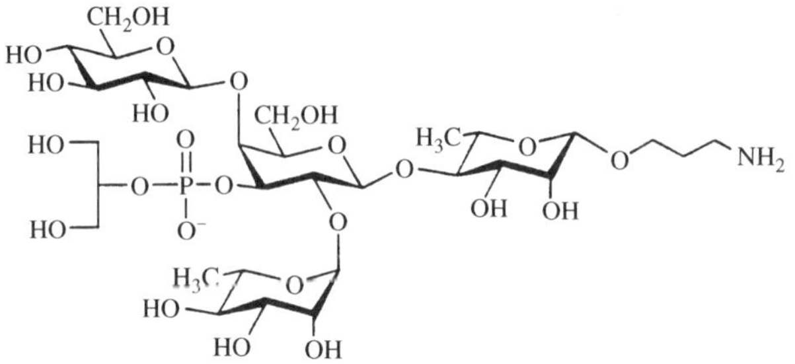

108

tetra-saccharide (**108**) per KLH molecule, respectively.[125] In rabbits, the tetrasaccharide conjugate elicited anti polysaccharide-specific antibodies when injected in Freund's complete adjuvant, whereas no antibody could be elicited in mice with the same conjugate. The weak immunogenicity of the trisaccharide conjugate in rabbits was attributed to the lack of the side-chain rhamnose epitope. That this unit was the immunodominant residue was demonstrated by inhibition experiments: The interaction between antibodies raised in rabbits against the native polysaccharide, its KLH conjugate, or the tetrasaccharide conjugate and the homologous polysaccharide were equally well inhibited by L-rhamnose regardless of the immunogen, indicating no major differences in the epitope specificity of antibodies raised against the native polysaccharide and the synthetic tetrasaccharide. In contrast to the rabbit immunesera, interaction between the native type 23F polysaccharide and human sera raised against this polysaccharide could only moderately be inhibited by oligosaccharides and not at all by monosaccharides, leading to the suggestion that human IgG possesses a larger antibody binding-site than the rabbit IgG. Alternatively, the human anti-polysaccharide antibody may recognize a conformational epitope present on the polymeric form of the repeating unit but not on its oligosaccharide fragments. This study was the first that documented the need for evaluating experimental vaccines in humans by demonstrating the differences in epitope specificity between the rabbit and the human anti-polysaccharide antibodies.[125]

g. *Klebsiella*—Type 2.—A series of monomer (**118**), dimer (**119**), and trimer (**120**) of the repeating unit of the capsular polysaccharide of type 2 *Klebsiella* were prepared by Geyer[128] by bacteriophage-associated *endo*-glycanase-catalyzed hydrolysis of the native polysaccharide. Oligosaccharides **118–120** were covalently linked to proteins by the diazo method (p. 8). The edestin conjugates so prepared contained an average of 40 tri-, 49 hepta-, and 10 undeca-saccharide chains per protein molecule, the reducing-end residue having been converted into a polyhydroxyalkyl moiety. In a similar manner the oligosaccharides were coupled to hemocyanin, yielding conjugates having 1700, 575, and 980 chains of **118, 119,** and **120,** respectively. Immunization of rabbits with the edestin conjugates in Freund's

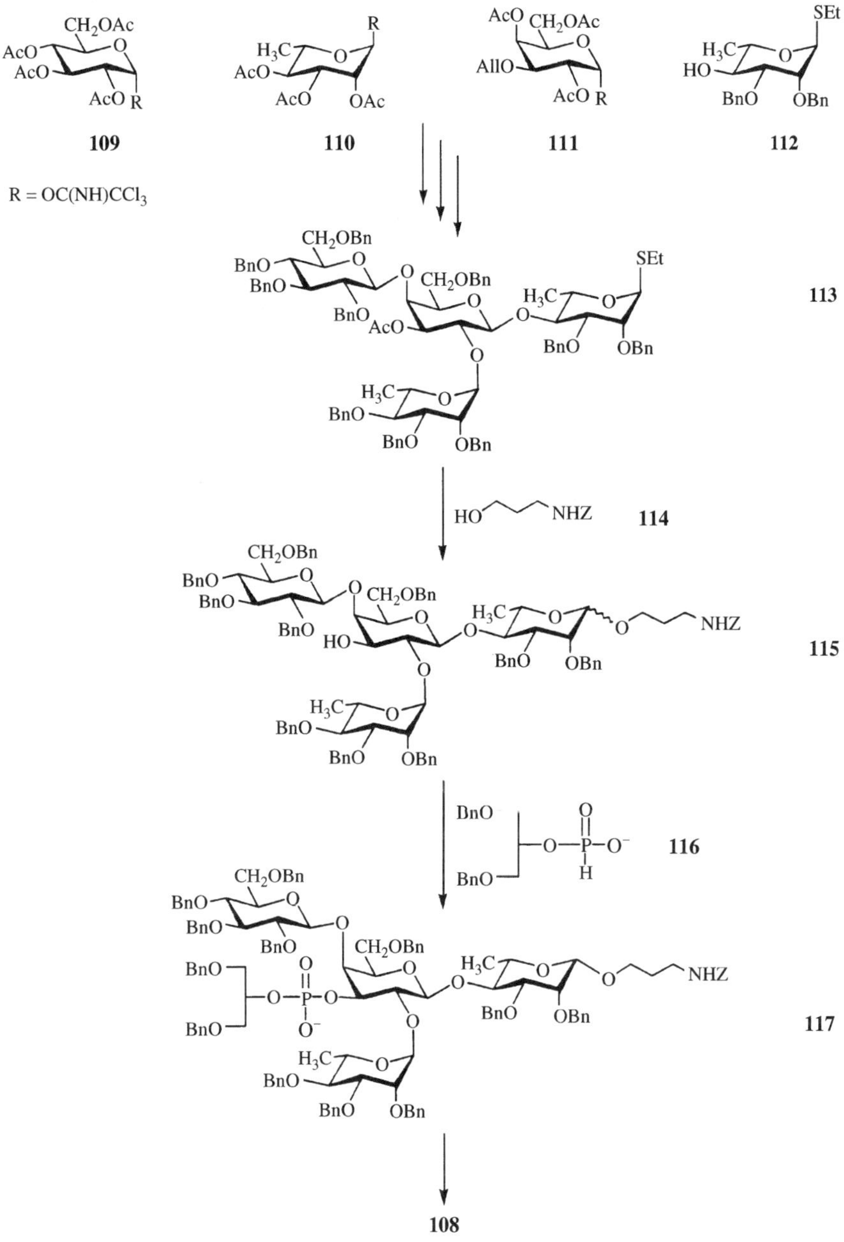

OLIGOSACCHARIDE–PROTEIN CONJUGATES

109 R = OC(NH)CCl₃

110

111

112

113

114 HO~~~NHZ

115

116

117

108

$$\begin{bmatrix} 4)\text{-}\beta\text{-}\text{D-Man}p\text{-}(1\to 4)\text{-}\alpha\text{-}\text{D-Glc}p\text{-}(1\to 3)\text{-}\beta\text{-}\text{D-Glc}p\text{-}(1\to \\ \uparrow 3 \\ 1 \\ \alpha\text{-}\text{D-Glc}p\text{A} \end{bmatrix}_n \quad \begin{array}{ll} \mathbf{118} & n=1 \\ \mathbf{119} & n=2 \\ \mathbf{120} & n=3 \end{array}$$

complete adjuvant three times at monthly intervals elicited polysaccharide-specific antibodies, as determined in passive hemagglutination assays. Both the octa- and the dodeca-saccharide conjugates elicited high titers after the first vaccination. The initially weak response to the conjugate of the monomeric repeating-unit could be boosted by a second injection. Interestingly, the hemagglutination titers for all conjugates were of comparable magnitude during the final phase of the immunization course. A major difference was found between the agglutinating ability of the antibodies raised against the different oligosaccharides: whereas the antibodies elicited by the octa- and the dodecasaccharide conjugates agglutinated *Klebsiella* type 2 well, the sera raised against the tetrasaccharide conjugate caused only very slight agglutination. In addition, the precipitin reaction between the native polysaccharide and antibacterial antibodies could be completely inhibited by the octa- **119** and the dodeca-saccharide **120**. In contrast, the tetrasaccharide **118** caused <80% inhibition at most. Rabbit sera raised against the oligosaccharide–edestin conjugates conferred passive protection in mice: the protection achieved by the octa- and the dodeca-saccharide conjugates was higher than that of the tetrasaccharide conjugate.[129] Similar results were obtained in the active immunization experiments. The hemocyanin conjugates of the octa- **119** and the dodeca-saccharide **120** offered significantly higher protection against challenge by the homologous organism than did the tetrasaccharide conjugate. It was concluded that the minimum saccharide length for a "substantial representation" of the serological specificity of the *Klebsiella* B5055 capsular polysaccharide corresponds to two contiguous repeating units and that protein conjugates of oligosaccharide fragments of bacterial polysaccharides can be as potent immunogens as the whole bacterium or its isolated polysaccharide.[128,129]

h. *Klebsiella*—Type 11.—A tetra- (**121**) and an octa-saccharide (**122**) corresponding to one and two repeating units, respectively, of the capsular polysaccharide of *Klebsiella* serotype 11 were isolated by depolymerization of the latter by a bacteriophage-associated glycanase.[130] The octasaccharide **122** was twice as efficient an inhibitor of the precipitin reaction between the polysaccharide and anti-polysaccharide antibodies than was the tetrasaccharide **121**. This finding led to the conclusion that the octasaccharide more closely resembles the antigenic determinant of the polysaccharide than does the tetrasaccharide. The octasaccharide **122** was attached to BSA and KLH using the method of Svenson and Lindberg[123]

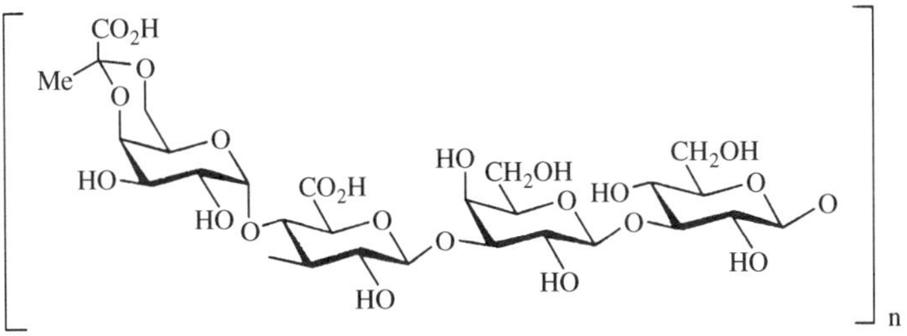

[Structure of compounds 121 (n=1) and 122 (n=2)]

(p. 15) to afford conjugates with BSA and KLH containing 5 and 18 octasaccharide chains, respectively, linked to one molecule of the protein. These loadings correspond to similar saccharide/protein weight-ratios for the two conjugates. The BSA conjugate was used in Freund's complete adjuvant, and the KLH conjugate was injected in combination with bentonite. In BALB/c mice both conjugates induced anti-*Klebsiella* type 11 polysaccharide IgG and IgM antibodies after primary immunization that could be boosted by a second injection to similar levels at 9 weeks, independent of the carrier used. This study confirmed the thymus dependence of the octasaccharide–protein conjugates and demonstrated the importance of the carrier protein in mice whose B cells carry X chromosome-linked immunodeficiency. It was concluded that two contiguous repeating units of the native polysaccharide are necessary to express the serologic specificity of the polysaccharide.[130]

3. Protein Conjugates of Oligosaccharides Related to Lipopolysaccharides

a. Oligosaccharide Fragments of O-Specific Polysaccharides.—

(i) *Salmonella*.—The O-specific polysaccharide of numerous *Salmonella* serogroups feature dideoxyhexose moieties **123–127** as side-chain residues that are the immunodominant monosaccharides defining the serogroup specificity. Albumin conjugates of dideoxyhexoses **123**, **124**, and **127** prepared as their *p*-aminophenyl glycosides[131] **128–130** (Table III) elicited antibodies in rabbits and

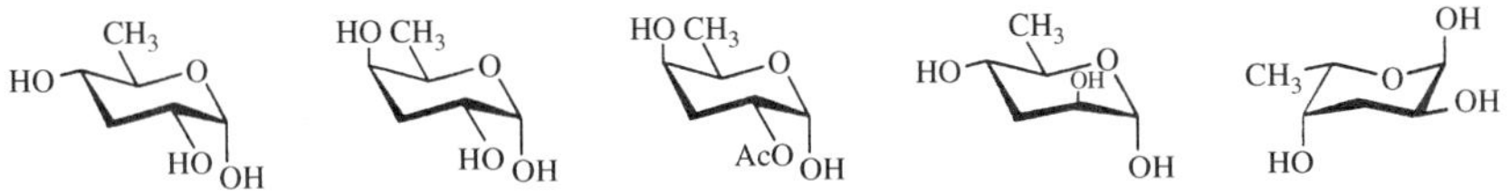

123	124	125	126	127
Paratose	Abequose	2-*O*-Acetyl-abequose	Tyvelose	Colitose
(Factor O:2)	(Factor O:4)	(Factor O:5)	(Factor O:9)	(Factor O:35,50)

goats specific to the haptenic residues.[132,133] The rabbit antibodies precipitated the synthetic immunogen against which they were raised, but not did not react with the homologous polysaccharide and lacked bactericidal activity. In contrast, the goat antibodies, although produced in low titers, agglutinated the homologous bacteria. Antisera raised in rabbits by the albumin conjugate of the 2-O-acetyl-abequose moiety 131 (O:5 specificity) were capable of precipitating the O-specific polysaccharide of *S. paratyphi* expressing the O:5 specificity. No cross-reaction was noted with the O:4 antigenic factor (abequose) that differs from the O:5 only by the absence of the acetyl group in the former. The rabbit anti-O:5 sera agglutinated *Salmonella* bacteria carrying the O:5 but not the O:4 factor.

(ii) *Salmonella.—S. typhi* **and** *S. typhimurium.*—Chemical synthesis of disaccharides (**132–135**, Table III) corresponding to the O-SPs of *S. typhi* (**136**) and *S. typhimurium* (**137**) and their conjugation to BSA were reported by several researchers.[43,86,135–141] The synthesis of the tyvelosyl-mannose disaccharide **133** employed the *xylo*-hexopyranosyl donor **138** and the *p*-nitrophenyl mannoside **139** as the acceptor.[86] Using $Hg(CN)_2$ as the promoter, the α-linked disaccharide **140** was obtained in a very high stereoselectivity, in spite of the participating group at O-2 of the donor. Removal of the protecting groups followed by reduction of the nitro function afforded the *p*-aminophenyl glycoside **133**. This and other disaccharide haptens listed in Table III were conjugated to BSA using the isothiocyanate procedure.[123] The conjugate of **133**, containing four disaccharide chains per conjugate molecule, elicited humoral antibodies specific to the O:9 determinant (**129**) of *S. typhi* in male mice when injected four times 1 week apart at 25 μg doses

TABLE III

Bovine Serum Albumin Conjugates of Mono- and Di-saccharides Corresponding to O-Specific Polysaccharides of *Salmonellae*[a]

Structure of Hapten[b]	Serogroup	Ref.[c]
α-Abe-(1 → O-C_6H_4pNH$_2$ (**128**)	O:4	131, *132, 133*
α-Tyv-(1 → O-C_6H_4pNH$_2$ (**129**)	O:9	131, *132, 133*
α-Col-(1 → O-C_6H_4pNH$_2$ (**130**)	O:35,50	131, *132, 133*
2-O-Ac-α-Abe-(1 → O-C_6H_4pNH$_2$ (**131**)	O:5	*137*, 142
α-Abe-(1 → 3)-α-D-Man*p*-(1 → O-C_6H_4pNH$_2$ (**132**)	O:4	136–138, *143*
α-Tyv-(1 → 3)-α-D-Man*p*-(1 → O-C_6H_4pNH$_2$ (**133**)	O:9	*42, 86*, 139, 140, *143*
α-Man*p*-(1 → 4)-α-L-Rha*p*-(1 → O-C_6H_4pNH$_2$ (**134**)	O:12	135, *143*
α-Par-(1 → 3)-α-D-Man*p*-(1 → O-C_6H_4pNH$_2$ (**135**)	O:2	140, 144, 145, *146*

[a] Conjugations were performed by the isothiocyanate procedure.

[b] Abe = 3,6-dideoxy-D-*xylo*-hexopyranosyl; Col = 3,6-dideoxy-L-*xylo*-hexopyranosyl; Par = 3,6-dideoxy-D-*ribo*-hexopyranosyl; Tyv = 3,6-dideoxy-D-*arabino*-hexopyranosyl.

[c] References in roman type are for synthesis; references in italic type are for conjugation.

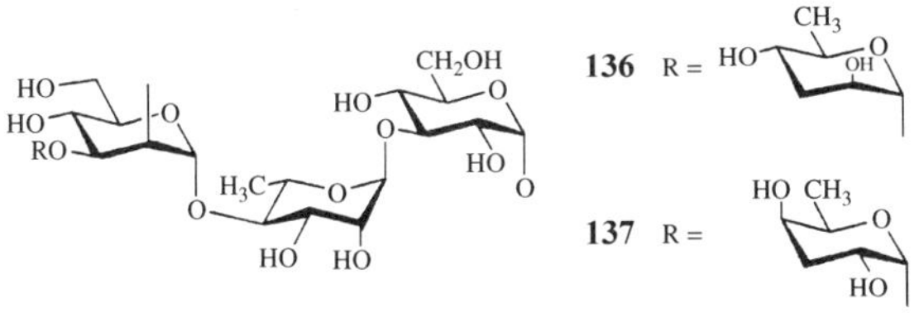

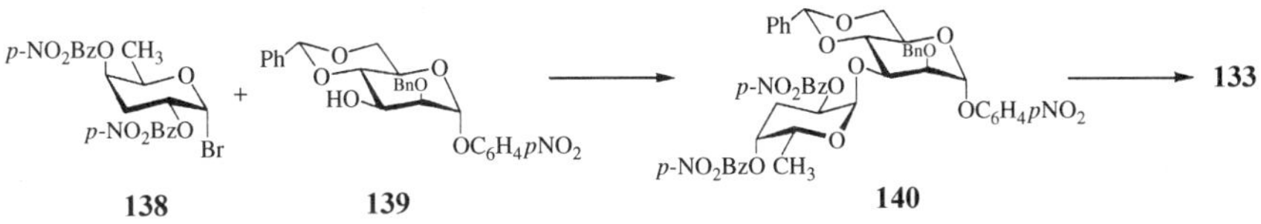

of the antigen, together with Freund's complete adjuvant.[40] Bactericidal activity was detected already 1 week after the initial immunization and reached a maximum in the sera taken 17 days after the last immunization. In a double-challenge experiment using the O:4 and O:9 strains having equal virulence, the growth of the homologous strain (O:9 specificity) was suppressed in comparison to the organism expressing the isomeric O:4-specific polysaccharide **137**. This study was the first to show that antibodies to a disaccharide component of an enteric bacterium may display specific bactericidal activity. In another study, the BSA conjugate of the Abe-Man disaccharide **132** expressing the O:4 specificity failed to induce detectable levels of anti-carbohydrate antibodies in mice.[147,148]

Rabbits immunized with BSA conjugates of disaccharides **132** and **134** (saccharide loading: 21 disaccharides per BSA in each) in Freund's complete adjuvant three times (10 μg of the synthetic antigen each time), elicited predominantly IgG antibodies against the haptenic disaccharides.[143] These sera had high complement-mediated bactericidal titers with no cross-reactivity detectable between serogroups O:4 and O:9. Although the anti-*Salmonella* specificity of the sera produced by the these immunogens exceeded that of those obtained after immunization with heat-killed bacteria, the serum hemagglutination and bactericidal titers elicited by the synthetic conjugates were lower than those obtained by whole bacteria. The immune response to the carrier protein BSA was approximately five-fold higher that the anti-hapten titer.[143] The high specificity of the antisera obtained after immunization with the synthetic immunogens has been used for the identification of *Salmonella* bacteria by immunofluorescence.[141,143,146]

A series of tetra- (**141**), octa- (**142**), and dodeca-saccharide fragments (**143**) of the O-specific polysaccharide (**139**) of *S. typhimurium* were prepared by partial degradation of the polysaccharide, using bacteriophage-associated endo-α-L-rhamnosidases.[123,149,150] BSA conjugates of the *S. typhimurium* octasaccharide

$$\begin{bmatrix} \text{3)-}\alpha\text{-D-Gal}p\text{-(1}\to\text{2)-}\alpha\text{-D-Man}p\text{-(1}\to\text{6)-}\alpha\text{-L-Rha}p\text{-(1}\to \\ \uparrow 6 \\ 1 \\ \alpha\text{-Abe} \end{bmatrix}_n \quad \begin{matrix} \mathbf{141} & n=1 \\ \mathbf{142} & n=2 \\ \mathbf{143} & n=3 \end{matrix}$$

142 were prepared by both the aldonate[151] and the isothiocyanate techniques (p. 9).[86] The former differs from the variation proposed by Kabat,[75] (p. 8) in that activation of the carboxyl group is achieved by a water-soluble carbodiimide. The octasaccharide conjugates, when injected in Freund's complete adjuvant in rabbits, elicited antibodies having O:4 (Abe-Man disaccharide) or O:12 (Man-Rha disaccharide) specificities. The titer for the latter was ~35 times higher than that of the O:4 specificity.[123,149] This is in contrast to the relative ratios of the O:4 and O:12 specificities of the serum raised against killed whole bacteria, for which the anti-O:4 titer was 1.6-fold higher than the O:12 titer.[149]

Precipitation of the *S. typhimurium* O-specific polysaccharide by homologous *Salmonella* antisera raised against the whole organism was inhibited by the *S. typhimurium*-derived octa- (**142**) and dodeca-saccharides, (**143**) but not by the tetrasaccharide **141**, indicating that the size of the polysaccharide determinant is larger than a tetrasaccharide and equal to or smaller than the octasaccharide **142**.[152] The specificities of the rabbit antibodies elicited by the octasaccharide conjugate were similar to those elicited by whole bacteria, whereas those elicited by the BSA conjugate of the disaccharide **132** carrying the immunodeterminant abequose residue (O:4 specificity) were different.[123] Reaction between the O-specific polysaccharide of *S. typhimurium* carrying the O:4 determinant and the antibodies against the O:4 determinant Abe-Man disaccharide **125** was more efficiently inhibited by the homologous Abe-Man disaccharide than by the octasaccharide **142**. This finding led to the assumption that the combining site of the antibodies raised against the synthetic disaccharide also accommodates the linking arm or portions thereof, which are absent in the native polysaccharide.[152]

In order to assess the effect of the protein carrier on anti-hapten immunogenicity, conjugates of the *S. typhimurium* octasaccharide **142** with diphtheria toxin (ds 13), edestin (ds 120), and the outer-membrane protein complex (porin) of *S. typhimurium* (ds 15) were prepared and evaluated in both mice and rabbits, with or without Freund's complete adjuvant.[153] The highest titers against either the native LPS or the octasaccharide **142** were obtained with the diphtheria toxin conjugate. The synthesis of this conjugate represents the first example of conjugational toxoiding whereby the toxic properties of a bacterial toxin are diminished by the modification of the lysine residues.[123] The resulting conjugate is immunogenic, and is capable of evoking anti-carbohydrate- and anti-toxin-specific antibodies, of which the latter can neutralize the native diphtheria toxin. When administered in Freund's complete adjuvant, all conjugates offered active protection in mice with a >10-fold increase in the LD_{50} value. The protective effect of the anti-octasaccharide antibodies was directly demonstrated by the finding that both the anti-octasaccharide–porin and the anti-octasaccharide–diphtheria toxin antibodies raised in rabbits conferred passive protection in mice against *S. typhimurium* challenge at up to 20-fold LD_{50} doses.[153]

A correlation between the oligosaccharide length and anti-saccharide immunogenicity was demonstrated by the finding that, in rabbits, BSA conjugates of the octa- (**142**) and the dodeca-saccharide (**143**) were about 20- and 200-fold more immunogenic, respectively, than the conjugate of the tetrasaccharide **141** conjugate when administered in Freund's complete adjuvant and assayed against the native LPS.[42] A corresponding trend in antibody response was also observed in mice. It should be noted that the actual size of the intact saccharide chain in the conjugates is shorter by one residue, because of the loss of the reducing-end terminus as a pyranose moiety during the conjugation process. This study also showed the importance of the saccharide loading on the immune response. Increase of the molar ratio of the octasaccharide **142** to BSA from 7 to 23 resulted in a 15- to 22-fold increase in the anti-carbohydrate antibody response in mice. Passive immunization experiments showed that the rabbit antiserum against the octasaccharide–BSA conjugate fully protected mice against challenge by *S. typhimurium* up to 100 times the LD_{50}, adding further support to the hypothesis that "humoral antibodies are important for protection."[42] It was demonstrated in both active and passive immunization experiments in mice and rabbits that the octa- and the dodeca-saccharide conjugates elicited opsonizing antibodies that enhance phagocytosis of O-antigenetically homologous organisms. A correlation of the bacterial clearance rate and the O-antigen-specific antibody titer was also established.[148] It was concluded that an artificial immunogen corresponding to the *S. typhimurium* O-antigen should contain about 20 octa- (**142**) or dodeca-saccharide (**143**) haptens per conjugate molecule.[42] The finding that a human serum albumin conjugate of the dodecasaccharide **143** was able to induce O-antigen-specific pyrogenic tolerance to the native LPS represents a further mechanism by which O-specific polysaccharide-based neoglycoproteins may contribute to antibacterial immunity.[154]

(iii) *Salmonella—S. illinois.*—The tetrasaccharide **144** corresponding to a complete chemical repeating unit of the O-specific polysaccharide of this bacterium was isolated by controlled acid hydrolysis of the O-specific polysaccharide, exploiting the weaker anomeric linkage of the rhamnose residue relative to the other interglycosidic linkages in this polysaccharide.[155] Coupling to edestin by the azo method[79] afforded a conjugate (**145**) in which the reducing-end rhamnose residues became a part of the linking moiety. According to the Kauffmann–White

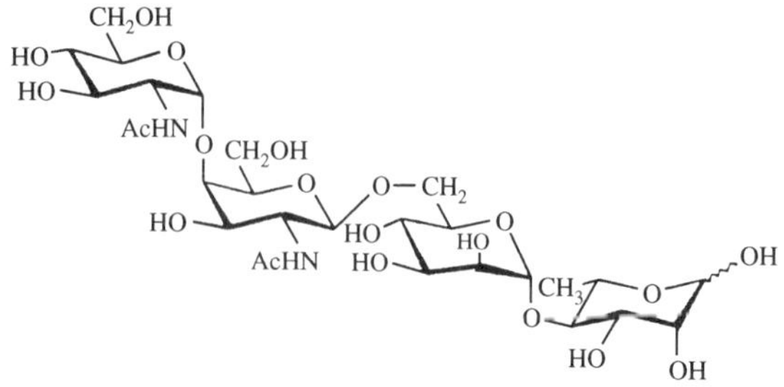

144

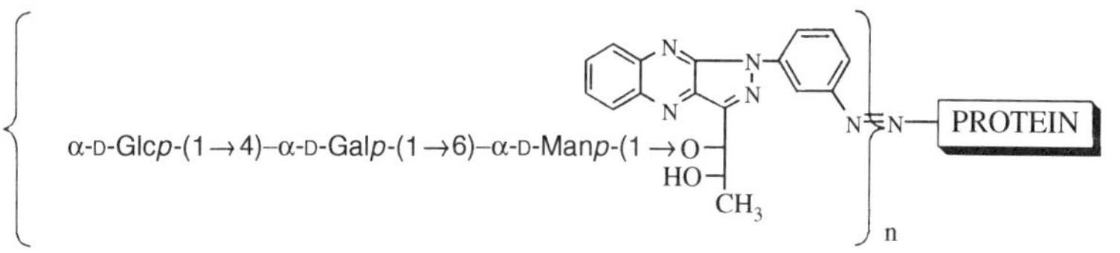

α-D-Glc$p$-(1→4)-α-D-Gal$p$-(1→6)-α-D-Man$p$-(1→O ... N=N—[PROTEIN] **145**

scheme of the classification of *Salmonellae, S. illinois* carries O-specificities 3, 15, and 34. In the antisera obtained by immunization of rabbits with 250 μg of the conjugate **145** in Freund's complete adjuvant most of the O-specificity was directed against the Glc—Gal—Man trisaccharide (O-factor 34), and only slightly against the Gal—Man (O-factor 15) and Man—Rha (O-factor 3) disaccharides. The high specificity of the antibodies was demonstrated by the lack of cross reaction with O-factor 12_2, which differs from O-factor 34 in the anomeric configuration of the Gal residue. This is the first report suggesting the use of protein conjugates of O-polysaccharide fragments for immunization against gram-negative pathogens.[155]

(iv) *Shigella dysenteriae—Sh. dysenteriae* **Type 1.**—The O-specific polysaccharide of *Sh. dysenteriae* type 1 consists of ~27 tetrasaccharide repeating-units **146**. Syntheses of oligosaccharide fragments corresponding up to six contiguous repeating-units with and without a spacer and their covalent attachment to human serum albumin was reported by several researchers.[156–164] The key intermediate in the synthesis of the spacer-equipped hexadecasaccharide **167** and its fragments

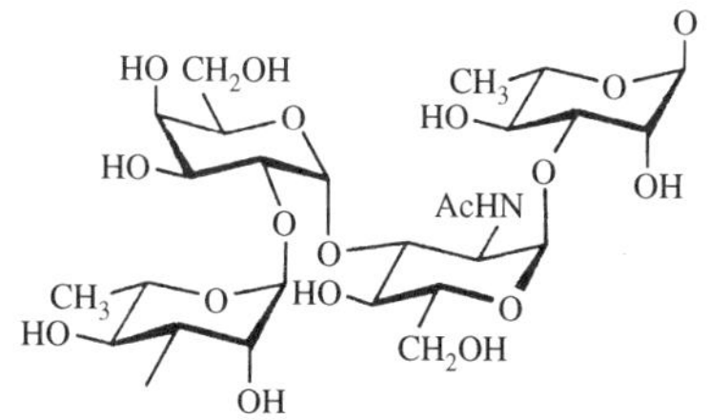

146

164, 165, and **166** was the thioglycoside building block **151,** which was assembled from four appropriately protected and activated monosaccharide precursors **147–150.**[162] The tetrasaccharide thioglycoside **151** was converted into the more reactive imidate **152,** which was linked to the spacer **153** to afford the intermediate **154.** Subsequent removal of the monochloroacetyl group of the rhamnose residue at the non-reducing-end afforded the acceptor **155.** Coupling of the tetrasaccharide donor **152** with the alcohol **155** furnished the dimeric repeating unit **156.** Two iterations of the deprotection–glycosylation sequence afforded the dodeca- (**157**) and hexadeca-saccharides (**158**). In the final stages of the synthesis, the protecting groups were removed followed by hydrazinolysis to afford the oligosaccharides **159–162** corresponding to 1 to 4 consecutive repeating-units of **146.** In order to increase the length of the connecting moiety, the oligosaccharide hydrazides were condensed with a heterobifunctional secondary spacer (**164**) that also introduced an aldehydo moiety in the saccharide constructs to afford constructs **164–167.** These were attached to human serum albumin using reductive amination.[165] The conjugates so prepared (Table IV) differ in two variables of oligosaccharide–protein vaccines: (*i*) the length of the saccharides and (*ii*) the saccharide/protein ratio.[163] The saccharide content of these conjugates was assayed by using MALDI-TOF mass spectrometry, which has a higher degree of accuracy than colorimetric methods. Except for the tetrasaccharide conjugate, all conjugates in Table IV were immunogenic in young outbred mice when injected subcutaneously three times at biweekly intervals at a dose of 2.5 µg of saccharide in the conjugates without an adjuvant.[166] Although the antibody response was poor 1 week after the first injection, almost all animals responded with antibody synthesis after the second injection and a booster response after the third. The anti O-SP IgG levels elicited by the synthetic oligosaccharide–HSA conjugates were significantly higher than those elicited by the conjugate of the native O-specific polysaccharide. If confirmed by clinical trials, it indicates a new approach for manufacturing saccharide-based vaccines. The highest anti-O-SP IgG levels were achieved by the hexadecasaccharide conjugate that contains nine saccharide chains per HSA (Fig. 1). Lower responses were obtained when the loading was either lower (item no. 7, 4 chains) or higher (item no. 9, 19 chains). A similar relationship between saccharide loading and immunogenicity was found for the dodecasaccharide conjugates (items nos. 4–6). The octasaccharide conjugate containing 20 chains (item no. 3) in contrast, was more immunogenic than the conjugate containing 11 chains (item no. 2) (Fig. 1). The complex nature of the influence on immunogenicity of the saccharide length and the loading was revealed by the finding that, at a higher loading, the octasaccharide conjugate containing 20 chains induced higher anti-polysaccharide IgG titers than did the dodecasaccharide conjugate at a lower loading (8 chains). It was concluded that at low levels of saccharide loading, receptor clustering

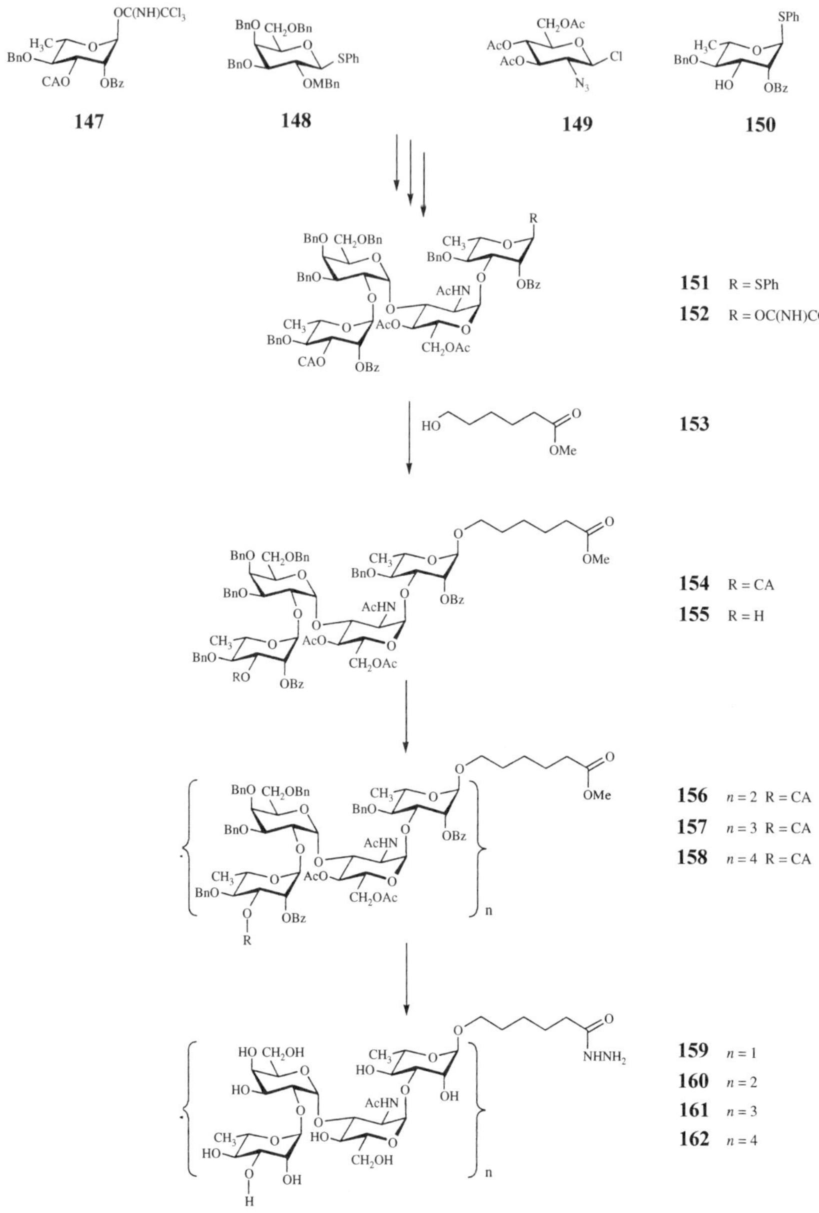

147	**148**	**149**	**150**

151 R = SPh
152 R = OC(NH)CCl$_3$

153

154 R = CA
155 R = H

156 n = 2 R = CA
157 n = 3 R = CA
158 n = 4 R = CA

159 n = 1
160 n = 2
161 n = 3
162 n = 4

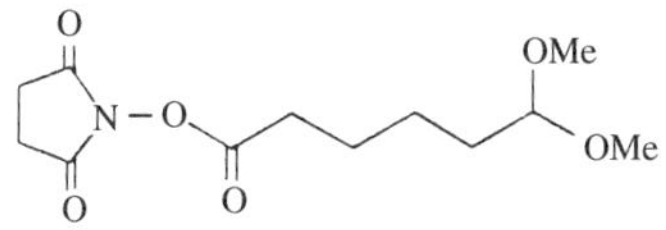

163

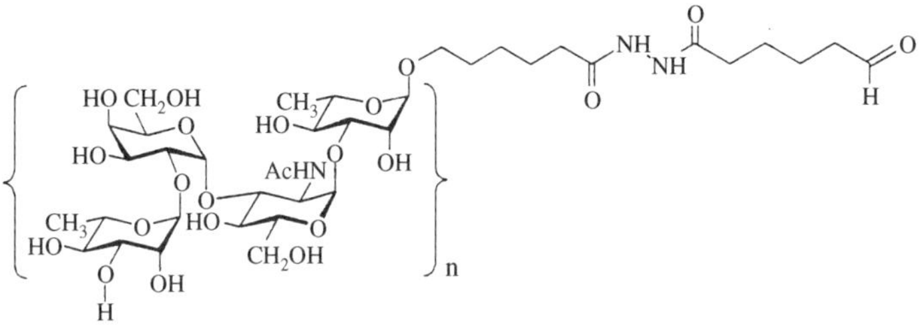

164 $n = 1$
165 $n = 2$
166 $n = 3$
167 $n = 4$

is insufficient for optimal immunogenicity, whereas at high saccharide loading, the T-cell epitopes of the carrier protein are shielded by the saccharide from antigen processing. Optimum immunogenicity of oligosaccharide–protein conjugates can thus be achieved at an intermediate loading defined by the saccharide length.[166]

(v) *Escherichia coli*—*E. coli* **Type O8.**—A heptasaccharide (**168**) was obtained from the O-specific polysaccharide of *E. coli* serotype O8 by bacteriophage-associated *endo*-mannosidase-mediated cleavage[167] and was attached to BSA by

TABLE IV

Human Serum Albumin Conjugates of Tetra-, Octa-, Dodeca-, and Hexadeca-saccharide Fragments of the O-Specific Polysaccharide of *Shigella dysenteriae* Type 1[163]

Item	Conjugate[a]	MW (Da)	Saccharide–Protein Ratio (mol/mol)
1	**IV**-HSA	77,970	13
2	**VIII**-HSA	83,160	11
3	**VIII**-HSA	98,000	20
4	**XII**-HSA	83,340	8
5	**XII**-HSA	89,200	10
6	**XII**-HSA	118,000	23
7	**XVI**-HSA	78,000	4
8	**XVI**-HSA	92,400	9
9	**XVI**-HSA	120,000	19

[a] The numbers **IV, VIII, XII,** and **XVI** stand for tetra-, octa, dodeca-, and hexadeca-saccharide, respectively.

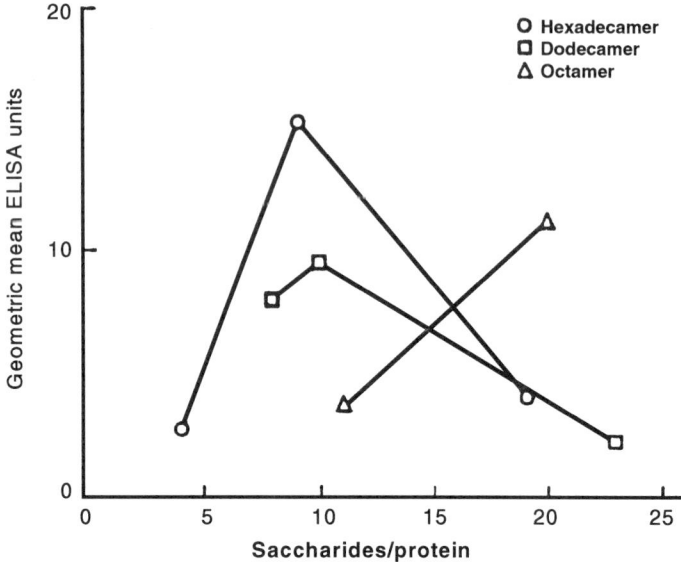

FIG. 1. Geometric mean of serum IgG anti-LPS antibodies (in ELISA units) elicited by HSA conjugates of synthetic oligosaccharide fragments of the O-SP of *Sh. dysenteriae* type 1 in mice 7 days after the their s.c. injection, 2 weeks apart, each injection containing 2.5 μg of saccharides. (Reproduced from *Proc. Natl. Acad. Sci. U.S.A.*, 96 (1999) 5194–5197, V. Pozsgay, C. Chu, L. Pannell, J. Wolfe, J. B. Robbins, and R. Schneerson, with permission from the publisher.)

the method of Svenson and Lindberg[123] to give a conjugate with an average of 20 saccharide chains per protein molecule.[168] When injected in rabbits in Freund's complete adjuvant at a dose of 20 μg of saccharide, the conjugate elicited anti-LPS and anti-oligosaccharide **168** IgG antibodies that could be boosted by repeated inoculations on days 14 and 21. Anti-polysaccharide IgG antibodies were elicited in rhesus monkeys when vaccinated with the conjugate without any adjuvant (100 μg) at monthly intervals. When an arbitrary titer of anti-polysaccharide IgG antibody was reached, the monkeys were infected by *E. coli* O8 inoculated in the kidney. Compared to the controls, the degree of renal scarring was significantly decreased in the vaccinated animals, although the duration of bacteriuria was the same. It was concluded that the decreased renal damage was due to interference of the polysaccharide-specific IgG antibodies with the endotoxin presentation to the host.[168]

$$\alpha\text{-D-Man}p\text{3Me-}(1\rightarrow 3)\text{-}\beta\text{-D-Man}p\text{-}(1\rightarrow 2)\text{-}\alpha\text{-D-Man}p\text{-}(1\rightarrow 2)\text{-}\alpha\text{-D-Man}p\text{-}(1\text{-}]_2$$

168

b. Oligosaccharide Fragments of the Core Region—Meningococcal oligosaccharides.—In contrast to the O-antigens that are specific for the different types of meningococcal bacteria, the inner core regions of the lipopolysaccharides are the same or similar for most immunotypes of *N. meningitidis*. This observation led to the assumption that antibodies directed against the core region might offer protection against several types simultaneously.[169] In order to identify the minimal oligosaccharide structures required for the induction of antibodies against the common meningococcal epitopes, Boons *et al.*[170-172] synthesized di- (**169, 170**) and tri-saccharides (**171, 172**) related to the inner core region of the most prevalent

β-D-Glc$p$-(1→4)-L-α-D-Hep$p$-(1→OCH$_2$CH$_2$CH$_2$NH$_2$ **169**

L-α-D-Hep$p$-(1→3)-L-α-D-Hep$p$-(1→OCH$_2$CH$_2$CH$_2$NH$_2$ **170**

α-D-GlcNAcp-(1→2)-L-α-D-Hep$p$-(1→3)-L-α-D-Hep$p$-(1→OCH$_2$CH$_2$CH$_2$NH$_2$ **171**

β-D-Glc$p$-(1→4)-[L-α-D-Hep$p$-(1→3)]-L-α-D-Hep$p$-(1→OCH$_2$CH$_2$CH$_2$NH$_2$ **172**

serotypes L1, L2, and L3,7,9. Starting material for the trisaccharide **171** was the *manno*-dialdehyde **173** that upon treatment with the reagent **174** afforded the silylated derivative **175** in which the silyl group is a masked hydroxyl group-equivalent.[170] Replacement of the allyloxy group by chlorine afforded the heptose donor **176** that was combined with the spacer **114** followed by selective deprotection at the site of the chain extension to afford the glycoside **177**. The latter was glycosylated with the heptosyl chloride **178** followed by deacetylation (→**179**). Subsequent stereoselective glycosylation with the glucosamine donor **180** furnished the intermediate **181**, from which the trisaccharide **171** was obtained after deprotection and functional group manipulations. Conjugation of the aminopropyl spacer-equipped saccharides **169–170** to the protein tetanus toxoid was achieved by the thioether method (p. 13).[105] The incorporation levels for the di- (**169** and **170**) and the trisaccharides (**171** and **172**) were 13, 26, 22, and 35 saccharide chains, respectively, per tetanus toxoid molecule.[169] Phosphoethanolamine-containing oligosaccharides **182, 183**, and **184** were isolated from the native polysaccharides of immunotypes L2 and L3,7,9, respectively. Compounds **183** and **184** were thiolated, and then coupled to bromoacetylated tetanus toxoid through a thioether linkage to afford conjugates having 20 and 15 chains of the L2 (**183**) and the L3,7,9 oligosaccharide (**184**) per tetanus toxoid, respectively.

In rabbits, all conjugates elicited substantial IgG antibody levels after repeated inoculations, while the IgM levels were low. The conjugate of the L2 immunotype-specific oligosaccharide **183** elicited both L2- and L-3,7,9-specific antibodies but not the L1 structure **182**, and the conjugate of the L3,7,9 oligosaccharide **184** induced antibodies that recognized the L1, L2, and the L3,7,9 structures. Thus, common epitopes are shared by the L2 and L3,7,9 types and by the L1 and L3,7,9 types, but not by the L1 and L2 serotypes. The finding that the conjugate of the branched trisaccharide **172,** but not the linear structures **170** and **171** were able to elicit IgG antibodies that recognized determinants of the L1 and L3,7,9 immunotypes led to the conclusion that the branched trisaccharide **172** represents the minimal structure necessary for inducing cross-reactive antibodies against

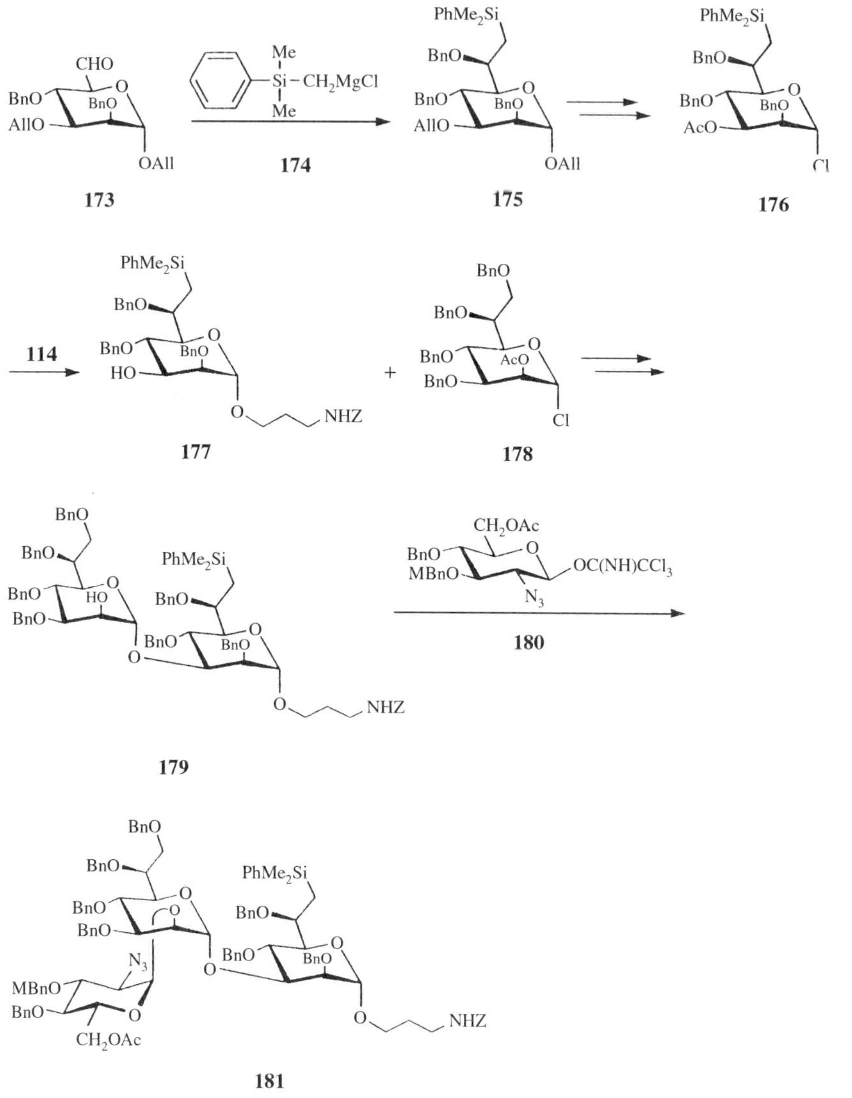

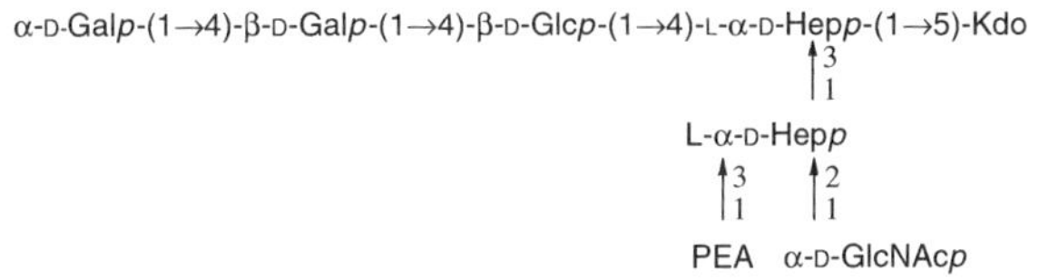

α-D-Gal*p*-(1→4)-β-D-Gal*p*-(1→4)-β-D-Glc*p*-(1→4)-L-α-D-Hep*p*-(1→5)-Kdo
↑3
|1
L-α-D-Hep*p*
↑3 ↑2
|1 |1
PEA α-D-GlcNAc*p*

182 (L1)

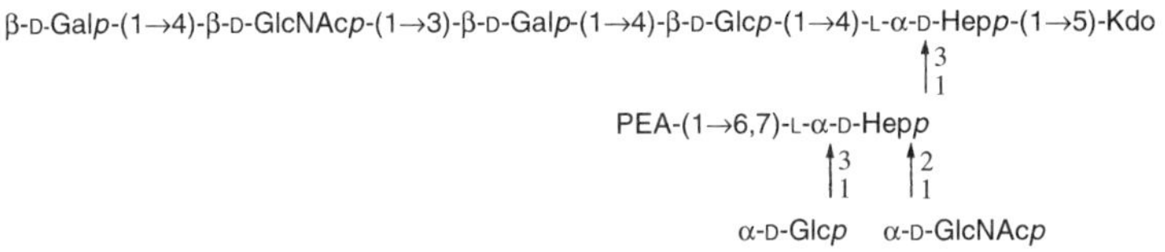

β-D-Gal*p*-(1→4)-β-D-GlcNAc*p*-(1→3)-β-D-Gal*p*-(1→4)-β-D-Glc*p*-(1→4)-L-α-D-Hep*p*-(1→5)-Kdo
↑3
|1
PEA-(1→6,7)-L-α-D-Hep*p*
↑3 ↑2
|1 |1
α-D-Glc*p* α-D-GlcNAc*p*

183 (L2)

β-D-Gal*p*-(1→4)-β-D-GlcNAc*p*-(1→3)-β-D-Gal*p*-(1→4)-β-D-Glc*p*-(1→4)-L-α-D-Hep*p*-(1→5)-Kdo
↑3
|1
L-α-D-Hep*p*
↑3 ↑2
|1 |1
PEA α-D-GlcNAc*p*

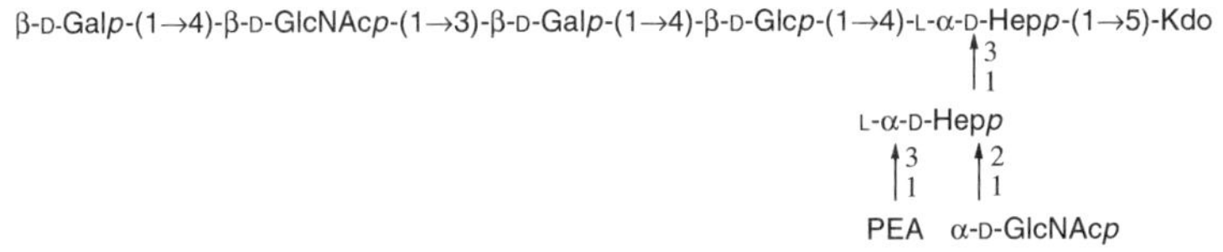

184 (L3,7,9)

several meningococcal immunotypes. Antibodies elicited by the conjugates of the oligosaccharides **169, 170,** and **171** were specific for the L2 immunotype.[169] In mice, the tetanus toxoid conjugates of the L2 (**183**) and the L3,7,9 oligosaccharide (**184**) elicited IgG antibodies only when added together with an adjuvant. These antibodies, although able to recognize epitopes on meningococcal lipopolysaccharide, were not bactericidal for group B meningococcal organisms, leading to the conclusion that it "will be difficult, if not impossible" to induce protective antibodies against multiple meningococcal serotypes simultaneously.[173]

c. Lipopolysaccharides of *Chlamydia*.—The genus-specific lipopolysaccharide of the intracellular gram-negative bacterium *Chlamydia* contains the unique oligosaccharide epitope **185** that is shared by all Chlamydiae.[174] In order to further prove the structure of the lipopolysaccharide and to study the epitope specificity of antibodies raised against the chlamydial lipopolysaccharide, Kosma synthesized di- (**186, 187**), tri- (**188**), tetra- (**189**), and penta-saccharides (**190**) related to **185** as allyl glycosides and conjugated to BSA.[175–178] A key precursor in the trisaccharide synthesis was the Kdo donor **191** that was condensed with the triol (**192**) to give, predominantly, the 6-*O*-linked disaccharide **193**. Subsequent transformations furnished the diol **194,** in which the HO-3 hydroxyl group was selectively glycosylated with the Kdo donor **191** to afford the intermediate **195**. Removal of the protecting group provided the trisaccharide **188**. Conjugation of the oligosaccharide allyl glycosides (**186–190**) to BSA was achieved by Lee's protocol.[179] Accordingly, ligands having the general formula **196** were combined with cysteamine (**197**) to afford amine **198** that was converted into the α-Kdo-(2→8)–α-Kdo-(2→4)–α-Kdo-(2→6)–α-D-GlcpNAc-(1→6)–β-D-GlcpNAc-(1→   **185**

α-Kdo-(2→4)–α-Kdo-(1→ OCH$_2$CH=CH$_2$ **186**

α-Kdo-(2→8)–α-Kdo-(1→ OCH$_2$CH=CH$_2$ **187**

α-Kdo-(2→8)–α-Kdo-(2→4)–α-Kdo-(1→ OCH$_2$CH=CH$_2$ **188**

α-Kdo-(2→8)–α-Kdo-(2→4)–α-Kdo-(2→6)–β-D-GlcpNAc-(1→ OCH$_2$CH=CH$_2$ **189**

α-Kdo-(2→8)–α-Kdo-(2→4)–α-Kdo-(2→6)–α-D-GlcpNAc-(1→6)–β-D-GlcpNAc-(1→ OCH$_2$CH=CH$_2$ **190**

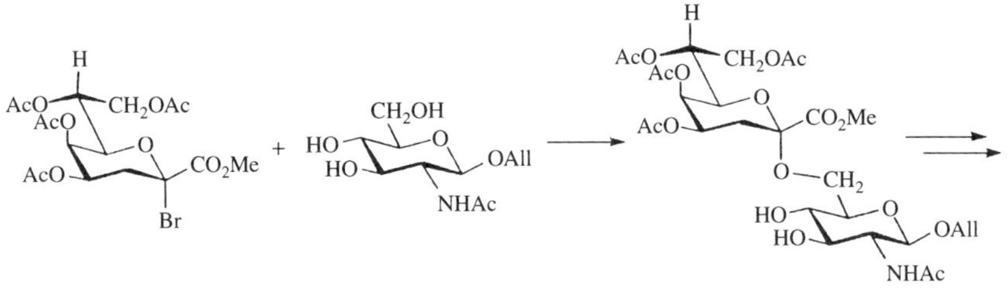

191   **192**   **193**

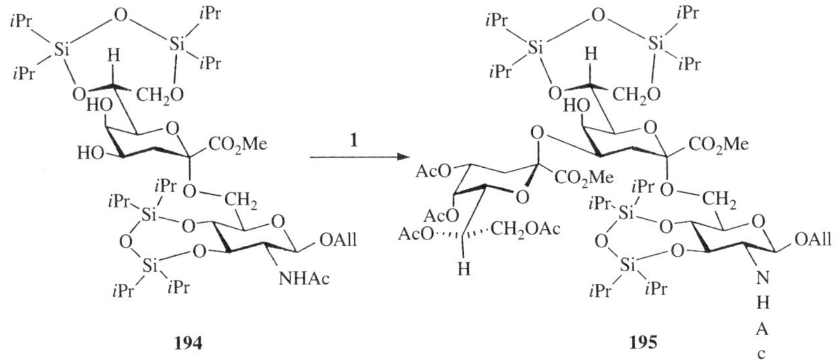

194   **195**

SUGAR-(1→OCH$_2$CH=CH$_2$) $\xrightarrow{\text{HS(CH}_2\text{)}_2\text{NH}_2}$ SUGAR-(1→O(CH$_2$)$_3$S(CH$_2$)$_2$NH$_2$
196   **197**   **198**

⟶ SUGAR-(1→ O(CH$_2$)$_3$S(CH$_2$)$_2$NCS

199

isothiocyanate-derivative **199** by reaction with thiophosgene followed by *in situ* coupling with BSA. The resulting conjugates contained an average of 2.5 to 6 oligo-saccharide chains per BSA molecule.[178] Both the tetra- (**189**) and the penta-saccharide (**190**) conjugate elicited lipopolysaccharide-specific antibodies in mice when injected with Freund's complete adjuvant.[178,180] Selected monoclonal IgG and IgM antibodies against the tetrasaccharide conjugate, prepared by the hybridoma technology, only recognized the complete Kdo–trisaccharide sequence, whereas others exhibited a more diverse antigen recognition pattern that also included Kdo monomer and Kdo-disaccharides.[178] Monoclonal antibodies against the BSA conjugate of the pentasaccharide **190** showed an absolute requirement for the terminal Kdo–trisaccharide segment.[180] One group of these conjugate-induced monoclonal antibodies exhibited 100-fold higher affinity toward chlamydia-specific recombinant lipopolysaccharide as compared to the antibodies elicited by chlamydial elementary bodies. This finding led to the conclusion that synthetic oligosaccharide-containing immunogens may be superior to their natural counterparts.[174]

III. Conclusion

The recognition that surface carbohydrates on the bacterial cell may be utilized as protective antigens made these macromolecules a target for vaccine developments many years ago. Current developments in carbohydrate chemistry may provide access to well-defined fragments of bacterial cell-surface polysaccharides either by specific degradation of the native polysaccharides or by chemical synthesis. As shown by the studies reviewed here, oligosaccharides related to bacterial extracellular polysaccharides hold great promise for the development of vaccines against infectious diseases. If successful in human trials, the synthetic oligosaccharide–protein conjugates may eliminate many of the problems associated with the use of native bacterial polysaccharides or their degradation products, including product heterogeneity and a potential for biological contamination, and may offer the advantage of improved immunogenicity. It has been demonstrated that protein conjugates of oligosaccharide fragments of cell-surface polysaccharides of pathogenic bacteria can elicit antibodies that are recognized by the native polysaccharide. In has also been shown that the IgG antibodies elicited by the oligosaccharide—protein conjugates may offer protection against the homologous organisms in animal models, even if the conjugates are used without an adjuvant.

A study on the *Shigella dysenteriae* type 1 system established the optimum range of saccharide loading in the conjugates. These and other studies have also demonstrated that oligosaccharides as small as a dimer or a timer of the repeating unit of O-specific or capsular polysaccharides might be sufficient for optimum antibody production in animal models. Although the requirements may show some variance for each individual polysaccharide, the findings may be regarded as the basis for the design of oligosaccharide-based conjugate antibacterial vaccines and are all the more encouraging because chemical synthesis of such fragments is already within the scope of numerous laboratories.

Although the efficacy of polysaccharide–protein conjugates in preventing enteric and respiratory infections has been established, it remains to be seen whether the antibody populations elicited by protein conjugates of oligosaccharide fragments of such polysaccharides share the same protective efficacy. It is expected that increasing availability of the experimental vaccine candidates will answer this question in the near future. Because of the lack of reliable animal models for most bacteria pathogenic to humans, human trials will be necessary in the future to verify the efficacy of synthetic carbohydrate—protein conjugates as vaccines. However, conclusions drawn from immunogenicity experiments in which dosage, adjuvant, and route of administration are not clinically acceptable msut be tempered when the experimental vaccines are considered for clinical use.

Acknowledgments

The author thanks Dr. John B. Robbins for his thoughtful review of the manuscript and many stimulating suggestions.

References

(1) L. Kenne and B. Lindberg, in G. O. Aspinall (Ed.), *The Polysaccharides, Vol. 2,* Academic Press, New York, 1983, pp. 287–363.
(2) V. N. Shibaev, *Adv. Carbohydr. Chem. Biochem.,* 44 (1986) 227–239.
(3) B. Lindberg, *Adv. Carbohydr. Chem. Biochem.,* 48 (1990) 279–318.
(4) F. Croakert, M. J. Lismont, M. P. van der Linden, and E. Yourassowky, *Rev. Med. Microbiol.,* 3 (1992) 241–248.
(5) M. Heidelberger and O. T. Avery, *J. Exp. Med.,* 38 (1923) 73–79.
(6) S. J. N. Devi, R. Schneerson, W. F. Vann, J. B. Robbins, and J. Shiloach, *Infect. Immun.,* 59 (1991) 732–736.
(7) C. Adlam, J. M. Knights, A. Mugridge, J. M. Williams, and J. C. Lindon, *FEMS Microbiol. Lett.,* 42 (1987) 23–25.
(8) Y. Wang, J. Huebner, A. O. Tzianabos, G. Martirosian, D. L. Kasper, and G. B. Pier, *Carbohydr. Res.,* 316 (1999) 155–160.
(9) D. N. Taylor, A. C. Trofa, J. Sadoff, C. Chu, D. Bryla, J. Shiloach, D. Cohen, S. Ashkenazi, Y. Lerman, W. Egan, R. Schneerson, and J. B. Robbins, *Infect. Immun.,* 61 (1993) 3678–3687.
(10) J. B. Robbins, R. Schneerson, and S. C. Szu, in M. M. Levine, G. C. Woodrow, J. B. Kaper, and G. S. Cobon (Eds.), *New Generation Vaccines,* Marcel Dekker, Inc., New York, 1997, pp. 803–815.
(11) T. Jr. Francis and W. S. Tillet, *J. Exp. Med.,* 52 (1930) 573–585.
(12) M. Heidelberger, M. M. Dilapi, M. Siegel, and A. W. Walter, *J. Immunol.,* 65 (1950) 535–541.
(13) C. M. MacLead, R. G. Hodges, M. Heidelberger, and W. G. Bernhard, *J. Exp. Med.,* 82 (1945) 445–465.
(14) M. Heidelberger, C. M. MacLoad, and M. M. Di Lapi, *J. Exp. Med.,* 88 (1949) 369–372.
(15) H. C. Neu, *Science,* 257 (1992) 1064–1073.
(16) C. Jones, *Carbohydrates in Europe,* No. 21 (1998) 10–16.
(17) H. J. Jennings, *Carbohydrates in Europe,* No. 21 (1998) 17–23.
(18) M. Cadoz, *Vaccine,* 16 (1998) 1391–1395.

(19) J. B. Robbins, R. Schneerson, and E. C. Gotschlich, *Lancet,* 350 (1997) 1709.
(20) R. M. Douglas, J. C. Paton, S. J. Duncan, and D. J. Hansman, *J. Inf. Dis.,* 148 (1983) 131–137.
(21) D. Wang and E. A. Kabat, in M. H. V. Van Regenmortel (Ed.), *Structure of Antigens, Vol. 3,* CRC Press, Baca Raton, 1996, pp. 247–276.
(22) M. J. Fine, M. A. Smith, C. A. Carson, F. Meffe, S. S. Sankey, L. A. Weissfeld, A. S. Detsky, and W. N. Kapoor, *Arch. Intern. Med.,* 154 (1994) 2666–2677.
(23) F. A. Wyle, M. S. Artenstein, B. L. Brandt, E. C. Tramont, D. L. Kasper, P. L. Altieri, S. L. Berman, and J. P. Lowenthal, *J. Inf. Dis.,* 126 (1972) 514–521.
(24) K. Landsteiner and H. Lampl, *Biochem. Zeitschr.,* 86 (1918) 343–394.
(25) W. F. Goebel and O. T. Avery, *J. Exp. Med.,* 54 (1931) 431–436.
(26) O. T. Avery and W. F. Goebel, *J. Exp. Med.,* 54 (1931) 437–447.
(27) R. Schneerson, O. Barrera, A. Sutton, and J. B. Robbins, *J. Exp. Med.,* 152 (1980) 361–376.
(28) J. B. Robbins, R. Schneerson, P. Anderson, and D. H. Smith, *JAMA,* 276 (1996) 1181–1185.
(29) D. L. Klein and R. W. Ellis, in M. M. Levine, G. C. Woodrow, J. B. Kaper, and G. S. Cobon (Eds.), *New Generation Vaccines,* Marcel Dekker, Inc., New York, 1997, pp. 503–525.
(30) W. D. Zollinger, in M. M. Levine, G. C. Woodrow, J. B. Kaper, and G. S. Cobon (Eds.), *New Generation Vaccines,* Marcel Dekker, Inc., New York, 1997, pp. 469–488.
(31) H. Peltola, *Drugs,* 55 (1998) 347–366.
(32) T. Lagergard, J. Shiloach, J. B. Robbins, and R. Schneerson, *Infect. Immun.,* 58 (1990) 687–694.
(33) M. R. Wessels, L. C. Paoletti, D. L. Kasper, J. L. DiFabio, F. Michon, K. Holme, and H. J. Jennings, *J. Clin. Invest.,* 86 (1990) 1428–1433.
(34) S. Szu, A. L. Stone, J. D. Robbins, R. Schneerson, and J. B. Robbins, *J. Exp. Med.,* 166 (1987) 1510–1524.
(35) A. Fattom and R. B. Naso, in M. M. Levine, G. C. Woodrow, J. B. Kaper, and G. S. Cobon (Eds.), *New Generation Vaccines,* Marcel Dekker, Inc., New York, 1997, pp. 979–988.
(36) R. J. Roantree, *Annu. Rev. Microbiol.,* 21 (1967) 443–466.
(37) R. J. Roantree, in G. Weinbaum, S. Kadis, and S. J. Ajl (Eds.), *Microbial Toxins, Vol. 5,* Academic Press, New York, 1971, pp. 1–37.
(38) U. Zähringer, B. Lindner, and E. Th. Rietschel, *Adv. Carbohydr. Chem. Biochem.,* 50 (1994) 211–276.
(39) U. Mamat, U. Seydel, D. Grimmecke, O. Holst, and E. T. Rietschel, in B. M. Pinto (Ed.), *Comprehensive Natural Product Chemistry, Vol. 3,* Elsevier, Amsterdam, 1999, pp. 179–239.
(40) A. A. Lindberg, L. T. Rosenberg, Å. Ljunggren, P. J. Garegg, S. Svensson, and N.-H. Wallin, *Infect. Immun.,* 10 (1974) 541–545.
(41) T. K. Eisenstein, *Infect. Immun.,* 12 (1975) 364–377.
(42) S. B. Svenson and A. A. Lindberg, *Infect. Immun.,* 32 (1981) 490–496.
(43) S. B. Svenson and A. A. Lindberg, *Prog. Allergy,* 33 (1983) 120–143.
(44) B. Kaijser and S. Ahlstedt, *Infect. Immun.,* 17 (1977) 286–289.
(45) J. B. Robbins and R. Schneerson, *J. Inf. Dis.,* 161 (1990) 821–832.
(46) J. B. Robbins, C. Chu, and R. Schneerson, *Clin. Infect. Dis.,* 15 (1992) 346–361.
(47) J. B. Robbins, R. Schneerson, S. Szu, and V. Pozsgay, in S. Plotkin and B. Fantini (Eds.), *Vaccinia, Vaccination and Vaccinology: Jenner, Pasteur and Their Successors,* Elsevier, Paris, 1996, pp. 135–143.
(48) M. Saxena and J. L. Di Fabio, *Vaccine,* 12 (1994) 879–884.
(49) S. J. Cryz, J. O. Que, A. S. Cross, and E. Fürer, *Vaccine,* 13 (1995) 449–453.
(50) I. Seppälä and O. Mäkelä, *J. Immunol.,* 143 (1989) 1259–1264.
(51) W. Richter and L. Kågedal, *Int. Arch. Allergy. Appl. Immunol.,* 42 (1972) 887–904.
(52) C. Fernandez and E. Sverremark, *Cell. Immunol.,* 153 (1994) 67–78.
(53) L. C. Paoletti, D. L. Kasper, F. Michon, J. L. DiFabio, K. Holme, H. J. Jennings, and M. Wessels, *J. Biol. Chem.,* 265 (1990) 18278–18283.

(54) P. W. Anderson, M. E. Pichichero, E. C. Stein, S. Porcelli, R. F. Betts, D. M. Connuck, D. Korones, R. A. Insel, J. M. Zahradnik, and R. Eby, *J. Immunol.,* 142 (1989) 2464–2468.
(55) C. A. Laferriere, R. K. Sood, J. M. deMuys, F. Michon, and H. J. Jennings, *Vaccine,* 15 (1997) 179–186.
(56) H. J. Jennings, *Adv. Carbohydr. Chem. Biochem.,* 41 (1983) 155–208.
(57) D. A. A. Ala'Aldeen and K. A. V. Cartwright, *J. Infect.,* 33 (1996) 153–157.
(58) W. E. Dick, Jr. and M. Beurret, *Contrib. Microbiol. Immunol.,* 10 (1989) 48–114.
(59) R. Z. Dintzis, *Pediatr. Res.,* 32 (1992) 376–385.
(60) W. Egan, *Ann. Rep. Med. Chem.,* 28 (1993) 257–265.
(61) D. Goldblatt, *J. Med. Microbiol.,* 47 (1998) 563–567.
(62) H. J. Jennings, *Curr. Top. Microbiol. Immunol.,* 150 (1990) 97–127.
(63) H. J. Jennings, *J. Inf. Dis.,* 165(suppl) (1992) S156–S159.
(64) H. J. Jennings and R. K. Sood, in Y. C. Lee and R. T. Lee (Eds.), *Neoglycoconjugates. Preparation and Applications,* Academic Press, New York, 1994, pp. 325–371.
(65) H. J. Jennings and R. A. Pon, in S. Dumitriu (Ed.), *Polysaccharides in Medicinal Applications,* Marcel Dekker, Inc., New York, 1996, pp. 443–479.
(66) C.-J. Lee, in S. Dumitriu (Ed.), *Polysaccharides in Medicinal Applications,* Marcel Dekker, Inc., New York, 1996, pp. 411–442.
(67) R. K. Sood, A. Fattom, V. Pavliak, and R. B. Naso, *Drug Disc. Today,* 1 (1996) 381–387.
(68) P. R. Paradiso and A. A. Lindberg, in F. Brown (Ed.), *New Approaches to Stabilisation of Vaccines Potency,* Karger, Basel, 1996, pp. 269–275.
(69) C. C. A. M. Peeters, P. R. Lagerman, O. de Weers, L. A. Oomen, P. Hoogerhout, M. Beurret, and J. T. Poolman, in A. Robinson, G. Farrar, and C. Wiblin (Eds.), *Vaccine Protocols,* Humana Press Inc., Totowa, NJ, 1996, pp. 111–133.
(70) G. O. Aspinall, G. Chatterjee, and P. J. Brennan, *Adv. Carbohydr. Chem. Biochem.,* 51 (1995) 169–242.
(71) A. Lipták, A. Borbás, and I. Bajza, *Med. Res. Rev.,* 14 (1994) 307–352.
(72) S. Müller-Loennies, U. Zähringer, U. Seydel, S. Kusumoto, A. J. Ulmer, and E. T. Rietschel, *Progr. Clin. Biol. Res.,* 397 (1998) 51–72.
(73) W. B. Neeley, *Adv. Carbohydr. Chem. Biochem.,* 15 (1960) 341–369.
(74) R. L. Sidebotham, *Adv. Carbohydr. Chem. Biochem.,* 30 (1974) 371–444.
(75) Y. Arakatsu, G. Ashwell, and E. A. Kabat, *J. Immunol.,* 97 (1966) 858–866.
(76) I. M. Outschoorn, G. Ashwell, F. Gruezo, and E. A. Kabat, *J. Immunol.,* 113 (1974) 896–903.
(77) E. Lai and E. A. Kabat, *Molec. Immun.,* 22 (1985) 1021–1037.
(78) A. W. Richter and R. Eby, *Molec. Immun.,* 22 (1985) 29–36.
(79) K. Himmelspach, O. Westphal, and B. Teichmann, *Eur. J. Immun.,* 1 (1971) 106–112.
(80) K. E. Stein, D. A. Zopf, B. M. Jonson, C. B. Miller, and W. E. Paul, *J. Immunol.,* 128 (1982) 1350–1354.
(81) H. R. Hanna and D. R. Bundle, *Can. J. Chem.,* 71 (1993) 125–134.
(82) R. Eby, *Carbohydr. Res.,* 70 (1979) 75–82.
(83) R. Eby and C. Schuerch, *Carbohydr. Res.,* 79 (1980) 53–62.
(84) R. Eby and C. Schuerch, *Carbohydr. Res.,* 102 (1982) 131–138.
(85) R. Eby and C. Schuerch, *Carbohydr. Res.,* 50 (1976) 203–214.
(86) G. Ekborg, P. J. Garegg, and B. Gotthammar, *Acta Chem. Scand. B,* 29 (1975) 765–771.
(87) M. Pittman, *J. Exp. Med.,* 53 (1931) 471–493.
(88) P. J. Garegg and B. Samuelsson, *Carbohydr. Res.,* 86 (1980) 293–296.
(89) P. J. Garegg, R. Johannson, I. Lindh, and B. Samuelsson, *Carbohydr. Res.,* 150 (1986) 285–289.
(90) P. Hoogerhout, D. Evenberg, C. A. A. van Boeckel, J. T. Poolman, E. C. Beuvery, G. A. van der Marel, and J. H. van Boom, *Tetrahedron Let.,* 28 (1987) 1553–1556.
(91) L. Chan and G. Just, *Tetrahedron Let.,* 29 (1988) 4049–4052.

(92) J. P. G. Hermans, L. Poot, M. Kloosterman, G. A. van der Marel, C. A. A. van Boeckel, D. Evenberg, J. T. Poolman, P. Hoogerhout, and J. H. van Boom, *Rec. Trav. Chim. Pays-Bas,* 106 (1987) 498–504.
(93) Z. Y. Wang and G. Just, *Tetrahedron Let.,* 29 (1988) 1525–1528.
(94) A. A. Kandil, N. Chan, M. Klein, and P. Chong, *Glycoconj. J.,* 14 (1997) 13–17.
(95) P. Hoogerhout, C. W. Funke, J. R. Mellema, G. N. Wagenaars, C. A. A. van Boeckel, D. Evenberg, J. T. Poolman, A. W. M. Lefeber, G. A. van der Marel, and J. H. van Boom, *J. Carbohydr. Chem.,* 7 (1988) 399–416.
(96) L. Chan and G. Just, *Tetrahedron,* 46 (1990) 151–162.
(97) A. A. Kandil, N. Chan, P. Chong, and M. Klein, *Synlett,* (1992) 555–557.
(98) S. Nilsson, M. Bengtsson, and T. Norberg, *J. Carbohydr. Chem.,* 11 (1992) 265–285.
(99) Verez Bencomo, V., Fernandez Santana, V., Figueroa Perez, I., Alan Padron, A., Lorenzo Acosta, Y., Mondelo Rodriguez, A., Castro Palomino, J., Sierra Gonzalez, G., and Barbera Morales, R., *XIXth Int. Carbohydr. Symp.,* Abstract BP 141, San Diego, 1998.
(100) C. J. J. Elie, H. J. Muntendam, H. Van Der Elst, G. A. van der Marel, P. Hoogerhout, and J. H. van Boom, *Rec. Trav. Chim. Pays-Bas,* 108 (1989) 219–223.
(101) T. Norberg and A. Chernyak, Unpublished results, ref. 3 in cit. 98 (1992).
(102) M. Nilsson and T. Norberg, *J. Carbohydr. Chem.,* 17 (1998) 305–316.
(103) P. Chong, N. Chan, A. Kandil, B. Tripet, O. James, Y. P. Yang, S. P. Shi, and M. Klein, *Infect. Immun.,* 65 (1997) 4918–4925.
(104) E. A. De Velasco, D. Merkus, S. Anderton, A. F. M. Verheul, E. F. Lizzio, R. van der Zee, W. van Eden, T. Hoffman, J. Verhoef, and H. Snippe, *Infect. Immun.,* 63 (1995) 961–968.
(105) C. C. A. M. Peeters, D. Evenberg, P. Hoogerhout, H. Käyhty, L. Saarinen, C. A. A. van Boeckel, G. A. van der Marel, J. H. van Boom, and J. T. Poolman, *Infect. Immun.,* 60 (1992) 1826–1833.
(106) D. Evenberg, P. Hoogerhout, C. A. A. van Boeckel, G. T. Rijkers, E. C. Beuvery, J. H. van Boom, and J. T. Poolman, *J. Inf. Dis.,* 165 (suppl 1) (1992) S152–S155.
(107) K. B. Reimer and B. M. Pinto, *J. Chem. Soc. Perkin 1,* (1988) 2103–2111.
(108) J. S. Andrews and B. M. Pinto, *J. Chem. Soc. Perkin 1,* (1990) 1785–1792.
(109) B. M. Pinto, K. B. Reimer, and A. Tixidre, *Carbohydr. Res.,* 210 (1991) 199–219.
(110) K. B. Reimer, S. L. Harris, V. Varma, and B. M. Pinto, *Carbohydr. Res.,* 228 (1992) 399–414.
(111) J. R. Marino-Albernas, S. L. Harris, V. Varma, and B. M. Pinto, *Carbohydr. Res.,* 245 (1993) 245–257.
(112) F. I. Auzanneau and B. M. Pinto, *Bioorg. Med. Chem.,* 4 (1996) 2003–2010.
(113) F. I. Auzanneau, F. Forooghian, and B. M. Pinto, *Carbohydr. Res.,* 291 (1996) 21–41.
(114) B. M. Pinto and D. R. Bundle, *Carbohydr. Res.,* 124 (1983) 313–318.
(115) K. B. Reimer, M. A. J. Gidney, D. R. Bundle, and B. M. Pinto, *Carbohydr. Res.,* 232 (1992) 131–142.
(116) L. Kenne, B. Lindberg, and S. Svensson, *Carbohydr. Res.,* 40 (1974) 69–75.
(117) W. F. Goebel, *J. Exp. Med.,* 72 (1940) 33–48.
(118) W. F. Goebel, *J. Exp. Med.,* 68 (1938) 469–484.
(119) W. F. Goebel, *J. Exp. Med.,* 69 (1939) 353–364.
(120) W. F. Goebel, *J. Biol. Chem.,* 110 (1935) 391.
(121) W. F. Goebel and R. E. Reeves, *J. Biol. Chem.,* 124 (1938) 207–220.
(122) H. Snippe, A.-J. van Houte, J. E. G. van Dam, M. J. De Reuver, M. Jansze, and J. M. N. Willers, *Infect. Immun.,* 40 (1983) 856–861.
(123) S. B. Svenson and A. A. Lindberg, *J. Immunol. Meth.,* 25 (1979) 323–335.
(124) G. H. Veeneman, L. J. F. Gomes, and J. H. van Boom, *Tetrahedron,* 45 (1989) 7433–7448.
(125) E. Alonso de Velasco, A. F. M. Verheul, A. M. P. van Steijn, H. A. T. Dekker, R. G. Feldman, I. M. Fernández, J. P. Kamerling, J. F. G. Vliegenthart, J. Verhoef, and H. Snippe, *Infect. Immun.,* 62 (1994) 799–808.

(126) A. M. P. van Steijn, J. P. Kamerling, and J. F. G. Vliegenthart, *Carbohydr. Res.,* 211 (1991) 261–277.
(127) A. M. P. van Steijn, J. P. Kamerling, and J. F. G. Vliegenthart, *J. Carbohydr. Chem.,* 11 (1992) 665–689.
(128) H. Geyer, S. Stirm, and K. Himmelspach, *Med. Microbiol. Immunol.,* 165 (1979) 271–288.
(129) A. Geyer, S. Schlecht, and K. Himmelspach, *Med. Microbiol Immunol.,* 171 (1982) 135–143.
(130) J. W. J. Zigterman, J. E. G. van Dam, H. Snippe, F. T. M. Rotteveel, M. Jansze, J. M. N. Willers, J. P. Kamerling, and J. F. G. Vliegenthart, *Infect. Immun.,* 47 (1985) 421–428.
(131) S. Stirm, O. Lüderitz, and O. Westphal, *Justus Liebigs Ann. Chem.,* 696 (1966) 180–193.
(132) A. M. Staub, S. Stirm, L. Le Minor, O. Lüderitz, and O. Westphal, *Ann. Inst. Pasteur,* 111 (Suppl. 5) (1966) 47–58.
(133) O. Lüderitz, A. M. Staub, and O. Westphal, *Bacteriol. Rev.,* 30 (1966) 192.
(134) K. Stellner, O. Lüderitz, O. Westphal, A. M. Staub, B. Leluc, C. Coynault, and L. Le Minor, *Ann. Inst. Pasteur,* 123 (1972) 43–54.
(135) G. Ekborg, J. Lönngren, and S. Svensson, *Acta Chem. Scand. B,* 29 (1975) 1031–1035.
(136) P. J. Garegg, H. Hultberg, and T. Norberg, *Carbohydr. Res.,* 96 (1981) 59–64.
(137) K. Eklind, P. J. Garegg, and B. Gotthammar, *Acta Chem. Scand. B,* 30 (1976) 305–308.
(138) K. Eklind, P. J. Garegg, and B. Gotthammar, *Acta Chem. Scand. B,* 30 (1976) 300–304.
(139) H. B. Borén, P. J. Garegg, and N.-H. Wallin, *Acta Chem. Scand.,* 26 (1972) 1082–1086.
(140) G. Ekborg, P. J. Garegg, and S. Josephson, *Carbohydr. Res.,* 65 (1978) 301–306.
(141) B. Svenungsson and A. A. Lindberg, *Med. Microbiol. Immunol.,* 163 (1977) 1–11.
(142) K. Stellner, O. Westphal, and H. Mayer, *Justus Liebigs Ann. Chem.,* 738 (1970) 179–191.
(143) G. Ekborg, K. Eklind, P. J. Garegg, B. Gotthammar, H. E. Carlsson, A. A. Lindberg, and B. Svenungsson, *Immunochemistry,* 14 (1977) 154–157.
(144) G. Alfredsson and P. J. Garegg, *Acta Chem. Scand.,* 27 (1973) 566–560 [correction:27 (1973) 1834].
(145) P. J. Garegg and B. Gotthammar, *Carbohydr. Res.,* 58 (1977) 345–352.
(146) B. Svenungsson and A. A. Lindberg, *Acta Path. Microbiol. Scand. B.,* 86 (1978) 35–40.
(147) S. B. Svenson and A. A. Lindberg, *Scand. J. Infect. Dis. Suppl.,* 24 (1980) 210–215.
(148) H. J. A. Jörbeck, S. B. Svenson, and A. A. Lindberg, *Infect. Immun.,* 32 (1981) 497–502.
(149) S. B. Svenson and A. A. Lindberg, *J. Immunol.,* 120 (1978) 1750–1757.
(150) U. Eriksson, S. B. Svenson, J. Lönngren, and A. A. Lindberg, *J. Gen. Virol.,* 43 (1979) 503–511.
(151) J. Lönngren, I. J. Goldstein, and J. E. Niederhuber, *Arch. Biochem. Biophys.,* 175 (1976) 661–669.
(152) H. J. A. Jörbeck, S. B. Svenson, and A. A. Lindberg, *J. Immunol.,* 123 (1979) 1376–1381.
(153) S. B. Svenson, M. Nurminen, and A. A. Lindberg, *Infect. Immun.,* 25 (1979) 863–872.
(154) A. A. Lindberg, S. E. Greisman, and S. B. Svenson, *Infect. Immun.,* 41 (1983) 888–895.
(155) G. Kleinhammer, K. Himmelspach, and O. Westphal, *Eur. J. Immunol.,* 3 (1973) 834–838.
(156) V. Pozsgay, C. P. J. Glaudemans, J. B. Robbins, and R. Schneerson, *Bioorg. Med. Chem. Let.,* 2 (1992) 255–260.
(157) V. Pozsgay, B. Coxon, and H. Yeh, *Bioorg. Med. Chem.,* 1 (1993) 237–257.
(158) V. Pozsgay and L. Pannell, *Carbohydr. Res.,* 258 (1994) 105–122.
(159) V. Pozsgay and B. Coxon, *Carbohydr. Res.,* 257 (1994) 189–215.
(160) V. Pozsgay, *J. Am. Chem. Soc.,* 117 (1995) 6673–6681.
(161) V. Pozsgay, *Angew. Chem. Int. Ed.,* 37 (1998) 138–142.
(162) V. Pozsgay, J. B. Robbins, and R. Schneerson, *Federal Register,* 63 (1998) 1117.
(163) V. Pozsgay, *J. Org. Chem.,* 63 (1998) 5983–5999.
(164) V. Pozsgay, *Tetrahedron Lett.,* Submitted for publication (1999).
(165) G. R. Gray, *Arch. Biochem. Biophys.,* 163 (1974) 426–428.

(166) V. Pozsgay, C. Chu, L. Pannell, J. Wolfe, J. B. Robbins, and R. Schneerson, *Proc. Natl. Acad. Sci. U.S.A.*, 96 (1999) 5194–5197.
(167) P.-E. Jansson, J. Lönngren, G. Widmalm, K. Leontein, K. Slettengren, S. B. Svenson, G. Wrangsell, A. Dell, and P. R. Tiller, *Carbohydr. Res.*, 154 (1985) 59–66.
(168) J. A. Roberts, M. B. Kaack, G. Baskin, and S. B. Svenson, *Infect. Immun.*, 61 (1993) 5214–5218.
(169) A. F. M. Verheul, G. J. P. H. Boons, G. A. van der Marel, J. H. van Boom, H. J. Jennings, H. Snippe, J. Verhoef, P. Hoogerhout, and J. T. Poolman, *Infect. Immun.*, 59 (1991) 3566–3573.
(170) G. J. P. H. Boons, M. Overhand, G. A. van der Marel, and J. H. van Boom, *Angew. Chem. Int. Ed.*, 28 (1989) 1504–1506.
(171) G. J. P. H. Boons, G. A. van der Marel, J. T. Poolman, and J. H. van Boom, *Rec. Trav. Chim. Pays-Bas,* 108 (1989) 339–343.
(172) G. J. P. H. Boons, F. L. van Delft, P. A. M. van der Klein, G. A. van der Marel, and J. H. van Boom, *Tetrahedron,* 48 (1992) 885–904.
(173) A. F. M. Verheul, J. A. M. van Gaans, E. J. H. Wiertz, and J. T. Poolman, *Infect. Immun.*, 61 (1993) 187–196.
(174) H. Brade, W. Brabetz, L. Brade, O. Holst, S. Löbau, M. Lukacova, U. Mamat, A. Rozalski, K. Zych, and P. Kosma, *Pure Appl. Chem.*, 67 (1995) 1617–1626.
(175) P. Kosma, J. Gass, G. Schultz, R. Christian, and F. M. Unger, *Carbohydr. Res.*, 167 (1987) 39–54.
(176) P. Kosma, G. Schultz, and H. Brade, *Carbohydr. Res.*, 183 (1988) 183–199.
(177) P. Kosma, R. Bahnmuller, G. Schultz, and H. Brade, *Carbohydr. Res.*, 208 (1990) 37–50.
(178) Y. Fu, M. Baumann, P. Kosma, L. Brade, and H. Brade, *Infect. Immun.*, 60 (1992) 1314–1321.
(179) Y. C. Lee and R. T. Lee, *Carbohydr. Res.*, 37 (1974) 193–201.
(180) L. Brade, K. Zych, A. Rozalski, P. Kosma, K. Bock, and H. Brade, *Glycobiology,* 7 (1997) 819–827.

MOLECULAR STRUCTURE OF THE CARBOHYDRATE–PROTEIN LINKAGE REGION FRAGMENTS FROM CONNECTIVE-TISSUE PROTEOGLYCANS

By N. Rama Krishna and Pawan K. Agrawal

Department of Biochemistry and Molecular Genetics, Cell Adhesion and Matrix Research Center, Comprehensive Cancer Center, The University of Alabama at Birmingham, Birmingham, AL 35294-2041, USA

I. Introduction	201
II. Isolation of Linkage-Region Oligosaccharides	205
III. Identification of Carbohydrate Spin Systems	206
IV. Identification of the Peptide Spin Systems	209
V. Conformational Analysis of the Interglycosidic Linkages	210
VI. Conformation of the Glycopeptide Linkage	211
VII. Influence of Glycosylation on Proteoglycan Core-Peptide Backbone Conformation	213
VIII. Structural Studies on the Carbohydrate–Protein Linkage Region	215
1. Dermatan Sulfate	215
2. Chondroitin Sulfate	216
3. Heparan Sulfate	219
4. Heparin	224
IX. NMR Spectral Properties of Linkage-Region Oligosaccharides	224
X. Biosynthesis of Carbohydrate–Protein Linkage-Region Oligosaccharides	228
References	229

I. Introduction

Proteoglycans are major structural and functional components of many connective tissues, wherein they influence mechanical properties, molecular transport, and morphogenesis. They modulate cell–cell and cell–matrix interactions, contribute to the maintenance of normal tissue architecture and function, and participate in cell adhesion and growth control. They are highly branched glycoproteins that contain a protein core to which large numbers of side chains of sulfate-substituted, negatively charged glycosaminoglycan (GAG) chains and N- and O-linked oligosaccharides are covalently attached.[1–12] Their high negative charge-density results from the presence of ester sulfate and uronic acid carboxylate groups in the repeating disaccharide units. Although the GAGs are alternating copolymers of uronic acid [D-glucuronic acid/L-iduronic acid (GlcA/IdoA)] and N-acetylhexosamine [2-acetamido-2-deoxy-D-glucose (2-acetamido-2-deoxy-D-galactose (GlcNAc/GalNAc)], the considerable complexity of the proteoglycan structure has been recognized. The GAGs are distinguished by their monosaccharide composition, the position and configuration of their glycosydic linkages, and the number and location of sulfate substituents. Based on the structure of the hexosamine, these are mainly classified into two categories—the galactosaminoglycans, dermatan sulfate and chondroitin sulfate containing N-acetylgalactosamine (GalNAc) and uronic acid, and the glucosaminoglycans, hyaluronan, heparan sulfate, and heparin, containing N-acetylglucosamine (GlcNAc) and uronic acid.

Dermatan sulfate (DS), an extracellular matrix component of fibrous connective tissues, is also present on cell surfaces as a proteoglycan. Several biological functions for it are known, such as extracellular matrix formation through interaction with several types of collagen,[12–14] inhibition of mitogenic activity to transforming growth factor-β,[15] and inhibition of the cell-attachment activity of fibronectin.[16] Dermatan sulfate chains also mediate such activities as anticoagulant activity, self-association activity, and antiproliferative activity. The anticoagulant activity is expressed by binding to heparin cofactor II.[17] Self-association via protein–protein interactions is known, but also occurs through carbohydrate–carbohydrate interactions.[18] Decorin, a common DS-bearing proteoglycan, has been shown to affect the kinetics of collagen fibril formation.[12] Structurally, the DS polymer possesses a repeating disaccharide region, composed of alternately arranged uronic acid [which may be either D-glucuronic acid (GlcA) or L-iduronic acid (IdoA)] and GalNAc residues. The uronic acid is always $(1 \to 3)$-linked to the GalNAc residue. Thus, there are two types of oligosaccharide units, -4GlcAβ3 GalNAcβ- and -4IdoAα3GalNAcβ-.* The disaccharide repeats are modified by sulfation. The former can be sulfated at C-4 and/or C-6 of the GalNAc unit, whereas the latter contains almost exclusively 4-sulfated GalNAc units, and a minor proportion of IdoA residues are sulfated at C-2. Combining sequential arrangements of IdoA residues and sulfate groups results in a wide variety of domain structures. Dermatan sulfate chains rich in iduronic acid have been demonstrated to be antiproliferative, and they also interact with spermine.[19]

Chondroitin sulfates (CS) are expressed on the cell surface and in extracellular matrices. Various amounts of CS proteoglycans are produced by most, if not all, vertebrate cells; the highest concentrations are found in such soft connective

---

* The IUPAC-IUBMB "short form" symbolism[18a] 2-Carb-38.5 is used here and elsewhere in this chapter for designation of oligosaccharide structures. Parent monosaccharide residues are the D enantiomers except for IdoA, which is the L enantiomer. Uronic acid residues modified by 4,5-unsaturation are prefixed by $\Delta^{4,5}$, and the associated anomeric symbol relates to the configuration at the 5-position of that residue *before* 4,5-elimination. The abbreviation HexA is used to denote either D-GlcA or L-IdoA.

tissues as cartilage.[20] These are increasingly implicated as important regulators of many biological processes, such as cell migration and recognition, extracellular matrix deposition, and morphogenesis.[21–23] CS has a linear polymer structure with repetitive sulfated disaccharide units containing alternate GlcA and GalNAc units (-4GlcAβ3GalNAcβ-).[24] The C-4 and/or C-6 positions of the GalNAc unit are usually sulfated. As has been reported,[24] the C-2 and C-3 positions of GlcA can be sulfated occasionally.[24] CS varies considerably in its size, in the number of GAG chains per core protein, and in the degree and profile of sulfation. This variability enables the storage of a massive amount of information and the encoding of a great variety of biological functions.[1,25]

Heparan sulfate (HS) has a ubiquitous distribution in the extracellular matrix and on cell surfaces of most mammalian cells.[26] HS on the host cell surface has been implicated in the adherence of numerous microbial agents,[27] including human immunodeficiency virus (HIV)[28] and herpes simplex virus.[29,30] HS exerts a variety of biological and biochemical activities, such as regulation of lipid metabolism,[31] control of blood fluidity at the endothelial surface,[32] modulation of cellular proliferation,[33,34] control of cell attachment to various proteins in extracellular matrices,[35–37] combination with acidic and basic fibroblast growth factors[38] and interleukin-3 and granulocyte–macrophage colony stimulating factor.[39–41] The GlcA and GlcNAc units are added, through β-(1 → 4) and α-(1 → 4) linkages, respectively, to the nonreducing termini of a nascent polysaccharide chain, in a process that appears to be catalyzed by a single enzyme.[42,43] The sequence -4GlcAβ4GlcNAcα1-is the basic repeat structure, which undergoes extensive biosynthetic modifications involving N-sulfation, O-sulfation, and uronate epimerization (GlcA→IdoA), and has a considerable degree of sequence variability. A single chain therefore contains sequences that are composed of high- and low-sulfated disaccharide units. The complex structure of HS appears to be tailored for specific interactions in diverse functional contexts.[44] The oligosaccharide sequences that are adapted for playing a fundamental role in the various activities are found in certain structural domains. Interactions between HS and proteins are being increasingly implicated in a variety of physiological processes, such as cytokine action, enzyme regulation, and cell adhesion.[45]

Heparin has the highest negative charge density of any known biological macromolecule and is thus prone to electrostatic interactions with a variety of proteins.[46] It exerts a variety of biological and biochemical activities, such as modulation of cellular proliferation,[33] potentiation of angiogenesis,[47] interactions with acidic and basic fibroblast growth-factors,[48–50] and is used in the clinic for its blood anticoagulant properties in preventing and treating thromboembolic disease[51] as well as for inhibiting replication of the human immunodeficiency virus type I (HIV-I) in cultures of CD4+ human cells.[52] It has been established that the anticoagulant activity depends on specific binding of heparin to antithrombin and that the antithrombin-binding site in heparin is a pentasaccharide with specific modifications by sulfation.[53] However, for other activities, the relation with the specific sequence of the HS is not well understood.[54] Interestingly, heparin is composed of the same alternating repeat sequence as mentioned for heparan sulfate (-4HexA4GlcNAcα1-), but contains more sulfate and IdoA and less N-acetyl groups and GlcA than HS. Although the major trisulfated disaccharide repeating unit of heparin is -4αIdoA(2S)4GlcNS(6S)α1-, substantial heterogeneity is observed in the fine structure of the polysaccharide. The structural variability of HS and heparin is the basis of the wide variety of their biological activities. Sulfate groups can be located at C-2 of the hexuronic acid and at C-2, C-3, and/or C-6 of the glucosamine (GlcN) residue. They add structural complexity to the carbohydrate backbone to form various active domain structures responsible for a number of biological activities. Structural studies have been reported on the binding domains of heparin to antithrombin III,[55] basic fibroblast growth factor,[56–58] and antiproliferative activity.[59] The polypeptide consensus sequences for heparin binding, based on clustered basic amino acids, show some variability,[60] other types of polysaccharide-binding regions are composed of multiple peptide loops (see, for example, ref.61). However, it is generally thought that the binding of heparin/HS sequences to proteins is largely electrostatic in nature and thus involves positively charged amino acid residues in the protein components.

From the foregoing it is evident that individual GAGs differ from each other in the type of hexuronic acid and glycosamine, or both, the position and configuration of the glycosidic linkages, and the degree and pattern of sulfation. More precisely, the following features should be noted: (a) The uronic acid is D-glucuronic acid for chondroitin 4- and 6-sulfates, mainly L-IdoA for DS and heparin (with D-GlcA as a minor constituent), and predominantly D-GlcA in HS; (b) The glycosamine is 2-amino-2-deoxy-D-glucose for HS and heparin and 2-amino-2-deoxy-D-galactose for CS and DS, (c) D-GlcA is β-(1 → 3)-linked to glycosamine in the CS, and β-(1 → 4)-linked in heparin and in HS. L-IdoA is linked α-(1 → 3) in DS and α-(1 → 4) in heparin; (d) The hexosamine is N-acetylated in DS, CS, and partially in HS, and is N-sulfated in heparin and (partially) in HS.

Despite significant structural variability and complexity in disaccharide repetitive sequences (GlcA/IdoA-GalNAc/GlcNAc), GAGs are O-glycosidically linked to L-serine residues of the protein cores via a common glycopeptide linkage-region consisting of the tetrasaccharide GlcAβ3Galβ3Galβ4XylβO-Ser, as previously reported by Rodén[1] and Lindahl and Rodén.[2] The monosaccharide residues have been labeled 1 to 4, starting with the serine-substituted xylopyranosyl moiety as Xyl-1, galactose substituted to xylose as Gal-2, Gal substituted with GlcA as Gal-3, and terminal GlcA as GlcA-4.

An important question in proteoglycan biosynthesis is how different GAGs (DS, CS, HS, and heparin) are assembled starting from the common tetrasaccharide structure. The precise molecular mechanisms, namely, whether it is the linkage-region sugars or the neighboring peptide sequence, or a combination thereof, that predetermines the biosynthesis of the different GAGs, remain unresolved. The structural characterization of the carbohydrate–protein linkage-region fragments from various sources undertaken by various investigators is an important step in unraveling these mechanisms. In this chapter, we present a brief overview of the current status concerning the ubiquity and structural variability of carbohydrate–protein linkage-regions so far reported for various kinds of proteoglycans. We also survey briefly NMR-spectroscopic and molecular-modeling approaches applicable to structural analysis of these complex carbohydrates. A compilation of ^{1}H NMR chemical shifts is also provided for comparative structural studies.

II. Isolation of Linkage-Region Oligosaccharides

The oligosaccharides representing the carbohydrate–protein linkage (CPL) region are usually isolated by exhaustive enzymatic digestion of commercially available heparin, heparan sulfate, chondroitin sulfate, and dermatan sulfate with such commercially available enzymes as protease, actinase, papain, pronase, β-galactosidase, heparinase, heparitinase I and II, $\Delta^{4,5}$-glycuronate-2-sulfatase, alkaline phosphatase, chondroitinase ACII, and/or chondroitinase ABC, followed by purification by gel chromatography, high-performance liquid chromatography (HPLC), and capillary electrophoresis.[62–68] Glycopeptide linkages involving the xylose–serine bond are more or less acid-sensitive (toward hydrolysis or anomerization) and are also alkali-sensitive (causing β-elimination or racemization). These drawbacks greatly complicate the isolation of glycopeptides from this region by chemical means. In the presence of alkaline sodium borohydride, the glycoprotein or proteoglycan undergoes reductive β-elimination with the release of an oligosaccharide-alditol.[69] Because the isolation of glycopeptide fragments containing carbohydrate chains O-glycosidically linked to only a short peptide chain is rather cumbersome, oligosaccharide-xylitols have been generally found appropriate for ^{1}H NMR structural studies. This also avoids the complicating effect of anomerization on the ^{1}H NMR spectrum. It is evident that, in all cases, the partial degradation techniques and subsequent chemical modifications applied do not generally affect the structure of the carbohydrate. A homogeneous sample is then subjected for NMR structural analysis.

In many instances, the linkage oligosaccharide is associated with a peptide moiety, and thus the amino acid sequence must be established and, in addition, the structural parameters of the CPL oligosaccharide determined. The latter comprise: (*i*) primary structure, (*ii*) type and location of noncarbohydrate substituents, and (*iii*) the linkage to the peptide backbone. For gaining insight into structure–function relations, it is important for the three-dimensional structure to be determined.

III. Identification of Carbohydrate Spin Systems

Despite the fact that the CPL-region oligosaccharides consist of few monosaccharide types, Xyl, Gal, and GlcA, they may contain GalNAc, GlcNAc, IdoA, and $\Delta^{4,5}$-HexA in the case of larger fragments. This results in severe overlapping in the ^{1}H NMR spectra, particularly in the 3.4–4.0 ppm region, whereas acetamido group protons at around δ 2.1, H-4 of $\Delta^{4,5}$-HexA at δ 6.0, and anomeric resonances (H-1) are generally well resolved in the 4.4–5.5 ppm region. The H-4 signals of 4-sulfated β-Gal and α-GalNAc also appear in the anomeric region (around 4.75 ppm for the former and 4.62 ppm for the latter) but may be readily distinguished from anomeric protons by their appearance as broad singlets or doublets with small couplings ($^3J_{45} = <1$ Hz).[70] Determination of complete ^{1}H NMR assignments for individual monosaccharide components facilitates the complete characterization of the conformation in solution. The sugar protons of these glycopeptides may be assigned by using standard methods for oligosaccharides.[71–77] Among these, the TOCSY technique is perhaps most useful because efficient coherence transfer through a linear spin-coupling network is generally possible for extended spin-lock times. Complete one-dimensional subspectra for each monosaccharide residue can potentially be extracted from cross sections through the two-dimensional TOCSY spectrum at the resonance frequency for each well-resolved anomeric proton signal. Resonance assignments in these subspectra are then usually made by inspection from prior knowledge of patterns of spin-coupling constants and the expected chemical shifts from data on model compounds. Although in principle, this method offers the possibility of obtaining all resonance assignments for a monosaccharide in a single experiment, in practice, TOCSY experiments with two or more spin-lock times may be needed to trace the sequential magnetization transfer from H-1 to all intra-ring protons. The combined use of TOCSY and phase-senstive COSY (PS-COSY) spectra can further augment the assignment process, besides yielding homonuclear coupling constants. By this approach, in most cases, the identification of all proton spin-systems belonging to the Xyl and GlcA residues is possible because of their large vicinal coupling constants ($^3J_{HH} => 7$ Hz), and up to H-4 for Gal residues. Small $^3J_{4,5}$ coupling constants (≤ 1 Hz) act as a bottleneck in the TOCSY coherence transfer step beyond H-4 for Gal residues. Characteristic TOCSY patterns are usually observed (Xyl and GlcA exhibit four cross-peaks due to partially superimposed H-5a + H-2 signals in the case of the former and H-3 + H-4 in the case of the latter; Gal shows three cross-peaks). The H-5 signal of the Gal residues can readily be assigned from the Gal-H-1/H-5 cross-peak in the NOESY spectrum.[78–80] The H$_2$-6 signal of the CH$_2$OH group of Gal and GalNAc residues can be assigned by the NOE between H-4 and H$_2$-6. In cases of ambiguity, however, triple-quantum-filtered (TQF)-COSY[81] can be employed, as it identifies mutually coupled three-spin systems (H-5 and H$_2$-6 of Gal, GalNAc, GlcNAc, and H$_2$-5 and H-4 of Xyl). The best way of identifying H_2-6 resonances of Gal residues is from one-bond $^1H-^{13}C$ heteronuclear correlation experiments (HMQC/HSQC), as the unsubstituted C-6 atoms of Gal moieties resonate at a characteristic ^{13}C NMR chemical shift ~61 ppm.[78–80,82] A similar approach can be adopted for assigning ^1H NMR resonances of other monosaccharide residues present in longer fragments.

Once the individual glycosyl residues are identified, sequential assignment of sugars in the carbohydrate chain can be accomplished from the interglycosidic NOE contacts between anomeric H-1 and the aglycone H-n protons.[78–80,83–85] Identification of the ^{13}C resonances of the glycosyl residues exhibiting glycosylation-induced shifts and/or observation of long-range heteronuclear interglycosidic connectivity in HMBC experiments[86] can also be used for the confirmation of sequential connectivity. Successful application of the NOE method requires a discrimination between interresidual and intraresidual NOEs and careful investigation for relevant protons in the case of coincident chemical shifts. In general, a NOE cross-peak between the glycone H-1 and aglycone-H-n, where n corresponds to H-3 for the (1 → 3) linkage and H-4 for the (1 → 4)-linkage, has always been observed.[79,80,83–85] Thus, sequence-specific resonance assignments may be made by identifying interresidue NOE connectivities in the NOESY spectrum. Additional weak NOE cross-peaks corresponding to H-1/H-n + 1e are also observed for β-(1 → 3) and β-(1 → 4) linkages. Hence, NOE cross-peaks at the chemical shift of the glycone H-1 identify the linkage, whereas the absence of an interresidue cross-peak for the anomeric proton identify the reducing-end monosaccharide. Thus, the primary structure (including monosaccharide chain length and interglycosidic linkage and sequence) may be deduced, from iterative comparison of TOCSY/PS-COSY/ NOESY spectral data in a systematic manner.

The first detailed ^1H NMR studies for a series of building blocks of the proteoglycan linkage region were reported by Van Halbeck and co-workers[87] in 1982. These authors checked and refined their assignments by iterative calculation of the theoretical spectra, using a computer simulation program. It is noteworthy that the assignments for the sugar residues, as suggested by 1D NMR methods, were found to be in complete agreement with the analysis of 2D homo- and hetero-nuclear studies[79,80] except for the chemical shift of H-α and one of the H-β resonances of the Ser residue, which were interchanged based on the one-bond $^1H-^{13}C$ correlation studies. Today, ^1H NMR assignments are routinely deduced from the application of diverse 2D NMR techniques. Occasionally, the increased spectral resolution in such homonuclear 3D-NMR techniques as 3D-NOESY-TOCSY are also exploited. The application of heteronuclear 3D-NMR techniques to the proteoglycans has been limited because of the difficulties associated with obtaining uniformly ^{13}C-labeled CPL region oligosaccharides.

According to accepted criteria and experimental results of conformational analysis of monosaccharides,[88] the pyranose ring of the Xyl, Gal, GalNAc, and GlcNAc residues adopt the $^4C_1$ (D) chair conformation.[78-80] The vicinal coupling constants ($^3J_{HH}$) and intraresidual NOE data are in good agreement. The conformations of the uronic acid residues are thought to play an important role in controlling the biological function of GAGs.[53,89] It is generally accepted that GlcA residues of CS are in the $^4C_1$ conformation, with an equatorially oriented carboxylate group. Consideration of coupling constants obtained either by 1D ^{1}H NMR or by DQF COSY indicates that sulfation at the 2- and 6-positions, as found in -2-deoxy-6-O-sulfo-2-sulfoamino-α-D-glucose also leaves unchanged the $^4C_1$ conformation.[90] The relative constancy of $^3J_{HH}$ coupling constants suggests that extension of the carbohydrate chain, either by unsaturated uronate or by β-D-GalNAc or α-D-GlcNAc residues, does not substantially modify the tetrasaccharide conformation.

The conformation of the L-IdoA residues of DS is also most commonly referred to as being $^4C_1$, involving an axial carboxylate group. Based on energy calculations and $^3J_{HH}$ coupling constants, a rapid conformational equilibrium among several ring conformations has been proposed,[54,91,92] and there is no simple way to predict which form dominates when it constitutes part of a polymer chain. It has been postualted that the flexible nature of the IdoA ring may be related to the protein-binding specificity of these componds.[93] X-Ray diffraction patterns of oriented fibers of DS have suggested that the L-IdoA residues adopt the $^4C_1$ (L) conformation.[94] NMR studies, including lineshape analysis and spectral density mapping, on 1,2,3,4,6-penta-O-acetyl-α-D-idopyranose, are consistent with a two-state equilibrum between the of $^4C_1$ and $^0S_2$ conformers.[95]

^{1}H NMR studies have indicated that the nonreducing ends of many CPL oligosaccharides possess a terminal uronate having 4,5-unsaturation for both di- and oligo-saccharides. Proton coupling constants for the terminal uronate indicate conformational flexibility.[96a] Crystallographic data show that this residue exists in two different forms ($^2H_1$ and $^1H_2$) within the same unit cell, indicating that these are of nearly equal energies.[96b] The nonterminal α-L-IdoA residue is likewise internally flexible, whereas internal α-D-GlcNAc, β-D-GalNAc, and β-D-GlcA residues are conformationally rigid in solution.

It is worth noting here that the ^{1}H NMR spectrum of a CPL oligosaccharide is unique: mere comparison of the ^{1}H NMR spectra of compounds (even without detailed interpretation) can often suggest the identity of compounds. When the spectrum does not match any of the spectra in existing databases, attempts can still be made to interpret the ^{1}H NMR spectrum in terms of (partial elements of) primary structure of the carbohydrate (including anomeric configuration and positions of the interglycosidic linkages), utilizing the strategy briefly described herein.

Although the information from resonances in the region between 3.3 and 4.1 ppm has been neglected in some studies in the literature because of resonance crowding, assignment of the individual resonances in this region is important in ascertaining information about additional intra- and inter-residual NOEs, which provide distance constraints in structure-refinement calculations.

In the case of longer oligosaccharide fragments obtained by enzymatic digestion of heparin and CS, the GlcNAc, GalNAc, GlcN, and IdoA residues are usually sulfated at the 2-, 4-, and/or 6-positions. The sulfate group has been so far reported at the 4- and 6-positions of GalNAc and GlcN residues and the 2-position of Δ^4-hexuronic acid. The amino group of GlcN may also be sulfated. Sulfation does not generally alter the multiplicity pattern of the protons directly bonded to the carbon atom substituted by the sulfate group, but causes significant changes in the ^{1}H and ^{13}C NMR chemical shifts for the nuclei either directly involved in sulfation or occupying adjacent positions. Downfield ^{1}H NMR shifts of 0.52–0.6 ppm for H-α and 0.4 ppm for H_2-α, together with 0.15–0.4 ppm downfield shifts of adjacent protons (H-β) are the usual consequences of sulfation.[97–101] However, N-sulfation of GlcN causes an upfield shift for H-2 and H-3 by approximately 0.63 and 0.15 ppm, respectively, and a downfield shift of ∼0.24 ppm for H-1.[98,100] The ^{13}C NMR data for sulfated CPL linkage oligosaccharides are somewhat scarce so far, but studies on related disaccharides suggest that sulfation induces a downfield shift of 7.6–7.9 ppm for Cα in addition to an upfield shift of 0.7–1.3 ppm for the Cβ resonances.[101]

The CPL oligosaccharides always have a β-(1 → 4) linkage between the Gal-2 and Xyl residues and a β-(1 → 3) linkage between Gal-3 and Gal-2, and HexA and Gal-3. The HexA is mostly β-D-GlcA but it has been identified as α-L-IdoA in dermatan sulfate proteoglycan isolated from the CPL region of bovine aorta after reductive β-elimination and subsequent digestion by chondroitinase ABC.[102]

IV. Identification of the Peptide Spin Systems

A peptide chain limited to Ser or a Ser-Gly dipeptide is usually present in oligosaccharide fragments derived from exhaustive enzymatic digestion.[103,104] The peptide resonances can be assigned in a sequence-specific manner following standard procedures.[105] However, in cases where the linkage oligosaccharide is linked to a longer peptide chain, the distinction between the resonances of peptide and carbohydrate moieties becomes an important step.[106] During our investigations of glycopeptides (synthesized by Rio et al.[107]), GlcAβ3Galβ3Galβ4XylβO-Ser-Gly (tetrasaccharide dipeptide, TSDP) and (GlcAβ3Galβ3Galβ4XylβO)-Ser-Gly-Ser-Gly-(GlcAβ3Galβ3Galβ4XylβO)-Ser-Gly (octasaccharide hexapeptide, OSHP), we found partial overlap of the Ser Hα the anomeric resonances between δ 4.38–4.71.[106] The characteristic multiplicity pattern [quartet (q) or double doublet (dd)] of Ser Hα was somewhat helpful in the tentative assignments of these resonances, as anomeric-H are generally doublets ($^3J_{1,2} = 7$–8 Hz). In these circumstances, once again, the TOCSY spectrum can reliably be used to identify serine residues, as Ser Hα exhibits only two cross-peaks corresponding to H-β and H-β′, whereas H-1 of the sugar moiety shows at least three or more cross-peaks. In the case of ambiguity, one-bond ^{1}H–^{13}C heteronuclear correlations observed in

HSQC/HMQC can be safely employed through Ser-Hα/Ser-Cα correlation in a well-defined region (51–53 ppm) that is well separated from the anomeric region (95–105 ppm).[106] The one-bond heteronuclear correlation of the ^{13}C resonances at 46–48 ppm provides a rapid way for the unambiguous ^{1}H NMR assignments of glycine residues.

V. Conformational Analysis of the Interglycosidic Linkages

Once complete ^{1}H NMR assignments have been ascertained, they can be used for establishing the solution conformation, using NOE and coupling-constant data and structure-refinement calculations. The primary tool for the generation of distance information in biomolecules is NOESY.[108] For small mixing times, the measured NOE is proportional to the inverse of the sixth power of the distance between relevant protons and is generally observable for protons separated by up to ~6 Å. For moderately small molecules with correlation times (τ_c) satisfying the condition $\omega_0\tau_c \approx 1.1$ (where ω_0 is the Larmor frequency in angular units), the NOESY peaks approach a null value, thus leading to a loss of distance information. A change in temperature and/or field strength recovers the NOE effects. An alternative technique, particularly attractive for moderately small molecules with null NOESY spectra, is ROESY,[93] which does not suffer from this null effect. It is thus relatively common to complement NOESY measurements on these molecules with ROESY experiments. In the ROESY spectrum, the ROE cross-peaks appear opposite in sign with respect to the diagonal, and for short mixing times the cross-peak intensity is proportional to the inverse of the sixth power of the internuclear distance. A relatively recent development in which magnetic-field-induced residual dipolar couplings (proportional to the inverse third power of distance) between nuclei can be directly measured in solution with the aid of such alignment agents as bicelles and rod-shaped virus particles,[110,111] holds considerable promise in obtaining longer range constraints. These distances derived from NOE/ROE data (or dipolar couplings) can be treated as distance constraints in a combined distance geometry and dynamical simulated annealing, first introduced to the field of complex carbohydrates by the author's laboratory.[80,112] This approach has been utilized by this laboratory to characterize the solution conformations of such CPL region fragments as GXS (Galβ4Xylβ O-L-Ser) [79], G'GXS (Galβ3Galβ4Xylβ O-L-Ser),[80] tetrasaccharide dipeptide (TSDP: (GlcAβ3Galβ3Galβ4Xylβ O-)Ser-Gly),[106] bis-glycosylated hexapeptide (BGH: (Galβ4XylβO)Ser-Gly-Ser-Gly-(Galβ4Xylβ O) Ser-Gly),[83] trisglycosylated hexapeptide (TGH: (Galβ4Xylβ O)Ser-Gly-(Galβ4 Xylβ O)Ser-Gly-(Galβ4Xylβ O)Ser-Gly),[83] and the octasacchaide hepapeptide (OSHP: (GlcA β3Galβ3Galβ4Xylβ O)Ser-Gly-Ser-Gly-(GlcAβ3Galβ3Galβ4 Xylβ O)Ser-Gly).[106]

During our studies toward unambiguous ^{1}H and ^{13}C NMR assignments for a synthetic glycopeptide, GlcAβ3Galβ3Galβ4XylβO-Ser-Gly (TSDP), we found that each interglycosidic linkage was characterized by the presence of two NOEs between the glyconic anomeric-H (H-1) and the aglyconic-H at the linkage site (H-n) and to its adjacent equatorial proton occupying the H-n + 1 position.[106] The H-n + 1 position corresponds to H-5e of Xyl and to H-4 of Gal. In the fragments studied by this laboratory (GXS, G'GXS, TSDP, and OSHP) NOESY cross-peak volumes corresponding to H-1/H-n were generally stronger (derived distances 2.4–2.5 Å) than H-1/H-n + 1 cross peaks (3.0–3.2 Å). However, it is noteworthy that some of the interresidue NOE cross-peaks may sometimes be partially superimposed with intraresidue NOE cross-peaks (as in TSDP.[106] In such circumstances, special care is needed to assess the interglycosidic distances accurately. Since the distance between 1,3-syn diaxial protons is relatively fixed (2.5 Å from X-ray diffraction data,[113] it is possible to calculate and subtract out the contribution of intraresidual cross-peaks to obtain an estimate of the interresidual NOE. We[106] and others[114] have found such a procedure of sufficient reliability. The derived interglycosidic distances were then used as distance constraints in distance geometry/simulated annealing calculations. These calculations generated a family of structures satisfying experimental constraints. The φ torsion angle (defined as O-5–C-1–O–C-n according to IUPAC convention[115] shows a strict preference for the -sc range, which is in accordance with the concept of the exo-anomeric effect.[116] The torsional angle ψ (defined as C-1–O–C-n–C-n-1), however, exhibits the -ac conformation for β-(1 → 3) but +ac for β-(1 → 4) interglycosidic linkages, respectively. The observed φ and ψ dihedral angles were found to be within 2° for φ and within 20° for ψ when compared with the published model disaccharide fragments GlcAβ3GalαOMe [101], GlcAβ3GalβO-benzyl[117] and Xylβ4XylβOMe.[118] The OSHP is one of the largest linkage-region glycopeptides fragments so far investigated for its solution conformation.[106] A comparison of the ^{1}H NMR chemical shifts for TSDP with those reported for larger fragments (see later) reveals sufficient similarity, suggesting that the conformation of the inner tetrasaccharide moiety GlcAβ3Galβ3Galβ4Xylβ-is virtually unaffected by the elongation of the saccharide chain. The set of glycosidic dihedral angles that define the overall conformation of the tetrasaccharide moiety of TSDP (from the energy-minimized average structure deduced from NMR and structure-refinement calculations[106]) is given in Table I, whereas the energy-minimized average structure for TSDP is shown in Fig. 1.

VI. Conformation of the Glycopeptide Linkage

The orientation of the peptide chain with respect to the oligosaccharide depends on the torsion angles defining the Xyl-Ser linkage as well as those defining the

TABLE I

Calculated Interglycosidic Torsion Angles[a] φ and ψ for the Energy-Minimized Average Structure of TSDP [GlcAβ3Galβ3Galβ4Xylβ-O]Ser-Gly, from NMR Studies[106]

Linkage	φ	Angle (Deg.)	ψ	Angle (deg.)
Xyl-Ser	ϕ_1 Xyl O-5–Xyl C-1–O–Ser C-β	−85 (−84)	ψ_1 Xyl C-1–O-Ser C-β–Ser C-α	+100 (+103)
Gal-Xyl	ϕ_2 Gal O-5–Gal C-1–O-Xyl C-4	−69 (−59)	ψ_2 Gal C-1–O-Xyl (C-4–Xyl C-3	+112 (+115)
Gal'-Gal	ϕ_3 Gal'O-5–Gal' C-1–O-Gal C-3	−63 (−57)	ψ_3 Gal' C-1–O-Gal C-3–Gal C-2	−150 (−149)
GlcA-Gal'	ϕ_4 GlcA O-5–GlcA C-1–O-Gal' C-3	−63 (−61)	ψ_4 GlcA C-1–O-Gal' C-3–Gal'C-2	147 (−148)

[a] The angles in parentheses are the average values of torsion angles within the family of structures.

peptide backbone. The $^3J_{\alpha,\beta}$ and $^3J_{\alpha,\beta'}$ couplings for Ser residues in CPL are typically about 4–5 Hz, suggesting that the predominant rotamer corresponds to χ = 60°.[79] We have observed in the case of OSHP that the two Xyl-Ser glycopeptide linkages are characterized by two NOEs (XylH-1/SerH-β and XylH-1/SerH-β'). Between these the more intense cross-peak corresponds to XylH-1-SerH-β' (upfield proton). Similar kinds of NOE observations were reported earlier by Paulsen et al.[119] for Xyl-Ser-containing glycopeptides. Use of NOE-derived distance-constraints in DG/SA calculations led to −sc and +ac conformations for the φ (Xyl O-5–Xyl C-1–O–Ser C-β) and ψ (Xyl C-1–O–Ser C-β–Ser C-α) torsional angles, respectively, for the Xyl-Ser linkage. The preference of φ for the -sc range is in accordance with the studies on O-glycopeptides containing a β-D-xylosyl group

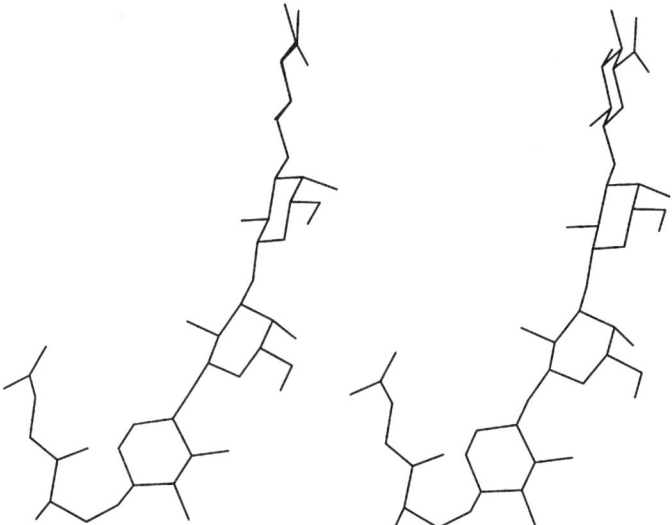

FIG. 1. Stereo view of the energy-minimized average structure of the tetrasaccharide moiety in TSDP([GlcAβ3Galβ3Galβ4XylβO]Ser-Gly) from NMR studies.

linked to an L-seryl residue[119,120] and glycoceramides incorporating β-linked glucose or galactose moieties.[121,122]

VII. Influence of Glycosylation on Proteoglycan Core-Peptide Backbone Conformation

Elucidation of the impact of N- and/or O-linked carbohydrate chains on the conformation and flexibility of peptides is a topic of current interest.[123] Glycosylated di- and tri-peptides have been used as model systems with which to look for interactions between attached sugars and the peptide, but the results are inconclusive.[124,125] Berman et al.[126] suggested that glycosylation has little effect on the dynamics and conformation of proteins. Measurements of the ^{13}C NMR relaxation times undertaken on bovine maxillary mucin have shown that O-linked carbohydrates alter the peptide–core conformation and decrease the mobility of the glycosylated residues.[127] Kessler et al.[128] concluded from NMR studies comparing cyclic hexapeptides and the corresponding glycosylated cyclic hexapeptides that there is no change in the backbone conformation of the peptide on glycosylation. In other glycoproteins, the impact of saccharide moieties is observed on the mobilities near the glycosylation sites, whereas the conformation of the protein is not affected.[129] From the measurements of amide exchange-rates, a decrease of conformational mobility for fucosylated proteinase inhibitor has been observed, and this result has been correlated with enhanced thermal stability of the protein on fucosylation.[130] Wyss et al.[131] assigned most of the protons of the adhesion domain of human CD_2, including resonances of the N-linked glycan chain, and identified several interactions between the carbohydrate chain and protein. Based on sequential amide proton NOEs, Andreotti and Kahne[132] and Liang and co-workers[133] were able to show that the conformation of the peptide backbone is radically different according to whether a mono- or disaccharide is attached. NMR studies on glycosylated recombinant human granulocyte-colony-stimulating factor suggest that the carbohydrate moiety decreases the local mobility around the glycosylation site, and this could be responsible for the stabilizing effect observed in the glycoprotein.[134] NMR studies on glycosylated analogs of the principal neutralizing determinant of gp 120 showed that covalent linking of a carbohydrate to the peptide has a major effect on the local conformation and imparts additional minor changes at more distant sites.[135]

NMR studies have been carried out for OSHP to determine the influence of tetrasaccharide substitution at Ser-1 and Ser-5 of (Ser-Gly)$_3$ hexapeptide. A comparison of ROESY data of OSHP and the unglycosylated hexapeptide (Ser-Gly)$_3$ was performed.[106] Under identical experimental conditions, both of these compounds exhibit strong intraresidue $d_{\alpha,N}$ (i, i) ROEs for glycines, $d_{\beta,N}$ (i, i) for Ser-3 and Ser-5, together with strong $d_{\alpha,N}$ $(i, i+1)$ ROEs. The backbone $d_{N,N}$

ROEs (sequential and/or nonsequential) were absent in case of the unglycosylated hexapeptide. The $d_{N,N}$ $(i, i+1)$ ROEs were also absent in the case of OSHP, except for Ser-5, which exhibits a weak d_{NN} $(i, i+1)$ ROE with Gly-6. Strong interresidue sequential $d_{\alpha,N}$ $(i, i+1)$ ROEs and the general absence of $d_{N,N}$ $(i, i+1)$ ROEs were consistent with a somewhat extended structure of the hexapeptide.[106,136] Thus a weak backbone ROE between Ser-5/Gly-6, which is present in OSHP but absent in unglycosylated hexapeptide, suggests that glycosylation at Ser-5 slightly alters the peptide backbone at this site, presumably because of its proximity to the C-terminal carboxylate groups, whereas glycosylation at Ser-1 has no effect on the peptide backbone.[106]

Studies on BGH and TGH showed that only trivial ROESY cross-peaks were observed in H_2O solutions at 298 K, but in 3:2 H_2O–(TFE) solvent mixture at 273 K there was an increase in the number of peaks in the NOESY spectra, and in spectral dispersion. The data in H_2O–TFE suggest that these glycopeptides exist in a conformational equilibrium.[83] Positive chemical-shift indices were observed for all the α protons of internal residues (Gly-2, Ser-3, Gly-4, Ser5) in BGH, TGH, and in OSHP, suggesting the presence of some specific conformations. The fact that Ser3, which is unglycosylated in both BGH and in OSHP, also shows a positive CSI indicates that these shifts are not a consequence of inductive effects asociated with glycosylation. In contrast, the serines residues in unglycosylated control peptide (Ser-Gly)$_7$ showed CSI indices of zero.[83] In these studies, no attempt was made to build a model for the peptide backbone because of uncertainties associated with the quantitative interpretation of NOESY data on a glycopeptide existing in a fast equilibrium among several conformations. Studies on longer glycopeptides (isotopically labeled if possible to take advantage of multidimensional NMR methods) might reduce the conformational flexibility problems associated with the shorter glycopeptides. It should also be noted that the findings in H_2O–TFE solvent mixture have no simple bearing to results in H_2O alone.

Experimental evidence based on NH-NH NOE data suggests that O-glycosylation of threonine (Thr) by α-D-GalNAc and β-D-Gal-(1 $\rightarrow$ 3)-α-D-GalNAc induces turns in the peptides Ac-Phe-Phe-D-Trp-Lys-Thr-Phe-NH$_2$ and Ac-Val-Thr-His-Pro-Gly-Tyr-NH$_2$[132,133]; however, our results provide evidence that attachment of xylose at serine residues in on the proteoglycan core proteins does not produce any dramatic changes, such as β-turns.[83,106] Steric hindrance between the acetamido group of α-D-GalNAc and the methyl group of Thr may be one of the factors responsible for the modification of the peptide backbone conformation observed in earlier studies.[132,133] This is supported, moreover, by a subsequent NMR study by Huang et al.[135] of glycosylated analogs of the principal neutralizing determinant of gp120 having O-glycosylation by α-D-GalNAc at Thr and Ser residues. These authors observed stronger NOEs around the Thr19 glycosylation site, indicative of more local rigidity as compared with the Ser6 site.

Structural studies have been reported for a GlcNAc-containing glycopeptide having GlcNAc linked to serine and/or threonine.[137] The authors observed that (*a*)

TABLE II
Oligosaccharide Fragments Isolated from the Carbohydrate–Protein Linkage-Region of Dermatan Sulfate Proteoglycans[a]

Source	Oligosaccharide Structure	Ref.
Pigskin		
1.	ΔHexA-GalNAc(S)-GlcA-Gal-Gal-Ser	138
Bovine aorta		
2.	$\Delta^{4,5}$ HexAα3 GalNAcβ4GlcAβ3Galβ3Galβ4Xyl-ol	102
3.	$\Delta^{4,5}$ HexAα3GalNAc(6S)β4GlcAβ3Galβ3Galβ4Xyl-ol	102
4.	$\Delta^{4,5}$ HexAα3GalNAc(4S)β4GlcAβ3Galβ3Galβ4Xyl-ol	102
5.	$\Delta^{4,5}$ HexAα3GalNAc(4S)β4IdoAα3Galβ3Galβ4Xyl-ol	102
6.	$\Delta^{4,5}$ HexAα3GalNAc(4S)β-4IdoAα3Gal(4S)β3Galβ4Xyl-ol	102

[a] $\Delta^{4,5}$ HexAα denotes the unsaturated residue derived from a β-L-IdoA residue by 4,5-elimination.

chemical-shift variations are restricted to the glycosylated amino acids, indicating that changes in peptide structure and dynamics are limited to the O-glycosylation, and carbohydrate moieties do not interfere with the peptide backbone (absence of ROEs between GlcNAc and amino acid residues other than Ser or Thr.[137]

VIII. STRUCTURAL STUDIES ON THE CARBOHYDRATE–PROTEIN LINKAGE REGION

In the following, we summarize briefly findings on the CPL fragments from dermatan sulfate, chondroitin sulfate, heparan sulfate, and heparin.

1. Dermatan Sulfate

Structural studies (Table II) have been reported for pigskin dermatan sulfates released from their parent proteoglycan.[102,138,139] The various products obtained by exhaustive proteolysis by both endo- and exo-peptidases were separated by gradient PAGE, detected by autoradiography, and quantified by videodensitometry. Complete digestion with chondroitin ABC lyase afforded a fragment, the structure of which was confirmed by sequential degradation from the nonreducing-end as $\Delta^{4,5}$HexA-GalNAc(S)-GlcA-Gal-Gal-Xyl-Ser. The Xyl residues in the CPL fragments from pigskin dermatan sulfate are not phosphorylated.[138]

The reductive β-elimination and subsequent chondroitinase ABC digestion of dermatan sulfate peptidoglycan from bovine aorta afforded five hexasaccharide alditols 2–6.[139] Three of these compounds 2–4 have the conventional hexasaccharide core $\Delta^{4,5}$-Hex3GalNAc(4S) β4GlcAβ3-Galβ3Galβ4Xyl-ol having either unsulfated (2) or 4/6 monosulfated GalNAc residues (3 and 4). The other two, 5 and 6, possess an internal IdoA instead of the usual GlcA residue ($\Delta^{4,5}$Hex3GalNAcβ4

IdoAα3Galβ3Galβ4Xyl-ol) having either C-4 of GalNAc monosulfated or disulfation on C-4 of GalNAc and Gal-3 residues. The latter two structures, **5** and **6**, account for 35% of the five hexasaccharide alditols isolated.[102] The conclusions of these studies are in striking contrast to the currently held notion that heparin, HS, and CS/DS share a common tetrasaccharide core in the linkage. These observations show that a structural feature characteristic of dermatan sulfate emerges at the fourth saccharide residue from the attachment site to the core protein. How general this structure is among dermatan sulfates of various cells and tissues and how variable it is during development, aging, or pathogenesis remains to be investigated.[102]

Sugahara et al.[102] also suggested that the dermatan sulfate chain, with iduronic acid closer to the linkage region, might be more flexible and mobile and thus may be able to swing around the core protein because of the specific conformational properties of IdoA. The conformation and flexibility of uronate residues have been the subjects of a number of studies.[90,140–142] These studies indicated that iduronic acid may be present in one of the three low-energy conformations, $^1C_4$, $^2S_0$, or $^4C_1$ or in all these three forms in rapid dynamic equilibrium. This equilibrium is highly sensitive to sulfation and carbohydrate sequences, as well as to such intermolecular factors as cation binding.[143]

2. Chondroitin Sulfate

The structural studies (Table III) on proteoglycans of shark cartilage, bearing predominantly chondroitin 6-sulfate, have been reported by Sugahara et al.[144] and De Waard et al.[145] Exhaustive protease digestion, reductive β-elimination, and subsequent digestion by chondroitinase ABC, gave 13 hexasacaharide alditols (**1–13**) having $\Delta^{4,5}$GlcAβ3GalNAcβ4GlcAβ3Galβ3Galβ4Xyl-ol* as the parent carbohydrate chain and that differ in the extent and sites of substitution of sulfation and/or phosphorylation. Compound 1 has no sulfate nor phosphate. Two of the monosulfated compounds have an O-sulfate group at C-6 (**2**) or C-4 (**4**) of the GalNAc residue. The third monosulfated compound (**3**) has a novel O-sulfate at C-6 of the Gal residue attached to xylitol. The two phosphorylated compounds (**5,6**) have O-phosphate at C-2 of Xyl-ol, and one of them (**6**) has an additional sulfate at C-6 of GalNAc.[144] These six compounds, containing 0 or 1 sulfate and/or phosphate residue, constitute ∼40% of the isolated linkage hexasaccharide alditols.[144] The other seven compounds (**7–13**), comprising ∼60% of the isolated linkage hexasaccharide alditols, were either disulfated or trisulfated. Two disulfated compounds have an O-sulfate at C-6 of the Gal residue attached to xylitol in combination with an O-sulfate at C-4 (**7**) or C-6 (**10**) of the GalNAc residue. The third disulfated compound (**9**) also has O-sulfate at C-6 of both Gal residues.

---

* Here and elsewhere, $\Delta^{4,5}$GlcAβ denotes 4-deoxy-α-L-*threo*-hex-4-enopyranosyluronic acid derived by 4,5-elimination from a β-D-GlcA residue.

TABLE III
Oligosaccharide Fragments Isolated from the Carbohydrate–Protein Linkage-Region of Chondroitin Sulfate Proteoglycans

Source	Oligosaccharide Structure	Ref.
Shark cartilage		
1. $\Delta^{4,5}$ GlcAβ3GalNAcβ4GlcAβ3Galβ3Galβ4Xyl-ol		144
2. $\Delta^{4,5}$ GlcAβ3GalNAc(6S)β4GlcAβ3Galβ3Galβ4Xyl-ol		144
3. $\Delta^{4,5}$ GlcAβ3GalNAcβ4GlcAβ3Galβ3Gal(6S)β4Xyl-ol		144
4. $\Delta^{4,5}$ GlcAβ3GalNAc(4S)β4GlcAβ3Galβ3Galβ4Xyl-ol		144
5. $\Delta^{4,5}$ GlcAβ3GalNAcβ4GlcAβ3Galβ3Galβ4Xyl(2P)-ol		144
6. $\Delta^{4,5}$ GlcAβ3GalNAc(6S)β4GlcAβ3Galβ3Galβ4Xyl(2P)-ol		144
7. $\Delta^{4,5}$ GlcAβ3 GalNAc(4S)β4GlcAβ3Galβ3Gal(6S)β4Xyl-ol		145
8. $\Delta^{4,5}$ GlcAβ3GalNAc(6S)β4GlcAβ3Gal(6S)β3Gal(6S)β4Xyl-ol		145
9. $\Delta^{4,5}$ GlcAβ3GalNAcβ4GlcAβ3Gal(6S)β3Gal(6S)β4Xyl-ol		145
10. $\Delta^{4,5}$ GlcAβ3 GalNAc(6S)β4GlcAβ3Galβ3Gal(6S)β4Xyl-ol		145
11. $\Delta^{4,5}$ GlcAβ3GalNAc(4S)β4GlcAβ3Gal(6S)β3Gal(6S)β4Xyl-ol		145
12. $\Delta^{4,5}$ GlcAβ3GalNAc(6S)β4GlcAβ3Gal(4S)β3Gal(6S)β4Xyl-ol		145
13. $\Delta^{4,5}$ GlcAβ3GalNAc(4S)β4GlcAβ3Gal(4S)β3Gal(6S)β4Xyl-ol		145
Human urinary trypsin inhibitor		
14. $\Delta^{4,5}$ HexAα3GalNAc(4S)β4GlcAβ3Gal(4S)β3Galβ4Xyl-ol		70
Inter-α-trypsin inhibitor in human plasma		
14'. $\Delta^{4,5}$ HexAα3GalNAc(4S)β4GlcAβ3Gal(4S)β3Galβ4Xyl-ol		146
Whale cartilage		
15. $\Delta^{4,5}$ GlcAβ3GalNAcβ4GlcAβ3Galβ3Galβ4Xyl-ol		147
16. $\Delta^{4,5}$ GlcAβ3GalNAc(6S)β4GlcAβ3Galβ3Galβ4Xyl-ol		147
17. $\Delta^{4,5}$ GlcAβ3GalNAc(4S)β4GlcAβ3Galβ3Galβ4Xyl-ol		147
18. $\Delta^{4,5}$ GlcAβ3GalNAc(4S)β4GlcAβ3Galβ3Gal(4S)β4Xyl-ol		147
Swarm rat chondrosarcoma		
19. $\Delta^{4,5}$ GlcAβ3GalNAcβ4GlcAβ3Galβ3Galβ4XylβO-Ser		66
20. $\Delta^{4,5}$ GlcAβ3GalNAc(4S)β4GlcAβ3Galβ3Galβ4XylβO-Ser		66
21. $\Delta^{4,5}$ GlcAβ3GalNAc(4S)β4GlcAβ3Gal(4S)β3Galβ4XylβO-Ser		66
22. $\Delta^{4,5}$ HexAβ3GalNAc(4S)β4GlcAβ3Galβ3Galβ4Xyl(2P)-ol		148
Engelbreth–Holmswarm mouse tumor		
23. $\Delta^{4,5}$ GlcAβ3GalNAcβ4GlcAβ3Galβ3Galβ4Xyl-ol		149
24. $\Delta^{4,5}$ GlcAβ3 GalNAc(6S)β4GlcAβ3Galβ3Galβ4Xyl-ol		149
25. $\Delta^{4,5}$ GlcAβ-3GalNAcβ4GlcAβ3Galβ3Galβ4Xyl(2P)-ol		149
26. $\Delta^{4,5}$ GlcAβ3GalNAc(4S)β4GlcAβ3Galβ3Galβ4Xyl-ol		149
27. $\Delta^{4,5}$ GlcAβ3GalNAc(6S)β4GlcAβ3Galβ4Xyl(2P)-ol		149
28. $\Delta^{4,5}$ GlcAβ3GalNAc(4S)β4GlcAβ3Galβ3Galβ4Xyl(2P)-ol		149
Bovine nasal septal cartilage		
29. $\Delta^{4,5}$ HexAα3GalNAcβ4GlcAβ3Galβ3Galβ4Xyl-ol		150
30. $\Delta^{4,5}$ HexAα3GalNAcβ4GlcAβ3Gal(4S)β3Galβ4Xyl-ol		150
31. $\Delta^{4,5}$ HexAα3GalNAc(6S)β4GlcAβ3Galβ3Galβ4Xyl-ol		150
32. $\Delta^{4,5}$ HexAα3GalNAc(4S)β4GlcAβ3Galβ3Galβ4Xyl-ol		150
33. $\Delta^{4,5}$ HexAα3GalNAc(4S)β4GlcAβ3Gal(4S)β3Galβ4Xyl-ol		150

Two of the trisulfated compounds have O-sulfate at C-6 of both Gal residues with the additional sulfate at C-6 (**8**) or C-4 (**11**) of GalNAc. The other two trisulfated compounds have O-sulfate at C-6 on Gal-2 and C-4 of Gal-3 in conjunction with sulfate at C-6 (**12**) or C-4 (**13**) of GalNAc.[145] The chondroitin 4-sulfate chain released from urinary trypsin inhibitor (UTI) by β-elimination using alkaline $NaBH_4$ followed by chondroitinase ABC digestion afforded[85] a single disulfated hexasaccharide alditol (**14**) possessing sulfate groups at C-4 of GalNAc and Gal-3. This disulfated structure **14** was also identified as the sole structural component in the linkage hexasaccharide alditol fraction isolated from inter-α-trypsin inhibitor (ITI) in human plasma.[146]

From the carbohydrate–protein linkage region of whale cartilage proteoglycans, which bear predominantly chondroitin 4-sulfate, four hexasaccharide alditols having the conventional structure $\Delta^{4,5}$GlcAα3GalNAcβ4GlcAβ3Galβ3Galβ4Xyl-ol were isolated.[147] These were nonsulfated (**15**), monosulfated either at C-6 (**16**) or at C-4 of the GalNAc residue (**17**), and disulfated having sulfate groups at C-4 of GalNAc and C-4 of Gal-3 (**18**).[147]

Based on ^{31}P NMR studies, the presence of 2-phosphoxylose (**22**) has been reported in the linkage region of chondroitin 4-sulfate proteoglycan of Swarm rat chondrosarcoma.[148] The authors also detected sulfate in the linkage region, but did not quantitate or localize it to a precise position.[148] The exhaustive enzymatic digestion of Swarm rat chondrosarcoma proteoglycans with chondroitinase ABC, papain, and pronase led to the isolation of nonsulfated, monosulfated, and disulfated glycopeptides containing the entire linkage region $\Delta^{4,5}$GlcAβ3GalNAcβ4GlcAβ-3Galβ3Galβ4Xyl$O$-Ser.[60] The monosulfated compounds have a sulfate at C-4 of the GalNAc residue and the disulfated component (**21**) has an additional sulfate at C-4 of the second galactose. The molar ratio of the isolated nonsulfated, monosulfated, and disulfated compounds was 53:37:10 based on the serine content.[66]

The linkage region fraction isolated from chondroitin sulfate chains attached to the hybrid proteoglycan of the Engelbreth–Holm Swarm mouse tumor shared the conventional hexasaccharide backbone structure $\Delta^{4,5}$GlcAβ3GalNAcβ4GlcAβ-3Galβ3Galβ4Xyl-ol. In addition to the presence of unsulfated and unphosphorylated hexasaccharide, monosulfated (sulfate at C-4 or C-6 of GalNAc) and/or phosphorylated (phosphate ester group at C-2 of xylose) fragments were isolated.[149] Five hexasaccharides (**29–33**) have been isolated from chondroitin 4-sulfate of bovine nasal septal cartilage.[150] These share the common core saccharide structure. One compound (**29**) does not contain sulfate. Two of the three monosulfated compounds (**31** and **32**) have O-sulfate at either C-6 or at C-4 of the GalNAc and the other monosulfated compound (**30**) has O-sulfate at C-4 of the Gal residue preceding the GlcA residue. The disulfated compound **33** has sulfate groups at C-4 of Gal and GalNAc residues.[150]

TABLE IV
Oligosaccharide Fragments Isolated from the Carbohydrate–Protein Linkage-Region of
Heparan Sulfate and Heparin Proteoglycans

Source	Oligosaccharide Structure	Ref.
Heparan sulfate		
Bovine lung		
1. -3Galβ3Galβ4Xyl(2P)-Ser		152
Bovine kidney		
2. $\Delta^{4,5}$ HexAα4GlcNAcα4GlcAβ3Galβ3Galβ4Xyl		67
Heparin		
Porcine intestinal		
1. $\Delta^{4,5}$ GlcAβ3Galβ3Galβ4XylβO-Ser		103
2. $\Delta^{4,5}$ GlcAα4GlcNAc(6S)β4GlcAβ3Galβ3Galβ4XylβO-Ser		103
3. ΔHexA(2S)α4GlcN(NS,6S)α4IdoAα4GklcNAcα4GlcAβ3Galβ3Galβ4XyklβO-Ser		68
4. ΔHexA(2S)α4GlcN(NS,6S)α4GlcAβ4GlcNAcα4GlcAβ3Galβ3Galβ4XylβO-Ser		68
5. ΔHexA(2S)α4GlcN(NS,6S)α4IdoAα4GlcNAcα4GlcAβ4GlcNAcα4GlcAβ3Galβ3Galβ4XylβO-Ser		68
6. ΔHexAα4GlcN(NS,6S)α4IdoAα4GlcNAc(6S)α4GlcAβ3Galβ3Galβ4XylβO-Ser		68
Porcine mucosal		
7. ΔHexβ3Galβ3Galβ4XylβO-Ser-NDBP		104
8. ΔHex(2S)4GlcNAcα4IdoAα4GlcNAc(6S)α4Glcβ3Galβ3Galβ4XylβO-Ser-NDBP		104

From the structural studies reported for the CPL region of chondroitin sulfate, it is evident that both galactose and GalNAc can bear sulfate group at positions 6 and/or 4. In most cases, position 4 of Gal-3 has been found to be sulfated. However, the Gal residues of the CPL region from porcine intestinal heparin are not sulfated. Thus, there seem to be distinct differences in the structure of the linkage region of CS and heparin. The structural differences in the linkage region may determine the character of the GAG species during its biosynthesis.

3. Heparan Sulfate

Although the linkage between heparan sulfate and protein has not been studied in detail, it has been known for long time that serine is the linkage amino acid and that serine, xylose, and galactose are present in the molar ration of 1:1:2.[151] Structural studies have been reported for heparan sulfate of bovine lung[152] and of bovine kidney (Table IV).[67] The heparan sulfate from bovine lung contains a monophosphate ester at C-2 of the xylose moiety, which is 4-substituted by a (1 → 3)-linked digalactose moiety. The unsulfated and unphosphorylated hexasaccharide $\Delta^{4,5}$HexA4GlcNAcα4GlcAβ3Galβ3Galβ4Xyl having both β- and α-anomeric forms of D-xylopyranose (**15** and **16**) (Table V) have been characterized from the heparan sulfate of bovine kidney after limited digestion with heparitinase I.[67]

TABLE V

^{1}H NMR Chemical Shifts for Monosaccharide Residues from the Carbohydrate–Protein Linkage-Region of Proteoglycans

Fragment	1	2	2	3	3	3	4	4	5	5	6
Reference	87	87	79	83	83 [TFE]	83 [TFE]	83 [TFE]	83 [TFE]	87	80	107
Ser H-α	3.99	3.99	3.92	4.29	4.50	4.81	4.48	4.80	4.04	3.98	
H-β	4.02	4.03	3.97	4.07	4.39	4.35	4.36	4.33	3.99	4.03	4.12
H-β′	4.25	4.25	4.22	4.24	4.19	3.95	4.17	3.97	4.26	4.25	4.28
Xyl H-1	4.43	4.46	4.42	4.49	4.55	4.49	4.54	4.49	4.47	4.46	4.51
H-2	3.32	3.38	3.38	3.38	3.48	3.44	3.47	3.44	3.38	3.38	3.41
H-3	3.45	3.61	3.57	3.62	3.70	3.70	3.69	3.69	3.62	3.62	
H-4	3.62	3.85	3.80	3.86	3.92	3.92	3.89	3.91	3.87	3.87	
H-5a	3.31	3.41	3.41	3.42	3.48	3.49	3.47	3.44	3.41	3.41	3.44
H-5e	3.97	4.11	4.08	4.12	4.19	4.18	4.17	4.17	4.12	4.12	4.15
Gal H-1		4.47	4.43	4.49	4.52	4.52	4.51	4.51	4.53	4.53	4.56
H-2		3.51	3.47	3.52	3.62	3.62	3.60	3.60	3.79	3.67	
H-3		3.65	3.61	3.65	3.72	3.72	3.71	3.71	3.82	3.82	
H-4		3.92	3.88	3.95	3.99	3.99	3.98	3.98	4.19	4.19	4.22
H-5		3.70	3.70	3.742	3.77	3.77	3.75	3.75	3.72	3.73	
H-6		3.80	3.80	3.82	3.92	3.92	3.91	3.91	3.76	3.75	
H-6′		3.75	3.75	3.76	3.83	3.83	3.82	3.82	3.68	3.72	
Gal H-1									4.61	4.61	4.64
H-2									3.61	3.60	
H-3									3.67	3.66	
H-4									3.93	3.92	3.95
H-5									3.69	3.69	
H-6									3.77	3.80	
H-6′									3.74	3.77	

Fragment	7	8	9	10	11
Reference	78	78	54	84	104
Xyl H-1	4.50	4.54	4.35	5.24	4.25
H-2	3.30	3.34	3.29	3.66	3.20
H-3	3.55	3.55	3.60	3.92	3.39
H-4	3.86	3.86	3.85	3.74	3.53
H-5a	3.38	3.38	3.41	3.92	3.14
H-5e	4.09	4.10	4.11	4.21	3.61
Gal H-1	4.51	4.51	4.53	4.60	4.01
H-2	3.64	3.61	3.68	3.72	3.40
H-3	3.81	3.81	3.83	3.83	3.24
H-4	4.17	4.17	4.18	4.22	3.79
H-5	3.70	3.71			
H-6	3.76	3.76			
H-6′	3.72	3.72			

TABLE V—Continued

Fragment Reference	7 78	8 78	9 54	10 84	11 104				
Gal' H-1	4.64	4.63	4.68	4.70	4.21				
H-2	3.73	3.73	3.42	3.75	3.50				
H-3	3.79	3.79	3.72	3.79	3.62				
H-4	4.17	4.17	4.18	4.18	3.95				
H-5	3.66	3.66							
H-6	3.76	3.76							
H-6'	3.72	3.72							
GlcA H-1	4.65	4.65	4.67	4.78					
H-2	3.40	3.39	3.76	3.37					
H-3	3.49	3.49		3.73					
H-4	3.49	3.49		3.86					
H-5	3.71	3.70							
HexA^a H-1				5.48	5.16				
H-2				4.17	3.81				
H-3				3.90	4.01				
H-4				4.02	5.76				
H-5				3.71					
NAc				2.08					
Fragment Reference	12 106	13 106	13 106	14 155	15 67	16 67	17 66	18 66	19 66	20 103
Xyl H-1	4.47	4.47	4.43	4.44	4.60	5.19	4.44	4.44	4.45	4.44
H-2	3.37	3.37	3.32	3.35	3.26		3.36	3.36	3.36	3.35
H-3	3.62	3.61	3.62	3.61	3.57		3.61	3.62		3.61
H-4	3.87	3.86	3.85	3.87	3.79			3.87		3.86
H-5a	3.41	3.40	3.39	3.41	3.41		3.40	3.40	3.41	3.40
H-5e	4.12	4.11	4.10	4.12	4.09		4.12	4.12	4.13	4.11
Gal H-1	4.53	4.52	4.52	4.54	4.53	4.52	4.53	4.53	4.53	4.53
H-2	3.67	3.67	3.67	3.68	3.68	3.68	3.67	3.68	3.68	3.67
H-3	3.82	3.81	3.81	3.83	3.82	3.82	3.83	3.83	3.83	3.83
H-4	4.19	4.18	4.18	4.19	4.19	4.19	4.19	4.19	4.18	4.19
H-5	3.72	3.72	3.72	b						
H-6	3.79	3.78	3.78	b						
H-6'	3.75	3.73	3.73	b						
Gal' H-1	4.66	4.66	4.66	4.67	4.66	4.66	4.66	4.66	4.69	4.66
H-2	3.76	3.72	3.72	3.74	3.74	3.74	3.75	3.74	3.78	3.74
H-3	3.82	3.81	3.81	3.79	3.79	3.79	3.80	3.80	4.01	3.78
H-4	4.19	4.18	4.18	4.16	4.15	4.15	4.16	4.16	4.75	4.16
H-5	3.68	3.67	3.67	b						
H-6	3.79	3.78	3.78	b						
H-6'	3.75	3.73	3.73	b						

(continued)

TABLE V—*Continued*

Fragment Reference	12 106	13 106	13 106	14 155	15 67	16 67	17 66	18 66	19 66	20 103
GlcA H-1	4.67	4.67	4.67	4.66	4.65	4.65	4.66	4.67	4.75	4.64
H-2	3.41	3.41	3.41	3.43	3.41	3.41	3.45	3.45	3.44	3.42
H-3	3.51	3.51	3.51	3.78	3.67	3.67	3.62	3.64	3.61	3.67
H-4	3.51	3.51	3.51	c	3.73	3.73				3.75
H-5	3.72	3.72	3.72	c						3.75
Glc-NAc H-1				5.45	5.39	5.39	4.54	4.62	4.63	5.41
H-2				4.17	3.92	3.92	4.00	4.07	4.07	3.94
H-3				3.88	3.83	3.83	3.90	4.15	4.15	3.80
H-4				3.99			4.10	4.62	4.63	3.85
H-5				b						4.02
H-6				b						4.45
H-6′				b						4.17
NAc				2.05	2.04	2.04	2.06	2.10	2.10	2.04
Uronic acid H-1					5.09	5.09	5.18	5.27	5.27	5.16
H-2					3.77	3.77	3.79	3.83	3.78	3.81
H-3					4.26	4.26	4.09	3.94	3.94	4.24
H-4					5.79	5.79	5.90	5.97	5.96	5.81

Fragment Reference	21 85	22 85	23 85	24 85			21 85	22 85	23 85	24 85
Xyl H-1	4.46	4.46	4.46	4.47	GlcN H-1		5.38	5.57	5.38	5.35
H-2	3.36	3.36	3.35	3.37	H-2		3.28	3.30	3.85	3.27
H-3	3.61	3.61	3.61	3.62	H-3		3.64	3.63		3.68
H-4	3.89	3.87	3.86	3.87	H-4		3.83	3.83	3.80	3.85
H-5a	3.41	3.41	3.39	3.41	H-5		3.99	3.98		4.03
H-5e	4.12	4.12	4.11	4.11	H-6		4.35	4.33		
					H-6′		4.20	4.18		
Gal H-1	4.53	4.53	4.53	4.53	NAc				2.03	
H-2	3.67	3.67	3.67	3.66						
H-3	3.83	3.83	3.82	3.83	IdoA					
H-4	4.19	4.18	4.19	4.19	H-1				4.98	
					H-2				3.75	
Gal′ H-1	4.66	4.66	4.66	4.66	H-3				4.11	
H-2	3.74	3.74	3.74	3.74	H-4				4.06	
H-3	3.77	3.77	3.76	3.79	H-5				4.78	
H-4	4.16	4.16	4.15	4.16						
					GlcN H-1				5.36	
GlcA H-1	4.65	4.65	4.65	4.65	H-2				3.28	
H-2	3.40	3.40	3.40	3.42	H-3				3.66	
H-3	3.68	3.68	3.67	3.68	H-4				3.84	
H-4	3.73	3.74	3.71	3.75	H-5				3.98	
					H-6				4.35	

TABLE V—Continued

Fragment Reference	21 85	22 85	23 85	24 85		21 85	22 85	23 85	24 85
GlcNAc H-1	5.34	5.36	5.35	5.36	H-6′			4.20	
H-2	3.89	3.86	3.89	3.91	ΔHexA				
H-3	3.77		3.81	3.78	H-1	5.49	5.50	5.51	5.14
H-4	3.82	3.82	3.80	3.73	H-2	4.61	4.61	4.61	3.76
H-5	3.71		3.74	4.00	H-3	4.32	4.32	4.33	4.27
					H-4	5.97	5.97	5.97	5.78
NAc	2.03	2.03	2.04	2.04					
Uronic acid H-1	4.95	4.51	4.49	5.02					
H-2	3.72	3.39	3.34	3.78					
H-3	4.11	3.79	3.37	4.13					
H-4	4.07	3.78	3.74	4.05					
H-5	4.75								

Structural formulas for CPL-region oligosaccharides in Table V.
1. XylβO-Ser
2. Galβ4XylβO-Ser
3. (Galβ4XylβO)Ser-Gly-Ser-Gly-(Galβ4XylβO)Ser-Gly
4. (Galβ4XylβO)Ser-Gly-(Galβ4XylβO)Ser-Gly-(Galβ4XylβO)Ser-Gly
5. Galβ3Galβ4XylβO-Ser
6. (GlcAβ3Galβ3Galβ4XylβO)Ser-Gly
7. GlcAβ3Gal-β3Galβ4XylβO-benzyl
8. GlcAβ3Gal-β3Galβ4XylβO-naphthymethanol
9. GlcAβ3Gal-β3Galβ4XylβOMe
10. GalNAcα4GlcAβ3Galβ3Galβ4XylβMU
11. ΔUronic3GalβGalβ4XylβO-Ser)-(AA)$_n$-NDBP
12. GlcAβ3Galβ3Galβ4XylβO Ser-Gly
13. (GlcAβ3Galβ3Galβ4XylβO)Ser-Gly-Ser-Gly-(GlcAβ3Galβ3Galβ4XylβO)Ser-Gly
14. GalNAcα4GlcAβ3Galβ3Galβ4XylβO-Ser
15. $\Delta^{4,5}$HexAα4GlcNAcα4GlcAβ3Galβ3Galβ4Xylβ
16. $\Delta^{4,5}$HexAα4GlcNAcα4GlcAβ3Galβ3Galβ4Xylα
17. $\Delta^{4,5}$GlcAβ3GalNAcβ4GlcAβ3Galβ3Galβ4XylβO-Ser
18. $\Delta^{4,5}$GlcAβ3GalNAc(4S)β4GlcAβ3Galβ3Galβ4XylβO-Ser
19. $\Delta^{4,5}$GlcAβ3GalNAc(4S)β4GlcAβ3Gal(4S)β3Galβ4XylβO-Ser
20. $\Delta^{4,5}$GlcAβ4GlcNAc(6S)α4GlcAβ3Galβ3Galβ4XylβO-Ser
21. ΔHexA(2S)α4GlcN(NS,6S)α4IdoAα4GlcNAcα4GlcAβ3Galβ3Galβ4XylβO-Ser
22. ΔHexA(2S)α4GlcN(NS,6S)α4GlcAβ4GlcNAcα4GlcAβ3Galβ3Galβ4XylβO-Ser
23. ΔHexA(2S)α4GlcN(NS,6S)α4IdoAα4GlcNAcα4GlcAβ4GlcNAcα4GlcAβ3Galβ3Galβ4XylβO-Ser
24. ΔHexAα4GlcN(NS,6S)α4IdoAα4GlcNAc(6S)α4GlcAβ3Galβ3Galβ4XylβO-Ser

[a] Hex corresponds to GalNAc in **10** and to $\Delta^{4,5}$HexA in **11**.
[b] 3.70–380.
[c] 3.70–375.

4. Heparin

Exhaustive digestion of porcine intestinal heparin with a mixture of bacterial heparinase and heparitinase I led to the isolation of two glycoserines (Table IV) in the molar ratio of 96:4.[103a] The major one is $\Delta^{4,5}$GlcAβ3Galβ3Galβ4Xylβ

O-Ser and the minor one $\Delta^{4,5}$GlcAβ4GlcNAcα4GlcAβ3Galβ3Galβ4Xylβ O-Ser having sulfate at C-6 of GlcNAc (**2**). Analysis of fractional precipitates of the acid hydrolyzate of heparin preparations (from pig intestinal mucosa, beef intestinal mucosa, and beef lung) reported by Iacomini *et al.* (103b) identified a major linkage region oligosaccharide with the prevalent structure GlcA-GlcNAc-GlcA-Gal-Gal-Xyl, where GlcNAc is partially 6-O-sulfated. The heparinase digestion led to the isolation of octa- and decasaccharides **3–6**.[85] In all of these oligosaccharides (**3–6**) the CPL tetrasaccharide GlcAβ3Galβ3Galβ4Xyl was neither sulfated nor phosphorylated. During structural studies on the linkage region of crude porcine mucosal heparin, Liu *et al.*[104] labelled heparin on the amino termini of these core protein remnants with hydrophobic, fluorescent [N-4(6-dimethylamino-2-benzofuranylophenyl (NDBP) isothiocyanate] and isolated tetrasaccharide **7** and octasaccharide **8** after treatment by a mixture of heparin lyase I and III, and heparin lyase I, respectively.

IX. NMR Spectral Properties of Linkage-Region Oligosaccharides

The ^{1}H NMR chemical shifts reported in the literature [66–69,78–80,83–85,87,102–104, 106,107,144–147,153,154] for the CPL-region oligosaccharides and oligosaccharide xylitols are compiled in Tables V and VI, respectively. The compilation of ^{1}H NMR chemical shifts (in D$_2$O solutions) reveals that anomeric resonances are usually well separated (between 4.5 and 5.8 ppm) from other signals in the crowded 3.6–4.1 ppm region. Among the anomeric resonances, H-1 of Xyl resonates at the highest field position (4.43–4.51 ppm), followed by H-1 of Gal-2 (4.53–4.55 ppm). The H-1 signals of Gal-3 and GlcA appear as partially overlapped resonances in the 4.65–4.67 ppm range; among these, the lower field resonance usually corresponds to the GlcA residue. The best way to assign these resonances unambiguously to Gal-3 H-1/GlcA H-1 is on the basis of COSY connectivity with H-2 as H-2 resonates at ∼3.41 ppm for GlcA and at ∼3.72 ppm for the Gal-3 residue. The chemical shift of the GlcA H-2 is marginally affected by GalNAc substitution at 4-position. The α-anomeric proton resonance of IdoA appears at 4.94–5.02 ppm, whereas the anomeric signals of α-GlcNAc and α-GlcNS, appear between 5.3 and 5.4 ppm. The H-1 signal of 2-sulfated α-HexA resonates at δ 5.49–5.51 ppm. The type of uronic acid can be judged by the chemical shift of the anomeric proton—a signal around 4.60 ppm ($^3J_{1,2}$ ∼8 Hz) indicate β-GlcA whereas an anomeric-proton signal at ∼5.0 ppm ($^3J_{1,2}$ 2.5–3.5 Hz) is characteristic of α-IdoA.[155,156] However, the anomeric-proton signal of a 2-sulfated α-IdoA residue is observed at ∼5.2 ppm, because sulfation of the uronic acid results in a downfield shift of the anomeric proton by 0.2 ppm.[67,98] The presence of $\Delta^{4,5}$GlcA can be deduced from the signals ∼5.18 and 5.85 ppm, which are characteristic for H-1 and H-4 atoms of $\Delta^{4,5}$ GlcA, respectively.[157,158]

TABLE VI
¹H NMR Chemical Shifts for Oligosaccharide Xylitols from the Carbohydrate–Protein Linkage-Region

Fragment Reference	1 87	2 87	3 147	4 69	5 102	6 102	7 102	8 102	9 102
Xyl H-1	3.76	3.76							
H-2	3.89	3.89							
H-3	3.78	3.79							
H-4	3.98	3.99	3.99	3.99	3.99	3.99	3.99	3.99	3.99
H-5a	3.85	3.85							
H-5e	3.79	3.79							
Gal H-1	4.56	4.62	4.62	4.62	4.62	4.62	4.62	4.62	4.62
H-2	3.56	3.78		3.72				3.73	3.74
H-3	3.67	3.84		3.86				3.84	3.85
H-4	3.93	4.20	4.19	4.19	4.20	4.19	4.20	4.21	4.19
H-5	3.70	4.72							
H-6	3.80	3.78							
H-6'	3.76	3.74							
Gal' H-1		4.62	4.70	4.70	4.67	4.66	4.67	4.69	4.74
H-2		3.61		3.79				3.73	3.76
H-3		3.67	4.01	4.01					4.02
H-4		3.93	4.75	4.75	4.16	4.16	4.16	4.16	4.66
H-5		3.69							
H-6		3.80							
H6'		3.74							
HexA H-1			4.75	4.75	4.67	4.67	4.68	5.04	5.15
H-2			3.44	3.44	3.46	3.47	3.46	3.70	3.71
H-3			3.62	3.62	3.63	3.63	3.63	4.00	3.96
H-4				3.81				4.10	4.13
H-5								4.54	4.74
GalNAc H-1			4.63	4.63	4.54	4.58	4.62	4.68	4.69
H-2			4.07	4.07	4.01	4.03	4.07	4.07	4.08
H-3			4.15	4.15	3.90	3.94	4.15	4.15	4.16
H-4			4.63	4.63	4.10	4.17	4.62	4.62	4.63
H-5									
H-6						4.23			
H-6'						4.22			
NAc			2.10	2.10	2.06	2.05	2.10	2.11	2.12
ΔHexA H-1			5.27	5.27	5.19	5.18	5.27	5.26	5.27
H-2				3.83				3.83	3.83
H-3			3.94	3.94	4.09	4.11	3.94	3.94	3.94
H-4			5.97	5.97	5.89	5.88	5.96	5.96	5.96

(continued)

TABLE VI—Continued

Fragment Reference	10 144	11 144	12 144	13 144	14 144	15 144	16 145	17 145	18 145	19 145	20 145	21 145	22 145
Xyl-ol H-1					3.79								
H-2					4.27	4.28							
H-3					3.89								
H-4	3.99	3.99	3.97	3.99	4.03	4.03	3.98	3.98	3.97	3.97	3.97	3.96	3.96
H-5					3.89								
H-5'					3.89								
Gal H-1	4.62	4.62	4.63	4.62	4.63	4.63	4.63	4.63	4.63	4.63	4.63	4.63	4.63
H-2	3.72	3.72	3.72		3.74		3.71	3.71	3.72	3.72	3.71	3.72	3.72
H-3	3.84	3.84	3.86		3.85		3.86	3.86	3.86	3.86	3.86	3.87	3.87
H-4	4.20	4.20	4.25	4.20	4.19	4.19	4.29	4.29	4.25	4.25	4.29	4.24	4.24
H-5			3.96				3.97	3.97	3.96	3.97	3.97	3.97	3.97
H-6			4.20				4.24	4.24	4.20	4.20	4.24	4.21	4.21
H-6'			4.19				4.20	4.20	4.19	4.19	4.20	4.19	4.19
Gal' H-1	4.67	4.67	4.67	4.67	4.67	4.67	4.68	4.68	4.67	4.67	4.68	4.70	4.70
H-2	3.75	3.75	3.75		3.74		3.76	3.75	3.75	3.75	3.76	3.78	3.78
H-3	3.80	3.80	3.80		3.81		3.83	3.83	3.81	3.81	3.83	4.01	4.01
H-4	4.16	4.16	4.16	4.17	4.16	4.16	4.21	4.21	4.16	4.16	4.21	4.75	4.75
H-5							3.93	3.93			3.93		
H-6							4.20	4.19			4.19		
H-6'							4.18	4.18			4.17		
GlcA H-1	4.67	4.68	4.67	4.67	4.67	4.67	4.68	4.68	4.67	4.68	4.68	4.75	4.75
H-2	3.46	3.47	3.45	3.46	3.46	3.47	3.46	3.47	3.46	3.47	3.45	3.45	3.44
H-3	3.62	3.64	3.62	3.64	3.63	3.64	3.63	3.64	3.63	3.64	3.63	3.62	3.62
H-4	3.78	3.74	3.78		3.78		3.74	3.74		3.74	3.78	3.76	3.80
H-5	3.72	3.74	3.72		3.72		3.78	3.74		3.74	3.72	3.70	3.70
GalNAc H-1	4.53	4.57	4.53	4.61	4.53	4.57	4.61	4.57	4.61	4.57	4.53	4.59	4.63
H-2	4.00	4.03	4.00	4.07	4.00	4.03	4.07	4.03	4.07	4.03	4.00	4.03	4.07
H-3	3.90	3.95	3.90	4.15	3.90	3.95	4.15	3.95	4.15	3.95	3.90	3.95	4.15
H-4	4.10	4.18	4.10	4.62	4.10	4.18	4.62	4.18	4.62	4.18	4.10	4.18	4.63
H-5		4.01				4.10				4.10		4.01	
H-6		4.23				4.23		4.23		4.23		4.22	
H-6'		4.22				4.22		4.22		4.22		4.22	
NAc	2.06	2.05	2.06	2.10	2.06	2.05	2.10	2.06	2.10	2.05	2.06	2.06	2.10

TABLE VI—Continued

Fragment	10	11	12	13	14	15	16	17	18	19	20	21	22
Reference	144	144	144	144	144	144	145	145	145	145	145	145	145
ΔGlcA H-1	5.18	5.18	5.18	5.27	5.18	5.18	5.27	5.18	5.27	5.18	5.19	5.18	5.27
H-2	3.79	3.78	3.79		3.80		3.83	3.78	3.84	3.79	3.79	3.78	3.83
H-3	4.09	4.11	4.10	3.94	4.10	4.11	3.94	4.11	3.94	4.11	4.09	4.11	3.94
H-4	5.90	5.88	5.90	5.97	5.90	5.88	5.97	5.88	5.97	5.88	5.89	5.88	5.97

Structural formulas for CPL oligosaccharide xylitols in Table VI.
1 Galβ4Xyl-ol
2 Galβ3Galβ4Xyl-ol
3 $\Delta^{4,5}$ GlcAβ3GalNAc(4S)β4GlcAβ3Galβ3Gal(4S)β4Xyl-ol
4 $\Delta^{4,5}$ HexAα3GalNAc(4S)β4GlcAβ3Gal(4S)β3Galβ4Xul-ol
5 $\Delta^{4,5}$ HexAα3GalNAcβ4GlcAβ3Galβ3Galβ4Xyl-ol
6 $\Delta^{4,5}$ HexAα3GalNAc(6S)β4GlcAβ3Galβ3Galβ4Xyl-ol
7 $\Delta^{4,5}$ HexAα3GalNAc(4S)β4GlcAβ3Galβ3Galβ4Xyl-ol
8 $\Delta^{4,5}$ HexAα3GalNAc(4S)β4IdoAα3Galβ3Galβ4Xyl-ol
9 $\Delta^{4,5}$ HexAα3GalNAc(4S)β4IdoAα3Gal(4S)β3Galβ4Xyl-ol
10 $\Delta^{4,5}$ GlcAβ3GalNAcβ4GlcAβ3Galβ3Galβ4Xyl-ol
11 $\Delta^{4,5}$ GlcAβ3GlaNAc(6S)β4GlcAβ3Galβ3Galβ4Xyl-ol
12 $\Delta^{4,5}$ GlcAβ3GalNAcβ4GlcAβ3Galβ3Gal(6S)β4Xyl-ol
13 $\Delta^{4,5}$ GlcAβ3GalNAc(4S)β4GlcAβ3Galβ3Galβ4Xyl-ol
14 $\Delta^{4,5}$ GlcAβ3GalNAcβ4GlcAβ3Galβ3Galβ4Xyl(2P)-ol
15 $\Delta^{4,5}$ GlcAβ3GalNAc(6S)β4GlcAβ3Galβ3Galβ4Xyl(2P)-ol
16 $\Delta^{4,5}$ GlcAβ3GalNAc(4S)β4GlcAβ3Gal(6S)β3Gal(6S)β4Xyl-ol
17 $\Delta^{4,5}$ GlcAβ3GalNAc(6S)β4GlcAβ3Gal(6S)β3Gal(6S)β4Xyl-ol
18 $\Delta^{4,5}$ GlcAβ3GalNAc(4S)β4GlcAβ3Galβ3Gal(6S)β4Xyl-ol
19 $\Delta^{4,5}$ GlcAβ3GalNAc(6S)β4GlcAβ3Galβ3Gal(6S)β4Xyl-ol
20 $\Delta^{4,5}$ GlcAβ3GalNAcβ4GlcAβ3Gal(6S)β3Gal(6S)β4Xyl-ol
21 $\Delta^{4,5}$ GlcAβ3GalNAc(6S)β4GlcAβ3Gal(4S)βGal(6S)β4Xyl-ol
22 $\Delta^{4,5}$ GlcAβ3GalNAc(4S)β4GlcAβ3Gal(4S)βGal(6S)β4Xyl-ol

The Xyl, Gal, GlcA, GalNAc, and $\Delta^{4,5}$ GlcA residues so far characterized in the CPL have β-D-anomeric configurations (formally α-L-*threo* for $\Delta^{4,5}$ GlcA), as evident from their $^3J_{1,2}$ coupling constants, which range between 7 and 8 Hz. The $^3J_{1,2}$ values of 2–3 Hz for GlcNAc, GlcN, IdoA, and $\Delta^{4,5}$HexA residues suggest that these monosaccharide residues have the α-anomeric configuration. A comparison of the ^{1}H NMR chemical shifts for **1** with those reported for serine suggest that the attachment of a β-D-Xyl*p* group to L-Ser produces changes in the chemical shifts (H-α = +0.14, H-β = 0.07, H-β′ = 0.29 ppm) for the Ser protons (for unsubstituted L-Ser: δ H-α = 3.84 and δ H-β = δ H-β′ = 3.96 ppm). Furthermore, the Ser H$_2$-β protons become nonequivalent in **1**. On substitution of the D-Xyl*p* residue in **1** at C-4 by a β-D-Gal*p* group, as in **2**, small changes in chemical shifts of the Xyl protons were observed. The largest shift increments appeared for H-4 and H-5*e*. The signal for H-5*e* is now found at a significantly downfield position (∼4.12 ppm) and separated from the overlapped proton signals. On extending compound **2** by a (1 → 3)-β-D-Gal*p* groups (resulting in compound **3**), the chemical shifts of the Xyl and Ser protons are little if any affected. Most protons of the internal Gal-2 undergo considerable downfield shifts, but H-4 resonates at a distinct position (~4.18 ppm). Substitution at C-3 of Gal-3 by β-D-GlcA also causes H-4 of Gal-3 to resonate around 4.18 ppm. The chemical shifts of H-4 of Gal thus seem to be highly characteristic for the β-(1 → 3) linkage. The chemical shift of H-2 of the GlcN residue is sensitive to the presence of N-acetyl on N-sulfate groups, as it resonates at 3.85 ppm in the former and ~3.24 ppm in the latter's case.[67]

X. BIOSYNTHESIS OF CARBOHYDRATE–PROTEIN LINKAGE-REGION OLIGOSACCHARIDES

Even though this chapter is concerned with the molecular structure of CPL-region oligosaccharides, a few words about the biosynthesis of these molecules is appropriate at this stage. The chain initiation of proteoglycans starts with the transfer by xylosyltransferase of the sugar residue from UDP-xylose to the hydroxyl group of serine on the core protein.[1,159] Starting from xylose, biosynthesis of the remaining linkage region and then the glycosaminoglycan proceeds by the stepwise addition of one sugar at at time to the growing carbohydrate chain under the influence of glycosyltransferases. The glycosyltransferases show a high degree of specificity with respect to the nucleotide sugar, the acceptor, the anomeric configuration, and position of the linkage formed.[1] The sequential addition of the two Gal residues in the CPL is directed by two separate galactosyltransferases,[160] with subsequent attachment of GlcA by a specific GlcA-transferase, to form the common tetrasaccharide CPL sequence GlcA-Gal-Gal-Xyl-Ser. Chain elongation then proceeds by subsequent addition to the nonreducing-terminal GlcA residue of either a GlcNAc or a GalNAc, followed by a GlcA, under the direction of the corresponding glycosyltransferases. Thus, the (-4GlcAβ4GlcNAcα1-) disaccharide repeat generates a heparin/heparan sulfate chain, whereas the (-4GlcAβ3GalNAcβ1-) disaccharide repeat results in a chondroitin sulfate sequence. The sequential action of enzymes that partially N-deacetylate, N-sulfate, C-5 epimerize, and O-sulfate the polymer chain gives GAG chains their complex structure and its microheterogeneity. After polymerization, the chains are modified into their final structures by glycosyluronic epimerases and various sulfotransferases.

The precise molecular signal that leads to a divergence in the biosynthesis, producing different GAGs starting from this common tetrasaccharide linkage region GlcA-Gal-Gal-Xyl-Ser, is not well understood and has been the subject of intense investigations in many laboratories. Initial studies suggested a serine-containing consensus core-protein sequence of the form acidic-acidic-X-Ser-Gly-X-Gly for formation of the xylose–serine linkage in chondroitin sulfate proteoglycan.[161,162] A significant role has been suggested for the acidic residues flanking the site of attachment, Ser-Gly, in the biosynthesis of heparan sulfate.[163–165] This biosynthesis is augmented by an adjacent tryptophan residue.[163] There is also evidence to suggest that the primary sequence at the site of attachment may be less important than the overall structure.[164,166]

NMR studies on different CPL glycopeptides containing different core-peptide sequences (and longer sequences), together with studies on complexes with purified glycosyltransferases (when they become available), might further elucidate the roles of core-protein conformation and its sequence in contributing to this divergence in the biosynthesis that results in the assembly of different proteoglycans.

ACKNOWLEDGMENTS

This work was supported in part by the Arthritis Foundation and by NCI Grant CA-13148 (NMR Facility). The authors thank Drs. John Baker and Ted Sakai for their critical reading of the manuscript and for their comments, and Dr. Derek Horton for his editorial corrections and comments on this manuscript.

REFERENCES

(1) L. Rodén, in W. J. Lennarz (Ed.), *The Biochemistry of Glycoproteins and Proteoglycans,* Plenum, 1980, pp. 267–371.
(2) U. Lindahl, and L. Rodén, in A. Gottschalk (Ed.), *Glycoproteins* 1972, pp. 491–517.
(3) J. F. Kennedy, in *Proteoglycans: Biological and Chemical Aspects of Human Life.,* Elsevier, Amsterdam, 1979, pp. 131–143.
(4) M. Hook, L. Kjellen, S. Johansson, and J. Robinson, *Annu. Rev. Bichem.,* 53 (1984) 847–869.
(5) L.-A. Fransson, (1985) in G. O. Aspinall, (Ed.), *The Polysaccharides,* Academic Press, Orlando, FL, Vol. 3, 1985, pp. 3378–3415.
(6) D. Heinegard, and Y. Sommarin, *Methods Enzymol.,* 144 (1987) 305–370.
(7) E. Ruoslahti, *Annu. Rev. Cell Biol.,* 4 (1988) 229–255.
(8) J. T. Gallagher, *Curr. Opin. Cell. Biol.,* 1 (1989) 1202–1218.
(9) L. Kjellen, and U. Lindahl, *Annu. Rev. Biochem.,* 60 (1991) 443–475.
(10) R. L. Jackson, S. J. Busch, and A. D. Cardin, *Physiol. Rev.,* 71 (1991) 481–539.
(11) T. E. Hardingham, and A. J. Fosang, *FASEB J.,* 6 (1992) 861–870.
(12) K. G. Vogel, M. Paulsson, and D. Heinegard, *Biochem. J.,* 223 (1984) 587–597.
(13) D. J. Bidanset, C. Guidry, L. C. Rosenberg, H. U. Choi, R. Timpl, and M. Hook, *J. Biol. Chem.,* 267 (1992) 5250–5256.
(14) B. Font, E. Aubert-Foucher, D. Goldschmidt, D. Eichenberger, and M. van der Rest, *J. Biol. Chem.,* 268 (1993) 25015–25018.
(15) Y. Yamaguchi, D. M. Mann, and E. Ruoslahti, *Nature,* 346 (1990) 281–284.
(16) K, Lewandowska, H. U. Choi, L. C. Rosenberg, L. Zardi, and L. A. Culp, *J. Cell. Biol.,* 105 (1987) 1443–1454.
(17) D. M. Tollefsen, C. A. Pestka, and W. J. Monafo, *J. Biol. Chem.,* 258 (1983) 6713–6716.
(18) L.-A. Fransson, I. A. Nieduszynski, C. F. Phelps, and J. K. Sheelan, *Biochim. Biophys. Acta,* 586 (1979) 179–188.
(18a) IUPAC-IUBMB *Nomenclature of Carbohydrates, Adv. Carbohydr. Chem. Biochem.,* 52 (1997) 43–177.
(19) M. Belting, and L.-A. Franson, *Glycoconjugate J.,* 10 (1993) 453–460.
(20) A. R. Poole, *Biochem J.,* 236 (1986) 1–14.
(21) L. A. Fransson, *Trends Biochem. Sci.,* 12 (1987) 406–411.
(22) E. Ruoslahti, *J. Biol. Chem.,* 264 (1989) 13369–13372.
(23) J. E. Silbert, and G. Sugumaran, *Biochim. Biophys. Acta.,* 1241 (1995) 371–384.

(24) A. Kinoshita, S. Yamada, S. M. Haslam, H. R. Morris, A. Dell, and K. Sugahara, *J. Biol. Chem.*, 272 (1997) 19656–19665.
(25) V. C. Hascall, and G. K. Hascall, in E. D. Hay, (Ed.), *Proteoglycans,* Plenum, New York, 1981, pp. 39–63.
(26) J. T. Gallagher, and M. Lyon, in *Heparin,* (D. A Lane, and U. Lindahl, Eds), Edward Arnold, London, 1989, pp. 135–158.
(27) K. S. Rostand, and J. D. Esko, *Infect. Immun* , 65 (1997) 1–8.
(28) M. Patel, M. Yanagishita, G. Rodriguez, D. C. Bou Habib, T. Oravecz, V. C. Hascall, and M. A. Norcross, *AIDS Res. Hum. Retroviruses,* 9 (1993) 167–174.
(29) E. Feyzi, E. Trybala, T. Bergstrom, U. Lindahl, and J. R. Spillmann, *Adv. Exp. Med. Biol.,* 313 (1992) 87–96.
(30) M. T. Sheih, D. WuDunn, R. I. Montgomery, J. D. Esko, and P. G. Spear, *J. Cell Biol.,* 116 (1992) 1273–1281.
(31) L. A. Cisar, A. J. Hoogewerf, M. Cupp, C. A. Rapport, and A. Bensadoun, *J. Biol. Chem.,* 264 (1989) 1767–1774.
(32) Z. M. Marcum, D. H. Atha, L. M. S. Fritz, P. Nawroth, D. Stern, and R. D. Rosenberg, *J. Biol. Chem.,* 261 (1986) 7507–7517.
(33) A. Clowes, and M. Karnovsky, *Nature,* 265 (1977) 625–626.
(34) S. C. Thornton, S. H. Mueller, and E. M. Levin, *Science,* 222 (1983) 623–625.
(35) J. Laterra, J. E. Silbert, and L. A. Culp, *J. Cell Biol.,* 96 (1983) 112–123.
(36) A. Woods, J. R. Couchman, S. Johansson, and M. Hook, *EMBO J.,* 5 (1983) 665–670
(37) R. G. LeBaron, J. D. Esko, A. Woods, S. Johansson, and M. Hook, *J. Cell Biol.,* 106 (1988) 945–952.
(38) O. Saksela, D. Moscateelli, A. Somer, and D. B. Rifkin, *J. Cell Biol.,* 107 (1988) 743–751.
(39) M. Y. Gordon, G. P. Riley, S. M. Watt, and M. F. Greaves, *Nature,* 326 (1987) 403–405.
(40) R. Roberts, J. T. Gallagher, E. Spooncer, T. D. Allen, F. Bloomfield, and M. Dexter, *Nature,* 332 (1988) 376–378.
(41) R. J. Linhardt, and T. Toida, in *Carbohydrates as Drugs,* (Z. B. Witczak, and K. A. Nieforth, Eds.) Dekker, New York, 1996.
(42) K. Lidholt, J. L. Weinke, C. S. Kiser, F. N. Lugemwa, K. J. Bame, S. Cheifetz, J. Massaque, U. Lindahl, and J. D. Esko, *Proc. Natl. Acad. Sci., U.S.A.* 89 (1992) 2267–2271.
(43) T. Lind, U. Lindahl, and K. Lidholt, *J. Biol. Chem.,* 268 (1993) 20705–20708.
(44) M. Salmivirta, K. Lidholt, and U. Lindahl, *FASEB J.,* 10 (1996) 1270–1279.
(45) U. Lindahl, K. Lidholt, D. Spillmann, and L. Kjellen, *Thromb. Res.,* 75 (1994) 1–32.
(46) U. Lindahl, *Pure Appl. Chem.,* 69 (1997) 1897–1902.
(47) J. Folkman, and D. E. Ingber, in *Heparin* (D. A. Lane, and U. Lindahl, Eds), 1989, pp. 317–333, Edward Arnold, London.
(48) T. Macig, T. Mehlman, R. Friesel, and A. B. Schreiber, *Science,* 225 (1984) 932–935.
(49) S. Faham, R. E. Hileman, J. R. Fromm, R. J. Linhardt, and D. C. Rees, *Science,* 271 (1986) 1116–1120.
(50) M. Klagsbrun and Y. Shing, *Proc. Natl. Acad. Sci. U.S.A.,* 82 (1985) 805–809.
(51) J. A. Marcum, and R. D. Rosenberg, in *Heparin* (D. A. Lane, and U. Lindahl, Eds), Edward Arnold, London, 1989, pp. 275–294.
(52) C. C. Rider, *Glycoconjugate J.,* 14 (1997) 639–642.
(53) U. Lindahl, D. S. Feingold, and L. Roden, *Trends Biochem. Sci.,* 11 (1986) 221–225.
(54) B. Casu, *Adv. Carbohydr. Chem. Biochem.,* 43 (1985) 51–134.
(55) U. Lindahl, in *Heparin* (D. A. Lane, and U. Lindahl, Eds), Edward Arnold, London, 1989, pp. 159–189.
(56) H. Habuchi, S. Suzuki, T. Saito, T. Tamura, T. Harada, K. Yoshida, and K. Kimata, *Biochem. J.,* 285 (1992) 805–813.

(57) D. J. Tyrrell, M. Ishihara, N. Rao, A. Horne, M. C. Keifer, G. B. Stauber, L. H. Lam, and R. J. Stack, *J. Biol. Chem.,* 268 (1993) 4684–4689.
(58) M. Maccarana, B. Casu, and U. Lindahl, *J. Biol. Chem.,* 268 (1993) 23898–23905.
(59) T. C. Wright Jr. J. J. Castellot Jr. M. Petitou, J.-C. Lormeau, J. Choay, and M. J. Karnovsky, *J. Biol. Chem.,* 264 (1989) 1534–1542.
(60) D. Spillmann, and U. Lindahl, *Curr. Opin. Struct. Biol.,* 4 (1994) 677–682.
(61) C. A. A. van Boeckel, P. D. J. Grootenhuis, and A. Visser, *Nature Struct. Biol.,* 1 (1994) 423–425.
(62) T. Helting and L. Roden *J. Biol. Chem.,* 244 (1969) 2799–2805.
(63) N. B. Schwartz, and L. Roden, *Carbohydr. Res.,* 37 (1974) 167–180.
(64) N. B. Schwartz, *J. Biol. Chem.,* 251 (1976) 285–291.
(65) A. Malmstrom, L. A. Farnsson, and U. Lindahl, *J. Biol. Chem.,* 250 (1975) 3419–3425.
(66) K. Sugahara, I. Yamshina, P. de Waard, H. Van Halbeek, and J. F. G. Vligenthart, *J. Biol. Chem.,* 263 (1988) 10168–10174.
(67) K. Sugahara, T. Tohno-oka, S. Yamada, K. H. Khoo, H. R. Morris, and A. Dell, *Glycobiology,* 4 (1994) 535–544.
(68) K. Sugahara, H. Tsuda, K. Yoshida, S. Yamada, T. De Beer, J. F. G. Vligenthart, *J. Biol. Chem.,* 270 (1995) 22914–22923.
(69) D. Aminoff, W. D. Gathmann, C. M. McLean, and T. Yadomae, *Anal. Biochem.,* 101 (1980) 44–53.
(70) S. Yamada, M. Oyama, Y. Yuki, K. Kato, and K. Sugahara, *Eur. J. Biochem.,* 233 (1995) 687–693.
(71) J. F. G. Vliegenthart, L. Dorland, and H. Van Halbeek, *Adv. Carbohydr. Chem. Biochem.,* 41 (1983) 209–374.
(72) C. A. Bush, *Bull. Magn. Reson.,* 10 (1989) 73–95.
(73) J. Dabrowski, *Methods Enzymol.,* 179 (1989) 122–156.
(74) S. W. Homans, *Progr. NMR Spectrosc.,* 22 (1990) 55–81.
(75) L. E. Lerner, in *NMR Spectroscopy and Its Application to Biomedical Research,* (S. K. Sarkar, Ed.), Elsevier, (1996), pp. 313–344.
(76) J. Dabrowski, in *Two Dimensional NMR Spectroscopy; Applications for Chemists and Biochemists.* 2nd Ed. edited by William R. Croasmun and R. M. K. Karlson, Verlag Chemic, (1994) 741–784.
(77) P. K. Agrawal, C. A. Bush, N. Qureshi, and K. Takayama, *Adv. Biophys. Chem.,* 4 (1994) 179–236.
(78) T. Fritz, P. K. Agrawal, J. D. Esko, and N. R. Krishna, *Glycobiology,* 7 (1997) 587–595.
(79) N. R. Krishna, B. Y. Choe, M. Prabhakaran, G. C. Ekborg, L. Rodén, and S. C. Harvey, *J. Biol. Chem.,* 265 (1990) 18256–18262.
(80) B. Y. Choe, G. C. Ekborg, L. Rodén, S. C. Harvey, and N. R. Krishna, *J. Am. Chem. Soc.,* 113 (1991) 3743–3749.
(81) S. W. Homans, R. A. Dwek, J. Boyd, M. Mahmoudian, W. G. Richards, and T. W. Rademacher, *Biochemistry,* 25 (1986) 6342–6350.
(82) P. K. Agrawal, D. C. Jain, R. K. Gupta, and R. S. Thakur, *Phytochemistry,* 24 (1985) 2479–2496.
(83) E. V. Curto, T. T. Sakai, M. J. Jablonsky, S. Rio-Anneheim, J. -C, Jacquinet, and N. R. Krishna, *Glycoconjugate J.,* 13 (1996) 599–607.
(84) A. Manzi, P. V. Salimath, R. C. Spiro, P. A. Keifer, and H. H. Freeze, *J. Biol. Chem.,* 270 (1995) 9154–9163.
(85) K. Sugahara, H. Tsuda, K. Yoshida, S. Yamada, T. de Beer, and J. F. G. Vliegenthart, *J. Biol. Chem.,* 270 (1995) 22914–22923.
(86) L. Lerner, and A. Bax, *Carbohydr. Res.,* 166 (1987) 35–46.

(87) H. Van Halbeek, L. Dorland, G. A. Veldink, J. F. G. Vliegenthart, P. J. Garegg, T. Norberg, and B. Lindberg, *Eur. J. Biochem.,* 127 (1982) 1–6.
(88) S. J. Angyal, *Angew. Chem. Int. Ed. Engl.,* 8 (1969) 157–226.
(89) D. A. Rees, *Int. Rev. Sci. Biochem. Ser.,* 1 (1975) 1–42.
(90) D. Mikhailov, K. H. Mayo, I. R. Vlahov, T. Toida, A. Pervin, and R. J. Linhardt, *Biochem. J.,* 318 (1996) 93–102.
(91) J. F. Stoddart, *MTP Int. Rev. Sci. Org. Chem. Ser. One,* 7 (1973) 1–30.
(92) J. R. Snyder and A. S. Serianni, *J. Org. Chem.,* 51 (1986) 2694–2702.
(93) B. Casu, M. Petitou, M. Provasoli, P. an Sinay, *TIBS,* 13 (1988) 221–225.
(94) S. Arnott, J. M. Guss, D. W. L. Huckins, I. C. M. Dean, and D. A. Rees, *J. Mol. Biol.,* 88 (1974) 175–184.
(95) D. A. Horita, P. J. Hajduk, and L. E. Lerner, *Glycoconjugate J.,* 14 (1997) 691–696.
(96a) M. Ragazzi, D. R. Ferro, P. Prevasoli, P. Pumilla, A. Cassinari, G. Torri, M. Guerrini, B. Casu, H. B. Nader, and C. P. Dietrich, *J. Carbohydr. Chem.,* 12 (1993) 523–535.
(96b) S. E. B. Gould, R. O. Gould, D. A. Rees, and A. W. Wight, *J. Chem. Soc. Perkin Trans II,* (1976) 392–402.
(97) R. R. Contreras, J. P. Kamerling, J. Breg, and J. F. G. Vligenthart, *Carbohydr. Res.,* 179 (1988) 411–418.
(98) S. Yamada, K. Yoshida, M. Segiura, and K. Sugahara, *J. Biochem.,* 112 (1992) 440–447.
(99) H. B. Nader, M. A. Porcianatto, I. L. S. Tersariol, M. A. S. Pinhal, F. W. Oliveira, C. T. Moraes, C. P. Dietrich, *J. Biol. Chem.,* 265 (1990) 16867–16813.
(100) A. Horne, and P. Gettins, *Carbohydr. Res.,* 225 (1991) 43–47.
(101) M. Zsiska, and B. Meyer, *Carbohydr. Res.,* 243 (1993) 225–258.
(102) K. Sugahara, Y. Ohkita, Y. Shibata, K. Yoshida, and A. Ikegami, *J. Biol. Chem.,* 270 (1995) 7204–7212.
(103a) K. Sugahara, S. Yamada, K. Yoshida, P. de Waard, and J. F. G. Vligenthart, *J. Biol. Chem.,* 267 (1992) 1528–1533.
(103b) M. Iacomini, B. Casu, M. Guerrini, A. Naggi, A. Pirola, and G. Torri, *Anal. Biochem.,* 274 (1999) 50–58.
(104) J. Liu, U. R. Desai, X.-J. Han, T. Toida, and R. J. Linhardt, *Glycobiology,* 5 (1995) 765–774.
(105) K. Wüthrich, *NMR of Proteins and Nucleic Acids,* Wiley, New York (1986).
(106) P. K. Agrawal, J. C. Jacquinet, and N. R. Krishna, *Glycobiology,* 9 (1999) 669–677.
(107) S. Rio, J.-M. Beau, and J.-C. Jacquinet, *Carbohydr. Res.,* 244 (1993) 295–313.
(108) J. Jeener, B. H. Meier, and R. R. Ernst, *J. Chem. Phys.,* 71 (1979) 4546.
(109) A. Bax, and D. G. Davis, *J. Magn. Reson.,* 63 (1985) 207–213.
(110) J. H. Prestegard, J. R. Tolman, H. M. Al-Hashimi, M. Andrec, in *Biological Magnetic Resonance,* Vol. 17, N. R. Krishna, and L. J. Berliner, Eds, Kluwer Academic/Plenum Publishers, 1999, pp. 311–356.
(111) M. R. Hansen, M. Rance, A. Pardi, *J. Amer. Chem. Soc.,* 120 (1998) 11210–11211.
(112) N. R. Krishna, B. Y. Choe, and S. C. Harvey, in *Computer Modelling of Carbohydrate Molecules,* A. French and J. W. Brady (Eds.) *Amr. Chem. Soc. Symp. Ser.,* (1990) 227–239.
(113) B. Sheldric, *Acta Crystallogr.,* B32 (1976) 1016–1020.
(114) S. M. A. Holmbeck, P. A. Petillo, and L. E. Lerner, *Biochemistry,* 33 (1994) 14246–14255.
(115) IUPAC Commission on the Nomenclature of Organic Chemistry (CNOC) Rules for the Nomenclature of Organic Chemistry: *Pure Appl. Chem.,* 55 (1983) 1269–172.
(116) R. U. Lemieux, S. Koto, and D. Voisin, in W. A. Szarek, and D. Horton, (Eds). *The Anomeric Effect, Origin and Consequences, ACS Symp. Ser.,* 87 (1979) 17–29.
(117) H. Baumann, B. Erbing, P.-E. Jansson, and L. Kenne, *J. Chem. Soc. Perkin Trans 1,* (1989) 2153–2165.

(118) M. Hricovini, I. Tvaroska, J. Mirsch, and A. J. Duben, *Carbohydr. Res.,* 198 (1990) 193–203.
(119) H. Paulsen, R. Busch, V. Sinnwell, and A. Pollex-Krüger, *Carbohydr. Res.,* 214 (1991) 227–234.
(120) H. Paulsen, *Angew. Chem. Int. Ed. Engl.,* 29 (1990) 823–839.
(121) K. P. Howard, and J. H. Prestegard, *J. Am. Chem. Soc.,* 117 (1995) 5031–5040.
(122) K. S. Bruzik, and P. G. Nyholm, *Biochemistry,* 36 (1997) 566–575.
(123) J. F. G. Vligenthart, *Pure Appl. Chem.,* 69 (1997) 1875–1878.
(124) M. C. Rose, W. A. Voter, H. Sage, C. F. Brown, and B. Kaufman, *J. Biol. Chem.,* 259 (1984) 3167–3172.
(125) H. Matsuura, T. Greene, and S. Hakomori, *J. Biol. Chem.,* 264 (1989) 10472–10476.
(126) E. Berman, A. Allerhand, and A. L. Davies, *J. Biol. Chem.,* 255 (1980) 4407–4410.
(127) T. A. Gerken, K. J. Butenhof, and R. Shogren, *Biochemistry,* 28 (1989) 5536–5543.
(128) H. Kessler, H. Matter, G. Gemmecker, M. Kottenhahn, and J. W. Bats, *J. Am. Chem. Soc.,* 114 (1992) 4805–4818.
(129) J. T. Davis, S. Hirani, C. Barlett, and B. R. Reid, *J. Biol. Chem.,* 269 (1995) 3331–3338.
(130) G. Mer, H. Hietter, and J.-F. Lefevre, *Nature Struct. Biol.,* 3 (1995) 45–53.
(131) D. F. Wyss, J. S. Choi, and G. Wagner, *Biochemistry,* 34 (1995) 1622–1634.
(132) A. H. Andreotti, and D. Kahne, *J. Am. Chem. Soc.,* 115 (1993) 3352–3353.
(133) R. Liang, A. H. Andreotti, and D. Kahne, *J. Am. Chem. Soc.,* 117 (1995) 10395–10396.
(134) V. Gervais, A. Zerial, and H. Oschkinat, *Eur. J. Biochem.,* 247 (1997) 386–395.
(135) X. Huang, Jr. J. J. Barch, F. T. Lung, P. P. Roller, P. L. Nora, J. Mushick, and R. R. Garrity, *Biochemistry,* 36 (1997) 10846–10856.
(136) H. J. Dyson, and P. E. Wright, *Annu. Rev. Biophys. Biophys. Chem.,* 50 (1991) 519–537.
(137) U. K. Saha, L. S. Griffith, J. Rodemann, A. Geyer, and R. R. Schmidt, *Carbohydr. Res.,* 304 (1997) 21–28.
(138) L. A. Fransson, B. Havsmark, and I. Silverberg, *Biochem. J.,* 269 (1990) 381–388.
(139) J. Glossel, W. Hoppe, and H. Kresse, *J. Biol. Chem.,* 261 (1986) 1920–1923.
(140) D. R. Ferro, A. Provasolli, M. Ragazzi, G. Torri, G. Gatti, J. C. Jacquinet, P. Sinaÿ, M. Petitou, and J. Choay, *J. Am. Chem. Soc.,* 108 (1986) 6773–6778.
(141) D. R. Ferro, A. Provasolli, M. Ragazzi, B. Casu, G. Torri, V. Bossennec, B. Perty, P. Sinaÿ, M. Petiou, and J. Choay, *Carbohydr. Res.,* 195 (1990) 157–167.
(142) T. N. Huckerby, P. N. Senderson, and A. I. Nieduszynski, *Carbohydr. Res.,* 138 (1985) 199–206.
(143) B. Casu, D. R. Ferro, M. Ragazzi, and G. Torri, in *Dermatan Sulfate Proteglycans* (J.E. Scott, Ed.), Portland Press, London, 1993, pp. 41–53.
(144) K. Sugahara, Y. Ohi, T. Harada, I. Yamashina, P. De Waard, J.F.G. Vliegenthart, *J. Biol. Chem.,* 267 (1992) 6027–6035.
(145) P. De Waard, J. F. G. Vligenthart, T. Harada, and K. Sugahara, *J. Biol. Chem.,* 267 (1992) 6036–6043.
(146) S. Yamada, M. Oyama, H. Kinugasa, T. Nakagawa, T. Kawasaki, S. Nagasawa, K.-H. Khoo, H. R. Morris, A. Dell, and K. Sugahara, *Glycobiology,* 5 (1995) 335–341.
(147) K. Sugahara, M. Masuda, T. Harada, I. Yamashina, P. De Waard, and J. F. G. Vligenthart, *Eur. J. Biochem.,* 202 (1991) 805–811.
(148) T. R. Oegema, Jr. E. L. Kraft, G. W. Jourdian, and T. R. Van Valen, *J. Biol. Chem.,* 259 (1984) 1720–1726.
(149) K. Sugahara, N. Mizuno, Y. Okumura, and T. Kawasaki, *Eur. J. Biochem.,* 204 (1992) 401–406.
(150) T. De Beer, A. Inui, H. Tsuda, K. Sugahara, and J. F. G. Vliegenthart, *Eur. J. Biochem.,* 240 (1996) 789–797.
(151) J. Knecht, J. A. Cifonelli, and A. Dorfman, *J. Biol. Chem.,* 242 (1967) 4652–4661.
(152) L. A. Fransson, I. Silverberg, and I. Carlstedt, *J. Biol. Chem.,* 260 (1985) 14722–14726.
(153) M. Nilssen, J. Westman, and C.-M. Svahn, *J. Carbohydr. Chem.,* 12 (1993) 23.
(154) K. W. Neumann, J. Tamura, and T. Ogawa, *J. Glycoconjugate,* 13 (1996) 933–936.

(155) R. J. Linhardt, K. G. Rice, Z. M. Merchant, Y. S. Kim, and D. L. Lohse, *J. Biol. Chem.,* 261 (1986) 14448–14454.
(156) R. J. Linhardt, H. Wang, D. Loganathan, and J. Bae, *J. Biol. Chem.,* 267 (1992) 2380–2387.
(157) S. Hirano, *Org. Magn. Res.,* 2 (1970) 577 (1970).
(158) A. S. Perlin, D. M. Mackie, and C. P. Dietrich, *Carbohydr. Res.,* 18 (1971) 185–194.
(159) L. Rodén, and N. B. Schwartz, *Biosynthesis of Connective Tissue Proteoglycans,* in MTP International Review of Science, Biochemistry, Ser. 1, Biochemistry of Carbohydrates, (W. Whelan, Ed.), N.T. Butterworths, London (1975).
(160) L. Rodén, *Metabolic Conjugation and Metabolic Hydrolysis,* (N.H. Fishman, Ed.), Academic Press, New York, 1970.
(161) M. A. Bourdon, T. Krusis, S. Campbell, N. B. Schwartz, and E. Ruoslahti, *Proc. Natl. Acad. Sci. U.S.A.,* 84 (1987) 3194–3198.
(162) A. E. Kearns, S. C. Campbell, J. Westley, N. B. Schwartz, *Biochemistry,* 30 (1991) 7477–7483.
(163) L. Zhang, and J. D. Esko, *J. Biol. Chem.,* 269 (1994) 19295–19299.
(164) J. D. Esko, *Curr. Opin. Cell Biol.,* 3 (1991) 805–816.
(165) J. D. Esko, and L. Zhang, *Curr. Opin. Struct. Biol.,* 6 (1996) 663–670.
(166) D. M. Mann, Y. Yamaguchi, M. A. Bourdon, E. Ruoslahti, *J. Biol. Chem.,* 265 (1990) 5317–5323.

THE CONFORMATION OF C-GLYCOSYL COMPOUNDS

By Jesús Jiménez-Barbero, Juan Félix Espinosa, Juan Luis Asensio, Francisco Javier Cañada, and Ana Poveda

Instituto de Química Orgánica, CSIC, Juan de la Cierva 3, 28006 Madrid, and Servicio Interdepartamental Investigación, Universidad Autónoma Madrid, 28048-Cantoblanco, Spain

I. Introduction	235
II. Nomenclature	236
III. C-Glycosyl Compounds in Chemistry and Biochemistry	237
IV. Conformation of Saccharides: An Overview on the Relative Role of the Exoanomeric Effect and Steric Interactions	239
V. The Use of NMR and Molecular Modeling as Tools in Conformational Analysis of C-Glycosyl Compounds in Solution	243
VI. Conformation of C-Glycosyl Compounds	246
1. Conformation in the Solid State	246
2. Conformation in Solution	247
VII. Conclusions	279
References	280

I. Introduction

Interactions between carbohydrates and proteins mediate a broad range of biological activities, starting from fertilization and extending to such pathological processes as tumor metastasis.[1] In all these processes, the three-dimensional structures of both molecular entities are of paramount importance. Because the carbohydrate ligands are subject to hydrolytic attack, C-glycosyl compounds that afford the possibility for improved chemical and biochemical stability have been developed.[2] In addition, C-glycosyl compounds also appear in Nature,[3] especially as C-nucleosides[4] and C-glycosyl arenes.[5] Some of these naturally occurring compounds show interesting pharmacological and biological properties. Certain natural products also contain fragments having C-glycosyl architecture,[6–8] which have presented synthetic targets for organic chemists. All of these facts stimulated important advances in synthetic methodologies for obtaining these structures, as reviewed by several authors.[2–4]

However, these methylene-bridged analogues do not behave simply as noncleavable glycosides, and some additional differences in the behaviors of C- and O-glycosyl compounds are to be expected. Apart from the expected variations in bond lengths and bond angles, the conformational similarity of the intersaccharide linkages has been the subject of debate: because the replacement of an oxygen atom by a methylene group results in a change in both the size and the electronic properties of the linkage, as the flexibility and the energy barriers to rotation around the interresidue torsion angles could be markedly changed. Thus, the exo-anomeric effect present in glycosides,[9] resulting from the presence of the acetal function, disappears in the C-glycosyl analogue, along with a consequent variation of steric interactions between both residues.

The group of Kishi[10] at Harvard has been very active in this field, and based on NMR data, have proposed that glycosides and their C-glycosyl analogues share the same conformational characteristics in the free state. Moreover, the finding that the conformation of "C-lactose" bound to peanut agglutinin is basically identical to the conformation of its parent O-linked lactose bound to the same protein has led this group to claim that the conformational similarity between O- and C-glycosyl compounds is a general phenomenon.[11] However, other groups have reported that the assumption of similar conformations for glycosides and their C-glycosyl analogues is open to question, at least in certain cases.[12]

This chapter overviews studies from different groups on the conformational behavior of C-glycosyl compounds, especially those containing C-glycopyranosyl moieties, both in solution and in the solid state, and either isolated or when they are bound to proteins. The aim of this article is to determine up to which point the different reported results may be generalized. Comparisons between C-glycosyl compounds and glycosides make it is clear that the former are key compounds for deducing the relative importance of steric and stereoelectronic effects in the conformational analysis of oligosaccharides.[13] Major attention is paid to the conformational preferences around the Φ, Ψ, and ω angles because these torsions define the three-dimensional structure of oligosaccharides. Nevertheless, it must be noted that some unusual conformations have been reported for the six-membered rings in some synthetic O-alkyl or -acyl-protected derivatives.[14] Because these anomalous conformations probably relate to the presence of nonnatural protecting groups and/or nonaqueous solvents, these reports are treated only marginally. Furthermore, several reports have dealt with the conformational analysis of naturally occurring and synthetically-prepared C-nucleosides,[15] both in solution and in the solid state,[16] but a detailed overview of these investigations is likewise beyond the scope of this article.

II. Nomenclature

Throughout this paper and for sake of brevity, C-glycosyl compounds are frequently named after their corresponding related glycoside and not according to their formal nomenclature.[16a] Thus, "C-lactose" denotes a lactose analogue having the interresidue oxygen atom replaced by —CH_2—. The definitions of torsion angles, these are given referenced to the corresponding protons, i.e., Φ_H as H-1′ –C-1′–X–C-Y, where X is the interglycosidic atom and Y is the attached carbon of the aglycon, and Ψ_H as C-1′–X–C-Y–H-Y.

III. C-GLYCOSYL COMPOUNDS IN CHEMISTRY AND BIOCHEMISTRY

As already mentioned, potential biological applications of C-glycosyl compounds as well as interest in their structures has stimulated several research groups to develop new synthetic strategies,[2,17] and thanks to these efforts, most of the C-analogues of common naturally occurring glycosides are now accessible. Several reviews and books have appeared dealing with this subject. Along with oligosaccharide analogues, a number of C-glycosylpeptides have also been prepared.[18] There have been numerous applications to chemical and biochemical problems, especially in the use of C-glycosyl compounds as enzyme inhibitors. These include in particular, C-glycosyl-based inhibitors of glycogen phosphorylase[19] (such as compound **1**) sweet almond β-glucosidase,[20,21] glucoamylase,[22] glucosidases, and

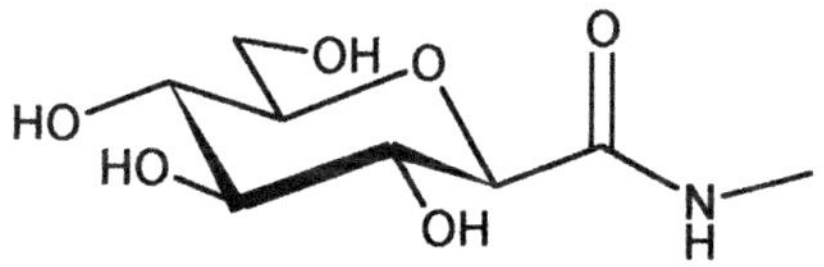

1

mannosidases,[23] and also glycosyltransferases[24] have been designed. Some analogues have been employed as biological probes.[25] Their potential uses as antiviral, anti-microbial, and as chemotherapeutic agents have also been documented.[26] Indeed, a polymer constituted by C-sialic acid moieties (**2**) has been shown to inhibit the attachment of virus particles to erythrocytes.[27] The integration of related

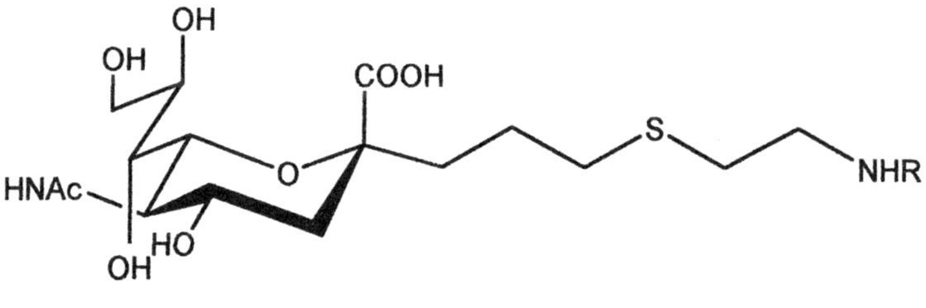

2

C-sialyl moieties into liposomes showed excellent inhibition of viral binding.[28] For practical applications, such moieties have also been integrated into molecular scaffolds for identifying GlcNAc 6-O-sulfotransferase activity specific to lymphoid tissue.[29]

C-Mannopyranosyl derivatives have been shown to inhibit the receptor-mediated adhesion of *Escherichia coli* to yeast cells.[30] Synthetically prepared C-glycosyl analogues of lipid A and lipid X are recognized by lipid A synthase and exhibit biological activities similar to those of their natural O-phosphorylated counterparts.[31] In a related context, and regarding the use of C-glycosyl compounds as glycoside isosteres, the relative affinity of glycosides and their C-glycosyl analogues toward mannose/glucose-specific lectins has been investigated.[32,33] In studies with a trimannoside analogues, it has been shown that the observed specificity of the binding to concanavalin A is slightly lower for the C-glycosyl compounds in comparison with their natural counterparts. However, the C-glycosyl analogues (**3**) of a linear

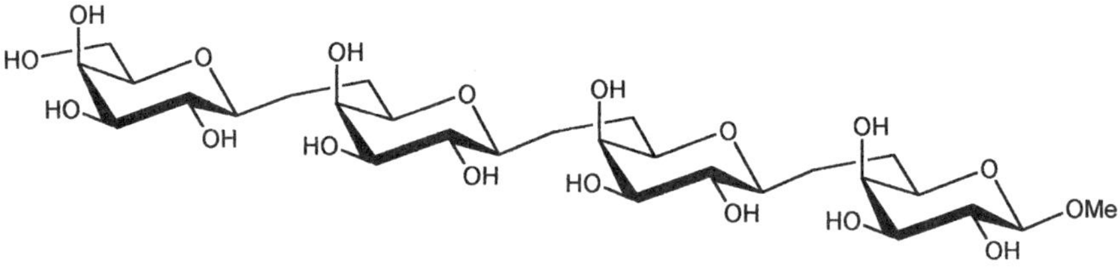

3

oligosaccharide containing (1→6)-linked β-D-galactopyranose residues have been shown to exhibit the same binding to monoclonal immunoglobulins as the natural O-linked determinants.[34] Likewise, the interaction of a mixed synthetic C/O-pentasaccharide (**4**) with antithrombin AT-III has shown that the introduction of one C-glycosyl linkage scarcely affects the affinity constant. The anti-Factor Xa activity is also similar for both pentasaccharide entities.[35]

C-Maltosyl fluoride analogues have been used as glycosyltransferase donors,[36] and a GlcNAc-CH_2-Asn-containing peptide has been used as an endo-glycosidase acceptor to form a high-mannose-type N-glycopeptide.[37] Also, phosphonate and phosphate mannose-containing analogues of the sialyl Le^X tetrasaccharide, having C-glycosyl linkages (**5**), have also shown moderate inhibition of the binding of that ligand to E-, P-, and L-selectins.[38]

Per-O-acetylated C-glycosyl conjugates may be very cytotoxic, as exemplified by studies on the affinity and cytotoxicity of hexose keto-C-glycosyl conjugates toward the glucose transporter Glut-1.[39] A C-glycosyl bromoketone derivative has been elegantly employed for the labeling and identification of the acid–base catalytic residue of *Saccharomyces cerevisiae* α-glucosidase.[40] The enhanced chemical stability of these analogues have also prompted other applications. Thus, for instance, spin-labeled *C*-glycosyl analogues have been prepared for use as paramagnetic probes,[41] a *C*-ribosyl coumarin has been prepared for use as a photophysical probe of oligonucleotide dynamics,[42] and water-soluble macrocyclic glycophanes have been prepared containing *C*-glycosyl linkages.[43]

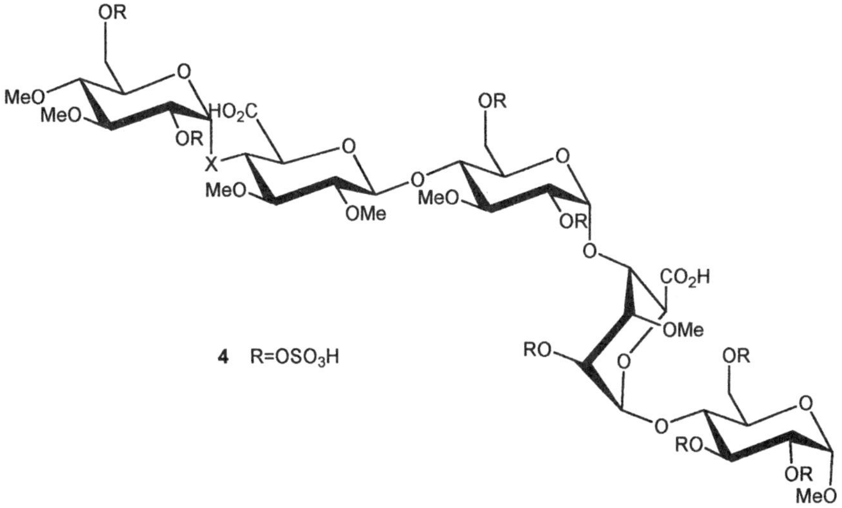

4   R=OSO$_3$H

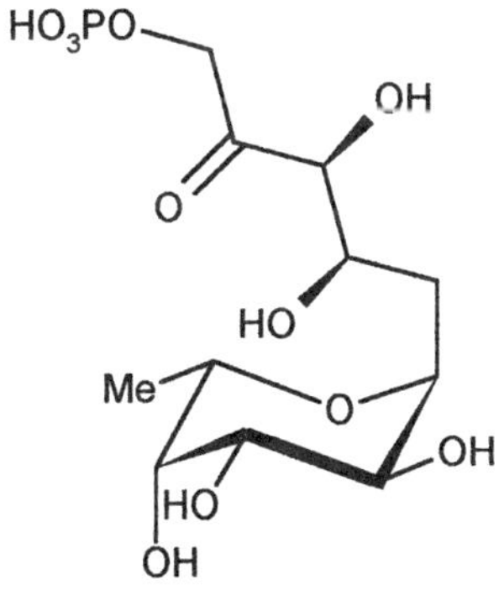

5

IV. Conformation of Saccharides: An Overview on the Relative Role of the Exoanomeric Effect and Steric Interactions

Despite a number of X-ray structural investigations as well as a variety of computational studies that unequivocally prove the existence of the exo-anomeric effect in glycosides,[9] a clear experimental estimation of the strength of the exo-anomeric (stereoelectronic) contribution in water solution is elusive. The so-called exo-anomeric effect refers to the preference for a gauche orientation of the aglycone O–R bond with respect to the endocyclic C–O bond. The most widely accepted theory of the exo-anomeric effect points to an interaction between one of the interglycosidic oxygen lone-pairs and the antibonding σ^* orbital of the C-1–O-5 bond of the corresponding pyranose ring.[9] *Ab initio* calculations *in vacuo* have estimated the strength of the exo-anomeric effect as being between 1.5 and 4.0 kcal/mol. With respect to the same fragment in the corresponding *C*-glycosyl analogue, Wiberg and Murcko,[44] and Houk and co-workers,[45] using *ab initio* calculations, have estimated the gauche O–C-1–C-2–C-3 preference to be between 0.3 and 0.8 kcal/mol in simple systems. It has also been suggested that these calculations, overestimate the preference for antiperiplanar conformations.[9]

When both *O*- and *C*-glycosyl compounds are considered, apart from torsional (gauche) interactions between vicinal bonds, 1,3-*syn*-diaxial interactions need also to be taken into account. In principle, if the three staggered conformations around the Φ angle of glycopyranosides are considered (Fig. 1), it is to be expected that the spatial orientation of the hydroxyl group at the 2-position of the pyranose ring should strongly influence the conformational equilibrium, in the absence of additional stereoelectronic (exo-anomeric) effects.

Two of these orientations are favored by the exo-anomeric effect (exo-*syn* and exo-*anti*), whereas one of them is not favored (non-exo). For the normal $^4C_1(D)$

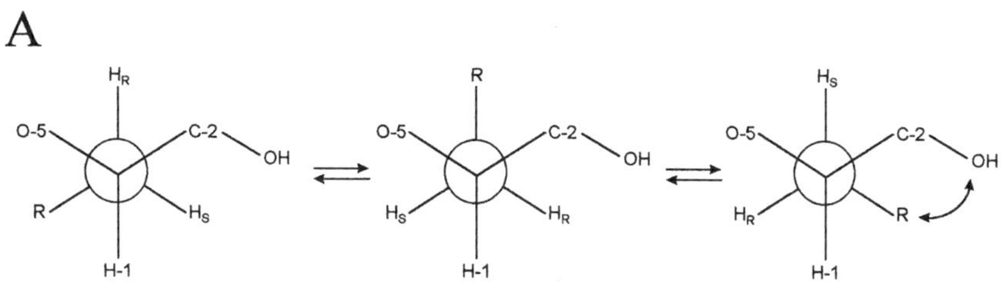

A

FIG. 1. Schematic view of the different orientations around the Φ angle of *C*-glycosyl compounds. The three possible staggered orientations are given for β-*C*-glycosyl (A, B), α-*C*-glycosyl (C, D), and α-L-*C*-sialyl-compounds (E). The three rotamers have been termed exo-*syn* (left), exo-*anti* (middle), and non-exo (right), in relation to those existing for normal glycosides, for which lone pairs of electrons would exist instead of the interglycosidic hydrogen atoms. The 1,3-*syn*-diaxial destabilizing interactions that can take place between the aglycone R and the hydroxyl group located at the C-2 position of the pyranose ring can also be observed. It may be noted that non-exo orientations (right) are destabilized for both β- and α-*C*-glycosyl compounds (A, C), which show equatorially oriented 2-OH groups, whereas exo-*anti* orientations (middle) show 1,3-*syn*-diaxial interactions for α-glycosides and for β-*C*-glycosyl compounds having an axially oriented 2-OH group. Exo-*syn* orientations do not show 1,3-*syn*-diaxial interactions except for α-L-*C*-sialyl compounds (E), for which the three orientations are basically equivalent.

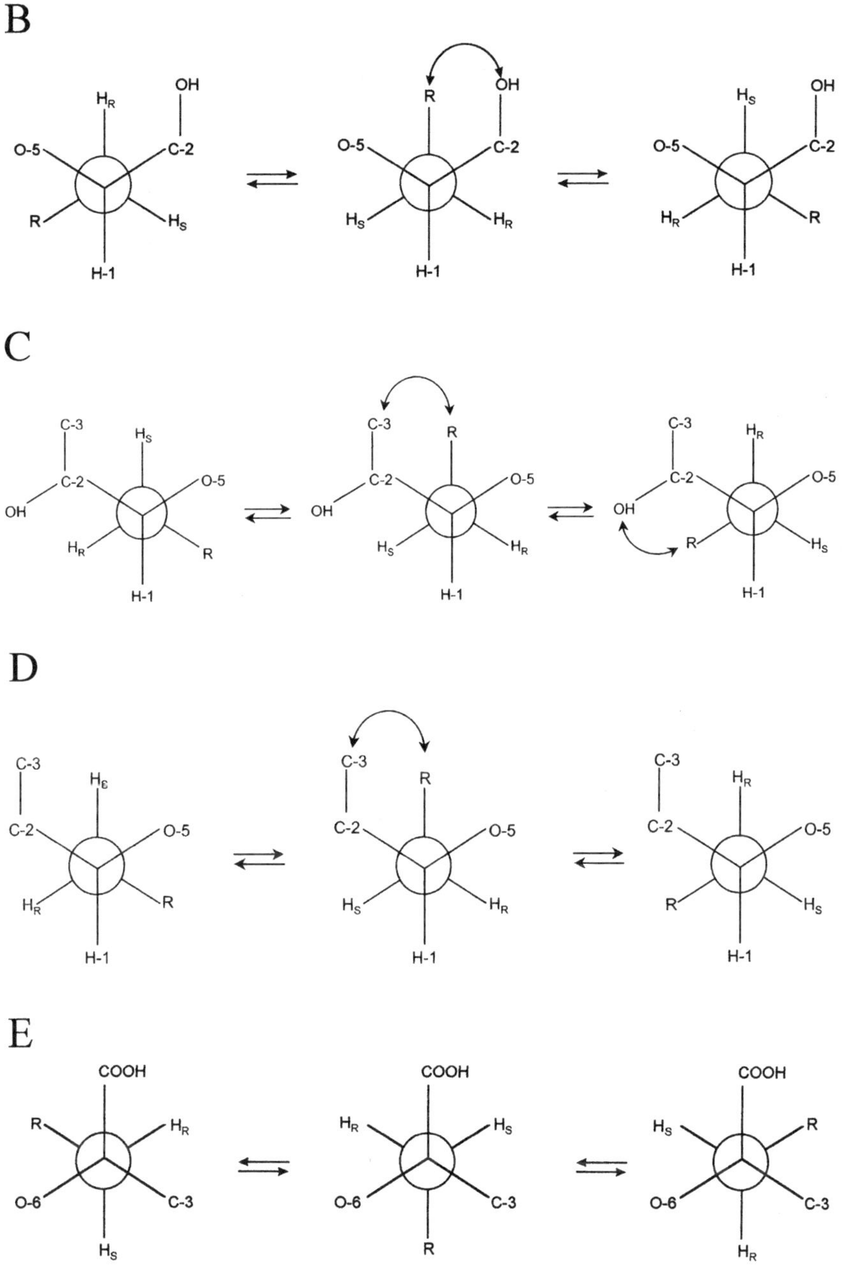

THE CONFORMATION OF C-GLYCOSYL COMPOUNDS

Fig. 1. *(Continued)*

chairs, there is a 1,3-*syn*-diaxial interaction between one equatorially substituted C-2 (*gluco* series) and the aglycone when the non-exo-anomeric (non-exo) conformation is considered. There are no steric interactions for the *syn*-exo-anomeric (*syn*-exo) orientation. The interactions for the *anti*-exo conformations depend on the α or β stereochemistry at the glycosidic linkage. In contrast, the 1,3-*syn*-diaxial interaction should be expected for one axially substituted C-2 (*manno* series) and the aglycone when the *anti* conformation of β anomers is considered. In this case, there are no 1,3-*syn*-diaxial interactions for either the *syn*-exo or the non-exo conformations. Interactions (except for the 1,3-type) occur for 2-deoxy sugars and for the alternative $^1C_4$ chair conformations. It is evident that, depending on the value of the stereoelectronic effect, the energy difference among the three rotamers will be different for glycosides and *C*-glycosyl compounds. Gauche-type torsional interactions should also be considered for the three rotamers. In fact, the separation and quantitative evaluation of the 1,3-*syn* diaxial interactions[45] can only be performed in the absence of the exo-anomeric effect. *Ab initio* calculations *in vacuo* by Houk *et al.*[45] showed that the relative stabilization of the gauche *vs.* anti forms rose from 0.7 kcal/mol to 2.2 kcal/mol upon adding a 3-hydroxyl group to 2-ethyltetrahydropyran to make 1,3-type interactions. According to these data, the 1,3-type destabilization should amount to about 1.5 kcal/mol.

Oligosaccharide conformations depend not only on Φ, but also on the aglyconic Ψ torsion angle (Fig. 2).

In the case of Ψ, no stereoelectronic effects have to be invoked (obvious exceptions are sucrose and trehalose-type compounds), and the existing conformational differences between C/O-pairs (if any) will arise strictly from the distinct van der Waals and torsional interactions attributable to the different bond lengths, bond angles, and atomic sizes in the corresponding analogues. Intuitively, it seems plausible that, because of the longer interglycosidic distances, the rotameric equilibrium around Ψ angles of *C*-glycosyl compounds should involve smaller energy barriers than those around Ψ of glycosides. It seems clear that the systematic

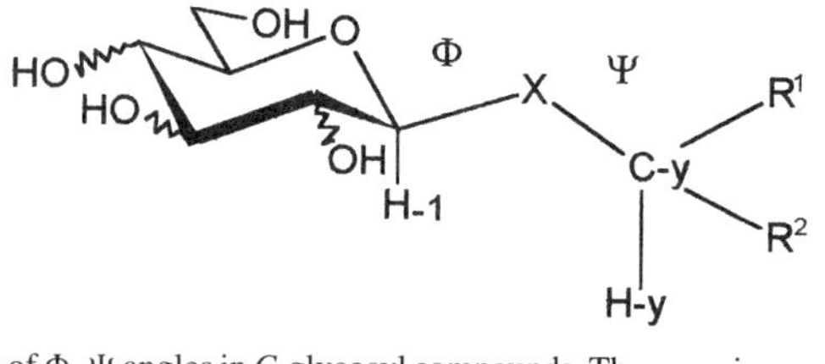

FIG. 2. Definition of Φ, Ψ angles in *C*-glycosyl compounds. They are given referenced to the corresponding protons: Φ_H as H-1'–C-1'–X–C-Y, where X is the interglycosidic atom and Y is the attached carbon of the aglycone, and Ψ_H as C-1'–X–C-Y–H-Y. The relationship with regard to heavy atoms is as follows: For β-D-glycosyl compounds, Φ_O is $\Phi_H - 120$, whereas for α-D-glycosyl analogues, Φ_O is $\Phi_H + 120$ and Ψ_C is $\Phi_H + 120$. Ψ(+) stands for +60 and Ψ(−) stands for −60.

comparison of *C*-glycosyl compounds having the *gluco* and *manno* configurations, as well as 2-deoxy analogues, should provide unambiguous data to verify the importance of steric effects in modulating the conformation around the Φ angle of oligosaccharides and analogues. Indeed, a global overview of all reported results should answer the question regarding the relative importance of the stereoelectronic and steric effects in oligosaccharide conformational analysis, and the controversy about similar or dissimilar conformations for *O*-glycosides and their C-analogues.

V. THE USE OF NMR AND MOLECULAR MODELING AS TOOLS IN CONFORMATIONAL ANALYSIS OF *C*-GLYCOSYL COMPOUNDS IN SOLUTION

It is obvious that a detailed knowledge of the structure of *C*-glycosyl compounds, both free and bound to proteins, is indeed relevant from both basic and applied scientific viewpoints. This information may be extracted by different means.[46,47] X-Ray crystallography has been widely employed for characterizing free and complexed carbohydrate-binding proteins.[48] Accordingly, examples of the application of X-ray to the study of these compounds are of prime interest. However, carbohydrates are often rather difficult to crystallize, probably because of their inherent flexibility. Furthermore X-ray only provides indirect information on the dynamics of the biomolecules and, moreover, for flexible structures, only one conformation may be analyzed. In consequence, NMR has been applied in this field because it provides both conformational and dynamic information.[13,49,50] Because of the particular characteristics of sugars, it is recognized that relaxation NMR parameters[51] should be complemented with such computational methods as molecular mechanics/dynamics[52] or Monte Carlo calculations[53] to define the structural features of the carbohydrate unambiguously. This task is commonly achieved by calculating potential-energy surfaces for the glycosidic linkages, using a force field.[54] When comparing such calculations to experimental data, especially in water solution, it should be kept in mind, that these surfaces provide just a first estimate of the conformational regions that are energetically accesible, and the possible presence of different conformational families. Protocols based on single-point conformers, restrained or unrestrained molecular dynamics,[55] or Monte Carlo calculations[53] may also be employed satisfactorily. With molecular mechanics calculations, care must be taken when considering the relative energy values provided by the force field. Nevertheless, the calculated geometries are usually very good approximations to those existing in solution and the solid state.

The existence of molecular motion around the glycosidic linkages of oligosaccharides has been firmly established.[56] In addition, later investigations have pointed out that the rates of overall and internal motions of small- and medium-size oligosaccharides may occur on similar time-scales.[57] Because NMR parameters are essentially time-averaged, the information deduced from NMR experiments generally corresponds to the time-averaged conformation in solution. Regarding the relaxation time-scale, ratios from transverse and longitudinal cross-relaxation rates obtained through off-resonance ROESY experiments[58] and/or through the comparison of data taken from individual NOESY, ROESY, or T-ROESY experiments[59] may be used to extract local correlation times for different pairs of protons in the oligosaccharide. These ratios are independent of interproton distances and may allow the estimation of specific correlation times. From these correlation times, proton–proton intra- and interresidue distances may be extracted. Because of the $\langle r^{-6} \rangle$ dependence of the NOE, minor populations of conformers can be detected, provided that they show exclusive interresidue proton–proton distances. For C-glycosyl compounds, distances between the interglycosidic protons and other protons on either ring may also be obtained. Intraresidue signals may also be taken as internal distance references as a first approximation. ^{13}C NMR relaxation parameters may also be employed to access the rates of overall and internal motions for saccharide molecules,[60] although to the best of our knowledge, this approach has not yet been applied to C-glycosyl compounds.

With regard to the monosaccharide moieties, the average shape of the pyranose rings may be deduced from the vicinal proton–proton coupling values.[61] The presence of two additional methylene protons at the pseudoglycosidic linkage affords more conformational information for C-glycosyl compounds than for the regular glycosides. Diastereotopic assignment of the prochiral H^R and H^S protons is an essential task that must be performed prior to any analysis. It is highly desirable for this assignment to be based exclusively on experimental spectroscopic and/or chemical data. A first indication of the relative orientation of the pyranose rings around the Φ and Ψ angles may be therefore obtained from the interglycosidic vicinal proton–proton coupling values, although it should always be remembered that the experimental values correspond to the time-averaged conformation in solution. Obviously, the best approach should combine both J and NOE data, and the postulated conformation or conformational equilibrium in solution should correlate with all of the NMR data in an unambiguous manner.

Considering the staggered basic conformations around the Φ angles for β- and α-linked C-glycosyl compounds (Fig. 1), the expected J values for the single rotamers may be deduced by applying the generalized Karplus[61a] equation proposed by Altona[61b] to the geometries provided by molecular mechanics calculations. These values (Table I) may serve as guide for comparison with the experimental values that are given later in this article for different C-glycosyl entities.

In an analogous manner, the expected J values for the basic conformations around Ψ angles for some equatorially and axially linked C-glycosyl compounds are given in Table II. In this case, additional eclipsed orientations may appear, especially when the glycosylation position is flanked by two equatorially oriented hydroxyl groups.

Returning to the question of relaxation parameters, the relationship between NOEs and proton–proton distances is well established[51] and can be worked out, at least semiquantitavely, when a full matrix-relaxation analysis is considered.

TABLE I

Expected *J* values (Hz) for the Basic Conformations Around Φ Angles for β- and α-Linked *C*-Glycosyl Compounds, Deduced by Applying the Generalized Karplus Equation Proposed by Altona to the Geometries Provided by MM3* Molecular-mechanics Calculations[a]

Proton Pair	syn-Exo-Φ	anti-Exo-Φ	non-Exo-Φ
β-H-1'/H$_{proS}$	1.0–1.4	5.0	11.2–11.5
β-H-1'/H$_{proR}$	10.8–11.3	1.9	3.2–4.0
α-H-1'/H$_{proS}$	10.5–11.6	No minimum	1.4–2.7
α-H-1'/H$_{proR}$	1.0–2.8	No minimum	11.0–11.6

[a] The range of variation accounts for the oscillations around the global or local minimum and to changes in the stereochemistry in the vicinity of both glyconic and aglyconic moieties.

In this case, it is obvious that the corresponding NOE intensities are sensitive to the respective conformer populations, and that therefore, an indication of the population distribution when these molecules are free in solution and even in the protein-bound state may be obtained by focusing on key NOEs. Regarding protein-bound conformations of *C*-glycosyl compounds, valuable information may be gained by X-ray crystallography. In addition, transferred (TR-NOE) experiments can be used for solution studies, provided that the exchange rate between the bound and the free state is fast.[62] In complexes of large molecules, cross-relaxation rates of the bound compound are opposite in sign to those of the free ligand and produce negative NOEs. Following this methodology, as pioneered by Bevilacqua *et al.*[62] for studying carbohydrate–protein interactions, several examples have been described.[63] Notably, the conditions required to monitor TR-NOEs appear to be satisfied frequently by sugar receptors.[63] The reason for this favorable situation probably rests in various facts: these interactions are not extremely strong, there is fast exchange between the free and the bound states of the ligand, and the perturbations of the conformational equilibrium of a given oligosaccharide upon binding to a protein are accessible to observation by TR-NOE. Different mixing times and protein/ligand molar ratios should be systematically used in order to

TABLE II

Expected *J* Values (Hz) for the Basic Conformations Around Ψ Angles for *C*-Glycosyl Compounds, Deduced by Applying the Generalized Karplus Equation Proposed by Altona to the Geometries Provided by MM3* Molecular-mechanics Calculations[a]

Proton Pair	syn-Ψ(+)	syn-Ψ(−)	syn-Ψ(Eclipsed)	anti-Ψ
H-X/H$_{proS}$	9.0–12.4	2.0–2.5	4.7	2.9–3.3
H-X/H$_{proR}$	1.4–2.5	12.3–12.4	4.3	3.8–4.1

[a] The range of variation accounts for the oscillations around the global or local minimum and to changes in the stereochemistry in the vicinity of both glyconic and aglyconic moieties.

gain quantitative conclusions. Comparison with TR-ROESY[64] and/or QUIET-TR-NOESY[65] experiments should also be performed to detect spin-diffusion effects.

Independently of the use of NOEs, the molecular alignment in a strong magnetic field induced by the use of a dilute liquid-crystalline medium has been applied to deduce the orientation of oligosaccharides, both free and within protein-binding sites.[66] Other NMR experiments, also independent of NOE, that are based on the transfer of cross-correlation may be used to deduce the bound conformation of biomolecules,[67] provided they are labeled with stable isotopes (such as ^{13}C). The application of these methodologies to C-glycosyl compounds should permit to access new information about bound conformations that can be used for rational ligand–inhibitor design.

VI. Conformation of C-Glycosyl Compounds

In this section, the examples reported in the literature on the conformation of different C-glycosyl compounds is described. Because very few examples of X-ray studies of these analogues have been reported, this section has been arbitrarily arranged in two parts, and the solid-state studies are discussed first. Next, the studies in solution have been arranged according to the various glycosidic linkages in the analogues, starting from simple C-glycosyl compounds to trisaccharide and glycopeptide analogues. The O/C-lactose case, which is obviously to be included among the $(1 \rightarrow 4)$-linked analogues, has an independent section because the conformational behavior of this molecule has been debated extensively and both the free and protein-bound (to four different proteins) conformations have been studied.

1. Conformation in the Solid State

Curiously, among those C/O-disaccharide pairs that have been studied by X-ray, the gentiobiose, isomaltose, and sucrose pairs each show completely different conformational behavior in the solid state. C-Lactose bound to peanut agglutinin[11] has been studied, and this case will be treated later.

a. β-C-Gentiobiose.—The solid-state conformation of the C-linked analogue (**6a**) of methyl β-gentiobioside (**6b**) has been analyzed by X-ray crystallography and its conformational behavior has also been explored using molecular mechanics calculations with the PFOS program.[68] Interestingly, the anti-Φ conformer is found around the glycosidic angle, in contrast to the value observed for gentiobiose itself, which presents the usual *syn* exo-anomeric conformation around Φ.[69] This 120° variation is also accompanied by a 30° shift around the ψ angle. The PFOS calculations were also in agreement with this experimental observation, indicating that the anti-Φ conformer is indeed the more-stable conformer. This result strongly contrasts with the report by Goekjian and co-workers[70] that explained the experimental solution NMR data in terms of a major *syn* exo-anomeric conformation (see later).

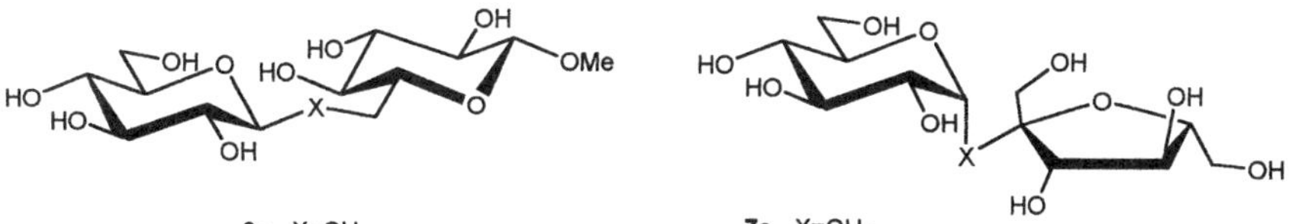

6a X=CH$_2$
6b X=O

7a X=CH$_2$
7b X=O

b. C-Sucrose.—The conformation of the C-analogue (**7a**) of sucrose octaacetate (**7b**) has been studied in the solid state[71] and compared with that described for the octaacetate of the natural disaccharide.[72] The existing differences are much more significant for Ψ [-75(C-) versus 22° (O-)] than for the Φ angles [66 (C-) versus 93° (O-)]. The shape of the five-membered ring is also different, $^3T_4$ (O-) versus $^2T_3$ (C-). Analysis of the bond distances and angles indicates that there is a partial compensation of the longer bond lengths (~0.1 Å) of the C-glycosyl derivative with a narrower bond angle (~6°).

c. C-Isomaltose.—The conformation of the hepta-*O*-acetylated C-analogue of isomaltose has been deduced for the crystal[14] and compared to that described for the corresponding glycoside. There are significant differences, especially around Ψ [64° (C-) versus 165° (O-)]. Although somehow different, the conformation around ω belongs to the same rotameric family [62° (C-) versus 80° (O-)]. The solid-state conformation around Φ is similar to that predicted by Goekjian *et al.*[70] for the deacetylated compound in solution.

2. Conformation in Solution

a. Simple C-Glycosyl Derivatives.—Several reports have dealt with conformational and configurational aspects on *C*-glycosyl analogues. For example, the assignment of either α or β-configurations to O-protected *C*-glycopyranosyl and *C*-glycofuranosyl analogues have been accomplished by using a combination of NMR data, including both ^{1}H and ^{13}C chemical shifts, proton–proton and proton–carbon coupling constants, and NOEs, which were shown to be the key parameters in this case.[73]

b. Conformation of β-C- and α-C-Glucosyl Derivatives.—An initial communication[74] described work on the conformation of simple β-D-glucopyranosyl (**8**) and 2-deoxy β-D-*arabino*-hexopyranosyl (**9**) derivatives C-linked to small alicyclic fragments. ^{1}H NMR coupling constants between the anomeric H-1 and the pseudoglycosidic CH$_2$ protons were measured in a variety of solvents. Molecules

TABLE III

Comparison Between the Experimental *J* Values (Hz) Described by Wu et al.[74] for the Glycosidic Torsion Φ of a Series of β- (8) and α-Linked (10) *C*-Glycosyl Compounds and Their 2-Deoxy Analogues (9, 11), and the Expected *J* Values (Hz) Given in Table I[a]

Proton Pair	Expected exo-Syn-Φ	Expected anti-Φ	Expected non-exo-Φ	Experimental (8, 10)	Experimental (9, 11)
β-H-1'/H$_{proS}$	1.0–1.4	5.0	11.2–11.5	1.5–2.7	5.0–5.4
β-H-1'/H$_{proR}$	10.8–11.3	1.9	3.2–4.0	8.1–10.7	7.5–7.7
α-H-1'/H$_{proS}$	10.5–11.6	No minimum	1.4–2.7	10.0–12.3	8.6–8.8
α-H-1'/H$_{proR}$	1.0–2.8	No minimum	11.0–11.6	2.7–4.7	5.6–6.0

[a] The range of *J* values measured for the different series is given in the columns with the experimental data.

specifically deuterated at the interglycosidic position were also studied to discriminate the H*s* and H*r* protons. One large (8.1–10.7) and one small (1.5–2.7 Hz) coupling were measured in a variety of solvents, namely D_2O, Me_2SO-d_6, CD_3OD, and C_5D_5N. The results were interpreted in terms of the existence of a major exo-anomeric conformation. The possibility of 1,3-*syn*-diaxial-type interactions between the equatorially oriented O-2 of Glc in **8** and the aglycone prompted the authors to study the 2-deoxy analogues (**9**). The couplings were significantly changed, becoming 7.5–7.7 Hz (large) and 5.0–5.4 (small). However, these variations were interpreted by the authors in terms of a conformation slightly deviated from the ideal exo-anomeric orientation so as to minimize interactions between the aglycone and the C-1–O-5 bond. The possible existence of additional conformations (non-exo) for the deoxy derivative through removal of O-2 and consequent elimination of the 1,3-*syn*-diaxial-type interaction was not considered. Inspection of Table III shows that the experimental coupling constants may be readily explained quantitatively by considering a significant contribution of non-exo-anomeric conformers, especially in the case of the deoxy derivatives. However, the authors explained the results by claiming that the exo-anomeric effect is not the major factor causing glycosides to adopt preferentially the exo-anomeric conformation. Nevertheless, they also stated that their results do not preclude the existence of stereoelectronic stabilization.

Regarding α-*C*-glycosyl compounds, a similar protocol was performed for the corresponding simple α-D-glucopyranosyl (**10**) and 2-deoxy-α-D-*arabino*-hexopyranosyl (**11**) derivatives, which were C-linked to similar alicyclic fragments.[74] The coupling constants are given in Tables III and IV. One large (10.0–12.3) and one small (2.7–4.7 Hz) coupling were recorded in the different solvents. Again, the results were interpreted in terms of a single conformer having the *syn*-exo-anomeric conformation, in agreement with data published for methyl α-glucopyranoside. The couplings for the 2-deoxy analogues (Table III) were significantly different, changing to 8.6–8.8 (large) and 5.6–6.0 Hz (small).

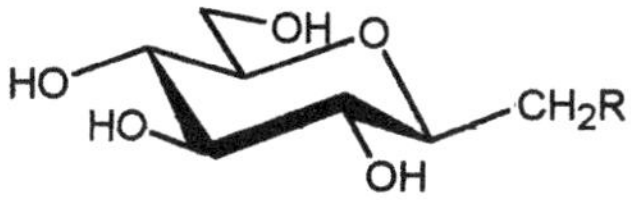

R= -CH$_2$-CH$_3$, -*CHOH-CH$_2$OH

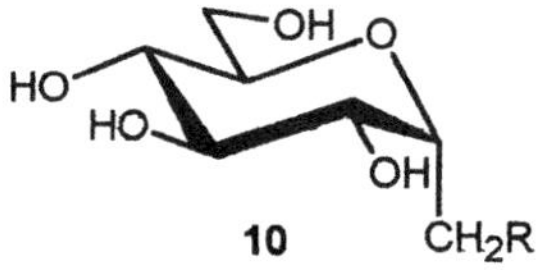

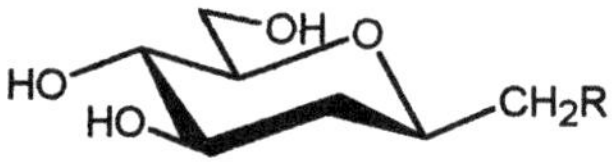

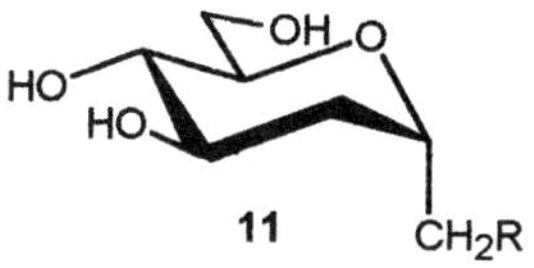

Again, and despite the major differences in experimental values, these variations were interpreted in terms of a conformation slightly deviated from the ideal exo-anomeric orientation; the possible existence of additional non-negligible contributions of non-exo-anomeric conformers for the deoxy derivatives was not considered. These results were extended in a more extensive publication.[75] Regarding the Ψ angle, the observed couplings (large, from 6.9–9.7, and small, from 3.1–5.3 Hz) were interpreted in terms of a unique zig-zag-like conformation, despite the clear scattering of the values, outside the limits of experimental uncertainty. Despite the aforementioned significant variation in couplings upon deoxygenation at C-2 (Tables III and IV), the authors stated that removal of the 2-hydroxyl group does not fundamentally alter the conformational behavior of C-glycosyl compounds. Moreover, based on these data, it was also claimed that *gauche* interactions and not 1,3-*syn*-diaxial ones are the primary factor in controlling the conformational behavior of these compounds, and that this behavior is independent of the structure and stereochemistry of the pyranose ring.

The conformational behavior in solvents other than water was also studied. In contrast with the aforementioned major conclusion, the presence of a modest temperature effect led the authors to conclude that these compounds exist as a

TABLE IV

Comparison Between the Experimental *J* Values (Hz) Described by Kishi *et al.*[74] for the Glycosidic Torsion Ψ of a Series of β- and α-Linked *C*-Glycosyl Compounds (8–11) and Their 2-Deoxy Analogues, and the Expected *J* Values (Hz) Given in Table II[a]

Proton Pair	Ψ(+)	Ψ(−)	Ψ(eclipsed)	*anti*-Ψ	Experimental
H-X/H$_{proS}$	9.0–12.4	2.0–2.5	4.7	2.9–3.3	3.1–5.3
H-X/H$_{proR}$	1.4–2.5	12.3–12.4	4.3	3.8–4.1	6.9–9.7

[a] The range of *J* values measured for the different series is given in the columns with the experimental data.

mixture of conformers in CD_3OD solution. Moreover, and along this reasoning, they analyzed their data for a simple two-state model. Indeed, as shown in Table 8 in that paper, it can be deduced that ~45% of non-exo-anomeric conformers exist for the deoxy derivatives, in contrast with 10–25% for the 2-hydroxyl analogue.

The results for the α analogues (**10, 11**) were also presented in the aforementioned paper,[75] with the same fundamental conclusions regarding the Φ and Ψ angles. Nevertheless, on the basis of the magnitudes of the couplings (usually >10 Hz for α-*C*-glycosyl compounds and usually ~9 Hz for the β analogues), they also claimed that axially linked α-*C*-glycosyl compounds should be slightly more rigid than the equatorially linked β-*C*-glycosyl analogues. No major solvent-polarity effect was deduced in either case. Again, the presence of a modest temperature effect led the authors to postulate that these α-*C*-glycosyl compounds could exist as a mixture of conformers in CD_3OD solution. In fact, analysis of the data for a simple two-state model for some of these compounds (Table 8 in that paper) also shows values above 40% for the population of non-exo-anomeric conformations.

A conformational study of these compounds was performed independently[76] in 1990 by using MM2 molecular mechanics calculations. The MM2-predicted conformational populations were used to deduce global average proton–proton vicinal coupling constants, and these were compared[76] with the vicinal coupling constants determined experimentally by Wa and co-workers[74] and Goekjian and co-workers[75] and shown in Tables III and IV. The MM2 program predicted significant populations of exo-*anti*-Φ (up to 16%) and non-exo (up to 18%) anomeric conformers, especially when 2-deoxy-*C*-glycosyl derivatives (**9, 11**) were considered. The agreement between the calculated and observed couplings was fair, thus providing a first indication of the importance of 1,3-*syn*-diaxial interactions. From the molecular mechanics results, the authors[76] built a virtual conformation around the different angles, which obviously does not have any physical meaning.

The importance of stereochemistry at position 2 of the pyranoid ring was also deduced for protected derivatives: these observations were experimentally supported for a series of 4,6-*O*-benzylidene β-*C*-glycosyl analogues[77] of the *gluco, manno, altro,* and *allo* series (**12**), bearing an acetamidomethyl group at the anomeric position. The couplings measured for the anomeric proton vary strongly (from 3.3 to 6.5 Hz) among the different series. No explicit mention of the occurrence of conformational equilibria was reported, and these heterogeneous values were explained by the authors as proof of considerable conformational distortion around Φ, depending on the substituents at the pyranose ring.

c. *C*-Glycosyl Analogues Having (1 → 4) Linkages.—Oligosaccharides having the commonly occurring (1 → 4) linkage have been exhaustively examined. A directly β-(1 → 4)-linked pseudo-*C*-disaccharide having no interglycosidic carbon atom, namely, the methyl cellobioside analogue **13** [methyl 4-*C*-2-deoxy-β-D-*arabino*-hexopyranosyl)-4-deoxy-α-D-*xylo*-hexopyranoside] has been studied through vicinal proton–proton couplings.[78] The H-1′–H-4 coupling for the unprotected compound and various protected derivatives were in all instances smaller than 2.5 Hz, indicating a preponderance of a *g*(-) orientation between these protons.

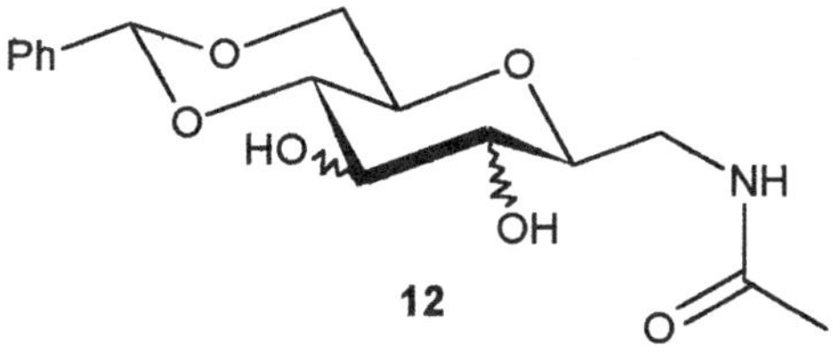

12

a *gluco*
b *manno*
c *altro*
d *allo*

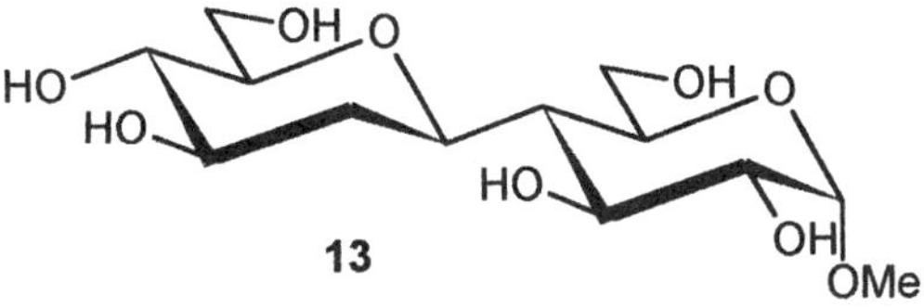

13

C-Glycosyl analogues of several common (1 → 4)-linked disaccharides have also been studied by Babirad's group.[79] In the first place, and regarding β-linked compounds, β-C-Glc-(1 → 4)-Glc (C-cellobiose, **14**), β-C-Man-(1 → 4)-Glc (**15**), 1,6-anhydro-C-cellobiose (**16**), and β-C-Man-(1 → 4)-1,6-anhydro-Glc (**17**) were analyzed. The couplings that define the Φ angle are (Table V) 8.7–10.0 (large) and 1.7–3.8 Hz (small). The corresponding couplings related to the ψ angle (Table VI) are 9.3–9.9 (large) and 4.5–5.4 Hz (small) for the 1,6-anhydro derivatives and 4.7–5.2 Hz (large) and 3.5–3.8 Hz (small) for those compounds that have the aglycone in the regular $^4C_1$ chair conformation (**14, 15**). It is noteworthy that the

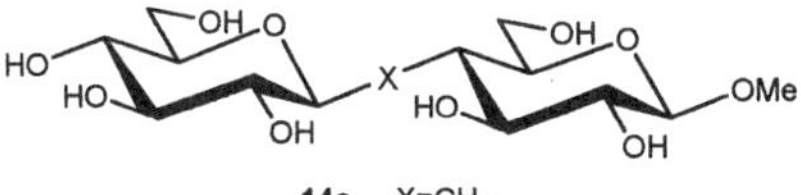

14a X=CH_2
14b X=O

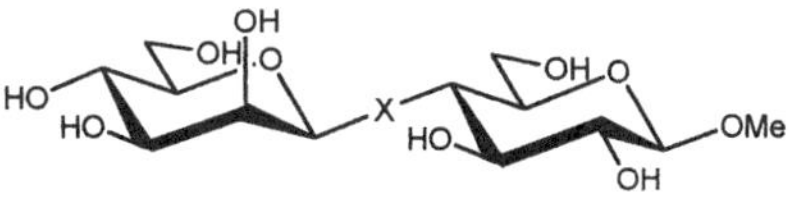

15a X=CH_2
15b X=O

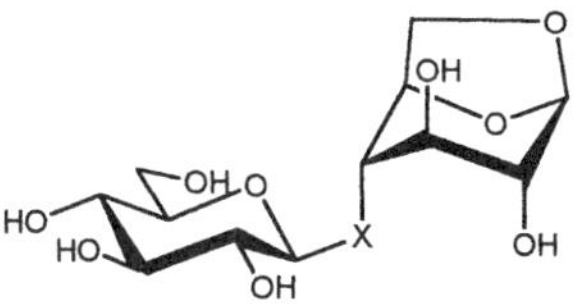

16a X=CH_2
16b X=O

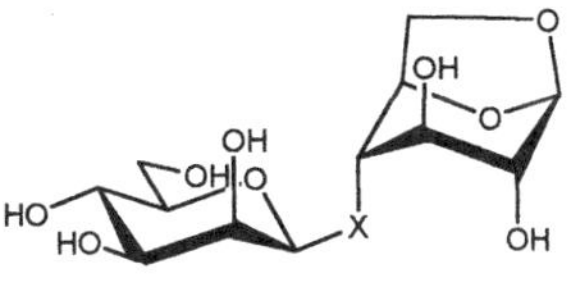

17a X=CH_2
17b X=O

TABLE V

Comparison Between the Experimental J Values (Hz) Described by Babirad et al.[79] for the Glycosidic Torsion Φ of a Series of β- (14–17) and α-(1 → 4)-Linked (18–21) C-Glycosyl Compounds, and the Expected J Values (Hz) Given in Table I[a]

Proton Pair	Expected Exo-syn-Φ	Expected Exo-anti-Φ	Expected non-Exo-Φ	Experimental (14–21)
β-H-1'/H$_{proS}$	1.0–1.4	5.0	11.2–11.5	1.7–3.8
β-H-1'/H$_{proR}$	10.8–11.3	1.9	3.2–4.0	8.7–10.0 Hz
α-H-1'/H$_{proS}$	10.5–11.6	No minimum	1.4–2.7	10.1–11.2
α-H-1'/H$_{proR}$	1.0–2.8	No minimum	11.0–11.6	3.1–4.6

[a] The range of J values measured for the different series is given in the columns with the experimental data.

aglycone adopts a different spatial orientation when the 1,6-anhydro ring is cleaved. In all these instances interpretation of the vicinal couplings was assisted with a diamond-lattice analysis. This analysis permitted the authors to conclude that the conformational preference around Φ is overwhelmingly for the *syn*-exo-anomeric orientation, whereas deviations from ideal staggered orientations are expected around the aglyconic bond to avoid steric interactions. However, the data gathered in Table V indicate that, for the β linkages, the observed couplings related to the Φ angle are in between those expected for the *syn*-exo and *anti*-exo (or non-exo) conformations, and that there is not any unique conformation present in solution. The values for Ψ may also be explained (Table VI) by conformational averaging around this angle. Nevertheless, the authors also claimed that the conformational behavior of these compounds agrees well with that of the corresponding disaccharides.

TABLE VI

Comparison Between the Experimental J Values (Hz) Described by Babirad et al.[79] for the Glycosidic Torsion Ψ of a Series of β- and α-Linked (1 → 4)-Linked C-Glycosyl Compounds, and the Expected J values (Hz) Given in Table II[a]

Proton Pair	Expected $\Psi(+)$	Expected $\Psi(-)$	Expected Ψ (eclipsed)	Expected anti-Ψ	Experimental 14–15, 16–17 18–19, 20–21
H-X/H$_{proS}$	9.0–12.4	2.0–2.5	4.7	2.9–3.3	3.5–3.8, 4.5–5.4 2.9–3.3, 5.3–5.5
H-X/H$_{proR}$	1.4–2.5	12.3–12.4	4.3	3.8–4.1	4.7–5.2, 9.3–9.9 4.6–6.0, 8.4–9.0

[a] The range of J values measured for the different series is given in the columns with the experimental data. For the experimental data, the first two values correspond to β linkages (14–17) and the third and fourth to α linkages (18–21). The first and third numbers correspond to aglycones having Glc (14–15, 18–19), and the second and fourth to 1,6-anhydro-Glc conformations (16–17, 20–21).

The α-analogues, α-C-Glc-(1 → 4)-Glc (**18,** C-maltose), α-C-Man-(1 → 4)-Glc (**19**), 1,6-anhydro-C-maltose (**20**), and α-C-Man-(1 → 4)-1,6-anhydro-Glc (**21**), all having (1 → 4) linkages, were investigated using similar methodology.[79] The couplings that define the Φ angle are (Table V) 10.1–11.2 (large) and 3.1–4.6 Hz (small). The corresponding couplings related to the ψ angle are (Table VI) 8.4–9.0 (large) and 4.6–6.0 Hz (small) for the 1,6-anhydro derivatives and 5.3–5.5 (large) and 2.9–3.3 Hz (small) for those compounds having the aglycone in the normal $^4C_1$ chair conformation. The diamond lattice-assisted analysis of the vicinal couplings permitted the authors to reach conclusions similar to those reached earlier. Indeed, the couplings listed in Table V indicate that the *syn*-exo-anomeric orientation dominates around Φ, whereas the scattered coupling values measured for H-4 show that some degree of conformational equilibrium is taking place around the aglyconic bond (Table VI).

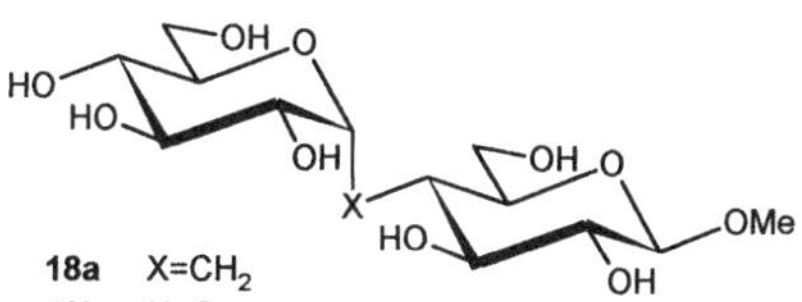

18a X=CH$_2$	**19a** X=CH$_2$
18b X=O	**19b** X=O
20a X=CH$_2$	**21a** X=CH$_2$
20b X=O	**21b** X=O

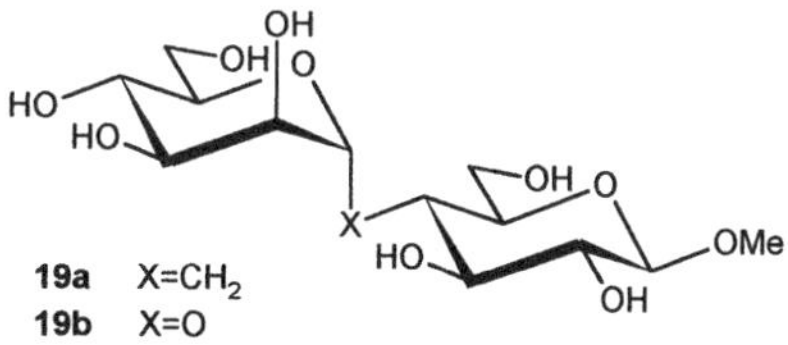

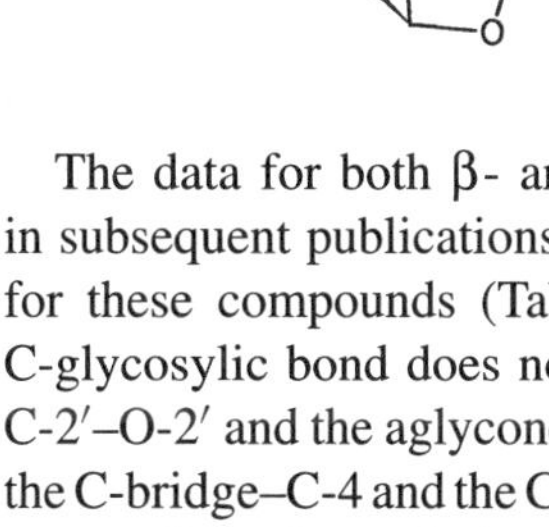

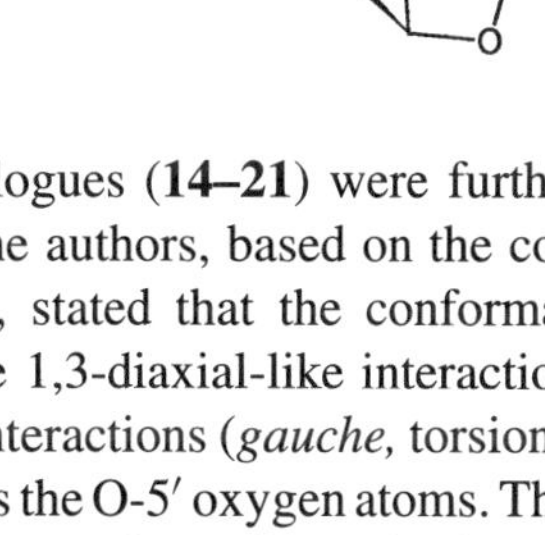

The data for both β- and α-linked analogues (**14–21**) were further analyzed in subsequent publications,[80,81] wherein the authors, based on the coupling data for these compounds (Tables V and VI), stated that the conformation at the C-glycosylic bond does not depend on the 1,3-diaxial-like interactions between C-2′–O-2′ and the aglycone, but on steric interactions (*gauche,* torsional) between the C-bridge–C-4 and the C-2′ carbon versus the O-5′ oxygen atoms. Three features emerge from the analysis: (*a*) the *syn*-exo-anomeric conformation is preponderant, (*b*) this favored orientation is independent on the configuration at C-2′, and (*c*) steric repulsions results in distortion around the aglyconic bond. The use of diamond lattice analysis to detect 1,3-*syn*-diaxial interactions of the various compounds allowed the impact of structural differences on the conformation around the C-aglyconic linkage to be predicted.

For β-linked analogues, this latter point was demonstrated by comparing the NMR data for C-cellobiose (**14**) and its 5-de(hydroxymethyl) analogue, Glcβ-$(1 \rightarrow 4)$-Xyl (**22**), for which the experimental couplings for H-4 changed from 3.5 and 4.7 Hz to 3.1 and 9.5 Hz, and thus this compound adopts preferentially an ideal extended conformation around the aglyconic linkage. In fact, according to Table VI, a higher percentage of the $\Psi(-)$ conformation should be present in this derivative **22,** although the values are still in accord with the presence of other conformers. In any case, this example demonstrates the existence of substantial averaging for Ψ of C-cellobiose (**14**) itself. This is also the case for C-lactose (see later). According to the authors' explanations, the conformational properties of natural glycosides may be explained, as a first approximation, by steric effects alone. After qualitatively comparing NOE and T_1 data for the pair of cellobiose analogues, C- and O-β-Glc-$(1 \rightarrow 4)$-Xyl (**22**), they concluded that the conformational analysis of carbon disaccharides may be extended directly to natural oligosaccharides. For α-linked compounds, the data for C-maltose (**18a**) and its 3-deoxy analogue (**23a**) were compared.[80,81] As regards the Ψ angle, the couplings for H-4 changed from 5.5 and 2.9 to 11.4 and 2.9 Hz, indicating that, in this α-analogue also, the deoxy derivative exists principally in an extended conformation around the aglyconic linkage. Again, according to these results, the conformational properties of natural glycosides may be explained, as a first approximation, by steric effects alone. A first-order analysis of NOE and T_1 data for the the C,O- pair of 3-deoxymaltose (**23**) also led to the conclusion that the conformational analysis of carbon disaccharides relates directly to that for natural oligosaccharides.

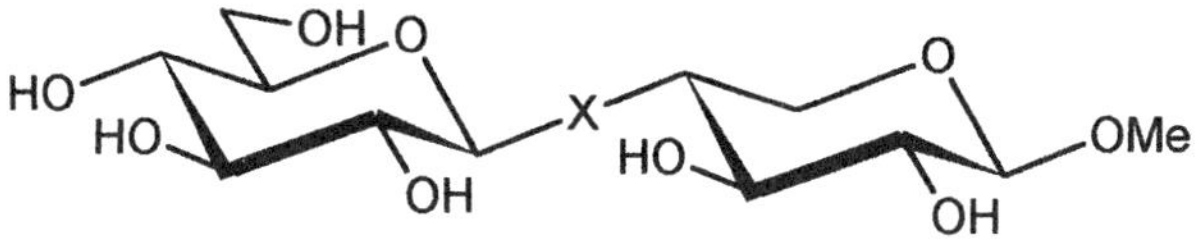

22a X=CH$_2$
22b X=O

At this point, and as a summary of their results, Wang and co-workers[82] then postulated that the analysis of steric interactions around the aglyconic bond may be performed by use of a diamond lattice. Because, according to their analysis, the C-glycosylic bond adopts the *syn*-exo-anomeric conformation, the conformational preferences of C-disaccharides may be predicted by focusing on the steric interactions around the aglyconic bond. In this context, an additional compound was studied, namely, the 3-deoxy derivative (**24a**) of C-cellobiose (**1a**). The couplings that define the Φ angle are 10.0 (large) and 2.0 Hz (small). The change in couplings related to the ψ angle upon deoxygenation at the C-3 position were significant. One of the couplings changes from 4.7 (in **14a**) to 9.5 Hz (in **24a**), which, according to the authors, indicates that 3-deoxy-C-cellobiose (**24a**) adopts a single conformation, in contrast to the results reported for C-cellobiose (**14a**), for which a single twisted conformer or a mixture of conformers was envisaged. Similar conclusions had been drawn for 3-deoxy-C-maltose (**23a**, with a change from 5.5 to 11.4 Hz, see preceding). According to Table VI, conformational averaging takes place in all instances, especially in the nondeoxygenated analogues (**14–17**), and also in the deoxy compounds, especially in the β-linked analogues (**22, 24**).

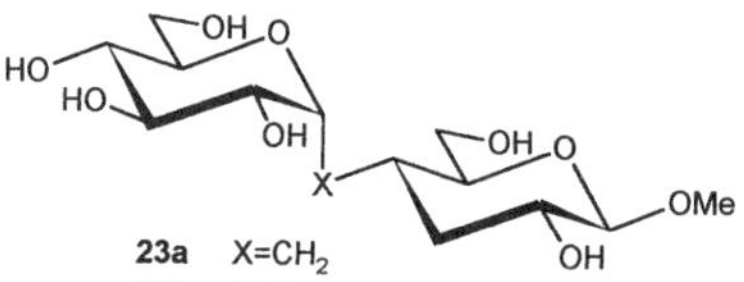

23a X=CH$_2$
23b X=O

24a X=CH$_2$
24b X=O

A direct, semiquantitative comparison between NMR data for glycosides and their C-glycosyl analogues was then performed.[83] In addition to the previously reported J data, NOE values and T$_1$ relaxation times were measured for pairs of C- and O-glycosyl compounds. Derivatives having CD$_2$ at the interglycosidic position were also studied to minimize multiple-spin effects. However, the authors explicitly mentioned that they did not use NOE and T$_1$ measurements as the principal means of conformational analysis. They first established the conformational preference of these C-glycosyl compounds by relying on the values of J couplings and subsequently compared the behavior of both types of compound through NOE and T$_1$ data For β linkages, the C- and O-Glcβ-(1 → 4)-Xyl pair (**22**) were further studied. The 1D-NOE data obtained by selective irradiation were compared qualitatively for both types of compounds, but the existence of strong overlapping precluded a clear analysis of the data. Although the authors explained their data (see earlier for the J-based analysis) in terms of a major *syn*-exo-anomeric conformation, a nonoverlapping H-2'–H-4 medium-to-strong NOE was reported for the C-glycosyl analogues (CH$_2$ and CD$_2$), which cannot be explained by the proposed conformer. Indeed, this NOE can only be explained by the presence of *anti*-exo-anomeric conformations. For the α analogues, the C/O-pair (**23**) of 3-deoxymaltose was further analyzed. In this case, although it is clear (Table V) that the major conformer in equilibrium should be the *syn*-exo-anomeric one, a H-1'–H-4 medium-to-strong NOE was reported for both 3-deoxy-C/O-maltose analogues, and this cannot be explained by the conformer described. In this case, two plausible explanations were suggested: either a slight deviation from the diamond analysis-type of ideal arrangement around the aglyconic bond, or the presence of a small proportion of another conformer in solution.

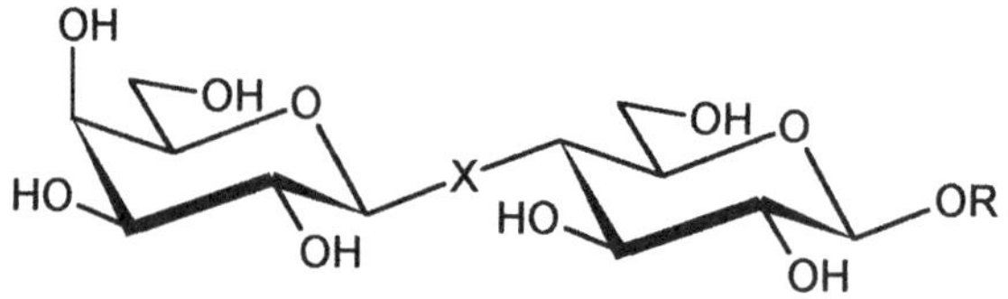

25a X=CH$_2$
25b X=O

d. The O/C-Lactose Case.—Most of the discussion on the conformational similarities between glycosides and their C-isosters has focused on this O/C-pair, and for this reason it is analyzed separately. The conformation of the O/C-lactose pair (**25**), both free in solution and bound to different proteins, has been analyzed in detail by NMR and molecular mechanics by Espinosa et al.[84,85] They concluded that the C/O-lactose pair primarily adopts the *syn*-exo-anomeric conformation around the Φ angle (Φ H-1'–C-1'–x–C-4, ~60°, *syn* conformation), although they also postulated that, in addition to this Φ value, a local minimum does exist for C-lactose only (Φ ~180°, *anti*-Φ conformation, also termed *gauche–gauche*.[85] Regarding the Ψ angle (Ψ, H-4–C-14–x–C-1'), they established for the first time that natural lactose is not monoconformational[86]: most of the population appears to be located in the *syn*-Ψ region (Ψ almost eclipsed, ~0°, >90%), but ~10% of the population is located in the *anti*-Ψ minimum (Ψ ~180, <10%). The situation is altered for C-lactose. In this case, the global minimum is shifted to the *syn*-exo-Φ/*anti*-Ψ region (Φ/Ψ, 60/180°, ~55% of the population), and the additional 45% of the population is divided between the *syn*-exo-Φ/*syn*-Ψ (eclipsed) conformer (40%) with Φ/Ψ, 60/0°, and the *exo*-*anti*-Φ/*syn*-Ψ conformation (5%) with Φ/Ψ, 180/0°.[12,84] The analysis was based on the interglycosidic coupling constants (Tables VII and VIII) and on the exclusive[87] interresidue NOEs that unequivocally characterize the *syn*, *anti*, and *gauche–gauche* regions of the conformational map. For C-lactose, these are H-1'–H-4, H-1'–H-3, and H-4–H-2', respectively. In addition, the two methylene protons present in C-lactose (**25a**)

TABLE VII

Comparison Between the Experimental *J* Values (Hz) for the Glycosidic Torsion Φ of C-Lactose (25) in Different Solvents, and the Expected *J* Values (Hz) Given in Table I

Proton Pair	Expected Exo-*syn*-Φ	Expected Exo-*anti*-Φ	Expected Non-exo-Φ	Experimental, Water	Experimental, Solvents with Increasing Polarity
H-1'/H$_{proS}$	1.0–1.4	5.0	11.2–11.5	1.2	2.4–1.6
H-1'/H$_{proR}$	10.8–11.3	1.9	3.2–4.0	10.3	8.7–10.3

TABLE VIII

Comparison Between the Experimental *J* Values in Different Solvents (Hz) Related to the Glycosidic Torsion Ψ of C-Lactose, and the Expected *J* Values (Hz) Given in Table II

Proton Pair	Expected Ψ(+)	Expected Ψ (eclipsed)	Expected anti-Ψ	Experimental
H-X/H$_{proS}$	8.0	4.7	2.9	3.6–5.2
H-X/H$_{proR}$	1.7	4.3	4.0	3.7–4.2

provide additional NOEs that can be correlated with a major orientation around either linkage (Φ or Ψ). In total, eight NOEs providing conformational information are observed in water solution[12,84]: three NOEs for both compounds and five more involving the interglycosidic methylene proton in **25a**. Because the corresponding NOE intensities are sensitive to their respective populations, at least qualitatively, an indication of the conformations present can be obtained by focusing on these key NOEs.[12,84,88] Indeed, all four *J* values and the eight NOEs are satisfactorily explained by the proposed conformational distributions. Studies on the behavior in nonaqueous solvents also showed an increase of the minor exo-*anti*-Φ conformer when the polarity of the medium was decreased, as deduced from the variations in *J* (Table VII) and NOE on going from water to Me$_2$SO, DMF, and pyridine.[12] Other C-lactose analogues bearing a hydroxyl group at the interglycosidic carbon atom (**26**) also show significant conformational equilibria, with substantial population of the *anti*-Φ conformation.[85]

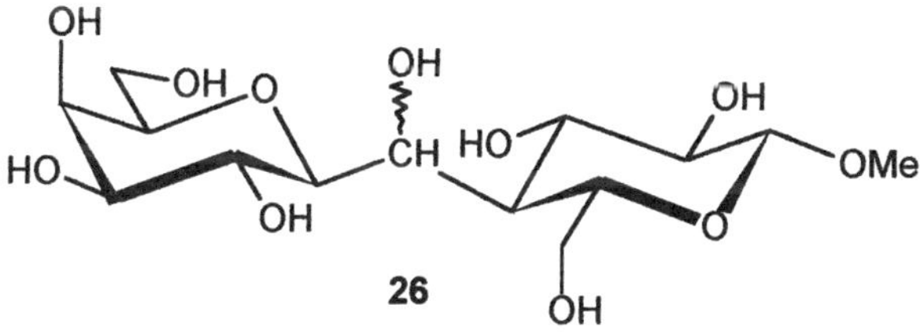

Independently, Rubinstenn et al.[89] quantitatively analyzed their NMR off-resonance ROESY data for methyl C-lactoside (**25a**, R = Me) in terms of a fixed Φ, two-state model for Ψ, proposing a 60:40 *syn:anti* conformational equilibrium.[89] However, neither a range of variation for the Φ/Ψ angles within the *syn* and *anti* families nor the presence of the *anti*-Φ conformation was included in that analysis. Indeed, the presence of the NOE for H-2′–H-4 can only be explained by this conformer. Nevertheless, and using this elegant off-resonance ROESY protocol, the existence of substantial conformational heterogeneity and internal mobility around the glycosidic linkages of C-lactose was demonstrated by the observation of distinct correlation times (from 0.12 to 0.20 ns) for different pairs of protons in the molecule. Interestingly, Espinosa et al.[12,84] had previously deduced an average correlation time of 0.15 ns for this molecule, in perfect agreement with this investigation.

Thus far, all of the reports have dealt with conformational studies and comparisons of conformational behavior of C/O-glycosyl compounds in the isolated state. Probably more important is their behavior when bound to biologically relevant proteins, including enzymes. It has been shown that, in general, for different protein-bound saccharides, the oligosaccharides are bound near their global minimum conformation, although there have also been several examples of either major or local conformational variations upon binding.[46–50]

The conformation of C-lactose (**25a**, R = H) bound to peanut lectin agglutinin has been studied through X-ray analysis.[11] The bound conformation, defined by Φ_O −63, Ψ_C 120°, is practically identical to that of the parent O-lactose (**25b**, R = H) bound to the same protein. (Φ_O −69, Ψ_C 118°). The authors were tempted to conclude from this particular observation that the conformational similarity in protein-bound states is a general phenomenon. Nevertheless, it is clear that the architecture of the binding site should also be analyzed (see later). In contrast, Espinosa et al.[12] and Asensio et al.[90] demonstrated that the three staggered conformations of C-lactose (*syn*-exo-Φ, *syn*-Ψ; *syn*-exo-Φ, *anti*-Ψ; and *anti*-exo-Φ, *syn*-Ψ) are recognized by three different proteins, and in some cases their behavior differs from that with natural lactose.

A protocol based on TR-NOESY experiments assisted by docking methods was performed. Regarding the C/O-lactose recognition by galectin-1,[90] a comparison between the NOESY/ROESY spectra of C-lactose recorded in the absence and in the presence of galectin-1 showed the disappearance of both H-1′–H-3 and H-2′–H-4 NOEs, thus providing evidence that neither the *anti*-Ψ nor the exo-*anti*-Φ geometries were recognized by this lectin. In contrast, the H-4–H-1′ NOE of moderate intensity for the free ligand corresponded now to the strongest interresidual contact in the spectrum. These findings indicate that the major bound conformation belongs to the exo-*syn*-Φ, *syn* Ψ family. Four additional cross-peaks involving the methylene protons also pointed to the recognition of this latter conformation. In order to assess specific binding of this analogue, competitive TRNOE experiments were performed: methyl β-lactoside was added to the NMR tube containing the C-lactose/galectin-1 solution. It was observed that the cross-peaks corresponding to the C-analogue changed their sign to positive, whereas the cross-peaks pertaining to the lactoside appear as negative, indicating that both ligands approach the same binding sites and that the affinity for lactose[91] is higher than for the C-analogue. The precise bound conformation was deduced from docking experiments, using the X-ray structure of the lectin, with the resulting protein–sugar complexes immersed within a 25 Å sphere containing water molecules. In all cases, major interactions take place between the galectin and the galactose moiety. However, the results showed that enthalpic differences between the complexes highly favor the exo-*syn*-Φ, *syn* Ψ family. Only in this situation is it possible for three hydrogen bonds, involving the remote Glc residue, to be formed between O-3 of this residue: two with Glu-71 (2.5 Å) and one with Arg-48 (2.6 Å). These additional interactions are no longer possible for the exo-*syn*-Φ, *anti*-Ψ or the exo-*anti*-Φ, *syn* Ψ families. The docking experiments thus clearly support the exclusive recognition of the *syn*-conformer. Obviously, the aforementioned interacions, which are defined by the shape of the protein-binding site, also take place for natural lactose, and in this case, the same conformations of lactose and its *C*-glycosyl analogue are bound by this lectin.

In contrast, and based on similar TR-NOE and docking studies, Espinosa *et al.*[12,92,93] and Rivera *et al.*[94] have shown that the conformations selected by other sugar-binding proteins differ between lactose and C-lactose. In the first place, application of this methodology to the binding of C-lactose by ricin-B,[64] a toxic galactose-binding lectin, showed the exclusive recognition of the *anti*-Ψ conformer[12,92] in contrast to the NOE and docking results for the binding of natural lactose,[64] which was in the *syn*-Ψ conformation. Regarding the three-dimensional structure of the lectin, ricin-B differs substantially from galectin-1 in both the folding topology and in the shape of the binding site. Binding studies have been carefully performed.[92] Indeed, probes at physiological pH show basically no interactions between the protein and the remote glucose residue apart from the van der Waals and hydrogen-bond stabilizing contacts between the toxin and the galactose moiety: only a weak hydrogen bond involving the 6-OH group of Glc is possible for ricin.[94] Therefore, the energetically favored global minimum of C-lactose (exo-*syn*-Φ, *anti*-Ψ) is well tolerated for binding. In contrast, because the global minimum of lactose is the exo-*syn*-Φ, *eclipsed*-Ψ conformer, this is the recognized geometry of the natural compound and, thus, in this case, different conformers of C- and O-lactose are bound by the same lectin.

Passing to hydrolytic enzymes, the same group has also demonstrated experimentally[93] that the β-galactosidase of *Escherichia coli* can even bind the high-energy exo-*anti*-Φ, *syn*-Ψ conformer of C-lactose (destabilized by 9 kJ/mol). This geometry of C-lactose, which is not detectable for natural O-lactose and which is only 5% populated for C-lactose, is selected by this glycosidase. Obviously, no comparison with the natural O-lactose is possible, because it is readily hydrolyzed by the enzyme.

The aforementioned conclusions were based upon TR-NOESY experiments, which were performed at different mixing times and a number of ligand–enzyme molar ratios. In all instance, strong and negative NOEs were observed, as expected for ligand binding. Indeed, the NOESY spectra of C-lactose (**25a**, R = H) recorded in the presence of the enzyme showed the disappearance of both H-1'–H-3 and H-1'–H-4 NOEs, indicating that neither the global minimum exo-*syn*-Φ, *anti*-Ψ conformation, nor the 40% populated exo-*syn*-Φ, *syn*-Ψ geometry were recognized by the enzyme. In turn, the H-4–H-2' NOE, which displayed weak intensity for the free ligand, was now the strongest interresidual one in the spectrum. These findings indicate that the bound conformation belongs to the high-energy exo-*anti*-Φ, *syn*-Ψ family. Also, the NOEs for the methylene protons, observed in the TR-NOESY spectral series, indicated the exclusive recognition of this conformation. In order to confirm that the binding of C-lactose was active-site selective, competitive TR-NOE experiments were also performed by adding increasing amounts of a known competitive inhibitor, namely isopropyl 1-thio β-D-galactopyranoside (**27**),

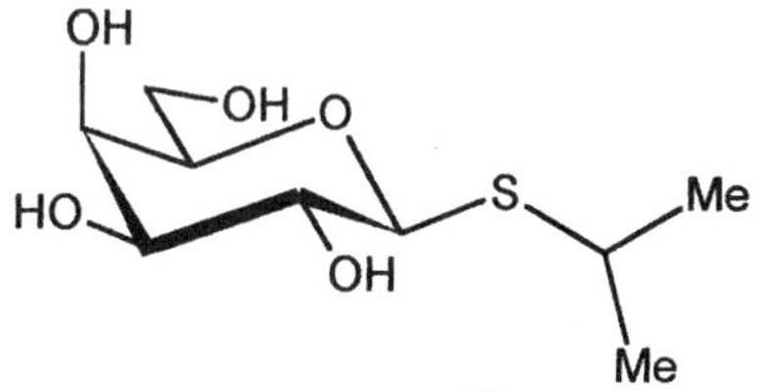

27

to a NMR tube containing C-lactose and the enzyme. It was observed that the TR-NOE signals of C-lactose disappear at equimolecular concentrations of **27** with concomitant appearance of strong negative NOEs for **27**. Docking experiments, using the binding site of the *E. coli* enzyme as deduced from the X-ray crystallographic analysis, were performed. A systematic search of the possible conformations around the glycosidic linkage, by varying φ and ψ angles, showed that only one conformational family was allowed within the binding site, and this corresponded to the exo-*anti*-Φ, *syn*-Ψ conformation. Because the binding site is located in a deep pocket of the three-dimensional structure, none of the other conformational families can occupy the binding site without severe steric conflicts, a conclusion in agreement with the experimental NMR results. In this case, the authors speculated that the recognition of this high-energy conformation of C-lactose could have implications for the catalytic mechanism, lowering the energy barrier necessary for hydrolysis. Results from this group have also demonstrated that the *anti*-Φ conformer of "S-lactose" (the lactose analogue having a thio-interglycosidic linkage) is also bound, and that therefore the recognition of the high-energy conformer of C-lactose is not an artifact due to the presence of the CH_2-group.[94a]

Regarding the binding of similar or disimilar conformations of C- and O-lactose by the different proteins, it thus seems clear that the features of the protein-binding sites are of paramount importance. They can restrict the accesible ligand conformations, and thus alter the conformational equilibria for the intrinsically flexible C-lactose. Therefore, and along this line of reasoning, the aforementioned discussion of the X-ray data for C-lactose–peanut agglutinnin lectin[11] can readily be reconciled with this interpretation. In this particular instance, sugar–lectin interactions identical to those described for the lactose–galectin-1 complex[90] take place within the binding site: three hydrogen bonds exist between O-3 of glucose and two amino acids, and these bonds are only possible if C-lactose adopts the *syn* conformation. In both cases (galectin-1 and peanut agglutinnin), the conformational equilibrium is merely shifted to select one conformation that is already 40% populated in solution. The hydrogen bonds to the Glc moiety may readily provide the energy required for selection of this conformer, and not the global minimum, which is populated to about 55%.

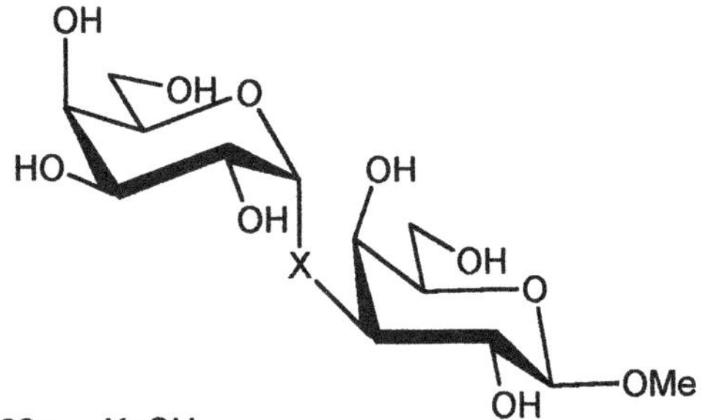

28a X=CH$_2$
28b X=O

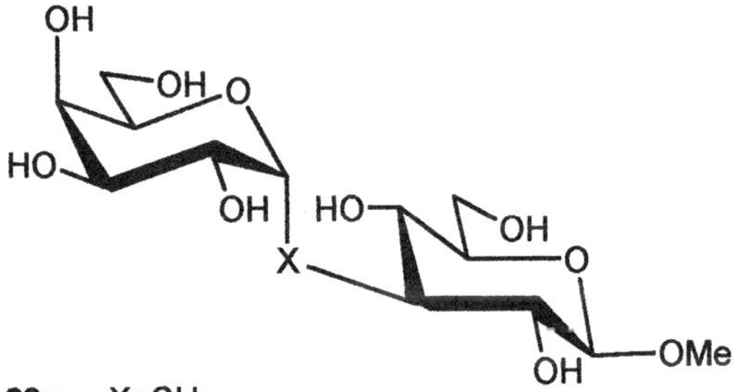

29a X=CH$_2$
29b X=O

e. *C*-Glycosyl Analogues with (1 → 3) Linkages.—Initial studies by Wang et al.[82] examined two different compounds, α-C-Gal-(1 → 3)-Gal (**28a**) and α-C-Gal-(1 → 3)-Glc (**29a**), which differ in configuration at C-4 of the aglycon. It is to be expected that the conformation around Ψ will be modulated to some extent by the orientation of HO-4 group. The HH couplings that define the Φ angle are (Table IX) 11.0–12.1 (large) and 3.1–3.3 Hz (small). Inspection of the data in Table IX thus indicates a high percentage of the *syn*-exo-conformation around Φ. The corresponding couplings related to the ψ angle are (Table X) 11.3 Hz (large) and 3.3 Hz (small) for the α-C-Gal-(1 → 3)-Gal derivative **28a** and 6.1 Hz (large) and 3.6 Hz (small) for the α-C-Gal-(1 → 3)-Glc analogue **29a**. The significant change in couplings upon configurational inversion at C-4 of the aglycone was interpreted in terms of a disturbance of the ideal staggered conformation. However,

TABLE IX

Comparison Between the Experimental J Values (Hz) Described for the Glycosidic Torsion Φ of a Series of α-(28–31) and β-(1 → 3)-Linked (32) C-Glycosyl Compounds, and the Expected J Values (Hz) Given in Table I[a]

Proton Pair	Expected Exo-syn-Φ	Expected Exo-anti-Φ	Expected non-Exo-Φ	Experimental 28, 29	Experimental
α-H-1'/H$_{proS}$	10.5–11.6	No minimum	1.4–2.7	11.0–12.1	10.2 (**30**)
α-H-1'/H$_{proR}$	1.0–2.8	No minimum	11.0–11.6	3.1–3.3	2.4 (**30**)
β-H-1'/H$_{proS}$	1.0–1.4	5.0	11.2–11.5	—	9.4–10.1 (**32**)
β-H-1'/H$_{proR}$	10.8–11.3	1.9	3.2–4.0	—	3.7–3.0 (**32**)

[a] The range of J values measured for the different series is given in the columns with the experimental data.

the changes can be readily explained in terms of a shift of the conformational equilibrium toward the Ψ(-) conformer when passing from the Glc (**29a**) to the Gal (**28a**) configuration in the aglycon.

A direct comparison for the C-/O-pairs was then performed.[83] The J data, NOEs, and T_1 relaxation times were measured for pairs of C- and O-glycosyl compounds, and deuterated analogues at the interglycosidic position were also synthesized and studied. The authors explicitly noted that they did not use NOEs and T_1 measurements as the principal means of conformational analysis. They first established the conformational preference of C-glycosyl compounds by relying on the values of J couplings, and then compared the behavior of both types of compounds through NOE and T_1 data. With respect to the C/O-pair of α-Gal-(1 → 3)-Gal (**28**), and beyond the J values, 1D-NOE data obtained by selective irradiation were qualitatively compared for both compounds. Although only qualitative, the H-1'–H-4 and H-1'–H-3 NOEs were reported to have basically the same intensities in the O-glycoside. However, H-1'–H-4 showed double the intensity of H-1'–H-3 for the C-analogue, suggesting some sort of conformational differences that were not mentioned by

TABLE X

Comparison Between the Experimental J Values in Different Solvents (Hz) Related to the Glycosidic Torsion Ψ of a Series of β- (32) and α-(1 → 3)-Linked (28–30) C-Glycosyl Compounds, and the Expected J Values (Hz) Given in Table II[a]

Proton Pair	Expected $\Psi(+)$	Expected $\Psi(-)$	Expected Ψ(eclipsed)	Expected anti-Ψ	Experimental 28, 29, 30*, 32
H-X/H$_{proS}$	9.0–12.4	2.0–2.5	4.7	2.9–3.3	3.3, 3.6, 3.9, 6.7
H-X/H$_{proR}$	1.4–2.5	12.3–12.4	4.3	3.8–4.1	11.3, 6.1, 4.4, 6.0

[a] For the experimental data, the values are given for aglycones having the Gal (**28**), Glc (**29**), and Man (**30***, **32**) configurations, in this order. The asterisk denotes the values for the peracetylated derivative.

the authors, who stated that both C/O-pairs show the same conformation around both glycosidic linkages.

Other α-Gal-(1 → 3) derivatives have been studied by Ferrito and Vogel.[95] Thus, the solution conformation of the C-glycosyl analogue (**30a**) of α-Gal-(1 → 3)-α-Man-OMe (**30b**) has also been studied by NMR spectroscopy. One large (10.2 Hz) and one small (2.4 Hz) coupling constant was found for H-1 of Gal, whereas a large and a medium coupling were measured for H-3 of Man. These data, which were also complemented by NOE measurements, were interpreted in terms of a major exo-anomeric conformation around Φ and a staggered conformation around Ψ. An interesting variation around Ψ was detected for its peracetylated analogue. In this case, similar large (11.6 Hz) and small (3.0 Hz) couplings were found for H-1', but two small values were measured for H-3 (3.9 and 4.4 Hz), indicating a different conformation around this aglyconic angle. These values were interpreted in terms of a nearly eclipsed Ψ torsion, although, according to Table X, a shift in the rotamer populations could also be expected.

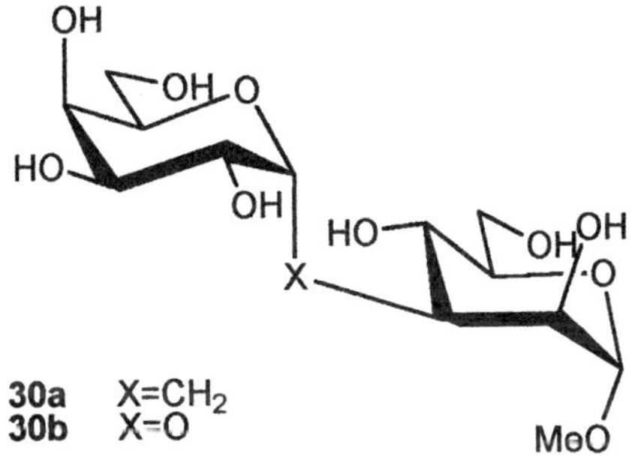

30a X=CH$_2$
30b X=O

The solution conformation of the C-glycosyl analogue (**31**) of the 6-deoxy derivative of β-Gal5N-(1 → 3)-α-Altf-OMe has also been studied[96] by using J and NOE data. The compound studied bore an (R)-hydroxyl group at the interglycosidic carbon atom, and thus only one coupling constant to the anomeric H-1

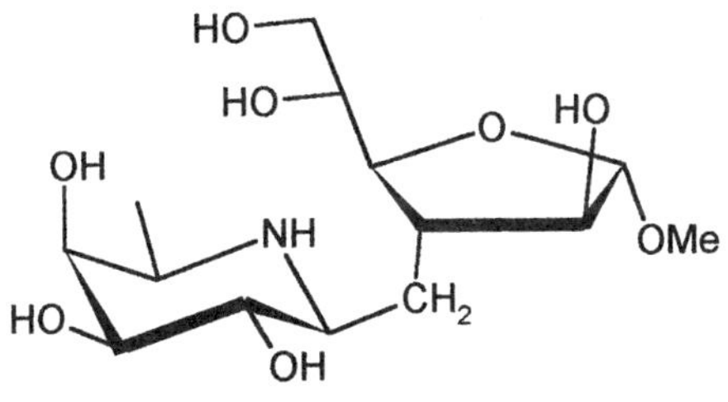

31

proton is available. The coupling value (1.3 Hz) again indicated a major *syn* exo-anomeric conformation for Φ. This conclusion was also corroborated by NOEs.

In general, therefore, all of the data from both the Kishi and Vogel groups[82,95] indicate that the *syn*-exo-anomeric conformation is primarily adopted for the *C*-glycosyl analogues of α-Gal-(1 → 3)-linked disaccharides (**28–31**). Minor populations, if any, of other conformers could be detected. In contrast, the conformational equilibrium around Ψ strongly depends on the arrangement around C-3 of the aglycon. These features vary drastically when β-C-Man-(1 → 3) linkages are considered.

Marquis and co-workers[97] have also analyzed the solution conformation of the *C*-glycosyl analogue (**32**) of β-C-Ara5N-(1 → 3)-α-Man-OMe (having an axially oriented hydroxyl group at C-2 of the nonreducing end) by *J* (Tables IX and X) and NOE NMR data. Couplings of medium size were observed for H-3 of the Man residue, thus indicating major conformational averaging around the Ψ torsion angles. Regarding Φ, the combination of *J* and nuclear Overhauser enhancements permitted stereospecific assignment of the interglycosidic protons, and the observed values lead to the proposal of a temperature-dependent conformational equilibrium between the *syn*-exo-anomeric and nonexomeric conformations around Φ. Indeed, a major non-exo-anomeric form was proposed for the β-Man-5N linkage. From this work, it may be deduced that the orientation of the hydroxyl group at position C-2 of the nonreducing pyranoid ring strongly influences the conformational behavior at the glycosidic torsion angle (see also later for trehalose-type compounds).

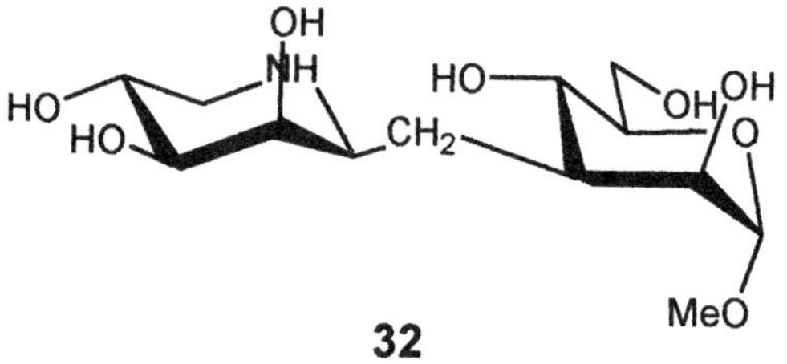

32

f. C-Sucrose.—The conformation of the C-analogue (**7a**) of sucrose (**7b**) has also been studied.[71,98] The only available *J* values are obviously, those related to the glycosidic Φ linkage of Glc (8.5, 3.6), and these were interpreted in terms of a 4:1 mixture of *syn*-exo-anomeric and non-exo-anomeric conformers. A qualitative comparison of NOEs between the *C*- and *O*-glycosyl compounds prompted the authors to conclude that both are flexible in solution, as explored in depth in previous reports for natural sucrose.[99] The similarity was not further investigated at this stage. Nevertheless, significant variations in the couplings (up to 2.3 Hz) for the five-membered ring of the *O*- and *C*-glycosyl compounds are evident. It was concluded that the stereoelectronic stabilization in sucrose would serve only to reinforce the thermodynamic preference for the exo-anomeric conformation, which already exists in the *C*-glycosyl analogue. In a subsequent report,[98] the authors demonstrated that the *C*-glycosyl analogue, but not the natural compound, is able to bind calcium ions in methanolic (but not in water) solution, and that this complexation process induces a conformational variation toward almost ideal exo-anomeric conformations for both glycosidic linkages, as evidenced by optical rotation, *J*, and NOE changes. Replacement of the 1'-hydroxymethyl group by a proton also abolished the cation-binding ability. Based on these data, the authors established that the inertness of sucrose toward calcium binding might be due to the presence of the lone pairs on the glycosidic oxygen atom, and that therefore, this observed difference should be considered when using *C*-glycosyl compounds as surrogates for natural products.

The behavior of C-sucrose prompted the analysis of other *C*-furanosyl analogues.[71,98] Thus, the *C*-arabinofuranosyl analogue (**33a**) was examined and compared with the corresponding O-glycoside **33b**. Data for **33** and its 2-epimer **34** were also examined. Deoxy derivatives were also synthesized and their NMR spectra registered. Marked changes were found between both series of compounds, with major differences in the conformational behavior of the five-membered rings for the C/O-pairs. In addition, the *C*-glycosyl analogues show no exo-anomeric preferences and have equatorial anomeric C-C bonds, in contrast to the O-furanosides, with axially oriented anomeric C-O bonds. A similar conclusion had been previously reported by other authors.[100] It was stated that this difference is probably due to the existence of a stereoelectronic effect for the normal glycosides. The presence of quaternary O- or C-linked anomeric centers as in the sucroses series appears to provide additional flexibility at the corresponding glycosidic linkages, and both types of compound are conformationally more similar. In fact, they demonstrated that both series are indeed flexible, although their flexibility characteristics were not further analyzed. C-Nucleoside derivatives from D-arabinose and 2-deoxy-D-erythro-pentose have been also analyzed, and it was shown that they also behave very differently from the analogous N-nucleosides. In particular, stronger *syn*-preferences were observed for the conformations of the N-series, a feature that does not sustain for the *C*-glycosyl analogues.[15,16]

33a X=CH$_2$
33b X=O

34a X=CH$_2$
34b X=O

The data for the sucrose pair (**7**) and its analogues (**33**, **34**), thus show a clear indication that the O/C-pairs of these derivatives have distinctly different conformational behavior. Nevertheless, these compounds present a special case from the conformational point of view, and the conclusions drawn from them should not be overemphasized.

g. C-Glycosyl Analogues of Trehalose and (1 → 1)-Linked Trehalose Analogues.—Trehalose derivatives have also been studied.[101–103] In these compounds, both bonds linking the glycosyl groups are glyconic, and therefore the corresponding glycosides show overlapping exo-anomeric effects. Many combinations have been studied, α,α (**35**), β,β (**36**), and α,β (**37**), with different Glc- or Man-type configurations on either side of the glycosidic linkage. The measured vicinal couplings for the α,α- and β,β-C-glycosyl derivatives with C_2-symmetry suggested that these compounds adopt an unique double syn-exo-anomeric conformation, although unsymmetrical, partially protected derivatives of these molecules first had to be studied to come to this conclusion. The couplings measured for these compounds were used as input to simulate the ^{1}H NMR spectra of the appropriate derivatives having C_2-symmetry. The conformation of the C-glycosyl analogue (**36a**) of β,β-trehalose (**36b**) has also been studied by Martin and Lai.[102] Also as mentioned before, the coupling pattern of the bridge methylenic protons had to be simulated to obtain the best values of individual coupling constants, giving large and small values of 9.6 and 2.4 Hz, respectively, indicating a major double syn-exo-anomeric conformation, in agreement with the conclusions by Wei and Kishi[101] and similar to those predicted for β,β-trehalose itself.[54a] The couplings (11.0, 3.0 Hz) for the per-O-acetyl derivative also accorded with these features. Derivatives monosubstituted by NO_2 or NHAc groups at the pseudoglycosidic carbon provided coupling values (8.2 or 7.1 Hz, respectively) that permitted consideration of a significant contribution of other conformers. Therefore, and according to this conclusion, the C/O-compounds both behave similarly from the conformational point of view. The NMR spectra of the α,β-C-glucosyl derivative (**37a**) could be analyzed directly [101] (Table XI). The couplings for the α-H-1 were 9.7 and 4.0, whereas those for the β-H-1 were 7.1 and 3.7 Hz. These values were interpreted by the authors as indicative of the presence of an ideal exo-anomeric conformation for the α linkage, along with the existence of a conformer either somewhat

TABLE XI

Comparison Between the Experimental J Values (Hz) Described for the Glycosidic Torsion Φ of a Series of β- and α-(1 → 1)-Linked C-Glycosyl Trehalose-like Compounds (36–39), and the Expected J Values (Hz) Given in Table I[a]

Proton Pair	Exo-syn-Φ	anti-Φ	non-Exo-Φ	Experimental (β,β, **36**) (α,β, **37**)	Experimental
β-H-1′/H$_{proS}$	1.0–1.4	5.0	11.2–11.5	3.7 (4.2, **36**)	8.5(β-Man, **39**)
β-H-1′/H$_{proR}$	10.8–11.3	1.9	3.2–4.0	7.1 (9.3, **36**)	5.5(β-Man, **39**)
α-H-1′/H$_{proS}$	10.5–11.6	No minimum	1.4–2.7	9.7	7.7(α-Man, **38**)
α-H-1′/H$_{proR}$	1.0–2.8	No minimum	11.0–11.6	4.0	7.0(α-Man, **38**)

[a] The range of J values measured for the different series is given in the columns with the experimental data.

distorted from the exo-anomeric position or a mixture of staggered conformers for the β linkage. Nevertheless, according to the data in Table XI, equilibria are probably taking place in both cases, although the existence of several rotamers is more evident for the β linkage.

From these data and quoting their previous reports already mentioned for C-maltose (**18**) and C-cellobiose (**14**), the authors generalized that axial *C*-glycosylic linkages are conformationally more rigid than the C-equatorial ones, and these in turn more rigid than the C-aglyconic carbon atoms. Solvent effects suggested a slight weakening in the *syn*-exo-anomeric preference of the α-glycosylic bond with varying solvent polarity (the large coupling changes from 10.2 to 9.7 Hz, whereas the small one varies from 3.2 to 4.0 Hz). The C-2 epimers for both α and β anomers were also studied (**38, 39**). For the α-Man case (**38**), the corresponding couplings change (from C- to O-compounds) from to 9.7 and 4.0 to 7.7 and 7.0 Hz, whereas for the β-Man linkage (**39**), they vary from 7.1 and 3.7 to 5.5 and 8.5 Hz. These significant changes were interpreted as indicative of deviations from the

35a X=CH$_2$
35b X=O

38a X=CH$_2$
38b X=O

36a X=CH$_2$
36b X=O

37a X=CH$_2$
37b X=O

39a X=CH$_2$
39b X=O

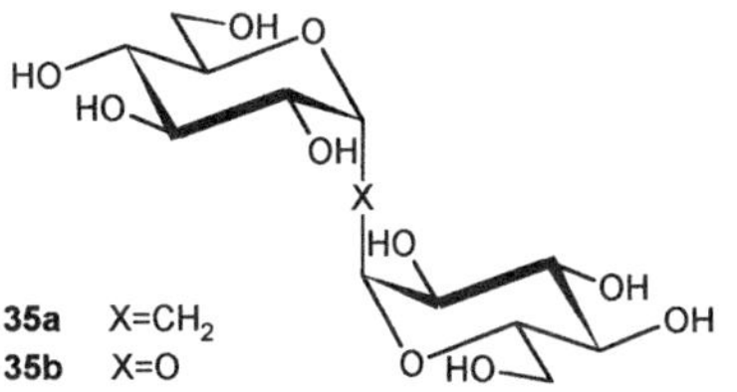

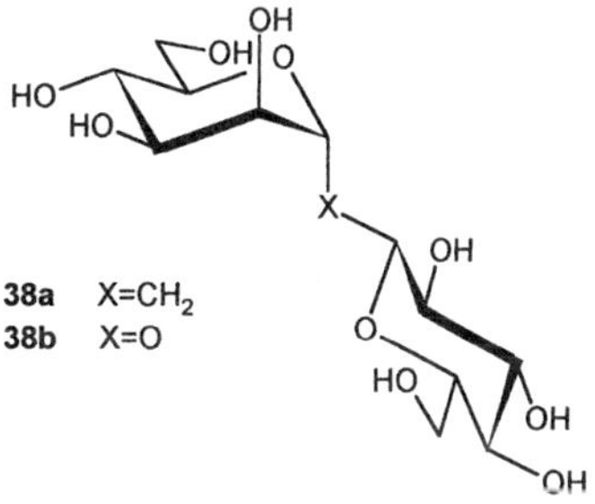

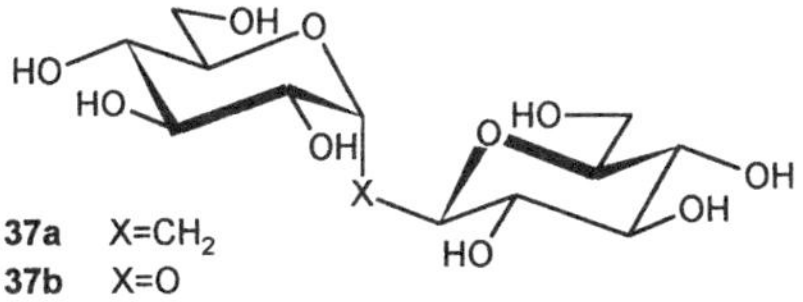

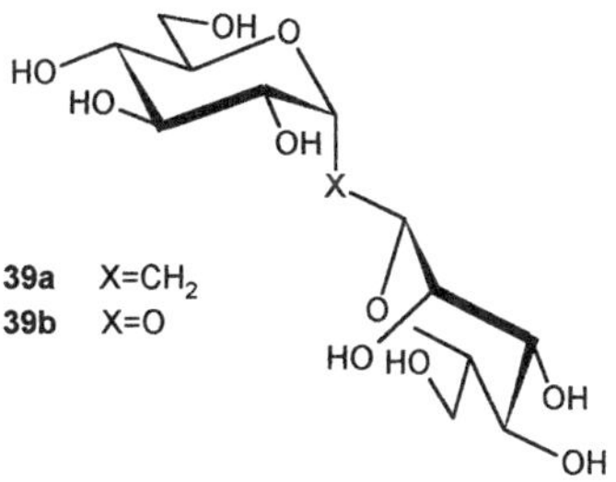

TABLE XII

Comparison Between the Experimental *J* Values (Hz) Described for the Glycosidic Torsion Φ of a Series of β- and α-(1 → 1)-Linked *C*-Glycosyl Trehalose-like Compounds, and the Expected *J* Values (Hz) Given in Table I

Proton Pair	Exo-*syn*-Φ	*anti*-Φ	non-Exo-Φ	Experimental 40^a	Experimental 38, 39^b
β-H-1'/H$_{proS}$	1.0–1.4	5.0	11.2–11.5	3.2 (β-Gal)	8.5 (39)
β-H-1'/H$_{proR}$	10.8–11.3	1.9	3.2–4.0	8.0 (β-Gal)	5.5 (39)
α-H-1'/H$_{proS}$	10.5–11.6	No minimum	1.4–2.7	7.4 (α-Man)	7.7 (38)
α-H-1'/H$_{proR}$	1.0–2.8	No minimum	11.0–11.6	7.1 (α-Man)	7.0 (38)

^{a} Ref. 103.
^{b} Ref. 101.

ideal *syn*-exo-anomeric conformation, but no mention of the possible presence of non-exo-anomeric conformers was made. Indeed, according to the data shown in Table XI, the non-exoanomeric conformers are probably predominant for both α- and β-Man linkages.

These conclusions have been interpreted quantitatively for similar trehalose-related *C*-glycosyl analogues. Asensio *et al.*[103] analyzed other α,β-*C*-(1 → 1)-glycosyl derivatives. In particular, the β-Gal-(1 → 1)-α Man O/*C*-glycosyl pair (**40**) has been carefully studied by NOE/*J* data and time-averaged restrained molecular dynamics (tar-MD). As the glycoside shows overlapping exo-anomeric effects, the exo-anomeric preference should be weakened with respect to regular (1 → 2, 3, 4, or 6)-linked glycosides. In this case, the conformational analysis was based on coupling-constant values and on the exclusive interresidue NOEs that characterize the different low-energy minima.

The couplings (Table XII) clearly indicate the existence of major averaging around both linkages. For the glycoside, the *best experimental* conformer distribution was obtained through tar-MD simulations[104] with the AMBER 5.0 force field,[105] and the experimental NOEs. The distribution indicated that the minimum showing two *syn*-exo-anomeric conformations was the major one (95%), followed by a local minimum in which the Man linkage showed the non-exo geometry (∼4–5%). In contrast to this behavior, the MM3* map for the *C*-glycosyl analogue predicted five conformational families. The *experimental* conformer distributions for **40** were again obtained from 15-ns tar-MD simulations, which included the *observed J*(4)/NOE(14) data with conformational information. The observed data were in full concordance with the values from the molecular dynamics study. A distinctly different conformational behavior of the *C*-glycosyl analogue was demonstrated as compared to the glycoside. Indeed, the Φ$_{Man}$ orientations were quite different: a 52% population of conformers having Φ$_{Man}$ 60°, were found for **40a** (non-exo-anomeric), as compared with only 5% for the glycoside **40b**. Although the conformational distribution about the β–Φ$_{Gal}$ linkage was more

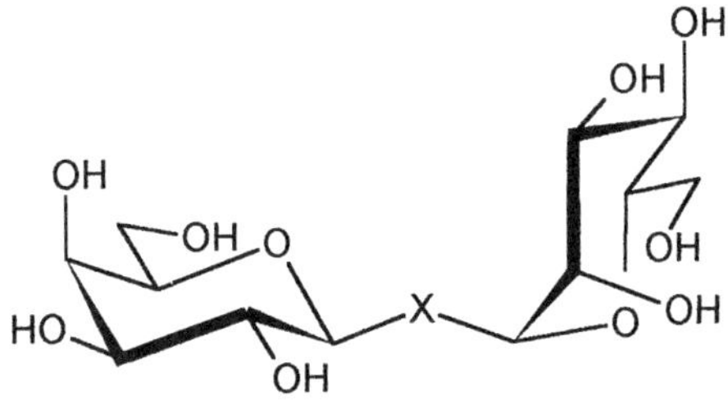

40a X=CH$_2$
40b X=O

similar to that of the glycoside, a 10% population of non-exo conformers was still deduced, whereas their presence in the glycoside was less than 1%. Probably, the equatorial orientation of O-2 of Gal destabilizes the non-exo conformers via 1,3-*syn*-diaxial interactions. That the conformer that shows a double *syn*-exo-anomeric orientation is strongly preferred in the glycoside **40b,** in contrast to a preference for the non-exo conformer around Φ_{Man} in the *C*-glycosyl analogue **40a,** is clearly consistent with the importance of the stereoelectronic effect. Therefore, the authors performed a comparison of the conformer ratios for this pair to give a lower limit for the stereoelectronic effect in water (for $\Phi_{\alpha-Man}$, ~1.9 kcal/mol). Because this system presents overlapping exo-anomeric effects, which are opposed, this value is probably smaller than the exo-anomeric effect for normal glycosides. This analysis has also been extended to β-Man-(1 → 1) linkages.

For C-β-Man5N-(1 → 1)-β-Man (**41**) and C-β-Glc5N-(1 → 1)-β-Glc (**42**), which are also trehalose-like systems and whose key feature is the different configuration at position C-2 of one saccharide, major conformational equilibria have

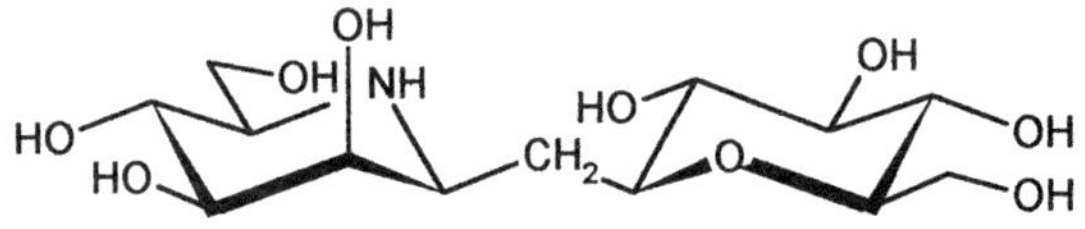

41

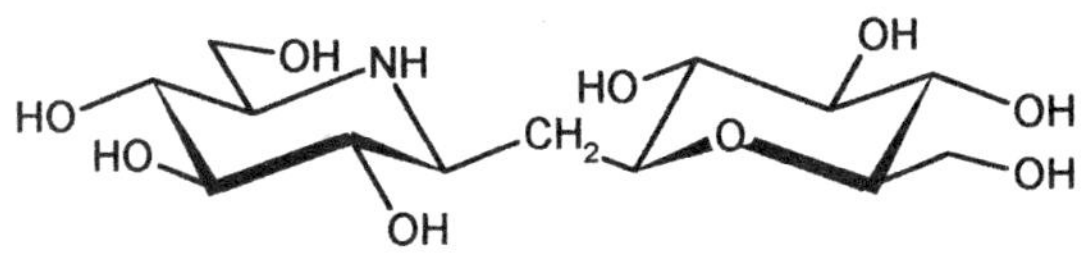

42

TABLE XIII

Comparison Between the Experimental *J* Values (Hz) Described for the Glycosidic Torsion Φ of of β-(1 → 1)-Linked *C*-Glycosyl Trehalose-like Compounds (41, 42), and the Expected *J* Values (Hz) Given in Table I[a]

Proton Pair	Expected Exo-*syn*-Φ	Expected Exo-*anti*-Φ	Expected non-exo-Φ	Experimental 41, 42
β Man-H-1'/H_S	1.0–1.4	5.0	11.2–11.5	6.8 (**41**)
β Man-H-1'/H_R	10.8–11.3	1.9	3.2–4.0	6.5 (**41**)
β Glc-H-1'/H_S	1.0–1.4	5.0	11.2–11.5	3.5 (**42**)
β Glc-H-1'/H_R	10.8–11.3	1.9	3.2–4.0	7.5 (**42**)

[a] The experimental *J* values are given for the 5N-residue of **41** and **42**.

been reported for both the C- and imino-C linkages.[106] The experimentally based population distributions for these compounds were also obtained by using NOE/*J* data (Table XIII) in combination with time-averaged restrained molecular dynamics (tar-MD). The data showed that the effect of the configuration at C-2 of the imino sugar and sugar residues on the conformational populations is noticeable at both basic and acid pH values.

The distribution around $Φ_{Man5N}$ of C-β-Man5N-(1 → 1)-β-Man (**41**) was found to depend on the pH of the solution. Nevertheless, in all cases, the minor conformer is that which presents a 1,3-type destabilizing interaction between the hydroxyl group at C-2 and the vicinal residue, as illustrated in Fig. 1 for β linkages. In particular, when OH-2 is equatorial (the Glc analogue, **42**, of **41**), the minor conformation is always non-exo-anomeric (4–12%). However, when OH-2 is axial (Man example, **41**), the *exo-anti* conformer is the less populated one (6–7%). As a further step, the free-energy values for the rotameric equilibria were estimated, which showed the key influence of the configuration at C-2. For equatorial OH-2 groups, ΔGexo-*syn*/non-exo amounts to 1.35 ±0.35 kcal/mol, whereas the ΔGexo-*syn*/exo-anti value is 0.65 ±0.25 kcal/mol. For axial OH-2 groups, the value of ΔGexo-*syn*/non-exo is only 0.45 ±0.35 kcal/mol, whereas the ΔGexo-*syn*/exo-anti amounts to 1.30 ±0.20 kcal/mol. A simple subtraction of the ΔG values gave a direct estimation of the ΔG contribution to the rotation around Φ that arises from the interaction between OH-2 and the vicinal residue R. Thus, the additional free-energy cost for the exo-*syn*/non-exo transition of Φ within a pyranose ring with an equatorial hydroxyl group at position C-2, in comparison to that for a ring with an axial OH at the same position, is 1.05 ±0.15 kcal/mol. Similarly, the corresponding relative energy cost for the exo-*syn*/exo-anti transition is −0.75 ±0.05 kcal/mol. Therefore, the exo-anti conformation is destabilized in β-Man compounds (such as **41**), and the non-exo conformation is also destabilized in β-Glc moieties. Obviously, in this values are contained both 1,3-*syn*-diaxial interactions as well as other steric (torsional) and polar effects (see a similar earlier case for **32**).

TABLE XIV

Comparison Between the Experimental *J* Values (Hz) Described for the Glycosidic Torsion Φ of a Series of β-Gentiobiose (6) and α-Isomaltose (43), (1 → 6)-Linked *C*-Glycosyl Compounds, and the Expected *J* Values (Hz) Given in Table I

Proton Pair	Expected Exo-*syn*-Φ	Expected *anti*-Φ	Expected non-Exo-Φ	Experimental 6, 7[a]
β-H-1'/H$_{proS}$	1.0–1.4	5.0	11.2–11.5	2.7 (**6**)
β-H-1'/H$_{proR}$	10.8–11.3	1.9	3.2–4.0	8.6 (**6**)
α-H-1'/H$_{proS}$	10.5–11.6	No minimum	1.4–2.7	11.7 (**43**)
α-H-1'/H$_{proR}$	1.0–2.8	No minimum	11.0–11.6	3.2 (**43**)

[a] Ref. 70.

h. *C*-Glycosyl Analogues with 1 → 6 Linkages.—*C*-Isomaltose (**43a**), with an α-(1 → 6) linkage, has also been analyzed by Goekjian and co-workers.[70] The observed couplings between the anomeric proton with the CH$_2$ protons are 3.2 and 11.7 Hz (Table XIV). Although the authors stated that, as with their natural O-linked counterparts, these *C*-glycosyl compounds are not conformationally rigid, especially around Ψ and ω, the data were again interpreted in terms of a strongly favored conformation, especially around Φ- −70°, as supported by the data in Table XIV. The couplings that define Ψ are 10.8, 10.8, 5.3, and 3.6, whereas those defining ω are 2.2 and 9.5. The Ψ and ω values were described as being around −75 and −40°. These data were extended in a subsequent publication[107] in which the authors established the presence of a predominant form along Φ/Ψ/ω, with the magnitude of the couplings precluding a major contribution of other conformers.

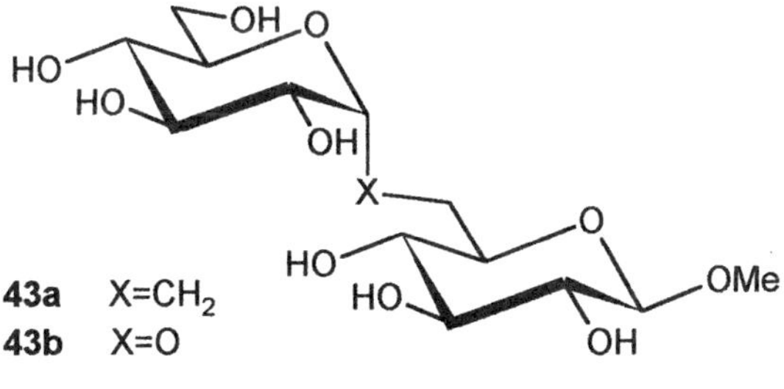

43a X=CH$_2$
43b X=O

β-*C*-Gentiobiose (**6**), having a β-(1 → 6) linkage, has been studied in a similar manner.[70] The observed couplings for the anomeric proton with the CH$_2$ protons were 2.7 and 8.6 Hz. Although the data were again interpreted in terms of a predominant *syn*-exo-anomeric conformation, the *J* values in Table XIV indicate that a contribution of the *anti*-exo or non-exo conformers should also be considered. The data given for Ψ, which were also elaborated in a later publication,[107] were 11.3 and 5.0 Hz, whereas those defining ω were 2.5 and 9.2 Hz. With these data at hand, Ψ and ω values were described around -60 and $-35°$. The authors therefore proposed the presence of a predominant form along Φ/Ψ, with a high participation (70–80%) of a unique conformer along ω. As noted by the authors, these values are not in agreement with the X-ray structure already discussed, which established a major anti-Φ conformation.[68] Other O-protected, $(1 \to 6)$-linked β-Glc analogues of **6** have been studied in solution by using proton vicinal coupling constants, and these also provided data in agreement with a major *syn*-exo-anomeric conformation for Φ, and a major anti Ψ conformation.[102]

The *C*-disaccharide analogues C-β-Man5N-$(1 \to 6)$-β-Man (**44**) and C-β-Glc5N-$(1 \to 6)$-β-Glc (**45**) have been studied by Asensio *et al.*[106] in order to isolate the effect of 1,3-*syn*-diaxial interactions, with different conclusions. Because these compounds have three bonds in the interresidue linkages, no important polar and/or steric interactions between the residues, apart from the 1,3-*syn*-diaxial ones, are likely to occur. It is well known that $(1 \to 6)$-linked oligosaccharides are rather flexible around the ω and Ψ linkages. In the first place, diastereotopic assignment of the prochiral H^R and H^S protons was performed. Next, the observed J and NOE values indicated that the non-exo-anomeric orientation around Φ_{MAN5N} of **44** is highly populated in solution. The experimental J values and the NOE-derived distances were included in a tar-MD protocol to deduce a 75% population of the non-exo conformer. In contrast, application of the same protocol to the C-β-Glc5N-$(1 \to 6)$-β-Glc (**45**) provided evidence of a major exo-anomeric *syn* orientation (>70%). Thus, the results obtained for these analogues were similar to those reported already for the $(1 \to 1)$ Glc/Man pair (**41, 42**), with an even higher percentage of non-exo conformers in the absence of interresidual interactions for the $(1 \to 6)$-linked compound. Thus, following the protocol already outlined,[106] there is an additional energy cost, from changing the orientation of the 2-OH group, of 1.7 ± 0.1 kcal/mol for the exo-*syn*/non-exo transition and of -0.7 ± 0.4 kcal/mol for the exo-*syn*-/exo-anti transition. The 1.7 ± 0.10 kcal/mol energy value for the 1,3-*syn*-diaxial interaction between one equatorial hydroxyl group and the aglycone is in excellent agreement with the *ab initio* results reported by Houk *et al.*[45] for the equivalent interaction in 2-ethyl-3-hydroxytetrahydropyran (1.5 kcal/mol) *in vacuo*.

Although the energy values for the 1,3-*syn*-diaxial interactions were obtained for C-(5-amino-5-deoxyhexopyranosyl) compounds, the corresponding values will be similar for regular C-analogues and somewhat larger for the corresponding O-glycosides, because the C-O bonds are shorter than the C-C bonds. In the same report,[106] the conformational behavior of a related β-linked mannopyranoside (**46**) was also studied to access the energetic contribution of the stereoelectronic effect. The β-mannoside **46** was chosen because there are basically no 1,3-*syn* diaxial interactions for either the *syn*-exo- or non-exo forms. Thus, the conformational equilibrium around Φ should be governed by the stereoelectronic contribution. According to both NMR and molecular-mechanics calculations, there was at least a ~99:1 ratio in favor of the exo-*syn* conformer versus the non-exo analogue, which represents a ΔG exo-*syn*/non-exo value of at least 2.75 kcal/mol. Because **41** showed a ΔGexo-*syn*/non-exo value of 0.45 kcal/mol and the analogue **44** presented a ΔGexo-*syn*/non-exo of −0.3 kcal/mol, even favoring the non-exo-anomeric conformer [assuming similar steric interactions around Φ_{Man}, for both exo-*syn* and non-exo conformers (very minor for these manno-compounds)], the additional stabilization of the exo-anomeric *syn* conformation of **46** in water provided by the stereoelectronic effect is at least 2.3 kcal/mol.

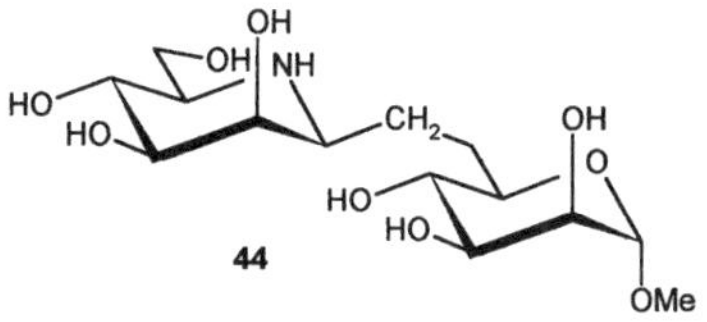

i. *C*-Glycosyl Analogues with (1 → 2) Linkages.—A combination of molecular mechanics and dynamics calculations with detailed NMR data (*J* couplings and NOE), has been applied to deduce the conformational behavior of the C-mannobiose/O-mannobiose pair (**47**) having the α-(1 → 2) linkage in water and other polar solvents. This study[108] has shown rather different conformational distributions for both compounds. In particular, the C-analogue **47a** shows a 20% population of conformers in the non-exoanomeric Φ region, in contrast to the natural compound **47b** for which no population is found within this area. The couplings in Table XV can only be explained by assuming an important contribution of the non-exo family. In addition, a medium-size interresidue H-2′/H-2 NOE is compatible only with the presence of this conformer to an appreciable extent. This behavior is also found for an analogue, **48** that has a hydroxyl group attached to the interglycosidic carbon atom.

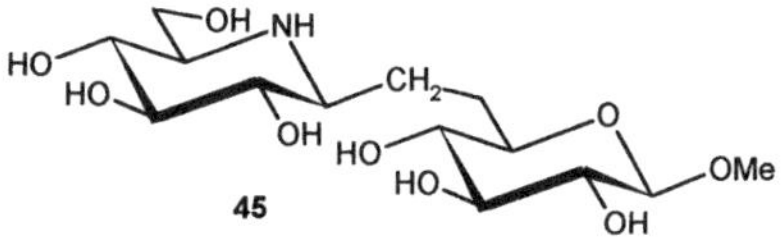

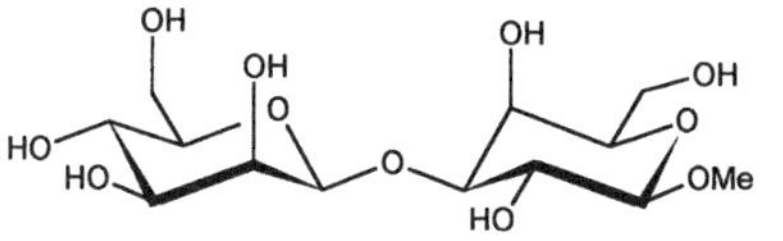

Later studies from the same group (Asensio *et al.*[94a]) have also shown that the α-Man-(1 → 3)-Man analogue (**49**) shows similar conformational behavior, with appreciable contributions of the non-exo-anomeric conformer.

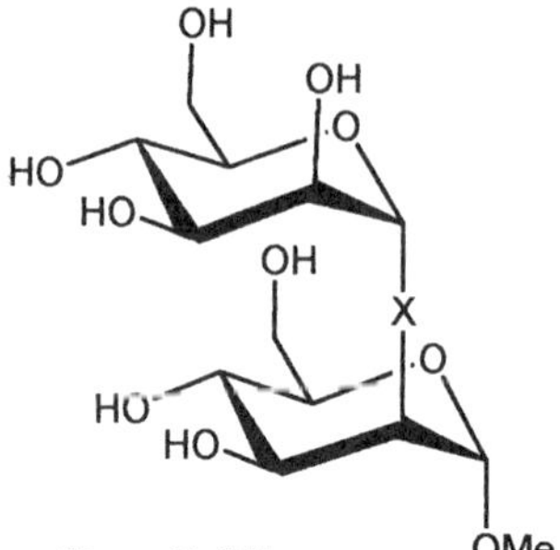

47a X=CH$_2$
47b X=O

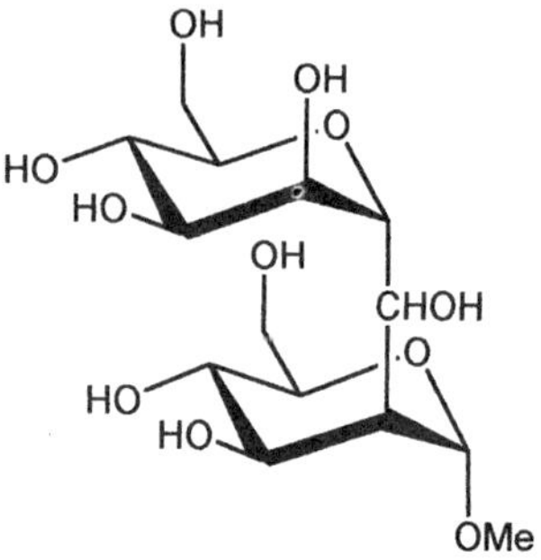

48

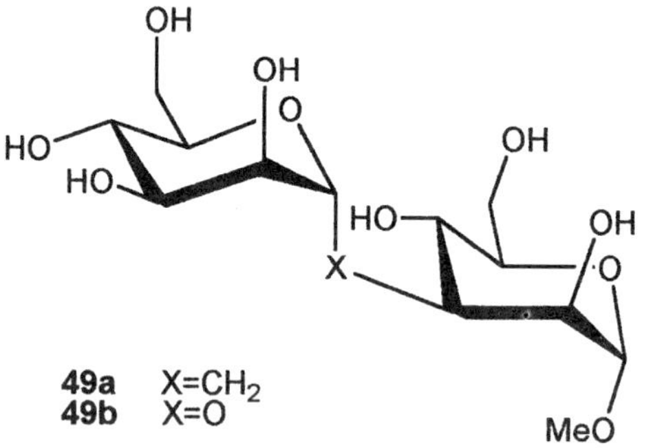

49a X=CH$_2$
49b X=O

j. C-Sialyl Compounds.—Sialyl-oligosaccharides are important compounds from the conformational standpoint. As with 2-deoxy aldoses, the lack of substitutent at C-3 renders the three basic staggered conformations around Φ essentially equivalent from the steric point of view (Fig. 1). None of them (exo-*syn*, exo-*anti*, and non-exo) show any 1,3-*syn*-diaxial-type interactions, as expected for regular glycosides when C-2 is substituted. Moreover, all of these orientations show two gauche-type interactions around the glycosidic linkage. Furthermore, the glycosidic carbon is quaternary, with the electron-withdrawing CO$_2$H group

TABLE XV

Comparison Between the Experimental *J* Values (Hz) Described for the Glycosidic Torsion Φ of α-(1 → 2)-*C*-Mannobiose (**47**) and the Expected *J* Values (Hz) of Table I

Proton Pair	Expected Exo-*syn*-Φ	Expected Exo-*anti*-Φ	Expected non-Exo-Φ	Experimental 47
α-H-1'/H$_{proS}$	10.5–11.6	No minimum	1.4–2.7	8.7
α-H-1'/H$_{proR}$	1.0–2.8	No minimum	11.0–11.6	3.5

attached to it. Therefore, in normal glycosides, this group should increase the participation of the exoanomeric effect from the interglycosidic oxygen atom.

For a C-glycosyl analogue (**50**) of αNeu5Ac-(2 → 3)-Gal, which bears a hydroxyl group at the interglycosidic carbon atom, the possibility of accessing up to 11 NOE and 1 J values, all providing conformational information, prompted the authors[109] to use time-averaged-restrained molecular dynamics (tar-MD) to get an ensemble average distribution of conformers in water solution. Thus, after conducting different tar-MD simulations using the NMR data as experimental constraints, with different starting geometries and dielectric constants, a very satisfactory match was obtained between the expected and the observed NMR parameters. With regard to use of the tar-MD method, the combination of 1 J and 11 NOE data to define only two dihedral angles gives confidence in the population estimates obtained. Indeed, all of the NOE-derived distances and J couplings were reproduced in a quantitative manner. The trajectories indicated the presence of the three different Φ values in the distribution (exo-*syn*, exo-*anti*, and non-exo), with a similar population for all of them. Regarding Ψ torsion, the surface is predominantly extended towards the staggered $g(-)$ conformation, although a participation of minima having an eclipsed Ψ angle is also present. Therefore, the results for the sialyl-O/C-galactose pair indicated that, in the absence of stereoelectronic stabilization, significant populations of conformers that are not consistent with the exo-anomeric disposition may be adopted around Φ.

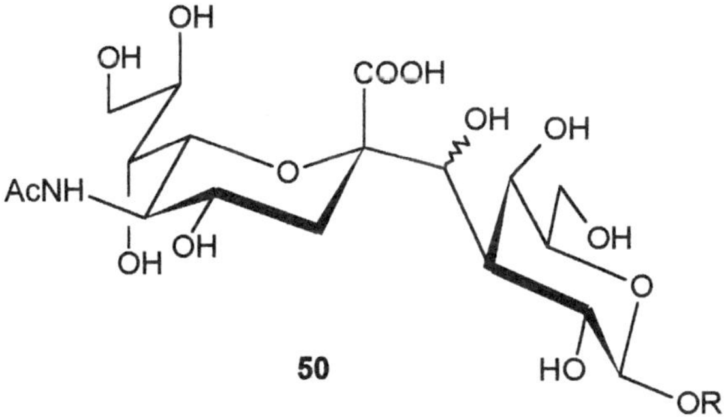

k. Trisaccharides.—The first pseudotrisaccharides to be analyzed were several C-glycosyl analogues of the Type II O(H) blood group determinant.[110] In particular, the authors studied the L-Fucα-(1 → 2)-Galβ-(1 → 4)-GlcβOMe (**51**), L-Fucα-(1 → 2)-Galβ-(1 → 4)-XylβOMe (**52**), L-Fucα-(1 → 2)-3-deoxyGalβ-(1 → 4)-GlcβOMe (**53**), and L-Fucα-(1 → 2)-3-deoxyGalβ-(1 → 4)-XylβOMe (**54**) pseudotrisaccharides. As an approximation, every compound was analyzed in terms of two independent disaccharide systems, using the diamond lattice and assuming

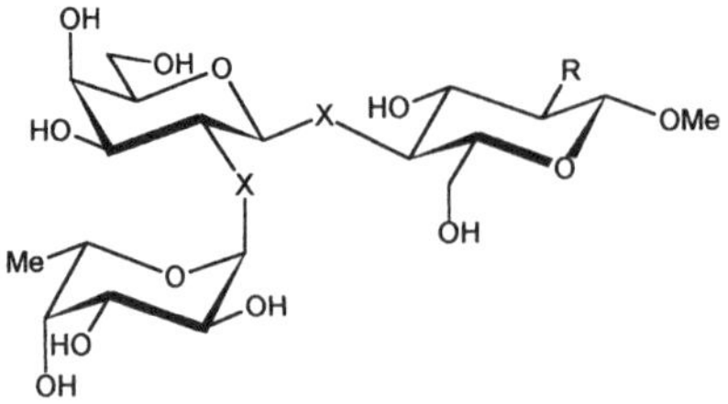

R= OH, NHAc

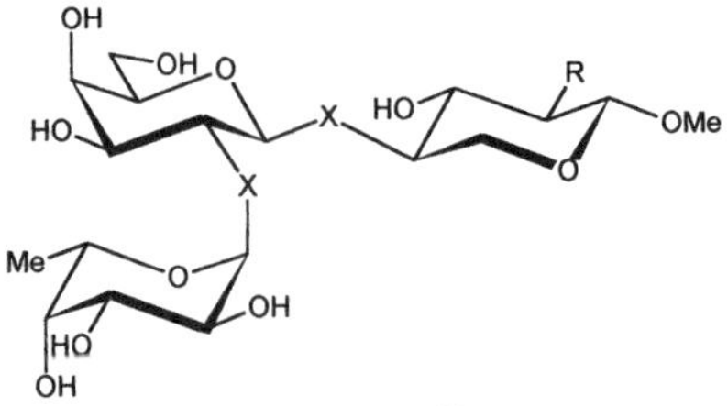

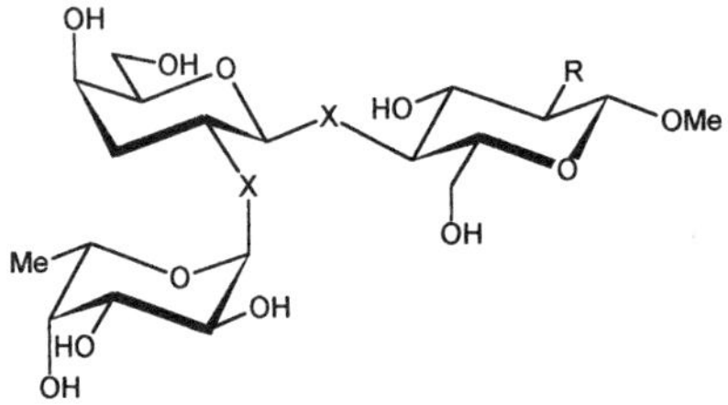

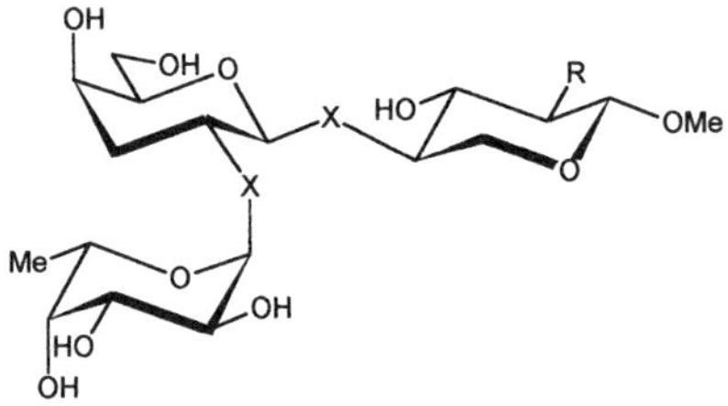

that every C-glycosylic linkage adopts the exo-anomeric conformation. For **51**, α-L-Fuc-(1 → 2)-β-Gal-(1 → 4)-β-Glc-OMe, the relevant couplings are 9.8, 4.7 Hz (Fuc H-1) and 9.5, 0.5 Hz (Gal H-1). For the aglyconic bonds, they are 5.5, 2.3 (Gal H-2) and 3.7, 4.0 (Glc H-4). These data (Tables XVI and XVII) were analyzed in terms of major exo-anomeric conformations around the glycosylic linkages and either a mixture of conformers or a single distorted one for the

TABLE XVI
Comparison Between the Experimental J Values (Hz) Described for the Glycosidic Torsions Φ of Trisaccharides, 51–54[a]

Proton Pair	Expected Exo-syn-Φ	Expected anti-Φ	Expected non-Exo-Φ	Experimental (51–54)
β-H-1′/H$_{proS}$	1.0–1.4	5.0	11.2–11.5	0.5
β-H-1′/H$_{proR}$	10.8–11.3	1.9	3.2–4.0	9.5
α-H-1′/H$_{proS}$	10.5–11.6	No minimum	1.4–2.7	9.8
α-H-1′/H$_{proR}$	1.0–2.8	No minimum	11.0–11.6	4.7

[a] The values for the acetamido analogues are very similar.

TABLE XVII
Comparison Between the Experimental *J* Values (Hz) Described for the Glycosidic
Torsions ψ of Trisaccharide 51[a]

Proton Pair	Expected $\Psi(+)$	Expected $\Psi(-)$	Expected Ψ(eclipsed)	Expected *anti*-Ψ	Experimental 51 (Gal, Glc)
H-X/H$_{proS}$	9.0–12.4	2.0–2.5	4.7	2.9–3.3	2.3, 3.7
H-X/H$_{proR}$	1.4–2.5	12.3–12.4	4.3	3.8–4.1	5.5, 4.0

[a] The couplings for H-2 of Gal and H-4 of Glc are given in this order. The values for the acetamido analogues are very similar.

aglyconic linkage. When the Xyl- or 3-deoxyGal-containing analogues were examined, as in the case of the disaccharides previously studied, the large couplings increase to about 10–11 Hz, indicating that the corresponding aglyconic linkages adopt a more ideal extended conformation around Ψ, as observed for the constitutent disaccharide entitites. The authors also stated that these structural modifications of the type II trisaccharide analogues should also be expected for the natural blood-group determinant. In a subsequent publication, Wei and co-workers[111] prepared and analyzed *C*-glycosyl analogues that contained the GlcNAc moiety of the H-type II determinant. Similar data to those for the Glc-containing compounds were obtained, and the authors concluded that the four GlcNAc- or XylNAc-containing pseudotrisaccharides possess a unique conformation in solution, and that the conformational preferences about the two interglycosidic linkages appear to be mutually independent. Some nonchair conformations were observed for certain synthetic intermediates.

The binding affinities of these analogues toward the lectin I of *Ulex europaeus* were investigated and compared to those of the corresponding O-glycosides, using a quantitative binding assay.[112] Both series provided identical results. Deoxygenation at C-3 of Gal causes a loss of affinity corresponding to 2.0 kcal/mol, whereas the loss due to the removal of the C-6 hydroxymethyl group of GlcNAc was 1.2 kcal/mol. The double substitution showed diminished affinity with a loss of 2.4 kcal/mol. The authors claimed that this behavior proves the conformational similarity of the corresponding *O*- and *C*-glycosyl derivatives. However, they also pointed out that the possibility of participation of O-3 of Gal and C-6 of GlcNAc in interactions with the lectin cannot be rigorously excluded. In fact, because the removal of these groups seems to induce the exclusive presentation of only one conformer for the corresponding aglyconic linkage, if the observed decrease in binding has a conformational origin, this fact could imply that the major conformation in solution of the XylNAc- or 3-deoxyGal-containing analogues is not the one recognized by the *Ulex*-I lectin, a conclusion that has not been yet demonstrated.

The conformational behavior of a C/O-analogue (**55**) of the Lewis X trisaccharide has also been thoroughly analyzed by off-resonance-ROESY NMR experiments and simulated annealing. The results indicated that both glycosidic linkages exhibit substantial conformational rigidity. In fact, very similar individual correlation times were deduced for different proton pairs of the pseudotrisaccharide.[113]

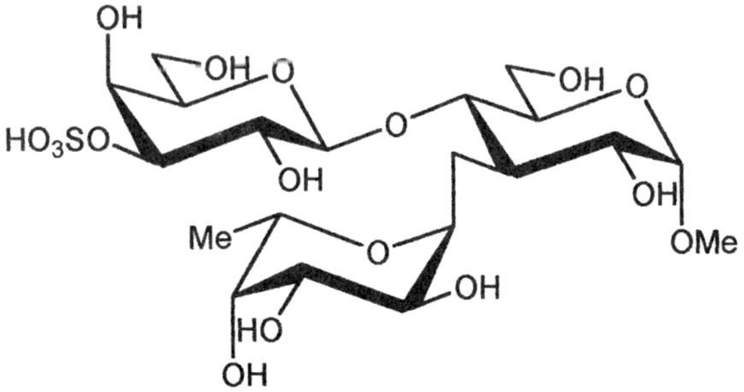

55

l. C-Glycopeptides.—C-Glycopeptides have also been used to explore different biologically relevant interactions and several reports have been presented on the conformational behavior of these *C*-glycosyl analogues. These include in particular a conformational study of a compound (**56**) containing a *C*-mannosyl moiety linked to the indole ring of Trp7 of human ribonuclease.[114] Through ROESY and coupling-constant NMR analysis it was demonstrated that the Man residue has the α configuration and that the six-membered pyranose ring in the glycopeptide

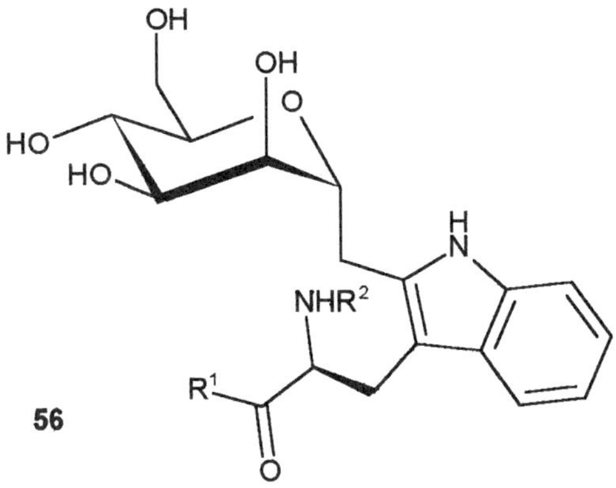

56

R^1 and R^2 are the vicinal amino acids of the polypeptide chain environment adopts several conformations on the NMR timescale. In addition, a C-glycopeptide analogue of an LH-RH agonist has been synthesized and its biological and conformational properties evaluated.[115] This peptidomimetic was shown to be four times less active than the agonist itself, and the NMR-based conformational analysis did not show any evidence of establishment of a β-turn, a feature that has been proposed to occur within the receptor binding site.

A conformational analysis, by NMR and molecular dynamics,[116] of a cyclic C-glycosyl analogue of β-GlcNAc-Asn demonstrated some variations in relation to the conformation of the regular N-glycosyl analogue.

VII. Conclusions

For several years, the conformation of C-glycosyl compounds has been assumed to be identical to that of glycosides. However, several examples described here have shown that this is not always the case. The conformation of a disaccharide or its C-glycosyl analogue is defined by the Φ and Ψ torsion angles. The C-glycosyl analogues seem to be more flexible than their parent compounds, showing different behavior especially around the Ψ angle. Regarding the degree of similarity around the Φ angle for C-*gluco* or C-*galacto* analogues having OH-2 equatorial, the conformational distribution around this angle indeed resembles the distribution deduced for glycosides having a major exo-anomeric *syn* orientation as concluded by Kishi,[10] although only in a qualitative manner. However, for the β-C-analogues, contributions of exo-anti (and, for α and β-analogues), minor, but detectable, participation of non-exo conformers are also manifested in many instances, in contrast to the almost exclusive participation of exo-*syn* conformers in normal glycosides. However, the *manno*-C-glycosyl compounds (having OH-2 axially oriented), deviate significantly from the behavior usually observed in glycosides by presenting high percentages of the non-exo rotamer.

Comparing the conformation of C-glycosyl compounds with that of the corresponding normal oligosaccharides, the addition of the exo-anomeric and 1,3-*syn*-diaxial effects in those glycosides having O-2 equatorial (the common natural β-*gluco* and β-*galacto* series) in the latter explains the unique existence of exo-anomeric conformations around Φ in these natural glycosides (>3.3 kcal/mol).

Regarding the use of C-glycosyl compounds as O-glycoside isosters, it is evident that, depending on the energy differences among energy minima, conformations different from the major one existing in solution could be bound by proteins without major energy conflicts. Evidently, topological features of the protein binding-site, restricting ligand mobility and demanding conformer selection, as well as the shifting of the inherent dynamic equilibrium of the flexible C-analogues, can contribute to the final results. When a lectin imposes a constraint by establishing interactions with several sugar residues, then the mobility decreases and only one topologically favored conformer will fit into the binding site. Alternatively, the possibility exists that intramolecular mobility could be maintained if the entropic penalty exceeds the enthalpic gain by weak sugar–protein interactions, or that the conformation of the global minimum reaches an optimal free-energy value. Although, the intrinsic flexibility of C-glycosyl compounds may be a problem from the entropic point of view in binding to the active site of proteins or enzymes, these compounds remain excellent probes for testing protein–carbohydrate interactions.

ACKNOWLEDGMENTS

We thank DGES for financial support (PB96-0833) and all the coauthors and coworkers whose names appear in the different publications. This review is dedicated to Prof. Dr. Manuel Bernabé on the occasion of his 65th birthday.

REFERENCES

(1) For leading and recent references, see (a) H. J. Gabius and S. Gabius, eds., *Glycosciences: Status and Perspectives*. Chapman & Hall, London, 1997; (b) W. I. Weis, *Curr. Opin. Struct. Biol.*, 7 (1997) 624–630; (c) H. Lis and N. Sharon, *Chem. Rev.*, 98 (1998) 637–674; (d) K. Drickamer, *Curr. Opin. Struct. Biol.*, 9 (1999) 585–590; (e) R. A. Dwek, *Chem. Rev.*, 96 (1996) 683–720; (f) P. R. Crocker and T. Feizi, *Curr. Opin. Struct. Biol.*, 6 (1996) 679–691; (g) A. Varki, *Glycobiology*, 3 (1993) 97–130.

(2) For reviews on chemistry of C-glycosyl compounds, see (a) Y. Du, R. J. Linhardt, and I. R. Vlahov, *Tetrahedron*, 54 (1998) 9913–9959; (b) M. D. Postema, *C-Glycoside Synthesis*, CRC Press, Boca Raton, 1995; (c) W. Levy, D. Chang, *Chemistry of C-Glycosides;* Elsevier: Cambridge, 1995; (d) F. Nicotra, *Topics Curr. Chem.*, 187 (1997) 55–83; (e) P. Sinaÿ, *Pure Appl. Chem.*, 69 (1997) 459–463; (f) J. M. Beau and T. Gallagher, *Topics Curr. Chem.*, 187 (1997) 1–54.

(3) L. J. Haynes, *Adv. Carbohydr. Chem. Biochem.*, 20 (1965) 357–369.

(4) S. Hanessian and A. G. Pernet, *Adv. Carbohydr. Chem. Biochem.*, 33 (1976) 111–188.

(5) C. Jaramillo and S. Knapp, *Synthesis* (1994) 1–20.

(6) M. D. Lewis, J. K. Cha, and Y. Kishi, *J. Am. Chem. Soc.*, 104 (1982) 4976–4978.

(7) L. Paterson and L. E. Keown, *Tetrahedron Lett.*, 38 (1997) 5727–5730.

(8) K. Horita, Y. Sakkurai, M. Nagasawa, S. Hachiya, and O. Yonemitsu, *Synlett*, (1994) 43–45.

(9) For leading references, see (a) R. U. Lemieux, S. Koto, and D. Voisin, *D. Am. Chem. Soc. Symp. Ser.*, 87 (1979) 17–29; (b) G. R. J. Thatcher, *The Anomeric Effect and Associated Stereoelectronic Effects;* American Chemical Society: Washington, DC, 1993; (c) A. J. Kirby, *The Anomeric Effect and Related Stereoelectronic Effects at Oxygen;* Springer-Verlag, Heidelberg, Germany, 1983; (d) H. Thögersen, R. U. Lemieux, K. Bock, and B. Meyer, *Can. J. Chem.*, 60 (1982) 44–65; (e) I. Tvaroška, and T. Bleha, *Adv. Carbohydr. Chem. Biochem.*, 47 (1989) 45–103; (f) I. Tvaroška and J. P. Carver, *J. Chem. Res. Synop.*, 1991, 6–7; (g) I. Tvaroška, and J. P. Carver, *J. Phys. Chem.*, 99 (1995) 6234–6241; (h) C. Cramer, D. G. Truhlar, and A. D. French, *Carbohydr. Res.*, 298 (1997) 1–14.

(10) Y. Kishi, *Pure Appl. Chem.*, 65 (1993) 771–778.

(11) R. Ravishankar, A. Surolia, M. Vijayan, S. Lim, and Y. Kishi, *J. Am. Chem. Soc.*, 120, (1998) 11297–11303.

(12) J. F. Espinosa, F. J. Cañada, J. L. Asensio, M. Martín-Pastor, H. Dietrich, M. Martin-Lomas, R. R. Schmidt, and J. Jiménez-Barbero, *J. Am. Chem. Soc.*, 118 (1996) 10862–10871.

(13) A general survey of conformation of carbohydrates is presented in: A. D. French and J. W. Brady, *Computer Modeling of Carbohydrate Molecules,* American Chemical Society, 1990.
(14) See, for instance (*a*) T. Skrydstrup, D. Mazéas, M. Elmouchir, G. Doisneau, C. Riche, A. Chiaroni, and J.-M. Beau, *Chem. Eur. J.,* 3 (1998) 1342–1356, (*b*) D. Urban, T. Skrydstrup, and J. M. Beau, *Chem. Comm.,* (1998) 955–956; (*c*) G. Rubinstenn, J. Esnault, J. M. Mallet, and P. Sinaÿ, *Tetrahedron: Asymmetry,* 8 (1997) 1327–1336.
(15) For a review, see J. G. Buchanan and R. H. Wightman, *Prog. Chem. Org. Nat. Prod.,* 44 (1984) 44, 243.
(16) See, for instance, (*a*) J. E. Abola, M. J. Sims, and D. J. Abraham, *J. Med. Chem.,* 17 (1974) 62–65; (*b*) B. M. Goldstein, D. T. Mao, and V. E. Marquez, *J. Med. Chem.,* 31 (1988) 1026–1031; (*c*) H. Li, W. H. Hallows, J. S. Punzi, K. W. Pankiewicz, K. A. Watanabe, and B. M. Goldstein, *Biochemistry,* 33 (1994) 11734–11744; (*d*) B. M. Goldstein, H. Li, W. H. Hallows, D. A. Langs, P. Franchetti, L. Capellacci, and M. Grifantini, *J. Med. Chem.,* 37 (1994) 1684–1688.
(16a) *IUPAC-IUBMB Nomenclature of Carbohydrates,* 2-Carb-33.7, 2-Carb-37.5, *Adv. Carbohydr. Chem. Biochem.,* 52 (1997) 43–177.
(17) For a review on the use of enzymes, see, for instance, W. D. Fessner, *Curr. Opin. Chem. Biol.,* 2 (1998) 85–97.
(18) For relevant examples, see (*a*) U. Tedebark, M. Meldal, L. Panza, and K. Bock, *Tetrahedron Lett.,* 39 (1998) 1815–1818; (*b*) F. Brukhart, M. Hoffmann, and H. Kessler, *Angew. Chem. Int. Ed. Engl.,* 36 (1997) 1191–1192.
(19) K. A. Watson, E. P. Mitchell, L. N. Johnson, J. C. Son, C. J. F. Bichard, M. G. Orchard, G. W. J. Fleet, N. G. Oikonomakos, D. D. Leonidas, M. Kontou, and A. Papageorgioui, *Biochemistry,* 33 (1994) 5745–5758.
(20) R. R. Schmidt and H. Dietrich, *Angew. Chem. Int. Ed. Engl.,* 30 (1991) 1328–1329.
(21) W. Lai and O. R. Martin, *Carbohydr. Res.,* 250 (1993) 185–193.
(22) B. A. Johns, Y. T. Pan, A. D. Elbein, and C. R. Johnson, *J. Am. Chem. Soc.,* 119 (1997) 4856–4865.
(23) M. A. Leeuwenburgh, S. Picasso, H. S. Overkleeft, G. A. van der Marel, P. Vogel, and J. H. van Boom, *Eur. J. Org. Chem.,* (1999) 1185–1189.
(24) For examples, see (*a*) R. R. Schmidt and K. Frische, *Bioorg. Med. Chem. Lett.,* 3 (1993) 1747–1750; (*b*) M. M. Vaghefi, R. J. Bernacki, N. K. Dalley, B. E. Wilson, and R. K. Robins, *J. Med. Chem.,* 30 (1987) 1383–1390.
(25) C. S. Kuhn, C. P. J. Glaudemans, and J. Lehmann, *Liebigs Ann. Chem.,* (1989) 357–366.
(26) For instance, (*a*) M. Mammen, G. Dahmann, and G. M. Whiteseides, *J. Med. Chem.,* 38 (1995) 4179–4190; (*b*) S.-K. Choi, M. Mammen, and G. M. Whitesides, *Chem. Biol.,* 3 (1996) 97–104.
(27) J. O. Nagy, P. Wang, J. H. Gilbert, M. E. Schaefer, T. G. Hill, M. R. Callstrom, and M. D. Bednarski, *J. Med. Chem.,* 35 (1992) 4591–4592.
(28) W. Spevak, J. O. Nagy, D. H. Charych, M. E. Schaefer, J. H. Gilbert, and M. D. Bernadski, *J. Am. Chem. Soc.,* 115 (1993) 1146–1147.
(29) K. G. Bowman, S. Hemmerich, S. Bhakta, M. S. Singer, A. Bistrup, S. D. Rosen, and C. R. Bertozzi, *Chem. Biol.,* 5 (1998) 447–460.
(30) C. R. Bertozzi and M. D. Bednarski, *Carbohydr. Res.,* 223 (1992) 243–253.
(31) H. Vyplel, D. Scholz, I. Macher, K. Schindlermaier, and E. Schutze, *J. Med. Chem.,* 34 (1991) 2759–2764.
(32) (*a*) R. V. Weatherman, K. H. Mortell, M. Chervenak, L. L. Kiessling, and E. J. Toone, *Biochemistry,* 35 (1996) 3619–3624. (*b*) R. V. Weatherman and L. L. Kiessling, *J. Org. Chem.,* 61 (1996) 534–538.
(33) O. Tsuruta, H. Yuasa, S. Kurono, and H. Hashimoto, *Bioorg. Med. Chem. Lett.,* 9 (1999) 807–810.

(34) (a) J. Wang, P. Kovac, P. Sinaÿ, and C. P. J. Glaudemans, *Carbohydr. Res.*, 308 (1998) 191–193.
(b) Y. C. Xin, Y. M. Zhang, J. M. Mallet, C. P. J. Glaudemans, and P. Sinaÿ, *Eur. J. Org. Chem.*, (1999) 471–476.

(35) (a) A. Hemboldt, M. Petitou, J. M. Mallet, J. P. Herault, J. C. Lormeau, P. A. Driguez, J. M. Herbert, and P. Sinaÿ, *Bioorg. Med. Chem. Lett.*, 7 (1997) 1506–1509; (b) M. Petitou, J. P. Herault, J. C. Lormeau, A. Hemboldt, J. M. Mallet, P. Sinaÿ, and J. M. Herbert, *Bioorg. Med. Chem.*, 6 (1998) 1509–1515.

(36) L. Bornaghi, J. P. Utille, D. F. Rekai, J. M. Mallet, P. Sinaÿ, and H. Driguez, *Carbohydr. Res.*, 305 (1997) 561–568.

(37) L. Wang, J. Fan, and Y. C. Lee, *Tetrahedron Lett.*, 37 (1996) 1975–1978.

(38) C. C. Lin, F. Moris-Vargas, G. Weitz-Schmidt, and C. H. Wong, *Bioorg. Med. Chem.*, 7 (1999) 425–433.

(39) C. Uriel, M. J. Egron, M. Santarromana, D. Scherman, K. Antonakis, and J. Herscovici, *Bioorg. Med. Chem.*, 4 (1996) 2081–2090.

(40) S. Howard and S. G. Withers, *Biochemistry*, 37 (1998) 3858–3864.

(41) J. M. J. Tronchet, M. Zsely, and M. Geoffrey, *Carbohydr. Res.*, 275 (1995) 245–258.

(42) R. S. Coleman and M. L. Madaras, *J. Org. Chem.*, 63 (1998) 5700–5703.

(43) C. S. Wilcox, and M. D. Coward, *Carbohydr. Res.*, 171 (1987) 141–160.

(44) K. B. Wiberg and M. A. Murcko, *J. Am. Chem. Soc.*, 111 (1989) 4821–4827.

(45) K. N. Houk, J. E. Eksterowicz, Y. Wu, C. D. Fuglesang, and D. R. Mitchell, *J. Am. Chem. Soc.*, 115 (1993) 4170–4177.

(46) D. R. Bundle and N. M. Young, *Curr. Opin. Struct. Biol.*, 2 (1992) 666–673.

(47) C. A. Bush, M. Martin-Pastor, and A. Imberty, *Ann. Rev. Biophys. Biomol. Struct.*, 28 (1999) 269–293.

(48) For relevant reviews, see (a) C. S. Wright, *Curr. Opin. Struct. Biol.*, 7 (1997) 631–636; (b) J. Bouckaert, T. Hamelryck, L. Wyns, and R. Loris, *Curr. Opin. Struct. Biol.*, 9 (1999) 572–577; (c) J. M. Rini and Y. D. Lobsanov, *Curr. Opin. Struct. Biol.*, 9 (1999) 578–584.

(49) For reviews on relevant applications of NMR data and calculations in conformational analysis of carbohydrates, see, for instance: (a) T. Peters and B. M. Pinto, *Curr. Opin. Struct. Biol.*, 6 (1996) 710–720; (b) A. Poveda, J. L. Asensio, J. F. Espinosa, F. J. Cañada, M. Martín-Pastor, and J. Jiménez-Barbero, *J. Mol. Graphics*, 15 (1997) 9–17.

(50) J. Jiménez-Barbero, J. L. Asensio, F. J. Cañada, and A. Poveda, *Curr. Opin. Struct. Biol.*, 9 (1999) 549–555.

(51) D. Neuhaus and M. P. Williamson, *The Nuclear Overhauser Effect in Structural and Conformational Analysis.* VCH Publishers, New York, 1989.

(52) For a review, see A. Imberty, *Curr. Opin. Struct. Biol.*, 7 (1997) 617–623.

(53) T. Peters and T. Weimar, *J. Biomol. NMR*, 4 (1994) 97–116.

(54) For leading references, see: (a) M. K. Dowd, P. J. Reilly, and A. D. French, *J. Comput. Chem.*, 13 (1992) 102–114; (b) S. Pérez, A. Imberty, S. B. Engelsen, J. Gruza, K. Mazeau, J. Jiménez-Barbero, A. Poveda, J. F. Espinosa, J. B. P. van Eick, G. Johnson, A. D. French, M. L. C. E. Kouwijzer, P. D. J. Grootenhuis, A. Bernardi, L. Raimondi, H. Senderowitz, V. Durier, G. Vergoten, and K. J. Rasmussen, *Carbohydr. Res.*, 314 (1998) 141–155. (c) J. L. Asensio, M. Martin-Pastor, and J. Jiménez-Barbero, *J. Mol. Struct. (Theochem)*, 395 (1997) 245–270.

(55) For instance, (a) T. J. Rutherford, J. Partridge, C. T. Weller, and S. W. Homans, *Biochemistry*, 32 (1993) 12715–12724; (b) T. J. Rutherford and S. W. Homans, *Biochemistry*, 33 (1994) 9606–9614; (c) T. J. Rutherford, D. C. A. Neville, and S. W. Homans, *Biochemistry*, 34 (1995) 14131–14137.

(56) M. Hricovini, R. N. Shah, and J. P. Carver, *Biochemistry*, 31 (1992) 10018–23.

(57) S. B. Engelsen, C. Herve du Penhoat, and S. Pérez, *J. Phys. Chem.*, 99 (1995) 13334–13351.

(58) For theory and applications of the off-resonance ROESY method to saccharide molecules, see (a) H. Desvaux, P. Berthault, N. Bilirakis, and M. Goldman, *J. Magn. Reson. A*, 108 (1994)

219–226; (b) G. Lippens, J. M. Wieruzeski, P. Talaga, J. P. Bohin, and H. Desvaux, *J. Am. Chem. Soc.,* 118 (1996) 7227–7228; (c) A. Poveda, M. Santamaria, M. Bernabe, A. Rivera, J. Corzo, and J. Jiménez-Barbero, *Carbohydr. Res.,* 304 (1997) 219–228.

(59) See, for instance, (a) D. G. Davis, *J. Am. Chem. Soc.,* 109 (1987) 6962–6963; (b) H. van Halbeek and L. Poppe, *Magn. Res. Chem.,* 30 (1992) S74–S86; (c) A. Poveda, J. L. Asensio, M. Martin-Pastor, and J. Jiménez-Barbero, *J. Biomol. NMR,* 10 (1997) 29–43; (d) A. Poveda, J. L. Asensio, M. Martin-Pastor, and J. Jiménez-Barbero, *Carbohydr, Res.,* 300 (1997) 3–10.

(60) For a review, see P. Dais, *Adv. Carbohydr. Chem. Biochem.,* 51 (1995) 63–161.

(61) For the relation between H/H couplings and conformation, see: (a) M. Karplus, *J. Chem. Phys.,* 30 (1959) 11–20; (b) C. A. G. Haasnoot, F. A. A. M. de Leeuw, and C. Altona, *Tetrahedron,* 36 (1980) 2783–2794.

(62) (a) V. L. Bevilacqua, D. S. Thomson, and J. H. Prestegard, *Biochemistry,* 29 (1990) 5529–5537. (b) V. L. Bevilacqua, Y. Kim, and J. H. Prestegard, *Biochemistry,* 31 (1992) 9339–9345.

(63) For a review, see A. Poveda and J. Jiménez-Barbero, *Chem. Soc. Rev.,* 27 (1998) 133–143.

(64) For instance, see J. L. Asensio, F. J. Cañada, and J. Jiménez-Barbero, *Eur. J. Biochem.,* 233 (1995) 618–630.

(65) For applications to the oligosaccharide field: T. Haselhorst, J. F. Espinosa, J. Jiménez-Barbero, T. Sokolowski, P. Kosma, H. Brade, L. Brade, and T. Peters, *Biochemistry,* 38 (1999) 6449–6459.

(66) For relevant examples, see (a) T. Rundlof, C. Landersjo, K. Lycknert, A. Maliniak, and G. Widmalm, *Magn. Reson. Chem.,* 36 (1998) 773–776; (b) G. R. Kiddle and S. W. Homans, *FEBS Lett.,* 436 (1998) 128–130; (c) P. J. Bolon and J. H. Prestegard, *J. Am. Chem. Soc.,* 120 (1998) 9366–9367; (d) H. Shimizu, A. Donohue-Rolfe, and S. W. Homans, *J. Am. Chem. Soc.,* 121 (1999) 5815–5816.

(67) For later applications, see (a) T. Carlomagno, I. C. Felli, M. Czech, R. Fischer, M. Sprinzl, and C. Griesinger, *J. Am. Chem. Soc.,* 121 (1999) 1945–1948; (b) M. J. J. Blommers, W. Stark, C. E. Jones, D. Head, C. E. Owen, and W. Jahnke, *J. Am. Chem. Soc.,* 121 (1999) 1949–1953.

(68) A. Neuman, F. Longchambon, O. Abbes, H. Gillier-Pandraud, S. Pérez, D. Rouzaud, and P. Sinaÿ, *Carbohydr. Res.,* 195 (1990) 187–197.

(69) D. C. Rohrer, A. Sarko, T. L. Bluhm, and Y. N. Lee, *Acta Crystallogr. Sect. B,* 36 (1980) 650–654.

(70) P. G. Goekjian, T.-C. Wu, H-Y. Kang, and Y. Kishi, *J. Org. Chem.,* 52 (1987) 4823–4825.

(71) D. J. O'Leary and Y. Kishi, *J. Org. Chem.,* 58 (1993) 304–306.

(72) J. D. Oliver and L. C. Strickland, *Acta Crystallogr. C,* 40 (1984) 820–824.

(73) M. Brakta, R. N. Farr, B. Chaguir, G. Massiot, C. Lavaud, W. R. Anderson Jr, D. Sinou, and G. D. Daves Jr, *J. Org. Chem.,* 58 (1993) 2992–2998.

(74) T. Wu, P. G. Goekjian, and Y. Kishi, *J. Org. Chem.,* 52 (1987) 4819–4823.

(75) P. G. Goekjian, T.-C. Wu, and Y. Kishi, *J. Org. Chem.,* 56 (1991) 6412–6422.

(76) F. J. Lopez-Herrera, M. S. Pino-González, and F. Planas-Ruiz, *Tetrahedron: Asymmetry,* 1 (1990) 465–475.

(77) X. Wang and P. H. Gross, *J. Org. Chem.,* 60 (1995) 1201–1208.

(78) R. W. Armstrong and B. R. Teergarden, *J. Org. Chem.,* 57 (1992) 915–922.

(79) S. A. Babirad, Y. Wang, P. G. Goekian, and Y. Kishi, *J. Org. Chem.,* 52 (1987) 4825–4827.

(80) Y. Wang, S. A. Babirad, and Y. Kishi, *J. Org. Chem.,* 57 (1992) 468–481.

(81) Y. Wang, P. G. Goekjian, R. V. Ryckman, W. H. Miller, S. A. Babirad, and Y. Kishi, *J. Org. Chem.,* 57 (1992) 482–489.

(82) Y. Wang, P. G. Goekian, D. M. Ryckman, and Y. Kishi, *J. Org. Chem.,* 53 (1988) 4151–4153.

(83) W. H. Miller, D. M. Ryckman, P. G. Goekjian, Y. Wang, and Y. Kishi, *J. Org. Chem.,* 53 (1988) 5580–5582.

(84) J. F. Espinosa, M. Manuel-Pastor, J. L. Asensio, H. Dietrich, M. Martin-Lomas, R. R. Schmidt, and J. Jiménez-Barbero, *Tetrahedron Lett.,* 36 (1995) 6329–6332.

(85) J. F. Espinosa, H. Dietrich, M. Martin-Lomas, R. R. Schmidt, and J. Jiménez-Barbero, *Tetrahedron Lett.,* 37 (1996) 1467–1470.

(86) J. L. Asensio and J. Jiménez-Barbero, *Biopolymers,* 35 (1995) 55–71.
(87) J. Dabrowski, T. Kozar, H. Grosskurth, and N. E. Nifant'ev, *J. Am. Chem. Soc.,* 117 (1995) 5534–5538.
(88) M. Martín-Pastor, J. F. Espinosa, J. L. Asensio, and J. Jiménez-Barbero, *Carbohydr. Res.,* 298 (1997) 15–47.
(89) G. Rubinstenn, P. Sinaÿ, and P. Berthault, *J. Phys. Chem. A,* 101 (1997) 2536–2540.
(90) J. L. Asensio, J. F. Espinosa, H. Dietrich, F. J. Cañada, R. R. Schmidt, M. Martín-Lomas, S. André, H.-J. Gabius, and J. Jiménez-Barbero, *J. Am. Chem. Soc.,* 121 (1999) 8995–9000.
(91) D. Solis, J. Jiménez-Barbero, M. Martin-Lomas, and T. Diaz-Mauriño, *Eur. J. Biochem.,* 223 (1994) 107–114.
(92) J. F. Espinosa, F. J. Cañada, J. L. Asensio, H. Dietrich, M. Martín-Lomas, R. R. Schmidt, and J. Jiménez-Barbero, *Angew. Chem. Int. Ed. Engl.,* 35 (1996) 303–306.
(93) J. F. Espinosa, E. Montero, A. Vian, J. L. Garcia, H. Dietrich, M. Martín-Lomas, R. R. Schmidt, A. Imberty, F. J. Cañada, and J. Jiménez-Barbero, *J. Am. Chem. Soc.,* 120 (1998) 10862–10871.
(94) A. Rivera, D. Solis, T. Diaz-Mauriño, J. Jiménez-Barbero, and M. Martin-Lomas, *Eur. J. Biochem.,* 197 (1991) 217–228; D. Solis, P. Fernandez, T. Diaz-Mauriño, J. Jiménez-Barbero, and M. Martin-Lomas, *Eur. J. Biochem.,* 214 (1993) 677–683.
(94a) Results in preparation.
(95) R. Ferrito and P. Vogel, *Tetrahedron: Asymmetry,* 5 (1994) 2077–2092.
(96) A. Baudat and P. Vogel, *J. Org. Chem.,* 62 (1997) 6252–6260.
(97) C. Marquis, S. Picasso, and P. Vogel, *Synthesis,* Special Issue, (1999) 1441–1452.
(98) D. J. O'Leary and Y. Kishi, *Tetrahedron Lett.,* 35 (1994) 5591–5594.
(99) C. Herve du Penhoat, A. Imberty, N. Roques, V. Michon, J. Mentech, G. Descotes, and S. Pérez, *J. Am. Chem. Soc.,* 113 (1991) 3720–3727.
(100) U. Ellervik and U. Magnusson, *J. Am. Chem. Soc.,* 116 (1994) 2340–2347.
(101) A. Wei and Y. Kishi, *J. Org. Chem.,* 59 (1994) 88–96.
(102) O. R. Martin and W. Lai, *J. Org. Chem.,* 58 (1993) 176–182.
(103) J. L. Asensio, F. J. Cañada, X. Cheng, N. Khan, D. R. Mootoo, and J. Jiménez-Barbero, *Chem. Eur. J.,* 6 (2000) 1035–1041.
(104) For instance, see D. A. Pearlman, *J. Biomol. NMR,* 4 (1994) 1–6.
(105) D. A. Pearlman, D. A. Case, J. W. Caldwell, W. S. Ross, T. E. Cheatham III, S. DeBolt, D. Ferguson, G. Siebal, and P. Kollman, *Comp. Phys. Commun.,* 91 (1995) 1–41.
(106) J. L. Asensio, F. J. Cañada, A. García, M. T. Murillo, A. Fernández-Mayoralas, B. A. Johns, J. Kozak, Z. Zhu, C. R. Johnson, and J. Jiménez-Barbero, *J. Am. Chem. Soc.,* 121 (1999) 11318–11329.
(107) P. G. Goejkian, T.-C. Wu, H.-Y. Kang, and Y. Kishi, *J. Org. Chem.,* 56 (1991) 6422–6434.
(108) J. F. Espinosa, M. Bruix, O. Jarreton, T. Skrydstrup, J.-M. Beau, and J. Jiménez-Barbero, *Chem. Eur. J.,* 5 (1999) 442–448.
(109) A. Poveda, J. L. Asensio, H. Bazin, T. Polat, R. J. Lindhardt, and J. Jiménez-Barbero, *Eur. J. Org. Chem.* (2000) 1805–1813.
(110) T. Haneda, P. G. Goekjian, S. H. Kim, and Y. Kishi, *J. Org. Chem.,* 57 (1992) 490–498.
(111) A. Wei, A. Haudrechy, C. Audin, H.-S. Jun, N. Haudrechy-Bretel, and Y. Kishi, *J. Org. Chem.,* 60 (1995) 2160–2169.
(112) A. Wei, K. M. Boy, and Y. Kishi, Y., *J. Am. Chem. Soc.,* 117 (1995) 9432–9437.
(113) P. Berthault, N. Birlirakis, G. Rubinstenn, P. Sinaÿ, and H. J. Desvaux, *J. Biomol. NMR,* 8 (1996) 23–31.
(114) T. de Beer, J. F. G. Vliegenthart, A. Loeffler, and J. Hofsteenge, *Biochemistry,* 34 (1995) 11785–11789.
(115) K. Michael, V. Wittmann, W. König, J. Sandow, and H. Kessler, *Int. J. Pept. Prot. Res.,* 48 (1996) 59–70.
(116) M. Hoffmann, F. Burkhardt, G. Hessler, and H. Kessler, *Helv. Chim. Acta,* 79 (1996) 1519–1528.

AUTHOR INDEX

A

Abbes, OI, 246, 272, 283(68)
Abola, J. E., 236, 237, 265, 281(16a)
Abraham, D. J., 236, 237, 265, 281(16a)
Adlam, C., 154, 194(7)
Adlington, R. M., 67, 104, 144(25)
Agnel, J.-P., 67, 144(32)
Agrawal, P. K., 206, 207, 208, 209, 210, 211, 212, 213, 214, 224, 231(77, 78, 82), 232(106)
Ahlstedt, S., 156, 195(44)
Ajl, S. J., 156, 195(37)
Ala'Aldeen, D. A. A., 157, 196(57)
Alan Padron, A., 161, 166, 197(99)
Albertı, A., 86, 145(86)
Alfredsson, G., 180, 198(144)
Al-Hashimi, H. M., 210, 232(110)
Allen, R. P., 94, 147(171)
Allen, T. D., 203, 230(40)
Allerhand, A., 213, 233(126)
Alonso de Velasco, E., 174, 176, 197(125)
Aloui, M., 13, 15, 16, 57(30), 58(49)
Altieri, P. L., 155, 195(23)
Altona, C., 244, 283(61b)
Alvarado, L., 20, 58(72)
Amatore, C., 11, 32, 57(17)
Aminoff, D., 205, 224, 231(69)
Anaya de Parrodi, C., 92, 93, 138, 147(168)
Anderson, P., 155, 195(28)
Anderson, P. W., 157, 166, 196(54)
Anderson, W. R., Jr., 247, 283(73)
Anderton, S., 165, 197(104)
André, S., 258, 260, 284(90)
Andrec, M., 210, 232(110)

Andreotti, A. H., 213, 214, 233(132, 133)
Andrews, C. W., 16, 58(54)
Andrews, J. S., 166, 197(108)
Angyal, S. J., 207, 232(88)
Antonakis, K., 238, 282(39)
Aoki, M., 35, 60(154, 155)
Aoki, Y., 32, 60(142)
Arakatsu, Y., 157, 158, 159, 182, 196(75)
Araki, F., 88, 146(112)
Araki, Y., 92, 147(159)
Arbusow, B. A., 40, 61(174)
Arimoto, M., 31, 60(135)
Armstrong, K. B., 22, 59(81)
Armstrong, R. W., 251, 283(78)
Arnott, S., 208, 232(94)
Artenstein, M. S., 155, 195(23)
Asensio, J. L., 236, 243, 244, 246, 256, 257, 258, 259, 260, 266, 268, 270, 272, 273, 275, 280(12), 282(49b, 50, 54c), 283(59c, 59d, 64, 84), 284(86, 88, 90, 92, 103, 106, 109)
Ashkenazi, S., 154, 194(9)
Ashwell, G., 157, 158, 159, 182, 196(75, 76)
Aspinall, G. O., 153, 157, 194(1), 196(70), 201, 229(5)
Atha, D. H., 203, 230(32)
Atherton, M. J., 56, 63(245)
Aubert-Foucher, E., 202, 229(14)
Audin, C., 277, 284(111)
Augé, J., 95, 147(178)
Auzanneau, F. I., 166, 197(112, 113)
Avalos, M., 97, 148(183)
Averett, D., 47, 48, 62(209)
Avery, O. T., 154, 155, 159, 160, 194(5), 195(25, 26)
Axon, J. A., 89, 146(126)

B

Babcock, B. W., 68, 129, 144(50, 52)
Babiano, R., 97, 148(183)
Babirad, S. A., 251, 252, 253, 254, 283(79–81)
Bae, J., 224, 234(156)
Baeschlin, D. K., 43, 61(194)
Baguley, P. A., 89, 143, 146(119)
Bahnmuller, R., 191, 199(177)
Bajza, I., 157, 196(71)
Baldwin, J. E., 67, 104, 144(25)
Ballell, L., 18, 58(64)
Ballestri, M., 90, 91, 147(152)
Bame, K. J., 203, 230(42)
Bamford, M. J., 95, 148(179)
Bamhaoud, T., 92, 147(163)
Bandaru, R., 47, 48, 62(209)
Banik, B. K., 25, 59(107)
Bär, H. P., 27, 60(120)
Barbera Morales, R., 161, 166, 197(99)
Barch, J. J., 213, 214, 233(135)
Barici, P., 23, 59(90)
Barrera, O., 155, 195(27)
Barrett, A. G. M., 53, 62(223)
Bartlett, C., 213, 233(129)
Bartlett, P. A., 50, 52, 55, 62(216–218, 220, 237)
Barton, D. H. R., 40, 41, 45, 61(175), 62(201), 66, 88, 90, 91, 92, 93, 108, 140, 143(16–18), 146(108–110), 147(145, 147, 155, 165), 148(214–216)
Basak, A., 67, 104, 144(25)
Baskin, G., 188, 199(168)
Batra, R., 79, 84, 143, 145(65), 150(299)
Bats, J. W., 213, 233(128)
Batta, G., 116, 149(240)
Battesti, C., 67, 144(32)
Baudat, A., 263, 284(96)
Baudry, M., 133, 150(275)
Baumann, H., 211, 233(117)
Baumann, M., 191, 192, 193, 199(178)
Baumberger, F., 67, 84, 118, 119, 120, 144(45)
Baumgartner, J., 101, 102, 106, 142, 148(197)
Bax, A., 207, 232(86, 109)
Bazin, H., 275, 284(109)
Beach, R.L., 27, 59(116, 117)
Beau, J.-M., 55, 62(233), 209, 224, 232(107), 235, 236, 237, 273, 280(2f), 281(14a, 14b), 284(108)
Beckwith, A. L. J., 86, 89, 92, 93, 102, 105, 130, 135, 138, 139, 145(83, 84),
146(126), 147(166, 168), 148(203), 150(270, 271, 280)
Bednarski, M. D., 67, 115, 144(39), 237, 238, 281(27, 30)
Beebe, X., 25, 59(104)
Begley, M. J., 102, 148(200)
Behrens, C. H., 52, 62(221)
Beigelman, L., 98, 99, 148(189)
Bellamy, F., 133, 150(275)
Belting, M., 202, 229(19)
Belucci, G., 23, 59(89, 90)
Bengtsson, M., 161, 197(98)
Benneche, T., 56, 63(244)
Bennek, J. A., 87, 146(96)
Benner, S. A., 43, 61(194)
Bensadoun, A., 203, 230(31)
Benson, S. W., 10, 45, 57(13)
Bergstrom, T., 203, 230(29)
Berliner, L. J., 210, 232(110)
Berman, E., 213, 233(126)
Berman, S. L., 155, 195(23)
Bernabe, M., 244, 283(58c)
Bernacki, R. J., 237, 281(24b)
Bernadski, M. D., 238, 281(28)
Bernardi, A., 243, 282(54b)
Bernasconi, C., 68, 129, 144(48)
Bernhard, W. G., 154, 194(13)
Berniger, E., 27, 59(114)
Berthault, P., 244, 257, 278, 282(58a), 284(89, 113)
Bertozzi, C. R., 67, 115, 116, 144(39), 149(232), 238, 281(29, 30)
Bertrand, P., 55, 62(236)
Betts, R. F., 157, 166, 196(54)
Beurret, M., 157, 196(58, 69)
Beuvery, E. C., 161, 165, 196(90, 106)
Bevilacqua, V. L., 245, 283(62a, 62b)
Beyer, G., 87, 146(103)
Bezuidenhoudt, B. C. B., 53, 62(223)
Bhakta, S., 238, 281(29)
Bichard, C. J. F., 237, 281(19)
Bidanset, D. J., 202, 229(13)
Bieber, M., 11, 57(23)
Bilirakis, N., 244, 282(58a)
Binkley, R. W., 87, 146(99, 100)
Birchenbach, L., 27, 59(114)
Birlirakis, N., 278, 284(113)
Bistrup, A., 238, 281(29)
Blackburne, I. D., 54, 62(230)
Blanton, J. R., 90, 146(131)

Blättler, M. O., 43, 61(194)
Bleha, T., 236, 240, 280(9e)
Blommers, M. J. J., 246, 283(67b)
Bloomfield, F., 203, 230(40)
Bluhm, T. L., 246, 283(69)
Bock, K., 193, 199(180), 236, 237, 240, 280(9d), 281(18a)
Bohin, J. P., 244, 283(58b)
Böhm, G., 12, 57(26, 27)
Boivin, J., 134, 150(277)
Bokelmann, C., 90, 146(132)
Bolon, P. J., 246, 283(66c)
Boojamra, C. G., 116, 149(232)
Boons, G.-J., 15, 16, 18, 58(43, 51, 53, 63), 189, 191, 199(169–172)
Borbás, A., 157, 196(71)
Borén, H. B., 180, 198(139)
Bornaghi, L., 238, 282(36)
Bose, A. K., 25, 59(107)
Bossennec, V., 216, 233(141)
Bouali, A., 117, 149(245)
Bouchu, M.-N., 133, 150(275)
Bouckaert, J., 243, 258, 282(48b)
Bou Habib, D. C., 203, 230(28)
Bourdon, M. A., 228, 234(161, 166)
Bowen, J. P., 16, 58(54)
Bowman, K. G., 238, 281(29)
Boy, K. M., 277, 284(112)
Boyd, J., 206, 231(81)
Brabetz, W., 191, 193, 199(174)
Braccini, I., 11, 57(18)
Brade, H., 191, 192, 193, 199(174, 176–178, 180), 246, 283(65)
Brade, L., 191, 192, 193, 199(174, 178, 180), 246, 283(65)
Brade, W., 119, 149(251)
Bradley, P. R., 20, 58(70)
Brady, J. W., 210, 232(112), 236, 243, 281(13)
Brakta, M., 247, 283(73)
Brånalt, J., 43, 61(192)
Brandt, B. L., 155, 195(23)
Brard, L., 82, 145(71)
Breg, J., 209, 232(97)
Brennan, P. J., 157, 196(70)
Breslow, R., 90, 147(139)
Brimacombe, J. S., 28, 60(121, 122)
Brown, C. F., 213, 233(124)
Brown, F., 157, 196(68)
Brown, R. S., 20, 58(72)
Bruix, M., 273, 284(108)

Brukhart, F., 237, 281(18b)
Brunckova, J., 22, 59(85), 89, 93, 111, 112, 113, 114, 146(127), 147(169), 149(228)
Brunings, K. J., 39, 61(172)
Bruzik, K. S., 213, 233(123)
Bryan, J. G. H., 28, 60(121)
Bryla, D., 154, 194(9)
Buchanan, J. G., 236, 265, 281(15)
Buckmelter, A. J., 69, 145(55)
Bull, D. H., 53, 62(226)
Buncel, E., 20, 58(70)
Bundle, D. R., 15, 58(45), 158, 168, 196(81), 197(114, 115), 243, 258, 282(46)
Bürger, H., 90, 91, 147(148)
Burgey, C. S., 88, 92, 146(111)
Burkhardt, F., 279, 284(116)
Burton, A., 15, 58(43)
Busch, R., 212, 213, 233(119)
Busch, S. J., 201, 229(10)
Bush, C. A., 206, 231(72, 77), 243, 258, 282(47)
Butenhof, K. J., 213, 233(127)
Byers, J. H., 116, 149(231)

C

Cadoz, M., 155, 194(18)
Caldwell, J. W., 268, 284(105)
Callstrom, M. R., 237, 281(27)
Calvo-Flores, F. G., 141, 150(287)
Campbell, S. C., 228, 234(161, 162)
Cañada, F. J., 236, 243, 246, 256, 257, 258, 259, 260, 266, 268, 270, 272, 273, 280(12), 282(49b, 50), 283(64), 284(90, 92, 93, 103, 106)
Cappa, A., 47, 62(210)
Card, P. J., 9, 57(4)
Cardin, A. D., 201, 229(10)
Caris, B. M. G., 44, 46, 62(200)
Carlomagno, T., 246, 283(67a)
Carlsson, H. E., 180, 181, 198(143)
Carlstedt, I., 219, 233(152)
Carr, C. J., 87, 146(91)
Carson, C. A., 155, 195(22)
Cartwright, K. A. V., 157, 196(57)
Carver, J. P., 236, 240, 243, 280(9f, 9g), 282(56)
Case, D. A., 268, 284(105)
Cassinari, A., 208, 232(96a)
Castellot, J. J., Jr., 204, 231(59)
Castro Palomino, J., 161, 166, 197(99)
Casu, B., 204, 208, 209, 210, 216, 219, 224,

230(54), 231(58), 232(93, 96a, 103b), 233(141)
Casu,B., 216, 233(143)
Catterall, H., 67, 144(33)
Cavallaro, C. L., 100, 101, 106, 142, 148(193, 194)
Cha, J. K., 235, 280(6)
Chaguir, B., 247, 283(73)
Chamberlin, A. R., 52, 62(222)
Chambers, R. D., 56, 63(245)
Chan, L., 161, 196(91), 197(96)
Chan, N., 161, 163, 164, 165, 197(94, 97, 103)
Chan, S. S. C., 22, 59(82)
Chandra, H., 69, 80, 145(59)
Chang, C.-T., 89, 114, 146(122)
Chang, D., 235, 237, 280(2c)
Chang, X., 55, 62(235)
Charych, D. H., 238, 281(28)
Chatgilialoglu, C., 69, 84, 85, 89, 90, 91, 113, 127, 128, 130, 134, 145(61), 146(116–118), 147(143, 149, 150, 152, 153), 150(264, 265)
Chatterjee, G., 157, 196(70)
Cheatham, T. E. III, 268, 284(105)
Cheifetz, S., 203, 230(42)
Chen, C., 92, 93, 138, 147(168)
Chen, G.-R., 116, 149(229)
Chénedé, A., 142, 150(295)
Cheng, X., 266, 268, 284(103)
Chern, C.-Y., 108, 148(215, 216)
Chernyak, A., 161, 197(101)
Chervenak, M., 238, 281(32a)
Chiappini, P. L., 117, 149(244)
Chiaroni, A., 236, 281(14a)
Chittenden, G. J. F., 10, 44, 46, 57(12), 62(200)
Choay, J., 204, 216, 231(59), 233(140, 141)
Choe, B. Y., 206, 207, 208, 210, 212, 224, 231(79, 80), 232(112)
Choi, H. U., 202, 229(13, 16)
Choi, J. S., 213, 233(131)
Choi, S.-K., 237, 281(26b)
Choi, S.-Y., 139, 150(283)
Chong, P., 161, 163, 164, 165, 197(94, 97, 103)
Christian, R., 116, 149(236), 191, 199(175)
Chu, C., 154, 156, 185, 187, 194(9), 195(46), 199(166)
Chua, J., 40, 61(179)
Cifonelli, J. A., 219, 233(151)

Cipolla, L., 121, 149(253, 254)
Cisar, L. A., 203, 230(31)
Clark, K. B., 90, 91, 147(152, 153)
Classon, B., 40, 41, 42, 43, 61(184, 190–192, 196)
Clive, D. L. J., 90, 147(140)
Clowes, A., 203, 230(33)
Cobon, G. S., 154, 155, 156, 157, 194(10), 195(29, 30, 35)
Coe, D. G., 38, 61(166)
Cohen, D., 154, 194(9)
Cole, S. J., 94, 95, 147(174)
Coleman, G. H., 10, 57(8)
Coleman, R. S., 239, 282(42)
Collins, P. M., 67, 87, 105, 106, 144(26), 146(104, 105)
Collum, D. B., 90, 147(138)
Connuck, D. M., 157, 166, 196(54)
Contreras, R. R., 209, 232(97)
Corey, E. J., 25, 47, 59(105)
Corzo, J., 244, 283(58c)
Cottier, L., 67, 68, 144(22)
Cotton, F. A., 45, 62(202)
Couchman, J. R., 203, 230(36)
Coward, M. D., 239, 282(43)
Coxon, B., 184, 198(157, 159)
Coynault, C., 198(134)
Cramer, C., 236, 240, 280(9h)
Cramer, F., 27, 60(120)
Crich, D., 66, 89, 92, 93, 95, 108, 109, 110, 111, 112, 113, 114, 118, 129, 130, 133, 135, 136, 137, 138, 139, 143, 143(6, 11), 146(124, 127, 168, 169, 176, 177), 148(214, 217–220), 149(221–223, 228, 246, 247), 150(270, 271, 278, 279, 281–283), 151(301, 302)
Croakert, F., 154, 194(4)
Croasmun, W. R., 206, 231(76)
Crocker, P. R., 235, 280(1f)
Cross, A. S., 156, 195(49)
Cruickshank, F. R., 10, 45, 57(13)
Cryz, S. J., 156, 195(49)
Csuk, R., 101, 142, 148(196)
Cuevas, G., 86, 145(82)
Cui, J., 116, 149(234)
Cullis, P. M., 143, 151(303)
Culp, L. A., 202, 203, 229(16), 230(35)
Cung, S.-K., 102, 148(202)
Cupp, M., 203, 230(31)
Curran, D. P., 66, 89, 90, 111, 112, 114, 115,

143(4, 5, 12, 14, 15), 146(122, 129), 147(142), 149(225, 226)
Curto, E. V., 207, 210, 214, 224, 231(83)
Czech, M., 246, 283(67a)
Czernecki, S., 42, 54, 61(188), 62(232), 143, 150(296)

D

Dabrowski, J., 206, 231(73, 76), 256, 284(87)
Dahmann, G., 237, 281(26a)
Dais, P., 244, 283(60)
Dak, K., 42, 61(189)
Dalley, N. K., 237, 281(24b)
Danishefsky, S. J., 23, 24, 25, 59(91, 92, 94–97, 99, 103, 104)
Da Rooge, M. A., 40, 61(179)
Date, V., 13, 57(31, 32)
Daube, M., 43, 61(194)
Daves, G. D., Jr., 247, 283(73)
David, S., 54, 62(229), 95, 147(178)
Davies, A. L., 213, 233(126)
Davies, M. J., 67, 144(33)
Davis, D. G., 232(109), 244, 283(59a)
Davis, J. T., 213, 233(129)
Davison, I. G. E., 92, 93, 138, 147(168)
Dean, I. C. M., 208, 232(94)
de Armas, P., 33, 37, 41, 60(147, 149)
de Beer, T., 205, 207, 217, 218, 219, 224, 231(68, 85), 233(150), 278, 284(114)
De Bolt, S., 268, 284(105)
Dekker, H. A. T., 174, 176, 197(125)
de Leeuw, F. A. A. M., 244, 283(61b)
de Lera, A. R., 36, 61(162)
Dell, A., 187, 199(167), 203, 205, 217, 218, 219, 224, 228, 230(24), 231(67), 233(146)
Della Bona, M. A., 86, 145(86)
Delmond, B., 90, 147(136)
de Muys, J. M., 157, 196(55)
Depew, M. C., 85, 145(81)
de Pouilly, P., 142, 150(295)
de Reuver, M. J., 172, 197(122)
Derouet, C., 11, 57(18)
Desai, U. R., 209, 219, 224, 232(104)
Descotes, G., 29, 60(126), 67, 68, 82, 88, 112, 117, 129, 133, 144(22, 23, 41, 42, 48), 145(71), 146(115), 149(245), 150(275), 264, 284(99)

Dess, D. B., 36, 60(157), 61(158)
Desvaux, H. J., 244, 278, 282(58a), 283(58b), 284(113)
Detsky, A. S., 155, 195(22)
De Velasco, E. A., 165, 197(104)
Devi, S. J. N., 154, 194(6)
de Waard, P., 205, 209, 216, 217, 218, 219, 223, 224, 231(66), 232(103a), 233(144, 145, 147)
de Weers, O., 157, 196(69)
Dexter, M., 203, 230(40)
Dezube, M., 52, 62(222)
Diaz-Mauriño, T., 258, 259, 260, 284(91, 94)
Di Cesare, P., 91, 147(157)
Dick, W. E., Jr., 157, 196(58)
Dietrich, C. P., 208, 209, 224, 232(96a, 99), 234(158)
Dietrich, H., 236, 237, 256, 257, 258, 259, 260, 280(12), 281(20), 283(84, 85, 90, 92, 93)
Di Fabio, J. L., 155, 156, 157, 195(33, 48, 53)
Dilapi, M. M., 154, 194(12)
Dimmel, R. D., 68, 129, 144(50)
Dintzis, R. Z., 157, 196(59)
Di Stèfano, C., 116, 129, 149(243)
Doddridge, Z. A., 143, 151(303)
Doisneau, G., 236, 281(14a)
Domann, S., 31, 60(136)
Donard, O., 90, 147(136)
Donna, A., 86, 134, 145(88)
Donohue-Rolfe, A., 246, 283(66d)
Dorfman, A., 219, 233(151)
Dorland, L., 206, 207, 224, 231(71), 232(87)
Dorta, R. L., 33, 37, 60(148, 150)
Douglas, R. M., 155, 195(20)
Dowd, M. K., 243, 266, 282(54a)
Dräger, G., 30, 31, 60(131, 136)
Drescher, M., 91, 100, 147(158)
Drickamer, K., 235, 280(1d)
Driguez, H., 238, 282(36)
Driguez, P. A., 238, 282(35a)
Du, Y., 235, 237, 280(2a)
Duben, A. J., 211, 233(118)
Duclos, A., 87, 146(92)
Duggan, P. J., 86, 130, 138, 139, 145(83, 84), 150(271, 280)
Dumartin, G., 90, 147(136)
Dumitriu, S., 157, 196(65, 66)
Duncan, S. J., 155, 195(20)
Duncan, S. M., 90, 147(141)

Dupuis, J., 67, 68, 69, 70, 71, 72, 73, 74, 79, 80, 81, 82, 83, 84, 85, 88, 94, 103, 105, 118, 130, 131, 144(24, 44, 47), 145(56, 67), 146(114), 147(170), 149(249)
Durier, V., 243, 282(54b)
Dussault, P., 52, 62(222)
Dussy, A., 143, 151(300)
Duynstee, H. I., 32, 60(146)
Dwek, R. A., 206, 231(81), 235, 280(1e)
Dygutsch, D. P., 90, 146(135)
Dyson, H. J., 233(136)

E

Eby, R., 157, 159, 161, 166, 196(54, 78, 82–85)
Egan, W., 154, 157, 194(9), 196(60)
Egron, M. J., 238, 282(39)
Eichenberger, D., 202, 229(14)
Eichenberger, E., 111, 112, 114, 115, 149(226)
Eisenstein, T. K., 156, 195(41)
Ekborg, G. C., 159, 180, 181, 182, 196(86), 198(135, 140, 143), 206, 207, 208, 210, 212, 224, 231(79, 80)
Eklind, K., 180, 181, 198(137, 138, 143)
Eksterowicz, J. E., 240, 242, 272, 282(45)
Elbein, A. D., 237, 281(22)
Elie, C. J. J., 161, 162, 197(100)
El Khadem, H. S., 67, 144(31)
El Kharraf, Z., 88, 146(115)
Ellervik, U., 265, 284(100)
Ellis, R. W., 155, 195(29)
Elmouchir, M., 236, 281(14a)
Emanuel, C. J., 69, 84, 85, 145(61)
Endo, T., 92, 147(159)
Engelsen, S. B., 243, 282(54b, 57)
Enholmm, E. J., 116, 149(230)
Erbing, B., 211, 233(117)
Eriksson, U., 181, 198(150)
Ernst, B., 12, 57(24)
Ernst, R. R., 210, 232(108)
Esko, J. D., 203, 206, 207, 208, 224, 228, 230(27, 30, 37, 42), 231(78), 234(163–165)
Esnault, J., 11, 57(18), 236, 281(14c)
Espinosa, J. F., 236, 243, 246, 256, 257, 258, 259, 260, 273, 280(12) 282(49b, 54b), 283(65, 84, 85), 284(88, 90, 92, 93, 108)

Evans, A., 31, 60(137)
Evans, W. L., 10, 57(9, 11)
Evenberg, D., 152, 161, 165, 166, 167, 173, 189, 196(90), 197(92, 95, 105, 106)
Ewing, D. F., 117, 149(245)

F

Fabian, M. A., 22, 59(83, 85)
Faham, S., 203, 230(49)
Fan, J., 238, 282(37)
Fantini, B., 156, 195(47)
Farkas, I., 116, 149(240)
Farnsson, L. A., 205, 231(65)
Farr, R. N., 247, 283(73)
Farrar, G., 157, 196(69)
Fattom, A., 155, 157, 195(35), 196(67)
Fayet, C., 87, 146(92)
Feingold, D. S., 204, 208, 230(53)
Feizi, T., 235, 280(1f)
Feldman, R. G., 174, 176, 197(125)
Felli, I. C., 246, 283(67a)
Ferguson, D., 268, 284(105)
Fernandes, A. C., 42, 61(188)
Fernandez, C., 156, 195(52)
Fernández, I. M., 174, 176, 197(125)
Fernandez, P., 259, 260, 284(94)
Fernández-Mayoralas, A., 140, 142, 150(286), 270, 272, 273, 284(106)
Fernandez-Santana, V., 161, 166, 197(99)
Ferreira, J. A., 66, 143(17)
Ferreri, C., 69, 84, 85, 145(61)
Ferretti, M., 23, 59(89)
Ferrier, R. J., 25, 26, 59(106, 112), 66, 67, 68, 82, 85, 105, 106, 112, 143(19), 144(21, 26), 149(227)
Ferrito, R., 263, 284(95)
Ferro, D. R., 208, 216, 232(96a), 233(140, 141, 143)
Fessner, W. D., 237, 281(17)
Fevig, T. L., 66, 143(12)
Feyzi, E., 203, 230(29)
Field, R. A., 13, 15, 16, 17, 18, 47, 48, 49, 57(29, 30), 58(49, 55, 56, 64), 62(208, 215)
Figueroa Perez, I., 161, 166, 197(99)
Filler, R., 55, 63(240)
Filzen, G. F., 130, 135, 137, 139, 150(270, 279, 281)

Fine, M. J., 155, 195(22)
Finn, J. M., 53, 62(227)
Fischer, H., 68, 69, 70, 105, 144(47)
Fischer, R., 246, 283(67a)
Fitchett, M., 85, 145(78, 79)
Fleet, G. W. J., 237, 281(19)
Fleming, I., 15, 58(48)
Fletcher, H. G., Jr., 11, 57(20), 87, 146(93)
Florent, J.-C., 87, 146(101)
Folkman, J., 203, 230(47)
Font, B., 202, 229(14)
Forooghian, F., 166, 197(113)
Fosang, A. J., 201, 229(11)
Franchetti, P., 236, 265, 281(16d)
Francis, T., Jr., 154, 194(11)
Fransen, C. T. M., 10, 57(12)
Fransisco, C. G., 33, 34, 37, 41, 60(147–150, 152, 153)
Fransson, L.-A., 201, 202, 203, 215, 219, 229(5, 18, 19, 21), 233(138, 152)
Fraser-Reid, B., 11, 13, 14, 15, 16, 19, 26, 32, 40, 49, 57(19, 31, 32, 36–41), 58(42, 54, 68), 59(109, 110), 61(176), 62(211), 88, 89, 92, 102, 103, 146(111, 125), 148(207)
Fréchet, J. M., 25, 59(102)
Fredericks, P. M., 54, 62(230)
Freeze, H. H., 207, 224, 231(84)
French, A. D., 210, 232(112), 236, 240, 243, 266, 280(9h), 281(13), 282(54a, 54b)
Friesel, R., 203, 230(48)
Friesen, R. W., 23, 24, 59(92, 94)
Frische, K., 237, 281(24a)
Fritz, L. M. S., 203, 230(32)
Fritz, T., 206, 207, 208, 224, 231(78)
Fromm, J. R., 203, 230(49)
Frost, J. W., 116, 149(235)
Fu, G. C., 89, 99, 146(128)
Fu, Y., 191, 192, 193, 199(178)
Fuglesang, C. D., 240, 242, 272, 282(45)
Fujioka, A., 17, 58(57)
Fujita, E., 31, 60(135)
Fukase, K., 32, 60(141, 143)
Funabashi, M., 87, 146(94)
Funke, C. W., 161, 197(95)
Fürer, E., 156, 195(49)
Furneaux, R. H., 66, 143(19), 144(20)
Fürst, A., 20, 58(73)
Fürstner, A., 101, 102, 106, 142, 148(196, 197)
Furusawa, Y., 17, 58(57)

G

Gabius, H.-J., 235, 258, 260, 280(1a), 284(90)
Gabius, S., 235, 280(1a)
Gallagher, J. T., 201, 203, 229(8), 230(26, 40)
Gallagher, T., 235, 237, 280(2f)
García, A., 270, 272, 273, 284(106)
Garcia, D. M., 53, 62(225)
Garcia, J. L., 259, 284(93)
García-Calvo-Flores, F., 97, 148(184)
García-Mendoza, P., 97, 141, 148(184), 150(287)
Garcia-Verdugo, C., 97, 148(183)
Garegg, P. J., 39, 40, 41, 42, 53, 61(171, 173, 181–184, 186), 62(224), 156, 159, 160, 161, 180, 181, 182, 195(40), 196(86, 88, 89), 198(136–140, 143–145), 207, 224, 232(87)
Garrity, R. R., 213, 214, 233(135)
Gass, J., 191, 199(175)
Gathmann, W. D., 205, 224, 231(69)
Gatti, G., 216, 233(140)
Gelas, J., 87, 146(92)
Gemmecker, G., 213, 233(128)
Geoffrey, M., 239, 282(41)
Geoffroy, M., 67, 144(27)
Gerigk, U., 90, 146(130)
Gerken, T. A., 213, 233(127)
Gerlach, M., 90, 146(130)
Géro, S. D., 40, 41, 61(175)
Gervais, V., 213, 233(134)
Gervay, J., 24, 25, 59(96, 99), 116, 149(237)
Gesson, J.-P., 55, 62(236)
Gettins, P., 209, 232(100)
Geurtsen, R., 16, 58(51)
Geyer, A., 178, 198(129), 214, 215, 233(137)
Geyer, H., 176, 178, 198(128)
Ghosez, A., 67, 141, 144(36), 150(292)
Gidney, M. A. J., 168, 197(115)
Giese, B., 66, 67, 68, 69, 70, 71, 72, 73, 74, 75, 78, 79, 80, 81, 82, 83, 84, 85, 86, 88, 89, 90, 91, 94, 96, 97, 100, 102, 103, 105, 115, 118, 122, 130, 131, 132, 133, 134, 135, 136, 137, 138, 139, 140, 141, 143(3, 9, 13–15), 144(24, 36, 40, 44, 46, 47), 145(56, 57, 62–65, 67–69, 72, 85), 146(114, 116), 147(152, 170), 148(187, 199, 209), 149(249), 150(272–274, 284, 292, 298, 299), 151(300)
Gilbert, B. C., 67, 84, 85, 144(33), 145(73–80)

Gilbert, J. H., 237, 238, 281(27), 281(28)
Gilges, S., 132, 133, 150(273)
Gillier-Pandraud, H., 246, 272, 283(68)
Gimisis, T., 69, 84, 85, 127, 128, 130, 145(61), 150(264, 265)
Girard, C., 116, 149(238)
Glänzer, B. I., 101, 142, 148(196)
Glaudemans, C. P. J., 184, 198(156), 237, 238, 281(25), 282(34a, 34b)
Glossel, J., 215, 233(139)
Göbel, T., 67, 141, 144(36), 150(292)
Goebel, W. F., 155, 159, 160, 169, 170, 171, 172, 195(25, 26), 197(117–121)
Goekjian, P. G., 246, 247, 248, 249, 250, 251, 252, 253, 254, 255, 261, 262, 271, 275, 283(70, 74, 75, 79, 81–83), 284(107, 110)
Gola, J., 116, 149(229)
Goldblatt, D., 157, 196(61)
Goldman, M., 244, 282(58a)
Goldschmid, H. R., 10, 57(10)
Goldschmidt, D., 202, 229(14)
Goldstein, B. M., 236, 265, 281(16b–d)
Goldstein, I. J., 182, 198(151)
Gomes, L. J. F., 173, 197(124)
Gómez, A. M., 89, 102, 103, 146(125)
González, C. G., 34, 60(152, 153), 98, 99, 148(189)
Goodman, B. K., 69, 145(60)
Goodman, L., 40, 61(178)
Gootz, R., 10, 57(14)
Gordon, M. Y., 203, 230(39)
Gore, J. L., 87, 146(98)
Görner, H., 87, 146(103)
Goto, M., 17, 58(57)
Gotschlich, E. C., 155, 195(19)
Gotthammar, B., 159, 180, 181, 182, 196(86), 198(137, 138, 143, 145)
Gottschalk, A., 201, 204, 229(2)
Goubeau, J., 27, 59(114)
Gould, R. O., 208, 232(96b)
Gould, S. E. B., 208, 232(96b)
Gravert, D. J., 25, 59(100)
Graves, D. P., Jr., 68, 129, 144(50)
Gray, G. R., 87, 146(96, 98), 185, 198(165)
Greaves, M. F., 203, 230(39)
Greenberg, M. M., 69, 145(60)
Greene, T., 213, 233(125)
Gregar, T. Q., 116, 149(237)
Greisman, S. E., 183, 198(154)
Grenier-Loustalot, M.-F., 67, 68, 112, 144(23)

Grice, P., 16, 58(53)
Griesinger, C., 246, 283(67a)
Griffith, D. A., 24, 25, 59(96, 97, 99)
Griffith, L. S., 214, 215, 233(137)
Griggs, J. L., 116, 149(232)
Griller, D., 90, 91, 147(146, 149, 150, 152, 153)
Grimmecke, D., 156, 157, 195(39)
Grindley, T. B., 143, 150(297)
Gröninger, K. S., 69, 70, 72, 74, 75, 79, 80, 81, 82, 83, 84, 85, 86, 97, 130, 131, 132, 133, 135, 137, 138, 145(57, 62, 67, 68, 72, 85), 150(272–274)
Grootenhuis, P. D. J., 204, 231(61), 243, 282(54b)
Gross, B., 91, 147(157)
Gross, P. H., 250, 283(77)
Grosskurth, H., 256, 284(87)
Grouiller, A., 117, 149(245)
Gruezo, F., 157, 158, 196(76)
Grunwald, E., 19, 58(69)
Gruza, J., 243, 282(54b)
Grynkiewicz, G., 29, 60(127)
Guerra, M., 69, 84, 85, 90, 91, 145(61), 147(153)
Guerrini, A., 90, 91, 147(153)
Guerrini, M., 86, 121, 141, 145(87), 149(254), 208, 209, 219, 224, 232(96a, 103b)
Gugger, A., 143, 150(299)
Guidry, C., 202, 229(13)
Gupta, R. K., 207, 231(82)
Guss, J. M., 208, 232(94)
Guthrie, R. D., 54, 62(230)

H

Haasnoot, C. A. G., 244, 283(61b)
Habuchi, H., 204, 231(56)
Hachiya, S., 235, 280(8)
Hadida, S., 90, 147(142)
Haight, A. R., 90, 147(141)
Hajduk, P. J., 208, 232(95)
Hakomori, S., 213, 233(125)
Hall, D. W., 112, 149(227)
Hallows, W. H., 236, 265, 281(16c, 16d)
Hamada, T., 25, 59(98)
Hamelryck, T., 243, 258, 282(48b)
Hammerschmidt, F., 91, 100, 147(158)
Hamor, T. A., 28, 60(121)
Han, X.-J., 209, 219, 224, 232(104)
Hanamoto, T., 34, 60(151)
Hanano, T., 49, 62(213)

Haneda, T., 275, 284(110)
Hanessian, S., 44, 54, 56, 61(197, 198), 62(229, 241), 66, 67, 115, 116, 143(17), 144(40), 149(238), 235, 280(4)
Hanna, H. R., 158, 196(81)
Hansen, M. R., 210, 232(111)
Hansman, D. J., 155, 195(20)
Hantsch, A., 27, 59(113)
Haque, M. B., 90, 95, 100, 147(154, 175)
Harada, T., 204, 216, 217, 218, 224, 231(56), 233(144, 145, 147)
Haraguchi, K., 127, 128, 130, 139, 150(266)
Harders, J., 30, 60(131)
Hardingham, T. E., 201, 229(11)
Harper, D. J., 36, 61(160)
Harris, S. L., 166, 197(110, 111)
Hart, D. J., 66, 143(2)
Hartung, J., 67, 144(36)
Hartwig, W., 66, 92, 143(1), 147(165)
Harvey, S. C., 206, 207, 208, 210, 212, 224, 231(79, 80), 232(112)
Harwig, C. W., 25, 59(100)
Harwood, J. S., 95, 133, 147(177)
Hary, C., 32, 60(140)
Hary, U., 32, 60(140)
Hascall, G. K., 203, 230(25)
Hascall, V. C., 203, 230(25, 28)
Hasegawa, A., 17, 58(60, 61)
Hasegawa, T., 87, 146(94)
Haselhorst, T., 246, 283(65)
Hashem, M. A., 32, 60(139)
Hashimoto, H., 238, 281(33)
Haslam, S. M., 203, 230(24)
Hasskerl, T., 94, 147(170)
Hassner, A., 27, 59(115)
Hasuoka, A., 32, 60(141–143)
Hatakeyama, S., 53, 62(225)
Haudrechy, A., 277, 284(111)
Haudrechy-Bretel, N., 277, 284(111)
Havsmark, B., 215, 233(138)
Hay, E. D., 203, 230(25)
Hayday, K., 68, 129, 144(49)
Haynes, L. J., 235, 280(3)
Hays, D. S., 89, 99, 146(128)
He, W., 86, 115, 145(89)
Head, D., 246, 283(67b)
Heidelberger, M., 154, 194(5, 12–14)
Heinegard, D., 201, 202, 229(6, 12)
Helferich, B., 10, 57(14)
Helting, T., 205, 231(62)
Hemboldt, A., 238, 282(35a, 35b)
Hemmerich, S., 238, 281(29)
Hendriks, K. B., 17, 58(59)
Herault, J. P., 238, 282(35a, 35b)
Herbert, J. M., 238, 282(35a, 35b)
Hermann, F., 110, 149(223)
Hermans, J. P. G., 152, 161, 197(92)
Hernández-Mateo, F., 97, 141, 148(184), 150(287)
Herpin, T. F., 122, 123, 124, 125, 126, 127, 130, 149(258, 260–262), 150(263)
Herscovici, J., 238, 282(39)
Hervé du Penhoat, C., 11, 57(18), 243, 264, 282(57), 284(99)
Hessler, G., 279, 284(116)
Hetzer, G., 116, 149(229)
Hicks, D. R., 40, 61(176)
Hiebl, J., 98, 148(188)
Hietter, H., 213, 233(130)
Hileman, R. E., 203, 230(49)
Hill, T. G., 237, 281(27)
Himmelspach, K., 158, 176, 178, 183, 184, 196(79), 198(128, 129, 155)
Hindsgaul, O., 12, 57(25)
Hirama, M., 55, 62(238)
Hirani, S., 213, 233(129)
Hirano, S., 224, 234(157)
Hodges, R. G., 154, 194(13)
Hoffman, T., 165, 197(104)
Hoffmann, M., 237, 279, 281(18b), 284(116)
Höfle, G., 90, 147(151)
Hofsteenge, J., 278, 284(114)
Hollósi, M., 116, 129, 149(243)
Holmbeck, S. M. A., 211, 232(114)
Holme, K., 155, 157, 195(33, 53)
Holmes, D., 16, 58(51)
Holst, O., 156, 157, 191, 193, 195(39), 199(174)
Homans, S. W., 206, 231(74, 81), 243, 246, 282(55a–c), 283(66b, 66d)
Hoogerhout, P., 152, 157, 161, 162, 165, 166, 167, 173, 189, 191, 196(69, 90), 197(92, 95, 100, 105, 106), 199(169)
Hoogewerf, A. J., 203, 230(31)
Hook, M., 201, 202, 203, 229(4, 13), 230(36, 37)
Hoppe, W., 215, 233(139)
Hori, H., 106, 148(212)
Horita, D. A., 208, 232(95)
Horita, K., 235, 280(8)
Horne, A., 204, 209, 231(57), 232(100)
Horner, J. H., 69, 84, 85, 139, 145(61), 150(283)

Horton, D., 22, 59(86–88), 116, 149(234), 211, 232(116)
Horwitz, J. P., 40, 61(179)
Hosokawa, H., 17, 58(57)
Hotta, K., 17, 58(61)
Houk, K. N., 240, 242, 272, 282(45)
Houlton, J. S., 123, 124, 125, 126, 127, 130, 149(261)
Houston, T. A., 25, 59(108)
Howard, K. P., 213, 233(121)
Howard, S., 238, 282(40)
Hoyer, F., 116, 149(242)
Hricovini, M., 211, 233(118), 243, 282(56)
Huang, J., 43, 61(193)
Huang, X., 139, 150(283)
Huang, X., Jr., 213, 214, 233(135)
Huckerby, T. N., 216, 233(142)
Huckins, D. W. L., 208, 232(94)
Hudson, C. S., 11, 57(20), 87, 146(91), 146(93)
Huebner, J., 154, 194(8)
Huheey, J. E., 15, 58(47)
Hultberg, H., 180, 198(136)
Hundey, F., 28, 60(122)
Husman, W., 95, 148(179)
Hüter, O., 67, 144(36)
Hwang, J.-T., 108, 109, 129, 148(217)

I

Iacomini, M., 209, 219, 224, 232(103b)
Ialongo, G., 127, 128, 130, 150(264, 265)
Ichikawa, Y., 111, 149(224)
Ikegami, A., 209, 215, 216, 224, 232(102)
Iley, D. E., 26, 59(110)
Imberty, A., 243, 258, 259, 264, 282(47, 52, 54b), 284(93, 99)
Inanaga, J., 34, 60(151)
Inazu, T., 17, 58(57), 58(58)
Ingber, D. E., 203, 230(47)
Ingold, K. U., 68, 89, 90, 113, 129, 144(52), 146(118)
Ingraham, L. L., 19, 58(69)
Ingrosso, G., 23, 59(89)
Innui, A., 217, 218, 233(150)
Insel, R. A., 157, 166, 196(54)
Ioannidis, P., 43, 61(196)
Ireland, R. E., 36, 61(159)
Isac-García, J., 97, 148(184)
Isac-Garcia, J., 141, 150(287)

Isbell, H. S., 67, 144(31)
Ishida, H., 17, 58(60), 58(61)
Ishida, N., 103, 148(208)
Ishihara, M., 204, 231(57)
Ison, E. R., 28, 45, 46, 60(124)
Itoh, K., 102, 148(205)
Itoh, Y., 127, 128, 130, 139, 150(266)
Iwamura, H., 68, 82, 129, 144(51, 53)
Izquierdo Cubero, I., 141, 150(291)

J

Jablonsky, M. J., 207, 210, 214, 224, 231(83)
Jackson, R. L., 201, 229(10)
Jacquinet, J.-C., 55, 62(233), 207, 209, 210, 214, 216, 224, 231(83), 232(107), 233(140)
Jäger, K. F., 80, 145(68)
Jahnke, W., 246, 283(67b)
Jain, D. C., 207, 231(82)
James, K., 17, 58(59)
James, O., 165, 197(103)
Janda, K. D., 25, 59(100)
Jang, D. O., 88, 90, 91, 100, 140, 146(108–110), 147(145, 147, 155), 148(191)
Jansson, P.-E., 187, 199(167), 211, 233(117)
Jansze, M., 172, 178, 179, 197(122), 198(130)
Jaramillo, C., 235, 280(5)
Jarosz, S., 40, 61(176)
Jarreton, O., 273, 284(108)
Jasperse, C. P., 66, 143(12)
Jaszberenyi, J. Cs., 66, 88, 90, 91, 108, 140, 143(17), 146(108–110), 147(145, 147, 155), 148(215, 216)
Jeanloz, R. W., 11, 57(21)
Jeener, J., 210, 232(108)
Jeffery, A., 87, 146(97)
Jenkins, I. D., 92, 147(167)
Jennings, H. J., 40, 61(185), 95, 148(181), 155, 157, 189, 191, 194(17), 195(33, 53), 196(55, 56, 62–65), 199(169)
Jewell, J. S., 26, 27, 28, 45, 59(111, 118), 60(123)
Jiao, X.-Y., 95, 133, 139, 147(177), 150(282)
Jiménez, J. L., 97, 148(183)
Jiménez-Barbero, J., 236, 243, 244, 245, 246, 256, 257, 258, 259, 260, 266, 268, 270, 272, 273, 275, 280(12), 282(49b, 50, 54b, 54c), 283(58c, 59c, 59d, 63–65, 84, 85), 284(86, 88, 90–94, 103, 106, 108, 109)

Johansson, R., 40, 41, 42, 61(183, 186), 160, 161, 196(89)
Johansson, S., 201, 203, 229(4), 230(36, 37)
Johns, B. A., 237, 270, 272, 273, 281(22), 284(106)
Johnson, C. R., 237, 270, 272, 273, 281(22), 284(106)
Johnson, G., 243, 282(54b)
Johnson, L. N., 237, 281(19)
Johnson, M. D., 36, 61(163)
Jones, C., 154, 194(16)
Jones, C. E., 246, 283(67b)
Jones, G. D. D., 143, 151(303)
Jones, J. K. N., 26, 27, 40, 59(111, 118), 61(185)
Jones, P. G., 22, 59(84)
Jonson, B. M., 158, 196(80)
Jörbeck, H. J. A., 181, 182, 183, 198(148, 152)
Jorgensen, W. L., 87, 93, 94, 146(106)
Josephson, S., 180, 198(140)
Jourdian, G. W., 217, 218, 233(148)
Juaristi, E., 86, 145(82)
Julina, R., 119, 149(250)
Jumbam, D. N., 101, 102, 106, 142, 148(197)
Jun, H.-S., 277, 284(111)
Jung, A., 32, 60(139)
Just, G., 161, 196(91), 197(93, 96)
Jütten, P., 87, 146(102)

K

Lacôte, E., 89, 146(123)
Kaack, M. B., 188, 199(168)
Kabat, E. A., 155, 157, 158, 159, 182, 195(21), 196(75–77)
Kadis, S., 156, 195(37)
Kågedal, L., 156, 195(51)
Kahlenberg, F., 121, 149(255)
Kahne, D., 107, 129, 148(213), 213, 214, 233(132, 133)
Kaijser, B., 156, 195(44)
Kaji, A., 118, 149(248)
Kakubo, K., 92, 147(161)
Kamerling, J. P., 174, 176, 178, 179, 197(125), 198(126, 127, 130), 209, 232(97)
Kanabus-Kaminska, J., 90, 91, 147(153)
Kandil, A., 165, 197(103)
Kandil, A. A., 161, 163, 164, 197(94, 97)
Kang, H.-Y., 246, 247, 271, 283(70), 284(107)

Kaper, J. B., 154, 155, 156, 157, 194(10), 195(29, 30, 35)
Kapoor, W. N., 155, 195(22)
Karl, H., 20, 58(75)
Karlson, R. M. K., 206, 231(76)
Karnovsky, M. J., 203, 204, 230(33), 231(59)
Karpeisky, A., 98, 99, 148(189)
Karplus, M., 244, 283(61a)
Kartha, K. P. R., 13, 15, 16, 17, 18, 44, 46, 47, 48, 49, 57(29, 30), 58(49, 55, 56, 64), 61(199), 62(208, 215), 95, 148(181)
Kasper, D. L., 154, 155, 157, 194(8), 195(23, 33, 53)
Katajima, T., 102, 148(205)
Kato, K., 206, 217, 231(70)
Katsuki, T., 52, 62(221)
Kaufman, B., 213, 233(124)
Kawahara, Y., 91, 147(156)
Kawamaura, S.-I., 51, 62(219)
Kawasaki, T., 217, 218, 224, 233(146, 149)
Käyhty, H., 165, 166, 167, 173, 189, 197(105)
Kearns, A. E., 228, 234(162)
Keck, G. E., 116, 149(230, 231)
Keifer, M. C., 204, 231(57)
Keifer, P. A., 207, 224, 231(84)
Keith, J., 47, 62(207)
Kenne, L., 153, 169, 194(1), 197(116), 211, 233(117)
Kennedy, J. F., 201, 229(3)
Kent, G. J., 27, 59(115)
Keown, L. E., 235, 280(7)
Kessler, H., 213, 233(128), 237, 279, 281(18b), 284(115, 116)
Khan, K. H., 54, 62(231)
Khan, N., 55, 62(235), 266, 268, 284(103)
Khan, S. H., 15, 57(41), 58(44), 67, 115, 144(39)
Khoo, K.-H., 205, 217, 218, 219, 224, 228, 231(67), 233(146)
Khuong-Huu, Q., 87, 146(101)
Kiddle, G. R., 246, 283(66b)
Kiessling, L. L., 238, 281(32a, 32b)
Kihlberg, J., 15, 58(45)
Kikuchi, N., 49, 62(213)
Kilian, W., 90, 91, 147(148)
Kim, D., 111, 149(225)
Kim, S. H., 275, 284(110)
Kim, Y., 224, 234(155), 245, 283(62b)
Kimata, K., 204, 231(56)
King, D. M., 84, 85, 145(73–77)
Kinoshita, A., 203, 230(24)

Kinoshita, I., 32, 60(141, 142)
Kinugasa, H., 217, 218, 224, 233(146)
Kirby, A. J., 22, 32, 59(84), 60(144), 236, 240, 280(9c)
Kirsching, A., 30, 31, 32, 60(130, 136, 139)
Kirschning, A., 30, 32, 60(131, 133, 134, 140)
Kirwan, J. N., 94, 95, 147(173, 174), 9594
Kiser, C. S., 203, 230(47)
Kishi, Y., 235, 236, 246, 247, 248, 249, 250, 251, 252, 253, 254, 255, 261, 262, 264, 265, 266, 271, 275, 277, 280(6, 10), 283(70, 71, 74, 75, 79–83), 284(98, 101, 107, 110–112)
Kiso, M., 17, 58(60, 61)
Kitagawa, O., 49, 62(213)
Kjellen, L., 201, 203, 229(4, 9), 230(45)
Klagsbrun, Y., 203, 230(50)
Klausener, A., 87, 146(103)
Klein, D. L., 155, 195(29)
Klein, M., 161, 163, 164, 165, 197(94, 97, 103)
Kleinhammer, G., 183, 184, 198(155)
Klemer, A., 11, 57(23)
Klemke, E., 25, 59(108)
Klich, G., 95, 148(180)
Kloosterman, M., 152, 161, 197(92)
Klundt, I. L., 40, 61(179)
Knapp, S., 235, 280(5)
Knecht, J., 219, 233(151)
Knights, J. M., 154, 194(7)
Koch, A., 67, 78, 83, 84, 97, 135, 136, 137, 139, 140, 144(36), 145(64), 148(187)
Kochetkov, N. K., 38, 61(168)
Kociensky, P., 89, 95, 146(120)
Kollman, P., 268, 284(105)
Komarov, I. V., 22, 59(84)
Komatsu, K., 53, 62(225)
Komatsu, M., 88, 146(112)
König, W., 279, 284(115)
Konradsson, P., 13, 14, 57(36, 37)
Kontou, M., 237, 281(19)
Köpper, S., 20, 58(77)
Kopping, B., 89, 90, 91, 134, 146(116), 147(152)
Kordish, R. J., 89, 146(121)
Koreeda, M., 25, 59(108)
Korones, D., 157, 166, 196(54)
Korth, H.-G., 68, 69, 70, 71, 72, 73, 74, 75, 76, 77, 78, 79, 80, 81, 82, 83, 84, 85, 86, 97, 105, 118, 130, 131, 132, 133, 135, 137, 138, 144(47), 145(56–59, 62, 66, 67, 72), 149(249), 150(272)

Koseki, K., 24, 25, 59(96), 59(99)
Koser, G. F., 30, 60(132)
Kosma, P., 191, 192, 193, 199(174–178, 180), 246, 283(65)
Koto, S., 211, 232(116), 236, 240, 280(9a)
Kottenhahn, M., 213, 233(128)
Kouwijzer, M. L. C. E., 243, 282(54b)
Kovac, P., 238, 282(34a)
Kozak, J., 270, 272, 273, 284(106)
Kozar, T., 256, 284(87)
Kozyrod, R. P., 67, 104, 144(25)
Kraft, E. L., 217, 218, 233(148)
Kresse, H., 215, 233(139)
Krishna, N. R., 206, 207, 208, 210, 212, 214, 224, 231(78–80, 83), 232(110, 112)
Kroder, K., 67, 144(36)
Kronzer, F. J., 10, 57(15)
Kropp, J. E., 27, 59(115)
Krusis, T., 228, 234(161)
Kuhn, C. S., 237, 281(25)
Kuhn, H., 90, 146(134)
Kumada, M., 103, 148(208)
Kuramochi, T., 35, 60(155)
Kurono, S., 238, 281(33)
Kurth, M. J., 49, 62(212)
Kusumoto, S., 32, 60(141–143), 157, 196(72)
Kuswik, G., 29, 60(127)
Kvarnström, I., 43, 61(192)

L

Lacôte, E., 89, 146(123)
Lacour, J., 31, 60(137, 138)
Ladlow, M., 102, 148(200, 201)
Laferriere, C. A., 157, 196(55)
Lafont, D., 29, 60(126)
Lagergard, T., 155, 195(32)
Lagerman, P. R., 157, 196(69)
Lai, E., 157, 196(77)
Lai, W., 237, 266, 272, 281(21), 284(102)
Laird, E. L., 87, 93, 94, 146(106)
Lam, L. H., 204, 231(57)
Lamberth, C., 78, 83, 84, 97, 132, 133, 135, 136, 137, 139, 140, 145(64), 150(273)
Lampl, H., 155, 195(24)
Lance, D. G., 27, 59(116, 117)
Landauer, S. R., 38, 61(165, 166)
Landersjo, C., 246, 283(66a)
Landsteiner, K., 155, 195(24)

Lane, D. A., 203, 204, 230(26, 47, 51, 55)
Langs, D. A., 236, 265, 281(16d)
Laterra, J., 203, 230(35)
Laupichler, L., 21, 58(78)
Lavaud, C., 247, 283(73)
Lavellée, P., 44, 61(197)
Law, F. C., 130, 150(269)
LeBaron, R. G., 203, 230(37)
Lee, A. W., 52, 62(221)
Lee, C.-J., 157, 196(66)
Lee, R. T., 157, 191, 196(64), 199(179)
Lee, W. W., 40, 61(178)
Lee, Y. C., 157, 191, 196(64), 199(179), 238, 282(37)
Lee, Y. N., 246, 283(69)
Leeuewburgh, M. A., 237, 281(23)
Lefeber, A. W. M., 161, 197(95)
Lefevre, J.-F., 213, 233(130)
Lefkidou, J., 117, 149(245)
Lehmann, J., 237, 281(25)
Leigh, D. A., 15, 58(45)
Leising, M., 88, 146(114)
Leismann, H., 87, 146(103)
Leisung, M., 69, 70, 72, 79, 80, 84, 86, 97, 130, 131, 133, 135, 137, 138, 145(62)
Leluc, B., 198(134)
Lemieux, R. U., 11, 17, 19, 20, 23, 57(16), 58(59, 67, 68, 71, 74), 211, 232(116), 236, 240, 280(9a, 9d)
Le Minor, L., 180, 198(132, 134)
Lennarz, W. J., 201, 203, 228, 229(1)
Lenz, R., 143, 150(298)
Leonidas, D. D., 237, 281(19)
Leontein, K., 187, 199(167)
LePage, T. J., 69, 82, 105, 145(54)
Lerman, Y., 154, 194(9)
Lerner, L., 206, 207, 208, 211, 231(75), 232(86, 95, 114)
Lesage, M., 90, 91, 147(146, 149, 150)
Leslie, R., 16, 58(53)
Lespinasse, A.-D., 117, 149(245)
Leteux, C., 54, 62(232)
Levin, E. M., 203, 230(34)
Levine, M. M., 154, 155, 156, 157, 194(10), 195(29, 30, 35)
Levine, S., 19, 20, 23, 58(67, 71)
Levy, D. E., 67, 115, 116, 144(37)
Levy, W., 235, 237, 280(2c)
Lewandowska, K., 202, 229(16)
Lewis, M. D., 235, 280(6)

Ley, S. V., 16, 58(53)
Li, H., 236, 265, 281(16c, 16d)
Li, P., 32, 60(145)
Li, Y.-L., 18, 58(62)
Liang, R., 213, 214, 233(133)
Lichtenthaler, F. W., 97, 116, 148(186), 149(242)
Lidholt, K., 203, 230(42–45)
Light, J., 90, 147(139)
Liguori, L., 121, 149(253)
Lim, J. J., 107, 129, 148(213)
Lim, L. B. L., 109, 118, 129, 148(218, 219), 149(246, 247)
Lim, S., 236, 258, 260, 280(11)
Lin, C. C., 238, 282(38)
Lin, T.-S., 140, 150(285)
Lind, T., 203, 230(43)
Lindahl, U., 201, 203, 204, 205, 208, 218, 229(2, 9), 230(26, 29, 42–47, 51, 53, 55), 231(58, 60, 65)
Lindberg, A. A., 156, 157, 172, 178, 180, 181, 182, 183, 188, 195(40, 42, 43), 196(68), 197(123), 198(141, 143, 146–150, 152–154)
Lindberg, B., 153, 169, 194(1, 3), 197(116), 207, 224, 232(87)
Lindh, I., 160, 161, 196(89)
Lindhardt, R. J., 275, 284(109)
Lindner, B., 156, 157, 195(38)
Lindner, H. J., 86, 88, 145(85), 146(114)
Lindon, J. C., 154, 194(7)
Lindsay Smith, J. R., 85, 145(80)
Linhardt, R. J., 116, 149(239), 203, 208, 209, 216, 219, 224, 230(41, 49), 232(90, 104), 234(155, 156), 235, 237, 280(2a)
Linker, T., 121, 149(255)
Lippens, g., 244, 283(58b)
Lipshutz, B. H., 47, 62(207)
Lipták, A., 157, 196(71)
Lis, H., 235, 280(1c)
Lismont, M. J., 154, 194(4)
Liu, H. T., 111, 149(225)
Liu, J., 209, 219, 224, 232(104)
Liu, L., 36, 61(159)
Liu, M.-C., 140, 150(285)
Liu, Z., 42, 61(190, 191)
Lizzio, E. F., 165, 197(104)
Ljunggren, Å., 156, 181, 195(40)
Löbau, S., 191, 193, 199(174)
Lobsanov, Y. D., 243, 258, 282(48c)
Loeffler, A., 278, 284(114)

Loganathan, D., 224, 234(156)
Lohse, D. L., 224, 234(155)
Longchambon, F., 246, 272, 283(68)
Longmore, R. W., 92, 93, 130, 135, 138, 147(168), 150(270)
Lönngren, J., 180, 181, 182, 187, 198(135, 150, 151), 199(167)
López, J. C., 15, 26, 58(42), 59(109), 89, 102, 103, 146(125), 148(207)
Lopez, R. M., 89, 99, 146(128)
Lopez-Herrera, F. J., 250, 283(76)
Lorenzo Acosta, Y., 161, 166, 197(99)
Loris, R., 243, 258, 282(48b)
Lormeau, J.-C., 204, 231(59), 238, 282(35a, 35b)
Lowenthal, J. P., 155, 195(23)
Lucarini, M., 69, 84, 85, 145(61)
Lüderitz, O., 179, 180, 198(131–134)
Lugemwa, F. N., 203, 230(42)
Lukacova, M., 191, 193, 199(174)
Lung, F. T., 213, 214, 233(135)
Luzzio, F. A., 40, 61(177)
Lycknert, K., 246, 283(66a)
Lyon, M., 203, 230(26)

M

Maccarana, M., 204, 231(58)
Macciantelli, D., 86, 145(86)
Macher, I., 238, 281(31)
Macig, T., 203, 230(48)
Mackenzie, G., 117, 149(245)
Mackie, D. M., 224, 234(158)
MacLoad, C. M., 154, 194(13, 14)
Madaj, J., 102, 142, 148(198)
Madara, M. L., 239, 282(42)
Madsen, R., 15, 49, 57(41), 62(211)
Magnus, P., 31, 60(137, 138)
Magnusson, G., 116, 149(233)
Magnusson, U., 265, 284(100)
Mahmood, K., 54, 62(231)
Mahmoudian, M., 206, 231(81)
Maitra, U., 90, 147(137)
Majer, Z., 116, 129, 149(243)
Mäkelä, O., 156, 195(50)
Malatesta, V., 68, 129, 144(52)
Malik, S., 54, 62(231)
Maliniak, A., 246, 283(66a)
Mallet, J.-M., 11, 32, 57(17), 57(18), 73, 81, 142, 145(63), 150(295), 236, 238, 281(14c), 282(34b, 35a, 35b, 36)
Malmstrom, A., 205, 231(65)
Mamat, U., 156, 157, 191, 193, 195(39), 199(174)
Mammen, M., 237, 281(26a, 26b)
Manhas, M. S., 25, 59(107)
Mann, D. M., 202, 228, 229(15), 234(166)
Manuel-Pastor, M., 256, 257, 283(84)
Manzi, A., 207, 224, 231(84)
Mao, D. T., 236, 265, 281(16b)
Marcantoni, E., 47, 62(210)
Marchiandi, M., 36, 60(156)
Marcum, J. A., 203, 230(51)
Marcum, Z. M., 203, 230(32)
Marino-Albernas, J. R., 166, 197(111)
Marioni, F., 23, 59(89, 90)
Marquez, V. E., 236, 265, 281(16b)
Marquis, C., 264, 284(97)
Marra, A., 11, 32, 57(17), 140, 142, 150(286)
Marsden, S. P., 25, 59(99)
Marsili, A., 23, 59(89)
Martin, J. C., 36, 60(157), 61(158)
Martin, O. R., 122, 149(256), 237, 266, 272, 281(21), 284(102)
Martin, V. S., 52, 62(221)
Martinez, F. N., 139, 150(283)
Martinho-Simões, J. A., 90, 91, 147(146, 153)
Martín-Lomas, M., 236, 256, 257, 258, 259, 260, 280(12), 283(84, 85), 284(90–94), 284(91)
Martín-Pastor, M., 236, 243, 244, 256, 257, 258, 259, 280(12), 282(47, 49b, 54c), 283(59c, 59d), 284(88)
Martirosian, G., 154, 194(8)
Massaque, J., 203, 230(42)
Massiot, G., 247, 283(73)
Masuda, M., 217, 218, 224, 233(147)
Masuda, S., 92, 147(161)
Matsumoto, K., 127, 128, 130, 139, 150(266)
Matsumoto, M., 102, 148(205)
Matsuura, H., 213, 233(125)
Matter, H., 213, 233(128)
Matthew, D., 141, 150(290)
Matulic-Adamic, J., 98, 99, 148(189)
Mayer, H., 180, 198(142)
Mayer, S., 92, 147(164)
Mayo, K. H., 208, 216, 232(90)
Mazac, C. J., 45, 62(205)
Mazéas, D., 236, 281(14a)
Mazeau, K., 243, 282(54b)

McCarthy, J. R., Jr., 40, 61(180)
McCloskey, C., 10, 57(8)
McClure, K. F., 25, 59(103)
McCombie, S. W., 66, 93, 143(16)
McDevitt, R. E., 14, 57(37)
Mcdonald, F. E., 24, 25, 59(95, 96, 99)
McElroy, E. B., 43, 61(193)
McKelvey, R. D., 68, 82, 129, 144(49–53)
McLean, C. M., 205, 224, 231(69)
McMills, M. C., 52, 62(222)
McNeil, M., 18, 58(64)
Medakovic, D., 106, 148(211)
Meffe, F., 155, 195(22)
Meggers, E., 143, 151(300)
Meguro, H., 106, 148(212)
Mehlman, T., 203, 230(48)
Meier, B. H., 210, 232(108)
Meixner, J., 94, 147(170)
Melcher, L. M., 53, 62(223)
Meldal, M., 237, 281(18a)
Mellema, J. R., 161, 197(95)
Menes, M. E., 40, 61(177)
Mentech, J., 264, 284(99)
Mer, G., 213, 233(130)
Merchant, Z. M., 224, 234(155)
Mereyala, H. B., 54, 62(228)
Merkus, D., 165, 197(104)
Merritt, J. R., 15, 57(39)
Metras, Γ., 67, 68, 112, 144(23)
Meuwly, R., 119, 120, 149(252)
Meyer, B., 11, 57(22), 209, 232(101), 236, 240, 280(9d)
Meyer, S. D., 36, 61(161)
Michael, K., 279, 284(115)
Michon, F., 155, 157, 195(33, 53), 196(55)
Michon, V., 11, 57(18), 264, 284(99)
Migita, T., 100, 148(192)
Mikhailopulo, I. A., 141, 150(288)
Mikhailov, D., 208, 216, 232(90)
Miller, C. B., 158, 196(80)
Miller, N. D., 99, 148(190)
Miller, R., 107, 129, 148(213)
Miller, W. H., 253, 254, 255, 262, 283(81, 83)
Minakata, S., 88, 146(112)
Minisci, F., 67, 86, 134, 144(36), 145(88)
Mirsch, J., 211, 233(118)
Mitchell, D. R., 240, 242, 272, 282(45)
Mitchell, E. P., 237, 281(19)
Miyasaka, T., 127, 128, 130, 139, 150(266)
Mizuno, N., 217, 218, 233(149)

Mizuta, H., 91, 147(156)
Mo, X.-S., 143, 151(302)
Moffatt, J. G., 38, 61(169, 170)
Monafo, W. J., 202, 229(17)
Mondelo Rodriguez, A., 161, 166, 197(99)
Monneret, C., 87, 146(101)
Montero, E., 259, 284(93)
Montgomery, R. I., 203, 230(30)
Monti, D., 117, 149(244)
Mootoo, D. R., 13, 14, 55, 57(31, 32, 36, 37), 62(234, 235), 266, 268, 284(103)
Moraes, C. T., 209, 232(99)
Morelli, I., 23, 59(89, 90)
Morgan, A. R., 20, 58(74)
Moriarty, R. M., 30, 60(128)
Moris-Vargas, F., 238, 282(38)
Morris, H. R., 203, 205, 217, 218, 219, 224, 228, 230(24), 231(67), 233(146)
Mortell, K. H., 238, 281(32a)
Moscateelli, D., 203, 230(38)
Motherwell, W. B., 45, 62(201), 66, 92, 108, 122, 123, 124, 125, 126, 127, 130, 143(11), 147(165), 148(214), 149(257–262), 150(263)
Moutel, S., 92, 147(162)
Mueller, S. H., 203, 230(34)
Mugridge, A., 154, 194(7)
Muhn, R., 86, 145(85)
Müller, E., 87, 146(103)
Müllor, I., 119, 149(250)
Müller, S. N., 73, 81, 145(63)
Müller-Loennies, S., 157, 196(72)
Muntendam, H. J., 161, 162, 197(100)
Muralidharan, P., 36, 61(164)
Murase, T., 17, 58(60)
Murcko, M. A., 240, 282(44)
Murillo, M. T., 270, 272, 273, 284(106)
Mushick, J., 213, 214, 233(135)
Mussini, P., 86, 141, 145(87)
Myerson, J., 50, 55, 62(216, 217, 237)
Myrvold, S., 116, 149(235)

N

Nader, H. B., 208, 209, 232(96a, 99)
Nadji, S., 49, 62(212)
Nagahama, T., 17, 58(61)
Nagao, Y., 31, 60(135)
Nagasawa, M., 235, 280(8)

Nagasawa, S., 217, 218, 224, 233(146)
Naggi, A., 86, 134, 145(88), 209, 219, 224, 232(103b)
Nagorski, W., 20, 58(72)
Nagy, J. O., 237, 238, 281(27, 28)
Nair, V., 87, 146(97)
Nakagawa, T., 217, 218, 224, 233(146)
Nakamura, K., 17, 58(57, 58), 97, 127, 128, 130, 139, 148(186), 150(266)
Naso, R. B., 155, 157, 195(35), 196(67)
Nawroth, P., 203, 230(32)
Neeley, W. B., 157, 196(73)
Németh, I., 142, 150(293)
Ness, R. K., 11, 57(20), 87, 146(93)
Neu, H. C., 154, 194(15)
Neuhaus, D., 243, 244, 282(51)
Neuman, A., 246, 272, 283(68)
Neumann, K. W., 224, 234(154)
Neumann, N. P., 66, 89, 143(8)
Neumann, W. P., 90, 146(130, 132–135)
Neville, D. C. A., 243, 282(55c)
Newcomb, M., 69, 84, 85, 102, 139, 145(61), 148(202), 150(283)
Newman-Evans, D. D., 89, 146(121)
Nicolaou, K. C., 99, 148(190)
Nicotra, F., 121, 149(253, 254), 235, 237, 280(2d)
Niederhuber, J. E., 182, 198(151)
Nieduszynski, I. A., 202, 216, 229(18), 233(142)
Nieforth, K. A., 203, 230(41)
Nifant'ev, N. E., 256, 284(87)
Niklasson, G., 43, 61(192)
Nilssen, M., 224, 234(153)
Nilsson, M., 161, 197(102)
Nilsson, S., 161, 197(98)
Nishida, A., 25, 59(98)
Nishida, Y., 106, 148(212)
Nishiyama, H., 102, 148(205)
Nishiyama, K., 91, 147(156)
Nishizawa, M., 53, 62(225)
Nix, M., 88, 146(114)
Noel, M., 40, 61(179)
Noguchi, Y., 53, 62(225)
Noirot, A.-M., 55, 62(233)
Nora, P. L., 213, 214, 233(135)
Norberg, T., 15, 58(44), 161, 180, 197(98, 101, 102), 198(136), 207, 224, 232(87)
Norcross, M. A., 203, 230(28)
Nurminen, M., 182, 183, 198(153)
Nyholm, P. G., 213, 233(122)

O

Oae, S., 43, 61(195)
Oba, M., 91, 147(156)
Ochiai, M., 31, 60(135)
Oegama, T. R., Jr., 217, 218, 233(148)
Oertle, K., 45, 62(204)
Ogasawara, K., 55, 62(238)
Ogawa, T., 224, 234(154)
Ohi, Y., 216, 217, 224, 233(144)
Ohki, H., 17, 58(61)
Ohkita, Y., 209, 215, 216, 224, 232(102)
Ohrui, H., 106, 148(212)
Ohta, B. K., 22, 59(85)
Oikonomakos, N. G., 237, 281(19)
Okamura, W. H., 36, 61(162)
Okumura, Y., 217, 218, 233(149)
O'Leary, D. J., 247, 264, 265, 283(71), 284(98)
Oliveira, F. W., 209, 232(99)
Oliver, D. J., 247, 283(72)
O'Neill, R. A., 15, 57(41), 58(44), 67, 115, 144(39)
Ono, N., 118, 149(248)
Oomen, L. A., 157, 196(69)
Oravecz, T., 203, 230(28)
Orchard, M. G., 237, 281(19)
Oriyama, T., 24, 25, 59(96, 99)
Ortega, C., 40, 61(183)
Oschkinat, H., 213, 233(134)
Ossowski, P., 20, 21, 53, 58(76), 59(79), 62(224)
Othme, K., 9, 45, 57(1), 62(205)
Ottosson, H., 14, 57(38)
Outschoorn, I. M., 157, 158, 196(76)
Overhand, M., 189, 199(170)
Overkleeft, H. S., 237, 281(23)
Owen, C. E., 246, 283(67b)
Oyama, M., 206, 217, 218, 224, 231(70), 233(146)

P

Paguaga, E., 107, 129, 148(213)
Painter, R. B., 67, 144(34)
Pakulski, Z., 41, 61(187)
Palacios, J. C., 97, 148(183)
Pan, Y. T., 237, 281(22)
Pankiewicz, K. W., 236, 265, 281(16c)
Pannell, L., 184, 185, 187, 198(158), 199(166)
Pant, C., 89, 95, 146(120)
Panza, L., 117, 149(244), 237, 281(18a)

Paoletti, L. C., 155, 157, 195(33, 53)
Papageorgioui, A., 237, 281(19)
Paradiso, P. R., 157, 196(68)
Pardi, A., 210, 232(111)
Park, S.-U., 102, 148(202)
Partridge, J., 243, 282(55a)
Pastori, N., 36, 60(156), 86, 134, 145(88)
Patchornik, A., 25, 59(101)
Patel, B., 95, 148(179)
Patel, M., 203, 230(28)
Paterson, J. M., 24, 59(96)
Paterson, L., 235, 280(7)
Paton, J. C., 155, 195(20)
Patroni, J. J., 141, 150(290)
Pattenden, G., 102, 148(200, 201)
Paul, W. E., 158, 196(80)
Paulsen, H., 13, 57(34), 87, 95, 146(95), 148(180), 212, 213, 233(119, 120)
Paulsson, M., 201, 202, 229(12)
Pavliak, V., 157, 196(67)
Pearlman, D. A., 268, 284(104, 105)
Pedulli, G. F., 69, 84, 85, 145(61)
Peeters, C. C. A. M., 157, 165, 166, 167, 173, 189, 196(69), 197(105)
Pelizzoni, F., 86, 145(86)
Peltola, H., 155, 195(31)
Penglis, A. A. E., 9, 57(2)
Peptitou, M., 238, 282(35a)
Perchyonok, V. T., 88, 146(113)
Pereyre, M., 66, 90, 143(7), 147(136)
Pérez, S., 243, 246, 264, 272, 282(54b, 57), 283(68), 284(99)
Pérez-Alvarez, M. D., 141, 150(287)
Perlin, A. S., 10, 57(10), 224, 234(158)
Pernet, A. G., 235, 280(4)
Perrin, C. L., 22, 59(81, 83, 85)
Perty, B., 216, 233(141)
Pervin, A., 208, 216, 232(90)
Pestka, C. A., 202, 229(17)
Pete, J.-P., 87, 146(101)
Peters, J. A., 67, 144(30)
Peters, T., 243, 246, 258, 282(49a, 53), 283(65)
Peterseim, M., 90, 146(132, 133, 135)
Petersen, P. M., 112, 149(227)
Peterson, J. M., 25, 59(99)
Petillo, P. A., 211, 232(114)
Petitou, M., 204, 208, 210, 216, 231(59), 232(93), 233(140, 141), 238, 282(35b)
Petretta, M., 143, 151(300)
Peukert, S., 79, 84, 145(65), 150(284)

Phelps, C. F., 202, 229(18)
Picasso, S., 237, 264, 281(23), 284(97)
Pichel, J. C., 95, 148(179)
Pichichero, M. E., 157, 166, 196(54)
Pier, G. B., 154, 194(8)
Pinhal, M. A. S., 209, 232(99)
Pino-González, M. S., 250, 283(76)
Pinto, B. M., 156, 157, 166, 168, 195(39), 197(107–115), 243, 258, 282(49a)
Pirkle, W. H., 53, 62(227)
Pirola, A., 209, 219, 224, 232(103b)
Pittman, M., 196(87)
Planas-Ruiz, F., 250, 283(76)
Plattner, P. A., 20, 58(73)
Plaza López-Espinosa, M. T., 141, 150(291)
Plotkin, S., 156, 195(47)
Plumb, J. B., 36, 61(160)
Poggozelski, W. K., 143, 151(304)
Polat, T., 275, 284(109)
Pollex-Krüger, A., 212, 213, 233(119)
Pompliano, D. L., 116, 149(235)
Pon, R. A., 157, 196(65)
Ponpipom, M. M., 44, 61(197, 198)
Pontén, F., 116, 149(233)
Poole, A. R., 203, 229(20)
Poolman, J. T., 152, 157, 161, 165, 166, 167, 173, 189, 191, 196(69, 90), 197(92, 95, 105, 106), 199(169, 171, 173)
Poopeiko, N. E., 141, 150(288)
Poot, L., 152, 161, 197(92)
Poppe, L., 244, 283(59b)
Porcelli, S., 157, 166, 196(54)
Porcianatto, M. A., 209, 232(99)
Portella, C., 87, 146(101)
Porter, N. A., 66, 143(14, 15)
Posner, G. H., 53, 62(226)
Postema, M. H. D., 67, 115, 116, 144(38), 235, 237, 280(2b)
Pourcel, M., 90, 147(136)
Poveda, A., 243, 244, 245, 258, 275, 282(49b, 50, 54b), 283(58c, 59c, 59d, 63), 284(109)
Powers, J. P., 69, 82, 105, 145(54, 55)
Pozsgay, V., 156, 184, 185, 187, 195(47), 198(156–164), 199(166)
Prabhakaran, M., 206, 207, 208, 212, 224, 231(79)
Praly, J.-P., 11, 57(16), 67, 68, 69, 75, 76, 77, 78, 79, 82, 83, 86, 88, 103, 105, 106, 112, 116, 129, 133, 144(23, 43), 145(58, 71), 146(115), 149(229, 243), 150(275)

Prandi, J., 92, 147(162–164)
Prasad, N., 25, 26, 59(106, 112)
Prestegard, J. H., 210, 213, 232(110), 233(121), 245, 246, 283(62a, 62b, 66c)
Prevasoli, P., 208, 232(96a)
Prévost, C., 19, 58(66)
Pria, S., 36, 61(164)
Pricota, T. I., 141, 150(288)
Priebe, W., 22, 59(86–88)
Provasoli, M., 208, 210, 232(93)
Provasolli, A., 216, 233(140, 141)
Pumilla, P., 208, 232(96a)
Punzi, J. S., 236, 265, 281(16c)
Pyle, R. E., 10, 57(8)

Q

Quaranta, L., 89, 146(123)
Que, J. O., 156, 195(49)
Quiclet-Sire, B., 134, 150(276, 277)
Quintard, J. P., 66, 143(7)
Quintero, L., 66, 143(6)
Quintero-Cortes, L., 92, 93, 138, 147(168)
Qureshi, N., 206, 231(77)

R

Rademacher, T. W., 206, 231(81)
Raffi, J., 67, 144(32)
Ragazzi, M., 208, 216, 232(96a), 233(140, 141, 143)
Rahm, A., 66, 143(7)
Rai, R., 90, 147(138)
Raimondi, L., 243, 282(54b)
Ramaiah, M., 66, 143(10)
Ramasamy, K. S., 47, 48, 62(209)
Ramos, L., 134, 150(277)
Rance, M., 210, 232(111)
Randolph, J. T., 25, 59(103)
Rao, H. S. P., 36, 61(164)
Rao, N., 204, 231(57)
Raphoz, C., 116, 149(229)
Rapport, C. A., 203, 230(31)
Rasmussen, J. R., 89, 146(121)
Rasmussen, K. J., 243, 282(54b)
Ratcliffe, A. J., 11, 32, 57(19)
Rauter, A. P., 42, 61(188), 61(189)
Ravishankar, R., 236, 258, 260, 280(11)
Rees, D. A., 208, 232(89), 232(94), 232(96b)

Rees, D. C., 203, 230(49)
Reeves, R. E., 171, 197(121)
Regitz, M., 66, 67, 68, 89, 94, 96, 143(13)
Reichlin, D., 49, 62(212)
Reid, B. R., 213, 233(129)
Reilly, P. J., 243, 266, 282(54a)
Reimer, K. B., 166, 168, 197(107, 109, 110, 115)
Reimer, L. M., 116, 149(235)
Rekai, D. E., 238, 282(36)
Rekaï, E. D., 73, 81, 145(63)
Remy, G., 67, 68, 144(22)
Renaud, P., 89, 146(123)
Renoux, R., 55, 62(236)
Reynolds, D. E., 10, 57(9)
Rhaese, H. J., 27, 60(120)
Rice, K. G., 224, 234(155)
Richards, W. G., 206, 231(81)
Richardson, A. C., 24, 40, 59(93)
Richardson, D. P., 50, 62(216)
Riche, C., 236, 281(14a)
Richert, C., 43, 61(194)
Richter, A. W., 157, 159, 161, 196(78)
Richter, W., 156, 195(51)
Richtmyer, N. K., 87, 146(91)
Rider, C. C., 203, 230(52)
Ries, M., 32, 60(139, 140)
Rietschel, E. T., 156, 157, 195(38, 39), 196(72)
Rifkin, D. B., 203, 230(38)
Rijkers, G. T., 165, 197(106)
Riley, G. P., 203, 230(39)
Rini, J. M., 243, 258, 282(48c)
Rio, S., 209, 224, 232(107)
Rio-Anneheim, S., 207, 210, 214, 224, 231(83)
Rist, G., 143, 150(299)
Ritchie, R. G. S., 26, 27, 29, 59(111, 119, 125)
Ritchie, T. J., 110, 129, 148(220), 149(221, 222)
Riva, S., 117, 149(244)
Rivera, A., 244, 259, 260, 283(58c), 284(94)
Roantree, R. J., 156, 195(36, 37)
Robbins, J. B., 154, 155, 156, 157, 161, 184, 185, 187, 194(6, 9, 10), 195(19, 27, 28, 32, 34, 45–47), 198(156, 162, 166)
Robbins, J. D., 155, 195(34)
Roberge, J. Y., 23, 25, 59(91, 104)
Roberts, B. P., 90, 94, 95, 100, 123, 124, 125, 126, 127, 130, 147(154, 171–175), 149(261, 262)
Roberts, C., 49, 62(211)
Roberts, D. H., 102, 148(203)
Roberts, J. A., 188, 199(168)

Roberts, R., 203, 230(40)
Robins, M. J., 36, 40, 61(163, 180), 141, 150(289)
Robins, R. K., 40, 61(180), 237, 281(24b)
Robinson, A., 157, 196(69)
Robinson, J., 201, 229(4)
Robles-Díaz, R., 97, 141, 148(184), 150(291)
Rodemann, J., 214, 215, 233(137)
Rodén, L., 201, 203, 204, 205, 206, 207, 208, 210, 212, 224, 228, 229(1, 2), 230(53), 231(62, 63, 79, 80), 234(159, 160)
Rodriguez, G., 203, 230(28)
Rodríguez Melgarejo, C., 141, 150(291)
Roe, B. A., 116, 149(232)
Rohrer, D. C., 246, 283(69)
Roland, S., 123, 124, 125, 126, 127, 130, 149(262)
Roller, P. P., 213, 214, 233(135)
Ronald, R. C., 45, 48, 62(206)
Rondinini, S., 86, 141, 145(87)
Roques, N., 264, 284(99)
Rose, L., 31, 32, 60(136, 140)
Rose, M. C., 213, 233(124)
Rosen, S. D., 238, 281(29)
Rosenberg, L. C., 202, 229(13, 16)
Rosenberg, L. T., 156, 181, 195(40)
Rosenberg, R. D., 203, 230(32, 51)
Ross, B. C., 122, 123, 126, 149(259)
Ross, W. S., 268, 284(105)
Rostand, K. S., 203, 230(27)
Roth, R. C., 87, 146(100)
Rotteveel, F. T. M., 178, 179, 198(130)
Rouzaud, D., 246, 272, 283(68)
Rozalski, A., 191, 193, 199(174, 180)
Rubinstenn, G., 73, 81, 145(63), 236, 257, 278, 281(14c), 284(89, 113)
Rückert, B., 69, 72, 74, 75, 79, 82, 86, 132, 145(57, 85)
Rüegge, D., 68, 69, 70, 105, 144(47)
Ruggeri, R. B., 25, 59(103)
Rundlof, T., 246, 283(66a)
Runsink, J., 87, 146(103)
Ruoslahti, E., 201, 202, 203, 228, 229(7, 15, 22), 234(161, 166)
Russo, G., 117, 149(244)
Rutherford, T. J., 243, 282(55a–c)
Rychnovsky, S. D., 50, 62(218), 69, 82, 105, 145(54, 55)
Ryckman, D. M., 255, 262, 283(83)
Ryckman, R. V., 253, 254, 261, 283(81, 82)
Rydon, H. N., 38, 61(165–167)
Ryu, I., 88, 146(112)
Rzadek, P., 143, 150(299)

S

Saarinen, L., 165, 166, 167, 173, 189, 197(105)
Sada, S., 17, 58(57)
Sadoff, J., 154, 194(9)
Sage, H., 213, 233(124)
Saha, U. K., 214, 215, 233(137)
Saito, T., 204, 231(56)
Sakai, T. T., 207, 210, 214, 224, 231(83)
Sakkurai, Y., 235, 280(8)
Saksela, O., 203, 230(38)
Salimath, P. V., 207, 224, 231(84)
Salley, J. M., 90, 146(131)
Salmivirta, M., 203, 230(44)
Samano, M. C., 141, 150(289)
Samano, V., 36, 61(163)
Samuelsson, B., 39, 40, 41, 42, 43, 61(171, 173, 181–184, 186, 190–192, 196), 160, 161, 196(88, 89)
Sandoval-Ramirez, J., 92, 93, 138, 147(168)
Sandow, J., 279, 284(115)
Sanford, G., 56, 63(245)
Sänger, W., 27, 60(120)
Sankey, S. S., 155, 195(22)
Sano, H., 100, 148(192)
Santamaria, M., 244, 283(58c)
Santarromana, M., 238, 282(39)
Santoyo-González, F., 97, 141, 148(184), 150(287)
Sarkar, S. K., 206, 231(75)
Sarko, A., 246, 283(69)
Sarma, K. D., 90, 147(137)
Sato, M., 99, 148(190)
Satoh, Y., 17, 58(57)
Saxena, M., 156, 195(48)
Scaiano, J. C., 89, 90, 113, 146(118)
Scartoni, V., 23, 59(90)
Schaefer, M. E., 237, 238, 281(27, 28)
Schäfer, A., 95, 148(180)
Scharf, H.-D., 87, 146(102, 103)
Scheit, K. H., 27, 60(120)
Scherman, D., 238, 282(39)
Schiesser, C. H., 88, 146(113)
Schilling, C. H., 9, 57(7)
Schindlermaier, K., 238, 281(31)

Schlecht, S., 178, 198(129)
Schmid, U., 13, 57(28)
Schmid, W., 116, 149(236)
Schmidt, R. R., 13, 57(33, 35), 214, 215, 233(137), 236, 237, 256, 257, 258, 259, 260, 280(12), 281(20, 24a), 283(84, 85), 284(90, 92, 93)
Schneerson, R., 154, 155, 156, 157, 161, 184, 185, 187, 194(6, 9, 10), 195(19, 27, 28, 32, 34, 45–47), 198(156, 162), 199(166)
Schneider, G., 27, 60(120)
Schnell, D., 87, 146(95)
Scholz, D., 238, 281(31)
Schreiber, A. B., 203, 230(48)
Schreiber, M., 95, 148(180)
Schreiber, S. L., 36, 61(161)
Schuerch, C., 10, 25, 57(15), 59(102), 159, 196(83–85)
Schulte-Frohlinde, D., 67, 144(34)
Schultz, G., 191, 199(175–177)
Schummer, D., 90, 147(151)
Schutze, E., 238, 281(31)
Schwartz, J., 100, 101, 106, 142, 148(193–195), 150(294)
Schwartz, N. B., 205, 228, 231(63, 64), 234(159, 161, 162)
Schwarz, V., 19, 58(65)
Schwenter, J., 20, 58(75, 77)
Schwidetzky, S., 97, 148(186)
Schwitter, U., 143, 151(300)
Seconi, G., 90, 91, 147(153)
Seepersaud, M., 55, 62(234)
Seepersaud, S., 55, 62(234)
Segiura, M., 209, 224, 232(98)
Sello, G., 86, 145(86)
Senderowitz, H., 243, 282(54b)
Senderson, P. N., 216, 233(142)
Seppälä, I., 156, 195(50)
Serianni, A. S., 208, 232(92)
Seydel, U., 156, 157, 195(39), 196(72)
Shah, P. M., 54, 62(231)
Shah, R. N., 243, 282(56)
Shalaby, M. A., 67, 144(31)
Sharon, N., 235, 280(1c)
Sharpless, K. B., 52, 62(221)
Sheelan, J. K., 202, 229(18)
Sheih, M. T., 203, 230(30)
Shekhani, M. S., 54, 62(231)
Sheldric, B., 211, 232(113)
Shen, G.-J., 111, 149(224)

Shen, W., 111, 149(225)
Shi, S. P., 165, 197(103)
Shibaev, V. N., 153, 194(2)
Shibata, Y., 209, 215, 216, 224, 232(102)
Shiloach, J., 154, 155, 194(6, 9), 195(32)
Shimizu, H., 246, 283(66d)
Shing, 203, 230(50)
Shkrob, I. A., 85, 145(81)
Shogren, R., 213, 233(127)
Shull, B. K., 25, 59(108)
Sidebotham, R. L., 157, 196(74)
Siebal, G., 268, 284(105)
Siegel, M., 154, 194(12)
Sierra Gonzalez, G., 161, 166, 197(99)
Silbert, J. E., 203, 230(23, 35)
Silverberg, I., 215, 219, 233(138, 152)
Sims, M. J., 236, 237, 265, 281(16a)
Sinaÿ, P., 11, 15, 32, 57(17, 18), 58(46), 73, 81, 122, 140, 142, 145(63, 70), 150(286, 295), 208, 210, 216, 232(93), 233(140, 141), 235, 236, 237, 238, 246, 257, 272, 278, 280(2e), 281(14c), 282(34a, 34b, 35a, 35b, 36), 283(68), 284(89, 113)
Singer, M. S., 238, 281(29)
Sinnott, M. L., 22, 59(83), 106, 148(210)
Sinnwell, V., 212, 213, 233(119)
Sinou, D., 247, 283(73)
Skelton, B. W., 141, 150(290)
Skrydstrup, T., 236, 273, 281(14a, 14b), 284(108)
Slebocka-Tilk, H., 20, 58(72)
Slettengren, K., 187, 199(167)
Sliedregt, L. A. J. M., 16, 58(52)
Slinger, C. J., 89, 146(121)
Smith, D. H., 155, 195(28)
Smith, M. A., 155, 195(22)
Snider, B. B., 25, 47, 59(105)
Snippe, H., 165, 172, 174, 176, 178, 179, 189, 191, 197(104, 122, 125), 198(130), 199(169)
Snyder, J. R., 208, 232(92)
Söderman, P., 43, 61(196)
Sofia, M. J., 102, 148(206)
Sokolowski, T., 246, 283(65)
Solis, D., 258, 259, 260, 284(91, 94)
Somer, A., 203, 230(38)
Sommarin, Y., 201, 229(6)
Sommermann, T., 121, 149(255)
Somsák, L., 66, 68, 69, 75, 76, 77, 78, 79, 82, 83, 85, 86, 102, 105, 106, 116, 129, 142,

144(21), 145(58), 148(198), 149(240, 241, 243), 150(293)
Son, J. C., 237, 281(19)
Sood, R. K., 157, 196(55, 64, 67)
Sowa, C. E., 21, 58(78)
Sparrowhawk, M. E., 56, 63(245)
Spear, P. G., 203, 230(30)
Spencer, R. P., 101, 142, 148(195), 150(294)
Spevak, W., 238, 281(28)
Spillmann, D., 203, 204, 218, 230(45), 231(60)
Spillmann, J. R., 203, 230(29)
Spiro, R. C., 207, 224, 231(84)
Spooncer, E., 203, 230(40)
Springer, R., 67, 144(36)
Sprinzl, M., 246, 283(67a)
Srinivasan, N. S., 56, 63(243)
Stacey, M., 28, 60(122)
Stack, R. J., 204, 231(57)
Staněk, J., 19, 58(65)
Stark, W., 246, 283(67b)
Staub, A. M., 180, 198(132–134)
Stauber, G. B., 204, 231(57)
Stauch, T., 18, 58(63)
Stein, E. C., 157, 166, 196(54)
Stein, K. E., 158, 196(80)
Stellner, K., 180, 198(134, 142)
Stenzel, W., 87, 146(95)
Stern, D., 203, 230(32)
Stick, R. V., 17, 58(59), 92, 141, 147(160), 150(290)
Stirm, S., 176, 178, 179, 180, 198(128, 131, 132)
Stoddart, J. F., 208, 232(91)
Stoffyn, P. J., 11, 57(21)
Stone, A. L., 155, 195(34)
Storer, R., 95, 148(179)
Stork, G., 102, 148(204, 206)
Strickland, L. C., 247, 283(72)
Strütz, A. E., 42, 61(189)
Suárez, E., 33, 34, 37, 41, 60(147–153)
Sugahara, K., 203, 205, 206, 207, 209, 215, 216, 217, 218, 219, 223, 224, 228, 230(24), 231(66–68, 70, 85), 232(98, 102, 103a), 233(144–150)
Sugai, T., 111, 149(224)
Sugawara, T., 68, 82, 129, 144(53)
Sugimoto, Y., 34, 60(151)
Sugumaran, G., 203, 230(23)
Sukkari, H. E., 55, 62(236)
Sun, L., 32, 60(145)
Sun, S., 89, 112, 113, 114, 146(124), 149(228)

Surolia, A., 236, 258, 260, 280(11)
Surzur, J.-M., 130, 150(267, 268)
Sustmann, R., 68, 69, 70, 71, 72, 73, 74, 75, 76, 77, 78, 79, 80, 81, 82, 83, 84, 85, 86, 97, 105, 118, 130, 131, 132, 133, 135, 137, 138, 144(47), 145(56–59, 62, 66, 67, 72), 149(249), 150(272)
Sutton, A., 155, 195(27)
Suzuki, S., 204, 231(56)
Svahn, C.-M., 224, 234(153)
Svenson, S. B., 156, 172, 178, 180, 181, 182, 183, 187, 188, 195(42, 43), 197(123), 198(147–150, 152–154), 199(167, 168)
Svensson, S., 156, 169, 180, 181, 195(40), 197(116), 198(135)
Svensson, S. C. T., 43, 61(192)
Svenungsson, B., 180, 181, 198(141, 143, 146)
Sverremark, E., 156, 195(52)
Symons, M. C. R., 69, 80, 145(59)
Szabó, M., 116, 129, 149(241)
Szarek, W. A., 9, 22, 26, 27, 28, 29, 45, 46, 48, 49, 57(6), 59(82, 111, 116–119), 60(123–125), 62(203, 214), 211, 232(116)
Szczerek, I., 26, 27, 59(111, 118)
Sznaidman, M., 22, 59(86)
Szu, S., 154, 155, 156, 157, 194(10), 195(34, 47)

T

Taguchi, T., 49, 62(213)
Tailler, D., 55, 62(233)
Takano, M., 17, 58(57)
Takano, S., 55, 62(238)
Takatani, M., 52, 62(221)
Takayama, K., 206, 231(77)
Takeda, T., 100, 148(192)
Takeyama, H., 17, 58(58)
Talaga, P., 244, 283(58b)
Talley, E. A., 10, 57(9)
Tamao, K., 103, 148(208)
Tamura, J., 224, 234(154)
Tamura, T., 204, 231(56)
Tanaka, H., 127, 128, 130, 139, 150(266)
Tang, C., 67, 115, 116, 144(37)
Tanner, D. D., 130, 150(269)
Taylor, D. N., 154, 194(9)
Taylor, P., 85, 145(80)
Tedebark, U., 237, 281(18a)
Teergarden, B. R., 251, 283(78)

Teichmann, B., 158, 183, 196(79)
Teissier, P., 130, 150(267, 268)
Tennigkeit, J., 27, 60(120)
Tersariol, I. L. S., 209, 232(99)
Thakur, R. S., 207, 231(82)
Thatcher, G. R. J., 22, 59(80, 82), 236, 240, 280(9b)
Theodorakis, E. A., 99, 148(190)
Therien, M., 56, 63(241)
Thiagarajan, V., 56, 63(242)
Thiem, J., 11, 20, 21, 57(22), 58(75–78), 59(79), 95, 148(180)
Thiéry, C., 67, 144(32)
Thøgersen, H., 236, 240, 280(9d)
Thomas, C. B., 84, 85, 145(73–77)
Thomson, D. S., 245, 283(62a)
Thornton, S. C., 203, 230(34)
Tilbrook, D. M. G., 92, 147(160)
Tilbrook, G., 141, 150(290)
Tiller, P. R., 187, 199(167)
Tillet, W. S., 154, 194(11)
Timpl, R., 202, 229(13)
Ting, P. C., 52, 62(220)
Tiwara, K. N., 28, 45, 46, 60(124)
Tixidre, A., 166, 197(109)
Tocher, D. A., 90, 95, 100, 147(154)
Togo, H., 35, 43, 60(154, 155), 61(195), 86, 115, 145(89)
Tohno-oka, T., 205, 219, 224, 228, 231(67)
Toida, T., 203, 208, 209, 216, 219, 224, 230(41), 232(90, 104)
Tollefsen, D. M., 202, 229(17)
Tolman, J. R., 210, 232(110)
Tomaru, Y., 51, 62(219)
Tomasik, P., 9, 57(7)
Tong, G. L., 40, 61(178)
Tonge, B. L., 38, 61(167)
Toone, E. J., 238, 281(32a)
Torregiana, E., 47, 62(210)
Torri, G., 36, 60(156), 86, 121, 134, 141, 145(86–88), 149(253), 208, 209, 216, 219, 224, 232(96a, 103b), 233(140, 141, 143)
Toupet, L., 82, 145(71)
Townsend, L. B., 40, 61(180)
Tozer, M. J., 122, 123, 124, 125, 126, 149(258–260)
Tramont, E. C., 155, 195(23)
Triolet, J., 67, 144(32)
Tripet, B., 165, 197(103)
Trofa, A. C., 154, 194(9)

Tronchet, J. M. J., 67, 144(27, 28), 239, 282(41)
Truhlar, D. G., 236, 240, 280(9h)
Trumtel, M., 140, 142, 150(286)
Trybala, E., 203, 230(29)
Tsuchiya, T., 9, 57(3)
Tsuda, H., 205, 207, 217, 218, 219, 224, 231(68, 85), 233(150)
Tsumta, O., 238, 281(33)
Tucker, L. C. N., 28, 60(121)
Tuinman, R. J., 25, 59(108)
Tullius, T. D., 143, 151(304)
Tvaros, I., 236, 240, 280(9e–g)
Tvaroska, I., 211, 233(118)
Tyrrell, D. J., 204, 231(57)
Tzianabos, A. O., 154, 194(8)

U

Uchiyama, T., 12, 57(25)
Udodong, U., 13, 14, 57(36, 38)
Ulmer, A. J., 157, 196(72)
Unger, F. M., 191, 199(175)
Urban, D., 236, 281(14b)
Uriel, C., 238, 282(39)
Usman, N., 98, 99, 148(189)
Usov, A. I., 38, 61(168)
Utille, J. P., 238, 282(36)

V

Vaghefi, M. M., 237, 281(24b)
Vaid, R. K., 30, 60(128)
Vaino, A. R., 22, 45, 48, 49, 59(82), 62(203, 214)
Valery, J.-M., 42, 61(188)
van Bekkum, H., 67, 144(30)
van Boeckel, C. A. A., 152, 161, 165, 166, 167, 173, 189, 196(90), 197(92, 95, 105, 106), 204, 231(61)
van Boom, J. H., 16, 32, 58(50, 52), 60(146), 152, 161, 162, 165, 166, 167, 173, 189, 191, 196(90), 197(92, 95, 100, 105, 106, 124), 199(169–172), 237, 281(23)
van Dam, J. E. G., 172, 178, 179, 197(122), 198(130)
van Delft, F. L., 189, 199(172)
van den Berg, R., 67, 144(30)
van der Elst, H., 161, 162, 197(100)
van der Klein, P. A. M., 189, 199(172)
van der Linden, M. P., 154, 194(4)

van der Marel, G. A., 16, 32, 58(52), 60(146), 152, 161, 162, 165, 166, 167, 173, 189, 191, 196(90), 197(92, 95, 100, 105), 199(169–172), 237, 281(23)
van der Rest, M., 202, 229(14)
van der Zee, R., 165, 197(104)
van Eden, W., 165, 197(104)
van Eick, J. B. P., 243, 282(54b)
van Gaans, J. A. M., 191, 199(173)
van Halbeek, H., 205, 206, 207, 217, 218, 224, 231(66, 71), 232(87), 244, 283(59b)
van Houte, A.-J., 172, 197(122)
Vann, W. F., 154, 194(6)
van Regenmortel, M. H. V., 155, 157, 195(21)
van Steijn, A. M. P., 174, 176, 197(125), 198(126, 127)
van Valen, T. R., 217, 218, 233(148)
Varela, O., 22, 59(87)
Varki, A., 235, 280(1g)
Varma, V., 166, 197(110, 111)
Vasella, A., 67, 84, 118, 119, 120, 144(45), 149(250–252)
Vedejs, E., 90, 147(141)
Veeneman, G. H., 16, 58(50), 173, 197(124)
Veldink, G. A., 207, 224, 232(87)
Venkataramanaiah, K. C., 54, 62(228)
Venkatasubramanian, N., 56, 63(242, 243)
Verez Bencomo, V., 161, 166, 197(99)
Vergoten, G., 243, 282(54b)
Verhart, C. G. J., 10, 44, 46, 57(12), 62(200)
Verheul, A. F. M., 165, 174, 176, 189, 191, 197(104, 125), 199(169, 173)
Verheyden, J. P. H., 38, 61(169, 170)
Verhoef, J., 165, 174, 176, 189, 191, 197(104, 125), 199(169)
Veyrières, A., 54, 62(232), 140, 142, 150(286)
Vian, A., 259, 284(93)
Viehe, H. G., 69, 81, 83, 84, 85, 145(67)
Vieler, R., 90, 146(130)
Vijayan, M., 236, 258, 260, 280(11)
Vincent, P., 67, 144(32)
Vismarra, E., 36, 60(156), 86, 121, 134, 141, 145(86–88), 149(253, 254)
Visser, A., 204, 231(61)
Viti, S. M., 52, 62(221)
Vlahov, I. R., 116, 149(239), 208, 216, 232(90), 235, 237, 280(2a)
Vlahova, P. I., 116, 149(239)
Vliegenthart, J. F. G., 174, 176, 178, 179, 197(125), 198(126, 127, 130), 205, 206, 207, 209, 213, 216, 217, 218, 219, 223, 224, 231(66, 68, 71, 85), 232(87, 97, 103a), 233(123, 144, 145, 147, 150), 278, 284(114)
Voelter, W., 116, 129, 149(243)
Vogel, K. G., 201, 202, 229(12)
Vogel, P., 237, 263, 264, 281(23), 284(95–97)
Voisin, D., 211, 232(116), 236, 240, 280(9a)
Vollerthun, R., 88, 92, 146(111)
von Sonntag, C., 67, 144(29, 34, 35)
Voter, W. A., 213, 233(124)
Vyplel, H., 238, 281(31)

W

Wagenaars, G. N., 161, 197(95)
Wagner, G., 213, 233(131)
Wahlstrom, J. L., 45, 48, 62(206)
Waki, Y., 86, 115, 145(89)
Waldemann, H., 12, 57(26, 27)
Waldmann, H., 13, 30, 57(28), 60(129)
Walker, F. J., 52, 62(221)
Wallace, S. S., 67, 144(34)
Wallin, N.-H., 156, 180, 181, 195(40), 198(139)
Walling, C., 88, 146(107)
Walter, A. W., 154, 194(12)
Walton, J. C., 89, 143, 146(119)
Wan, J. K. S., 85, 145(81)
Wang, D., 155, 157, 195(21)
Wang, H., 224, 234(156)
Wang, J., 238, 282(34a)
Wang, L., 238, 282(37)
Wang, P., 237, 281(27)
Wang, X., 250, 283(77)
Wang, Y., 154, 194(8), 251, 252, 253, 254, 255, 261, 262, 283(79–83)
Wang, Z. Y., 161, 197(93)
Ward, S. R., 85, 145(80)
Warner, J. L., 143, 151(303)
Watanabe, K. A., 236, 265, 281(16c)
Watson, K. A., 237, 281(19)
Watt, S. M., 203, 230(39)
Weatherman, R. V., 238, 281(32a, 32b)
Wei, A., 266, 277, 284(101, 111, 112)
Weibel, J.-M., 123, 124, 125, 126, 127, 130, 149(261, 262), 150(263)
Weidmann, H., 42, 61(189), 101, 142, 148(196)
Weimar, T., 243, 282(53)
Weinbaum, G., 156, 195(37)

Weinke, J. L., 203, 230(42)
Weintritt, V., 90, 146(130)
Weir, N. G., 95, 148(179)
Weiss, W. I., 235, 280(1b)
Weissfeld, L. A., 155, 195(22)
Welch, J. T., 9, 57(5)
Weller, C. T., 243, 282(55a)
Wessels, M., 155, 157, 195(33, 53)
Westley, J., 228, 234(162)
Westman, J., 224, 234(153)
Westphal, O., 158, 179, 180, 183, 184, 196(79), 198(131–134, 142, 155)
Wettach, R. H., 30, 60(132)
Wetter, H., 45, 62(204)
Wetterich, F., 78, 83, 84, 97, 135, 136, 137, 139, 140, 145(64)
Whelan, N. T., 228, 234(159)
White, A. H., 141, 150(290)
White, F. H., 55, 62(239)
Whitesides, G. M., 237, 281(26a, 26b)
Whitton, B. R., 87, 106, 146(104, 105)
Whitwood, A. C., 85, 145(80)
Wiberg, K. B., 240, 282(44)
Wiblin, C., 157, 196(69)
Widlanksi, T. S., 43, 61(193)
Widmalm, G., 187, 199(167), 246, 283(66a)
Wiertz, E. J. H., 191, 199(173)
Wieruzeski, J. M., 244, 283(58b)
Wietz-Schmidt, G., 238, 282(38)
Wight, A. W., 208, 232(96b)
Wightman, R. H., 236, 265, 281(15)
Wilcox, C. S., 239, 282(43)
Wiley, M. R., 116, 149(230)
Wilkinson, G., 45, 62(202)
Willers, J. M. N., 172, 178, 179, 197(122), 198(130)
Williams, D. R., 55, 62(239)
Williams, J. M., 154, 194(7)
Williams, S. J., 92, 147(160)
Williamson, M. P., 243, 244, 282(51)
Willis, C. R., 94, 95, 147(171, 173, 174)
Wilson, B. E., 237, 281(24b)
Wilson, R. L., 85, 145(79)
Wink, D. J., 139, 150(283)
Winkler, T., 12, 57(24)
Winstein, S., 19, 58(69)
Wisniewski, A., 102, 142, 148(198)
Witczak, Z. B., 203, 230(41)
Witczak, Z. J., 96, 97, 143, 148(182, 185, 296)
Withers, S. G., 238, 282(40)

Wittmann, V., 279, 284(115)
Witzel, T., 69, 81, 83, 84, 85, 132, 133, 145(67, 72), 150(272, 273)
Wolfe, J., 185, 187, 199(166)
Wong, C.-H., 111, 149(224), 238, 282(38)
Wong, C.-L., 90, 147(144)
Wong, D. H.-C., 56, 63(241)
Woodrow, G. C., 154, 155, 156, 157, 194(10), 195(29, 30, 35)
Woods, A., 203, 230(36, 37)
Woodward, S. S., 52, 62(221)
Wothers, P. D., 22, 59(84)
Woytek, A. J., 9, 57(1)
Wrangsell, G., 187, 199(167)
Wright, C. S., 243, 258, 282(48a)
Wright, P. E., 233(136)
Wright, T. C., Jr., 204, 231(59)
Wu, T., 248, 249, 250, 283(74)
Wu, T.-C., 246, 247, 249, 250, 271, 283(70, 75), 284(107)
Wu, Y., 240, 242, 272, 282(45)
Wu, Y.-D., 90, 147(144)
Wu, Y.-L., 18, 58(62)
Wu, Z., 14, 57(38)
WuDunn, D., 203, 230(30)
Wüthrich, K., 209, 232(105)
Wyatt, P., 15, 58(43)
Wyle, F. A., 155, 195(23)
Wyler, R., 119, 149(250)
Wyns, L., 243, 258, 282(48b)
Wyss, D. F., 213, 233(131)

X

Xiao, H., 55, 62(235)
Xie, F., 122, 149(256)
Xin, Y. C., 238, 282(34b)

Y

Yadomae, T., 205, 224, 231(69)
Yamada, H., 53, 62(225)
Yamada, R., 91, 147(156)
Yamada, S., 203, 205, 206, 207, 209, 217, 218, 219, 223, 224, 228, 230(24), 231(67, 68, 70, 85), 232(98, 103a), 233(146)
Yamago, S., 92, 147(161)
Yamaguchi, H., 31, 60(135)
Yamaguchi, Y., 202, 228, 229(15), 234(166)

Yamanoi, T., 17, 58(57), 58(58)
Yamashina, I., 216, 217, 218, 224, 233(144, 147)
Yamazaki, N., 111, 112, 114, 115, 149(226)
Yamshina, I., 205, 217, 218, 224, 231(66)
Yanagihara, K., 17, 58(57, 58)
Yanagishita, M., 203, 230(28)
Yang, D., 107, 129, 148(213)
Yang, J.-H., 140, 150(285)
Yang, W., 90, 147(140)
Yang, Y. P., 165, 197(103)
Yao, Q., 89, 92, 93, 95, 111, 112, 113, 130, 133, 135, 136, 137, 138, 139, 143, 146(127), 147(168, 176, 177), 150(271, 278, 279, 283), 151(301)
Yates, J. B., 116, 149(230)
Yeh, H., 184, 198(157)
Yeung, L. L., 16, 58(53)
Yishi, Y., 236, 258, 260, 280(11)
Yokoyama, M., 35, 60(154, 155), 86, 115, 145(89)
Yokoyama, Y., 34, 60(151)
Yonemitsu, O., 25, 59(98), 235, 280(8)
Yoshida, J.-I., 92, 147(161)
Yoshida, K., 204, 205, 207, 209, 215, 216, 218, 219, 223, 224, 231(56, 58, 85), 232(98, 102, 103a)
Yoshida, Z.-I., 51, 62(219)
Young, N. M., 243, 258, 282(46)
Yourassowky, E., 154, 194(4)
Yuan, H., 108, 109, 129, 148(217)
Yuasa, H., 238, 281(33)
Yuki, Y., 206, 217, 231(70)

Z

Zahradnik, J. M., 157, 166, 196(54)
Zähringer, U., 156, 157, 195(38), 196(72)
Zamboni, M., 127, 128, 130, 150(264)
Zamojski, A., 28, 41, 45, 46, 60(124), 61(187)
Zard, S. Z., 134, 150(276, 277)
Zardi, L., 202, 229(16)
Zbiral, E., 98, 116, 148(188), 149(236)
Zegelaar-Jaarsveld, K., 16, 32, 58(52), 60(146)
Zehavi, U., 25, 59(101)
Zeitz, H.-G., 67, 115, 144(40)
Zerial, A., 213, 233(134)
Zervas, L., 87, 145(90)
Zhang, B., 55, 62(235)
Zhang, H., 55, 62(234)
Zhang, L., 228, 234(163, 165)
Zhang, Y. M., 238, 282(34b)
Zhao, K., 32, 60(145)
Zhdanov, R. I., 67, 144(28)
Zheng, C. Y., 20, 58(72)
Zheng, D., 40, 41, 61(175)
Zheng, J., 87, 146(98)
Zhu, J.-L., 140, 150(285)
Zhu, Z., 270, 272, 273, 284(106)
Zigterman, J. W. J., 178, 179, 198(130)
Zollinger, W. D., 155, 195(30)
Zopf, D. A., 158, 196(80)
Zsély, M., 67, 144(27), 239, 282(41)
Zsiska, M., 209, 232(101)
Zwanenburg, B., 10, 44, 46, 57(12), 62(200)
Zych, K., 191, 193, 199(174, 180)

SUBJECT INDEX

A

Ab initio calculations, saccharide exo-anomeric effect, 240, 242
Acetal radicals, carbon-centered, structure, 68–69
Acetate protecting groups, addition, iodine effect, 49
Acetylated glycosyl bromides, in glycoamino acid synthesis, 18
2,1-Acyloxy group, rearrangements for 2-deoxy sugar synthesis, 131–135
AIBN, *see* Azobisisobutyronitrile
Alcohols, oxidation with NIS, 55–56
Alkenes
  nitryl iodide addition, 27
  from vicinal *trans*-diols, with GSR, 39–40
Allyl alcohol, transfer via nucleoside, 43–44
Anhydroalditols, from glycosyl derivatives
  fused and branched, 102–103
  properties, 93–95
  reduction, 87–92
  by single-electron transfer, 100–102
  stereoselectivity, 103–106
  synthesis, 95–100
  via radicals, 86–87
1,4-Anhydroalditols, 98–99
1,4-Anhydro-D-alditols, 98
1,5-Anhydroalditols, 97
1,4-Anhydro-1-deuterio-2,3:5,6-di-*O*-isopropylidene-D-mannitol, 106
1,6-Anhydro-3,4-di-*O*-acetyl-2-deoxy-2-iodo-β-D-glucopyranose, 54
1,5-Anhydrohexitols, 96–97
Anomeric nitro sugars, in *C*-glycosyl compound synthesis, 118–120

Anomeric oxidation, with organoselenium(IV) compounds, 33
Anomeric radicals
  diastereoselective hydrogen quenching, 107–111
  C-glycosyl compound synthesis
    addition to exo-glycals, 120–122, 126–127
    from anomeric nitro sugars, 118–120
    from glyculosonic acid derivatives, 118
    from glyculosyl halides and analogs, 116
    nucleosides, 127–128
    substituted anomeric radicals, 129–130
    from thio- and seleno-glyculosides, 117
  H-atom transfer, stereoselectivity, 103–106
  mannopyranoside inversion, 111–115
Antibacterial vaccines, from polysaccharides, 153–157
1-Azido-2-deoxy sugar, 29–30
Azobisisobutyronitrile, for anhydroalditols, 88

B

Bacterial polysaccharides, oligosaccharide–protein conjugates, 157–161
Barton decarboxylation method, for anomeric radicals, 108–110
Benzenesulfonamide, in glycal additions, 24–25
Benzyl 2-benzyloxycarbonylamino-2,3,4,6-tetradeoxy-6-iodo-α-D-*erythro*-hex-3-enopyranoside, 41–42
4,6-*O*-Benzylidene-2,3-*O*-isopropylidene-D-glucopyranos-1-yl radical, 80
4,6-*O*-Benzylidene-2-*O*-methyl-α-D-altropyranoside, 28–29

4,6-*O*-Benzylidene-3-*O*-methyl-α-
D-glucopyranoside, 28–29
β fragmentation, with organoselenium(IV)
compounds, 33
Biosynthesis, CPL oligosaccharides, 228–229
Bis-tetrahydrofuran acetogenin, precursor
synthesis, 55
[Bis(trifluoroacetoxy)iodo]benzene, 35–36
N Bromosuccinimide, in NPG activation, 13
3′-*O*-*tert*-Butyldimethylsilyl-5′-deoxy-5′-
iodothymidine, 43
tert-Butyl(2,3,4,6-tetra-*O*-acetyl-β-D-
glucopyranosyl)ketone, 69

C

Carbohydrate–protein linkage region
  oligosaccharides, biosynthesis, 228–229
  structural studies
    chondroitin sulfate, 216–219
    DS, 215–216
    heparan sulfate, 219
    heparin, 223–224
Carbohydrate radicals, ESR studies
  C-1 substituted pyranos-1-yl radicals, 82–83
  2-deoxyglucopyranos-1-yl radicals, 81–82
  free sugar-derived radicals, 84–85
  glycofuranos-1-yl radicals, 84
  hexo- and pento-pyranos-1-yl radicals, 69,
    80–81
  non-anomeric radicals, 83–84
Carbohydrate spin systems, 206–209
Cerium(III) chloride heptahydrate, in PMB ether
  cleavage, 47
Chlamydia, lipopolysaccharides, 191–193
Chondroitin sulfate, in CPL structural studies,
  216–219
Conformation
  C-1 substituted pyranos-1-yl radicals, 82–83
  2-deoxyglucopyranos-1-yl radicals, 81–82
  glycals, 21–22
  glycopeptide linkage, 211–213
  C-glycosyl compounds, 243–246
  C-glycosyl compounds in solid state
    β-*C*-gentiobiose, 246–247
    C-isomaltose, 247
    C-sucrose, 247
  C-glycosyl compounds in solution
    α-*C*-glucosyl derivatives, 247–250

    β-*C*-glucosyl derivatives, 247–250
    C-glycopeptides, 278–279
    O/C-lactose, 256–261
    with (1→3) linkages, 261–264
    with (1→4) linkages, 250–255
    with (1→6) linkages, 271–273
    and (1→1)-linked trehalose analogs,
      266–270
    C-sialyl compounds, 274–275
    simple derivatives, 247
    C-sucrose, 264–265
    trisaccharides, 275–278
  interglycosidic linkages, 210–211
  proteoglycan core-peptide backbone, 213–215
  saccharides, 239–243
CPL, *see* Carbohydrate–protein linkage region
Cyclic acetals, cleavage with iodine solution, 45
Cyclic thioacetals, removal with iodine
  solution, 45
Cyclizations
  electrophile-mediated reactions, 51
  glucopyranosides, 51–52
Cyclohexyl 3,4,6-tri-*O*-acetyl-2-deoxy-2-iodo-
  α-D-mannopyranoside, 20

D

DBU, *see* 1,8-Diazabicyclo[5.4.0]undec-7-ene
DDQ, in glycosyl iodide preparation, 13
Decarboxylation, Barton method, *see* Barton
  decarboxylation method
(7*S*,9*S*)-4-Demethoxy-7-*O*-(2,6-dideoxy-2-iodo-
  α-L-mannopyranosyl)adriamycinone,
  22–23
(7*S*,9*S*)-4-Demethoxy-7-*O*-(2,6-dideoxy-2-iodo-
  α-L-mannopyranosyl)daunomycinone,
  22–23
Deoxygenation, α-hydroxy ketone,
  GSR effects, 42
2-Deoxyglucopyranos-1-yl radicals,
  conformational effects, 81–82
3-Deoxyglycals, reaction with
  iododiazidobenzene, 31
2-Deoxy-α-*N*-glycopeptides, 21
2-Deoxyiodoglycosyl phosphoramidates, 29–30
2-Deoxy sugars
  by 2,1-acyloxy group rearrangements,
    131–135
  general preparation, 20

by 2,1-migrations of ester groups, 138–139
by 2,1-phosphatoxy group rearrangements, 135–138
Dermatan sulfate, in CPL structural studies, 215–216
Dess–Martin reagent
　advantages, 56
　synthesis and applications, 36
Dextran, related oligosaccharide–protein conjugates, 157–161
Diacetoxyiodobenzene
　in carbohydrate C-1–C-2 cleavage, 33
　in lactone formation, 34–35
Diastereoselectivity, substituted anomeric radicals, 129–130
1,8-Diazabicyclo[5.4.0]undec-7-ene, 89
DIB, *see* Diacetoxyiodobenzene
3,4-Di-*O-tert*-butyldimethylsilyl-6-*O*-tosyl-D-glucal, 30–31
Difluromethylene phosphonates, from phosphonyl radicals, 126–127
Dimethoxytrityl ethers, cleavage with iodine solutions, 45–47
Dimethyl(methylthio)sulfonium triflate, in glycosylation, 17–18
Dimethylphosphinothioate glycoside, as stable glycoside acceptor, 17
Diphenyliodonium triflate, in thioglycoside activation, 32
Disaccharides
　additions, iodine role, 44–45
　synthesis, 16–17, 20
3,6-Di-*O*-tributylstannyl-D-glucal, 54–55
DMTST, *see* Dimethyl(methylthio)sulfonium triflate
DS, *see* Dermatan sulfate
Dutton, Guy Gordon Studdy
　education, 1–2
　family life, 5–6
　fellowships and awards, 3–4
　Max Planck Institute stints, 3–4
　non-academic activities, 6–7
　polysaccharide structural studies, 2–3
　professional service, 4–5
　research programs, 6

E

Electron spin resonance
　C-1 substituted pyranos-1-yl radicals, 82–83
　2-deoxyglucopyranos-1-yl radicals, 81–82
　free sugar-derived radicals, 84–85
　glycofuranos-1-yl radicals, 84
　hexo- and pento-pyranos-1-yl radicals, 69, 80–81
　non-anomeric carbohydrate radicals, 83–84
Electrophiles
　mediated cyclizations, 51
　in thiol-catalyzed radical reductions, 93–95
Eliminations, radical-mediated, for glycal formation, 139–142
Epoxidation, Sharpless asymmetric epoxidation, 52–53
Escherichia coli
　adhesion to yeast cells, 238
　E. coli type O8, O-specific polysaccharides, 187–188
ESR, *see* Electron spin resonance
Ester groups, 2,1-migrations mechanism, 138–139
Exo-anomeric effect, in saccharide conformation, 239–243
Exo-glycals, radical addition, 120–122, 126–127

F

Ferrier reaction, in glycosidic bond formation, 25
FMO theory, *see* Frontier molecular orbital theory
Free sugars, derived radicals, ESR studies, 84–85
Frontier molecular orbital theory, for sugar-derived radicals, 85–86
Furanose, diastereoselective hydrogen quenching, 110–111

G

Garegg–Samuelsson reagent, 39–44
β-C-Gentiobiose, solid-state conformation, 246–247
Glucopyranosides, cyclization, 51–52
D-Glucopyranos-1-yl radicals, peracetylated, ESR studies, 80
α-C-Glucosyl derivatives, solution conformation, 247–250

β-C-Glucosyl derivatives, solution conformation, 247–250
Glycals
  activation, 19–26
  addition mechanisms, 22–23
  –iodine method, in oligosaccharide synthesis, 23–24
  by radical-mediated eliminations, 139–142
Glycoamino acids, 18
Glycofuranos-1-yl radicals, 84
Glycogen phosphorylase, C-glycosyl-based inhibitors, 237–238
Glycopeptide linkage, 211–213
C-Glycopeptides, solution conformation, 278–279
Glycosides
  as glycosyl donors, 15
  NMR studies, 255
  synthesis with NPGs, 13
O-Glycosides, and analogues
  diastereoselective hydrogen quenching, 107–111
  mannopyranoside inversion, 111–115
Glycosidic bonds, 20, 25
C-Glycosyl analogs
  with (1→3) linkages, 261–264
  with (1→4) linkages, 250–255
  NMR studies, 255
Glycosylation
  with DMTST, 17–18
  effect on proteoglycan backbone conformation, 213–215
  glycosyl iodide use, 10–13
  iodine use, 10
C-Glycosyl-based inhibitors, glycogen phosphorylase, 237–238
C-Glycosyl compounds
  biological applications, 237–239
  conformation in solid state
    β-C-gentiobiose, 246–247
    C-isomaltose, 247
    C-sucrose, 247
  conformation in solution
    α-C-glucosyl derivatives, 247–250
    β-C-glucosyl derivatives, 247–250
    C-glycopeptides, 278–279
    O/C-lactose, 256–261
    with (1→3) linkages, 261–264
    with (1→4) linkages, 250–255
    with (1→6) linkages, 271–273

  and (1→1)-linked trehalose analogs, 266–270
  C-sialyl compounds, 274–275
  simple derivatives, 247
  C-sucrose, 264–265
  trisaccharides, 275–278
  molecular modeling, 243–246
  NMR, 243–246
  nomenclature, 236–237
  per-O-acetylated conjugates, 238–239
  sterocontrolled synthesis
    from anomeric nitro sugars, 118–120
    from glyculosonic acid derivatives, 118
    from glyculosyl halides and analogs, 116
    radical addition to exo-glycals, 120–122, 126–127
    from thio- and seleno-glyculosides, 117
  structure, 237–239
Glycosyl derivatives, to anhydroalditols
  fused and branched anhydroalditols, 102–103
  properties, 93–95
  reduction, 87–92
  by single-electron transfer, 100–102
  stereoselectivity, 103–106
  synthesis, 95–100
  via radicals, 86–87
Glycosyl donors, thioglycosides as, 15
Glycosyl fluorides, 56
Glycosyl iodides
  conversion, 11–12
  in glycosylation, 10–13
  instability, 12
  preparation, 11
Glycosyl phosphates, 12–13
Glycosyltransferase donors, C-maltosyl fluoride analogs as, 238
Glyculosonic acid derivatives
  in C-glycosyl compound and analog synthesis, 118
  for stereocontrolled synthesis, 109
Glyculosyl halides, in C-glycosyl compound and analog synthesis, 116
GSR, see Garegg–Samuelsson reagent

H

Haemophilus influenzae type b, oligosaccharide–protein conjugates, 161–166

Heparan sulfate, in CPL structural
   studies, 219
Heparin, in CPL structural studies,
   223–224
Hexamethyldisilazane, in peracetylated sugar
   reaction, 13
Hexopyranos-1-yl radicals, 69, 80–81
Hydrogen
   silanes as source, 99–100
   transfer at anomeric radicals, 103–106
Hydrogen quenching, diastereoselective,
   anomeric radicals, 107–111
1,6-Hydrogen transfer, for α- to
   β-*O*-mannopyranoside inversion, 114
α-Hydroxy ketone, deoxygenation,
   GSR effects, 42
[Hydroxy(tosyloxy)iodo]benzene, *see* Koser
   reagent

I

IDCP, *see* Iodonium dicollidine perchlorate
Interglycosidic linkages, conformational
   analysis, 210–211
Iodination, with GSR, 39–40
Iodine
   addition to glycals, 19–20
   in glycosylation, 10
   in glycosyl fluoride synthesis, 56
   hypervalent species, applications, 30–37
   in lactone formation, 50
   in methyl thioglycoside activation,
      16–17
   molecular, in thioglycoside activation, 15
   in protecting group modification,
      44–49
   in thioglycoside activation, 18
Iodine azide, in 1-azido-2-deoxy sugar
   formation, 29–30
Iodine–glycal method, in oligosaccharide
   synthesis, 23–24
Iodine trifluoroacetate, in carbohydrate
   synthesis, 27–28
Iodocyclization, various reactions, 50–55
Iododiazidobenzene, reaction with
   3-deoxyglycals, 31
Iodoetherification
   in fused 6,6-ring synthesis, 52
   in linalyl oxide synthesis, 50–51

in sucrose total synthesis, 53
in THF acetogenin precursor synthesis, 55
Iodolactonization, *vs.* Sharpless asymmetric
   epoxidation, 52–53
Iodonium dicollidine perchlorate
   in 1,6-anhydro-3,4-di-*O*-acetyl-2-deoxy-2-
      iodo-β-D-glucopyranose formation,
      54
   in glucopyranoside cyclizations, 51–52
   in glycal activation, 20
   in thioglycoside inactivation, 16
N-Iodosuccinimide
   activation method, 24
   in alcohol oxidation, 55–56
   in α-D-mannoside formation, 20
   in 3,4,6-tri-*O*-acetyl-D-glucal reaction, 21
N-Iodosuccinimide–trimethylsilyl triflate, 15
Isocyanide method, in 1,4-anhydroalditol
   preparation, 98–99
C-Isomaltose, solid-state conformation, 247

K

Klebsiella type 2, protein–oligosaccharide
   conjugates, 176–178
Klebsiella type 11, protein–oligosaccharide
   conjugates, 178–179
Koser reagent, in glycal reactions, 30–31

L

Lactone formation
   with DIB/I$_2$, 34–35
   with iodine, 50
O/C-Lactose, solution conformation, 256–261
Lemieux's halide-ion glycosylation method, 18
cis-Linalyl oxide, 50–51
trans-Linalyl oxide, 50–51
Lipopolysaccharides, related oligosaccharides,
   protein conjugates
   Chlamydia, 191–193
   Escherichia coli type O8, 187–188
   meningococcal, core region, 189–191
   Salmonella, 179–180
   Salmonella illinois, 183–184
   Salmonella typhi, 180–183
   Salmonella typhimurium, 180–183
   Shigella dysenteriae type 1, 184–187

M

C-Maltosyl fluoride analogs, as glycosyltransferase donors, 238
α-O-Mannopyranoside, inversion, 111–115
β-O-Mannopyranoside, inversion, 111–115
C-Mannopyranosyl derivatives, effect on *Escherichia coli* adhesions, 238
α D Mannosides, 20
Max Planck Institute, stints of Guy Dutton, 3–4
Meningococcal oligosaccharides, core region, 189–191
Methyl 4-deoxy-2,6-di-O-pivaloyl-α-D-*erythro*-hexopyranosid-3-ulose, 42
Methyl 6-deoxy-6-iodo-α-D-glucopyranoside, from GSR treatments, 40
Methyl thioglycosides, activation with iodine, 16–17
2,1-Migrations, ester groups, mechanism, 138–139
Monosaccharides
  additions, iodine role, 44–45
  moieties, conformational analysis, 244
Mono-tetrahydrofuran acetogenin, precursor synthesis, 55

N

NIS, *see* N-Iodosuccinimide
Nitryl iodide
  addition to glycal, 26
  alkene addition, 27
  reaction with epoxide, 28
NPGs, *see* n-Pentenyl glycosides
Nuclear magnetic resonance
  glycosides and C-glycosyl analogs, 255
  C-glycosyl compounds, 243–246
  linkage-region oligosaccharides, 224, 227–228
Nucleophiles, in thiol-catalyzed radical reductions, 93–95
Nucleosides
  C-nucleoside-like compounds, 35–36
  radical routes, 127–128
  Rydon reagent applications, 38
  through allyl alcohol transfer, 43–44

O

Oligosaccharide–protein conjugates
  capsular polysaccharides
    Haemophilus influenzae type b, 161–166
    Klebsiella type 2, 176–178
    Klebsiella type 11, 178–179
    Streptococcus Group A, 166–168
    Streptococcus pneumoniae type 2, 168–171
    Streptococcus pneumoniae type 3, 171–173
    Streptococcus pneumoniae type 17F, 173–174
    Streptococcus pneumoniae type 23F, 174–176
  dextran-related conjugates, 157–161
  lipopolysaccharide-related
    Chlamydia, 191–193
    Escherichia coli type O8, 187–188
    meningococcal, core region, 189–191
    Salmonella, 179–180
    Salmonella illinois, 183–184
    Salmonella typhi, 180–183
    Salmonella typhimurium, 180–183
    Shigella dysenteriae type 1, 184–187
Oligosaccharides
  CPL, biosynthesis, 228–229
  linkage-region
    isolation, 205
    NMR spectral properties, 224, 227–228
  meningococcal, core region, 189–191
  solid-phase synthesis, 25
  synthesis
    glycosyl iodide use, 12
    by iodine–glycal method, 23–24
Organoselenium(IV) compounds, in anomeric oxidation, 33
Oxidation
  alcohols, with NIS, 55–56
  anomeric, with organoselenium(IV) compounds, 33
Oxygen–carbon β-bond effect, anhydroalditols, 92–93
β-Oxygen effect, *see* Oxygen–carbon β-bond effect

P

n-Pent-4-enoyl amide, 49
n-Pentenyl glycosides, 13–15
Pentopyranos-1-yl radicals, 69, 80–81

Peptides, core backbone of proteoglycan, 213–215
Peptide spin systems, 209–210
Peracetylated D-glucopyranos-1-yl radicals, 80
Peracetylated sugars
　reaction with iodine, 13
　in thioglycoside synthesis, 17
Phenyl 2,3,4,6-tetra-O-acetyl-1-seleno-β-D-glucopyranoside, 69
2,1-Phosphatoxy group, rearrangements for 2-deoxy sugar synthesis, 135–138
Phosphonyl radicals, for anomeric carbohydrate difluromethylene phosphonates, 126–127
Phosphorus reactions
　Garegg–Samuelsson reagent, 39–44
　Rydon reagents, 38
PMB ethers, cleavage, 47
Polysaccharides
　antibacterial vaccines based on, overview, 153–157
　bacterial, oligosaccharide–protein conjugates, 157–161
　capsular, oligosaccharide–protein conjugates
　　Haemophilus influenzae type b, 161–166
　　Klebsiella type 2, 176–178
　　Klebsiella type 11, 178–179
　　Streptococcus Group A, 166–168
　　Streptococcus pneumoniae type 2, 168–171
　　Streptococcus pneumoniae type 3, 171–173
　　Streptococcus pneumoniae type 17F, 173–174
　　Streptococcus pneumoniae type 23F, 174–176
　O-specific, oligosaccharide fragments
　　Escherichia coli type O8, 187–188
　　Salmonella, 179–180
　　Salmonella illinois, 183–184
　　Salmonella typhi, 180–183
　　Salmonella typhimurium, 180–183
　　Shigella dysenteriae type 1, 184–187
　structural studies by Guy Dutton, 2–3
Potassium fluoride, for anhydroalditols from glycosyl derivatives, 89
Protecting groups
　modification with iodine, 44–49
　in NPG activation, 13–15
Protein conjugates
　oligosaccharide components, dextran-related conjugates, 157–161
　oligosaccharide fragments
　　capsular polysaccharides
　　　Haemophilus influenzae type b, 161–166
　　　Klebsiella type 2, 176–178
　　　Klebsiella type 11, 178–179
　　　Streptococcus Group A, 166–168
　　　Streptococcus pneumoniae type 2, 168–171
　　　Streptococcus pneumoniae type 3, 171–173
　　　Streptococcus pneumoniae type 17F, 173–174
　　　Streptococcus pneumoniae type 23F, 174–176
　　lipopolysaccharide-related
　　　Chlamydia, 191–193
　　　Escherichia coli type O8, 187–188
　　　meningococcal, core region, 189–191
　　　Salmonella, 179–180
　　　Salmonella illinois, 183–184
　　　Salmonella typhi, 180–183
　　　Salmonella typhimurium, 180–183
　　　Shigella dysenteriae type 1, 184–187
Proteoglycan, core-peptide backbone conformation, 213–215
Pseudohalogens, class description, 26–30
Pyranose, in carbohydrate, constraint, 16
Pyranos-1-yl radicals, C-1 substituted, conformations, 82–83

R

Radicals
　acetal, carbon-centered, structure, 68–69
　addition, to exo-glycals, for C-glycosyl compound synthesis, 120–122, 126–127
　for anhydroalditols from glycosyl derivatives
　　nucleophilic and electrophilic properties, 93–95
　　overview, 86–87
　　reduction with tin and silicon hydrides, 87–92
　　synthesis, 95–100
　anomeric, see Anomeric radicals
　4,6-O-benzylidene-2,3-O-isopropylidene-D-glucopyranos-1-yl radical, 80
　C-1 substituted pyranos-1-yl radicals, 82–83
　carbohydrate, see Carbohydrate radicals
　2-deoxyglucopyranos-1-yl radicals, 81–82
　glycofuranos-1-yl radicals, 84

Radicals (Cont.)
  hexo-pyranos-1-yl radicals, 69, 80–81
  induced rearrangements, for 2-deoxy sugar synthesis
    by 2,1-acyloxy group rearrangements, 131–135
    by 2,1-migrations of ester groups, 138–139
    by 2,1-phosphatoxy group rearrangements, 135–138
  pento-pyranos-1-yl radicals, 69, 80–81
  peracetylated D-glucopyranos-1-yl radicals, 80
  phosphonyl radicals, 126–127
  reductions
    diastereoselective
      O-glycosides and analogs, 107–115
      C-glycosyl compound and analogs, 115–127
      nucleosides, 127–128
      substituted anomeric radicals, 129–130
    glycosyl derivatives to anhydroalditols
      nucleophilic and electrophilic properties, 93–95
      with tin and silicon hydrides, 87–92
  sugar-derived, see Sugar-derived radicals
  2,3,4,6-tetra-O-methyl-D-glucopyranos-1-yl radical, 69
  translocation, for α- to β-O-mannopyranoside inversion, 111–115
RAE, see Reverse anomeric effect
Reverse anomeric effect, in glycals, 22
6,6-Rings, fused, synthesis via iodoetherification, 52
Rydon reagents, applications, 38

S

Saccharides, conformation, 239–243
Salmonella, O-specific polysaccharides, 179–180
Salmonella illinois, O-specific polysaccharides, 183–184
Salmonella typhi, O-specific polysaccharides, 180–183
Salmonella typhimurium, O-specific polysaccharides, 180–183
Seleno-glyculosides, in C-glycosyl compound and analog synthesis, 117

Sharpless asymmetric epoxidation, vs. iodolactonization, 52–53
Shigella dysenteriae type 1, O-specific polysaccharides, 184–187
C-Sialyl compounds, solution conformation, 274–275
Silanes, as hydrogen source, 99–100
Silicon hydrides, for anhydroalditols from glycosyl derivatives, 87–92
Single-electron transfer, for anhydroalditol synthesis, 100–102
Solid-phase oligosaccharide synthesis, technique development, 25
Solid state, C-glycosyl compound conformation
  β-C-gentiobiose, 246–247
  C-isomaltose, 247
  C-sucrose, 247
Solution, C-glycosyl compound conformation analysis, 243–246
  α-C-glucosyl derivatives, 247–250
  β-C-glucosyl derivatives, 247–250
  C-glycopeptides, 278–279
  O/C-lactose, 256–261
  with (1->3) linkages, 261–264
  with (1->4) linkages, 250–255
  with (1->6) linkages, 271–273
  and (1->1)-linked trehalose analogs, 266–270
  C-sialyl compounds, 274–275
  simple derivatives, 247
  C-sucrose, 264–265
  trisaccharides, 275–278
Spin systems
  carbohydrate, identification, 206–209
  peptide, identification, 209–210
Stannylene ether, formation, 55
Stereoselectivity, H-atom transfer at anomeric radicals, 103–106
Sterics, in saccharide conformation, 239–243
Streptococcus, protein–oligosaccharide conjugates
  Group A, capsular polysaccharides, 166–168
  type 2, 168–171
  type 3, 171–173
  type 17F, 173–174
  type 23F, 174–176
Streptococcus pneumoniae, protein–oligosaccharide conjugates
  type 2, 168–171
  type 3, 171–173

type 17F, 173–174
type 23F, 174–176
C-Sucrose
  solid-state conformation, 247
  solution conformation, 264–265
Sucrose, total synthesis, by iodoetherification, 53
Sugar-derived radicals, structure
  C-1 substituted pyranos-1-yl radicals, 82–83
  carbon-centered acetal radicals, 68–69
  2-deoxyglucopyranos-1-yl radicals, 81–82
  FMO theory interpretations, 85–86
  free sugar-derived radicals, 84–85
  glycofuranos-1-yl radicals, 84
  hexo- and pento-pyranos-1-yl radicals, 69, 80–81
  non-anomeric carbohydrate radicals, 83–84
Sugars, free, derived radicals, ESR studies, 84–85
Sugar xanthates, radical deoxygenation, 91–92

T

TBDMS ethers, cleavage, with iodine solutions, 45
2,3,4,6-Tetra-O-acetyl-α-D-glucopyranosyl bromide, 69
2,3,4,6-Tetra-O-acetyl-α-D-glucopyranosyl iodide, 10–11
2,3,4,6-Tetra-O-acetyl-β-D-glucosyl 3,4,6-tri-O-acetyl-2-deoxy-α-D-*arabino*-hexopyranoside, 20
2,3,4,6-Tetra-O-acetyl-β-D-glucosyl 3,4,6-tri-O-acetyl-2-deoxy-2-iodo-α-D-mannopyranoside, 20
Tetrahydropyranyl ethers, cleavage with iodine solutions, 47
2,3,4,6-Tetra-O-methyl-D-glucopyranos-1-yl radical, 69

Thiocarbonyl derivatives, sugar xanthates, radical deoxygenation, 91–92
Thioglycosides, 15–18, 32
Thio-glyculosides, in C-glycosyl compound and analogue synthesis, 117
Thiol, catalyzed radical reductions, 93–95
THP, *see* Tetrahydropyranyl ethers
Tin hydrides, for anhydroalditols from glycosyl derivatives, 87–92
Trehalose, C-glycosyl and (1→1)-linked analogs, 266–270
β,β-Trehalose octaacetate, 10
1,1,1-Triacetoxy-1,1-dihydro-1,2-benziodoxol-3(1H)-one, *see* Dess–Martin reagent
3,4,6-Tri-O-acetyl-O-benzoyl-2-deoxy-2-iodo-D-glucopyranose, 19
2,3,4-Tri-O-acetyl-6-deoxy-6-iodo-α-D-glucopyranoside, 11–12
Trisaccharides, solution conformation, 275–278
Tris(trimethylsilyl)silane, 91
Trityl ethers, cleavage with iodine solutions, 45–47

V

Vaccines, antibacterial, *see* Antibacterial vaccines
Vicinal *trans*-diols, in alkene formation, 39–40
Vinyl glycosides, 15

X

Xyloside, fragmentation, 33–34

Y

Yeast cells, *Escherichia coli* adhesion, 238

ISBN 0-12-007256-4